Source Rocks in a Sequence Stratigraphic Framework

Edited by

Barry J. Katz

and

Lisa M. Pratt

AAPG Studies in Geology #37

Published by
The American Association of Petroleum Geologists
Tulsa, Oklahoma, U.S.A.
Printed in the U.S.A.

Published 1993

ISBN: 0-89181-045-5

Association Editor: Kevin T. Biddle
Science Director: Gary D. Howell
Publications Manager: Cathleen P. Williams
Special Projects Editor: Anne H. Thomas
Production: Custom Editorial Productions, Inc., Cincinnati, Ohio

About the Editors

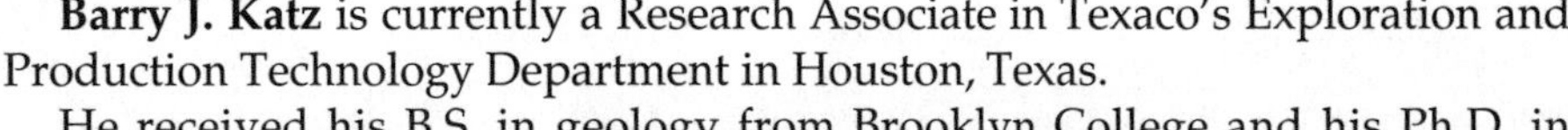

Barry J. Katz is currently a Research Associate in Texaco's Exploration and Production Technology Department in Houston, Texas.

He received his B.S. in geology from Brooklyn College and his Ph.D. in marine geology and geophysics from the Rosenstiel School of Marine and Atmospheric Sciences, University of Miami, in 1979. He was the recipient of the F. G. Walton Smith Prize for his dissertation, "An Application of Amino Acid Racemization—The Determination of Paleoheat Flow." He joined Texaco's Bellaire Research Laboratories in 1979 and has held various positions within Texaco's research organization.

Dr. Katz is currently an Associate Editor of the *AAPG Bulletin*, Chairman of the AAPG Research Activities Subcommittee, and a member of the AAPG Marine Geology Committee. He is also a member of the JOIDES's (Joint Oceanographic Institutions for Deep Earth Sampling) Pollution Prevention and Safety Panel.

The major themes of his research have been related to the processes controlling deposition of sedimentary organic matter and the characterization of organic facies and how this information may be applied to petroleum exploration.

As a consequence of his research interests, Dr. Katz has co-convened AAPG research conferences on lacustrine basin exploration and on organic facies variations within the Monterey Formation.

Lisa M. Pratt is Associate Professor of Geological Sciences at Indiana University. She is one of three senior scientists responsible for the funding and management of the Biogeochemical Laboratories at Indiana University.

Dr. Pratt obtained a B.A. in botany from the University of North Carolina in 1972, an M.S. in botany from the University of North Carolina in 1978, followed by a Ph.D. in geology from Princeton University in 1982. She was awarded a National Research Council Post-Doctoral Fellowship with the U.S. Geological Survey in 1982 and 1983. From 1983 through 1987, Dr. Pratt was a Research Geologist in Denver with the Branch of Petroleum Geology, U.S. Geological Survey. In 1987, she joined the faculty of Indiana University.

Dr. Pratt was an AAPG Distinguished Lecturer in 1990–1991, received the AAPG Matson Award in 1987, and served on the AAPG Publications Committee from 1986 to 1988. She was the elected SEPM Research Counselor in 1990–1992. She is currently a member of the Editorial Board for Geology and serves on the Board of Directors for the Geochemical Society. She is also a member of the Ocean History Panel for JOIDES (Joint Oceanographic Institutions for Deep Earth Sampling).

Dr. Pratt and her graduate students are involved in geochemical, stratigraphic, and sedimentologic studies of fine-grained sediments and sedimentary rocks, with particular emphasis on reconstruction of paleoclimatic and paleoceanographic conditions during accumulation of black shale. Recent studies have focused on strata ranging in age from the Precambrian Nonesuch Formation to the Ordovician Maquoketa Shale, the Cretaceous La Luna Formation, the Miocene Monterey Formation, and to modern sediments in the California Borderland Basins.

Table of Contents

Chapter 1

Introduction

Barry J. Katz
Exploration and Production Technology Department
Texaco Inc.
Houston, Texas, USA

Lisa M. Pratt
Department of Geological Sciences
Indiana University
Bloomington, Indiana, USA

Petroleum exploration has relied historically on the identification of structural anomalies through the use of geologic mapping combined with various geophysical methods (e.g., wire-line logging, seismic, magnetic and gravity). An exploration program of this type is viable when exploratory costs are low and numerous large structural features are untested. However, costs associated with exploration are increasing and most large structural anomalies have been tested so this strategy is no longer as viable.

A more viable exploration program in these times requires an examination of the diverse elements necessary for hydrocarbon accumulations to exist. In addition to an evaluation of the area under structural closure, necessary information includes hydrocarbon source rock quality and quantity, reservoir porosity and permeability, seal, migration network, and level of thermal maturity impacting both generation and preservation. These geologic elements must be evaluated in light of their temporal and spatial distributions (Magoon, 1988). Unfortunately, although many of these elements are considered, qualitative information all too often is presented independently of any geologic framework. Commonly, large volumes of data are gathered and are presented as only a mean value and/or range of values. Such information is of only limited usefulness, however, because it provides no insight into either the spatial or the stratigraphic distribution of geologic features. Recently there has been a major emphasis on assessment of quantitative data within a sequence stratigraphic framework. This integrated approach permits a better understanding of the net versus gross source and reservoir thickness in a given sampling locality. A more realistic understanding of the distribution of these facies provides some degree of predictability away from the actual sampling site and can aid in explaining the distribution of hydrocarbon reserves.

As a consequence of the growing interest in integrated sequence stratigraphy, several oral and poster technical sessions were organized at the 1991 Annual Convention of the American Association of Petroleum Geologists in Dallas. The session papers that dealt with petroleum source rocks in a sequence stratigraphic framework comprise the bulk of this volume. The papers describe various aspects of exploration and development problems currently facing petroleum geochemists. These problems include variability within source rock systems, geochemical variability of sedimentary organic matter and hydrocarbon products as a function of depositional setting, and the relationship between the accumulation of organic-rich strata and sea level. This volume covers a broad geographic and stratigraphic range, with the order of presentation progressing from general topics to specific case studies.

Chapter 2 by Katz et al. describes internal variability for three source rock systems with respect to richness, type, and nature of generated hydrocarbons, including variability at the molecular level. This chapter describes why these internal differences should be taken into consideration when designing a geochemical sampling program and how the nature of the sampling program may limit the applicability of the data generated. Results of this study raise questions about the appropriateness and/or correctness of many oil-to-source-rock correlations proposed in the literature.

Chapter 3 by Mello et al. examines the influence of depositional environment on organic geochemical character. The authors utilize lithologic and paleontologic data to support their environmental interpretations. Data are presented from several Brazilian basins (Paraná, Sergipe–Alagoas, Potiguar, and Ceará) covering a few of the depositional settings under which source rocks are deposited, hypersaline marine and hypersaline lacustrine settings. They conclude that there are distinct geochemical signatures for organic facies associated with subenvironments of hypersaline settings.

The following two chapters describe relationships between sea level and deposition of organic matter. Wignall and Maynard state that many thick organic-rich black shales form in distal marine settings where sedimentation rates are low and bottom waters are poorly oxygenated. They further conclude, however, that such a model does not adequately explain thinner, shallow-water black shales which are more commonly associated with potential hydrocarbon reservoir facies. Wignall and Maynard propose two depositional models for transgressive, shallow-water black shales. One model is associated with maximum flooding and is represented by the marine bands in the Upper Carboniferous of northern England which display marked lateral variations in facies. It is inferred that the most oxygen-depleted facies occur in the more basinal portions of the unit. The second depositional model is associated with initial phases of a marine transgression and is represented by the Jet Rock (lower Toarcian) of northern England which displays virtually no variability in organic facies and passes proximally into either sediment-starved marginal facies or nondepositional surfaces.

Steffen and Gorin describe a series of organic-facies changes that occur within a Berriasian sequence of southeastern France. They note that isolated kerogens associated with lowstand systems tracts are enriched in degraded terrestrial organic matter. This kerogen distribution contrasts with transgressive systems tracts which display increasing concentrations of autochthonous marine organic matter culminating at the time of maximum flooding. The regressive systems tract appears as a mirror image of the transgressive systems tract. Although the studied carbonate system is organic lean (organic carbon <0.2 wt.%) and is not of source rock quality, Steffen and Gorin believe that their conclusions are transferable to settings where preservation of organic matter would be more favorable and petroleum source rocks could develop.

The remaining nine chapters constitute the bulk of this volume and present a series of case studies. The first of these chapters was prepared by Herbin et al. This chapter describes the lateral and vertical distribution in the Jurassic Kimmeridge Clay, the major source rock in the North Sea region. The authors note that the Kimmeridge Clay was deposited as part of a highstand systems tract. An increase in organic carbon content was observed in a basinward direction and is inferred to result from changes in water column oxygen content and burial rate. Herbin et al. suggest that the vertical cyclicity in the Kimmeridge Clay has primary frequencies at 30,000 and 280,000 years. The consistency of organic matter type between alternating layers richer and poorer in organic matter suggests that the cyclicity is a combined result of interplay between preservation and productivity. Within this overall framework, maximum contents of organic matter were observed near the middle of the transgressive systems tracts.

The next chapter, written by Robison and Engel, describes petroleum source rocks deposited during the Late Cretaceous transgression in northeastern Africa. This source unit is responsible for much of the hydrocarbon accumulation in the Red Sea region. Robison and Engel suggest that third-order eustatic cycles are an important control on source rock deposition. They observe that the highest quality oil-prone source material was deposited within the condensed sections of the transgressive systems tract in partial response to reduced detrital input. They further note that the highstand systems tract is, in general, organic lean and dominated by gas-prone terrestrial organic matter.

The third of the case studies is by Meyers and Snowdon and describes a relationship between deposition of organic matter and the subsidence history during the Early Cretaceous of the northwestern Australian margin (Exmouth Plateau). Using samples recovered by the Ocean Drilling Program, their study suggests a progressive decrease in the amount of organic matter being introduced into the system as the basin subsided. Meyers and Snowdon conclude that the paleoenvironment was terrestrially dominated, the amount of terrestrial input decreased as the delta system retreated, and marine input provided little more than background throughout the time interval studied.

The last of the international case studies was prepared by Chandra et al. They describe the development of organic-rich shales within the Cauvery basin and note five sea level cycles. Source rock development was associated with the transgressive phases of each of these cycles. The authors note, however, that the highest quality source rocks developed during one cycle associated with the Aptian–Albian oceanic anoxic event. They conclude that sea level is but one of the several factors controlling the development of source rocks and that other global as well as local conditions will influence the composition and distribution of organic-rich sediments.

The last five chapters are case studies from North America. Mancini et al. examine the hydrocarbon source rock potential of several Mesozoic condensed sections from southwest Alabama. Of the several condensed sections examined, only two appear to exhibit source rock potential, namely the Upper Jurassic Smackover Formation and the Upper Cretaceous Tuscaloosa Group. The authors conclude that although condensed sections often represent the most favorable portion of a depositional sequence for source rock development, specific local environmental conditions are still required for deposition of sediment rich in marine organic matter.

The second North American case study was prepared by Lambert and describes the stratigraphic framework of the organic-rich Devonian Chattanooga (Woodford) Shale in Oklahoma and Kansas. The Chattanooga and its equivalents have been a major source for petroleum within the North American mid-continent region. Lambert notes that the Chattanooga Shale was deposited within a third-order depositional sequence, and three shale members were deposited within the overall transgression. The highest contents of organic carbon and marine organic matter were found in the middle shale member which represents deposition during the maximum extent of the transgression. The lower shale member, which represents the initial phases of the transgressive systems tract, appears to be dominated by oxidized organic matter, and the upper shale member, which represents the highstand systems tract, contains abundant terrestrial material, possibly indicating deltaic progradation into the depositional system.

The following two chapters deal with the Miocene Monterey Formation of southern and central California. The Monterey Formation has acted as both source and reservoir within the region. The first chapter is by Bohacs and relates the stratigraphic variability of source rock properties to systematic patterns associated with depositional sequences and sequence sets. He notes that source rock properties are best developed near the mid-sequence downlap position. Moderate source rock potential exists within the highstand systems tract. Bohacs further notes that deposition of the most organically enriched horizons and associated lithofacies (phosphatic shales and carbonaceous marls) are diachronous across the depositional basins. His work shows that placing organic geochemical data into a sequence stratigraphic framework allows better definition of depositional processes and construction of predictive models.

This chapter is followed by the work of Zaback and Pratt, which examines aspects of the inorganic and isotope geochemistry of the Monterey related to changing oceanographic conditions. The authors note that sediments display elevated Mn/Fe ratios when deposited during relatively high sea level and display a shift toward heavier sulfur isotopic compositions when deposited during relatively low sea level. Zaback and Pratt suggest that elevated Mn/Fe ratios may be associated with mobilization of Mn following an expansion of the oxygen-minimum zone, and that changes in sulfur isotopic ratios could result from high rates of sulfate reduction promoted by an increase in deposition of marine organic matter during the onset of intense coastal upwelling.

The final chapter of this volume is by Pasley et al. and examines organic facies variations within the Upper Cretaceous Mancos Shale of the San Juan basin, New Mexico. These authors note that the amount and type of organic matter preserved on an epicontinental shelf behaves in a predictable fashion. For example, they observe that the transgressive systems tract contains organic-rich shale with elevated pyrolytic yields. Within the transgressive systems tract, the maximum flooding surface contains the least amount of terrestrial organic matter and, therefore, may contain the highest proportion of oil-prone material. In contrast, both the lowstand and highstand systems tracts have lower contents of organic matter and lesser yields of pyrolyzable hydrocarbons. Maceral examination reveals that progradational systems also contain more terrestrial material than the transgressive systems tract, with terrestrial phytoclasts being particularly well-preserved in the lowstand tract.

One of the important contributions of the two sessions at the AAPG meeting in Dallas and this volume is an increasing awareness that source rock development has occurred under a broad spectrum of depositional conditions. The internal stratigraphy of many gray to black shales reveals that physical and chemical conditions were widely fluctuating during the deposition of any given source rock unit. Although recent literature has emphasized source rock development during condensed portions of stratigraphic sequences (Loutit et al., 1988), the works presented here reveal that source rocks develop within other portions of a stratigraphic sequence.

Moreover, there is continuing controversy concerning the possible effect of enhanced marine productivity during the transgressive systems tract. Detailed stratigraphic data are needed on type of organic matter and concentration of nutrients, like phosphorous, through the transgressive and condensed portions of marine sequences.

We hope that the chapters in this volume will serve as a focus for provocative discussion and a stimulus for further interdisciplinary research.

ACKNOWLEDGMENTS

This volume reflects not only the efforts of the authors but also those of the following reviewers, who provided constructive comments and suggestions: W.M. Ahr, J. Comer, J.S. Compton, J. Curiale, G. Demaison, R.W. Gallois, G. Grabowski, Jr., R. Haddad, W.E. Harrison, P. Heckel, M.A. Kruge, L.M. Liro, J. Palacas, K.E. Peters, D.D. Rice, V.D. Robison, R. Sassen, J. Schmoker, R. Tyson, and R. Witmer.

The editors also would like to thank Dr. R. Slatt for asking us to organize the sessions on "source rocks in a stratigraphic framework" and G. Howell for suggesting that a volume be prepared, based on the two sessions.

REFERENCES CITED

Loutit, T. S., J. Hardenbol, P.R. Vail, and G.R. Baum, 1988, Condensed sections: The key to age determination and correlation of continental margin sequences, *in* C.K. Wilgus, B.S. Hastings, H. Posamentier, J. Van Wagoner, C.A. Ross, and G.S.C. Kendall, eds., Sea-Level Changes: An Integrated Approach: SEPM Special Publication 42:183–213.

Magoon, L. B., 1988, The petroleum system—a classification scheme for research, exploration, and resource assessment, *in* L.B. Magoon, ed., Petroleum Systems of the United States: USGS Bulletin 1870:2–15.

Chapter 2

Implications of Stratigraphic Variability of Source Rocks

B. J. Katz, T. M. Breaux, E. L. Colling,[1] L. M. Darnell, L. W. Elrod, T. Jorjorian, R. A. Royle, V. D. Robison, H. M. Szymczyk, J. L. Trostle, and J. P. Wicks

Texaco Inc.
Exploration and Production Department
Houston, Texas, USA

ABSTRACT

There has been substantial progress in organic geochemistry, over the past two decades, in the areas of geochemical modeling and analytical methods. However, lesser attention has been paid by most geochemists to the natural variability of geologic systems. An examination of selected stratigraphic units from North America, South America, and southeast Asia suggests that variations in depositional conditions control how source rock intervals are stratigraphically distributed, as well as the character of the organic matter and the nature of the hydrocarbons generated upon thermal maturation.

The distribution of organic-rich intervals within a source sequence impacts the quantity of hydrocarbons generated as well as expulsion efficiency. Variations in the nature of the organic matter within the organic-rich intervals directly impact both the quantity and character of the generated products. Variations in geochemical character of many source rock systems can be substantial enough to raise questions on proposed oil-to-source rock correlations. Available data clearly indicate the need for more geologically controlled sampling programs that take into consideration the ultimate goal of the study.

INTRODUCTION

Organic geochemistry serves several roles in petroleum exploration and exploitation. Organic geochemistry is used to identify hydrocarbon source rocks and to correlate identified sources with oil accumulations. Correlation studies are aimed at establishing migration directions and pathways and the potential for down-dip hydrocarbon accumulations. Studies aimed at establishing source rock richness and organic matter type are used, in conjunction with various numeric models, to estimate the volumes and types of hydrocarbons which may have been generated.

Petroleum geochemistry has progressed substantially over the past two decades, largely as a result of analytical advances and increased computer capabilities. During this period there was an emphasis on the chemical aspects of geochemistry. Many publications have emphasized analytical techniques, some of

[1]*Current address: Saudi Aramco, Dhahran, Saudia Arabia*

which are now considered routine, while others are still viewed as experimental (e.g., Espitalié et al., 1977; Larter and Douglas, 1980; Hayes et al., 1990). Other publications noted the occurrence of specific compound classes or individual compounds (e.g., Brassell et al., 1981; Moldowan, 1984; Simoneit et al., 1986). Although attempts were made to relate the occurrence of these compounds to organic input and geologic conditions, the necessary supporting data were commonly not available to the various investigators. In many cases, conclusions were drawn based on limited sampling or when the necessary supporting data were unavailable.

An example of this problem can be found in the paper by Seifert et al. (1978). In this paper, the authors identify bisnorhopane in Monterey Formation-derived oils. They concluded, based largely on the available natural product literature, that there were three possible origins for this compound. One required a strongly oxic depositional/diagenetic environment. The other two assumed that ferns provided an important precursor. Available data now indicates that neither strongly oxic conditions nor significant fern input were available during the deposition of the Monterey Formation. Later work (Tornabene et al., 1979; Katz and Elrod, 1983; Schoell et al., 1992) has suggested that bisnorhopane is probably derived from sulfate-reducing bacteria.

Concurrently, advances in computer capability have led to the development of 1-, 2-, and 3-D models which simulate the generation, expulsion, and migration of hydrocarbons. Although these models are numeric and have been developed using chemical and physical laws, they require geochemical input. This input includes organic richness, type, and the areal and stratigraphic distribution of the effective source units.

With the development of these analytical tools and the identification of a wide spectrum of compounds, numerous studies have begun emphasizing geologic questions using these various chemical tools and mathematical models. These studies have attempted to establish available resources, depositional environments, levels of thermal maturity, and the interrelationships among oils and possible source rocks. Unfortunately, although a few studies have examined the variability within source rock systems (Macko and Quick, 1986; Wenger and Baker, 1986; Curiale and Odermatt, 1989), most studies have relied upon only limited sample populations with little regard to the observed variability within the lithostratigraphic column, including bedding type, mineralogy, and faunal, floral, and trace fossil assemblages. The potential geochemical variability associated with the same processes that varied the lithostratigraphic record could alter many of the conclusions drawn by these various investigators, with potentially significant economic implications. Variability in organic-richness and kerogen type could alter resource assessments, while variability in isotopic and molecular composition could complicate correlation studies.

This chapter will examine intrabasinal and interbasinal variations in source rock geochemistry. Bulk and molecular data are presented from three source rock systems: the Monterey Formation (California), the Oriente Group (Peru), and the Pematang Brown Shale (Indonesia). The internal geochemical variability of these systems is documented and the implications to petroleum exploration are discussed. This chapter will also describe sampling strategies in light of the observed geochemical variability of source rock systems and in the context of the study.

ANALYTICAL METHODS

Each sample was ground to ~44 μm prior to analysis. Organic carbon and total sulfur were determined on all samples using a LECO analyzer following the removal of carbonate minerals with dilute HCl. Most samples were subjected to "Rock-Eval" pyrolysis as described by Espitalié et al. (1977). Pyrolysis-gas chromatography was performed on the Monterey sample suite as outlined by Colling et al. (1986). Bitumen was extracted from several samples using a high-pressure flow-through system and an azeotropic mixture of chloroform, methanol, and acetone (70:15:15 by volume) as the solvent. The bitumens were subjected to HPLC to initially separate the saturated hydrocarbons, aromatic hydrocarbons, resins, and asphaltenes, and then the saturated hydrocarbons were further separated into normal paraffins, isoprenoids, and naphthenes (Nolte and Colling, 1989; and Nolte, 1991). The isoprenoid and naphthene fractions were recombined and analyzed using a Hewlett-Packard 5890 gas chromatograph with a 5970B mass selective detector at 70 eV electron impact ionization. Other samples underwent HCl and HF digestion to isolate kerogens for visual and/or chemical analysis. Elemental composition of the isolated kerogens was determined on a Carlo-Erba elemental analyzer model 1106. Visual assessment was performed in transmitted and fluorescent light using standard organic petrographic techniques. Stable carbon isotopic analysis was performed using a Finnigan MAT Delta E isotope ratio mass spectrometer. These data are presented relative to the Peedee belemnite standard.

ANALYTICAL RESULTS AND IMPLICATIONS

Monterey Formation

The Monterey Formation or its equivalent is present in many of the onshore and offshore basins of central and southern California (Figure 1). Approximately 13 billion barrels of recoverable oil have been attributed to this unit (Dolton et al., 1981). It is mid-Miocene in age and generally considered to have been deposited under dysaerobic marine conditions (Pisciotto and Garrison, 1981). Dysaerobic rather

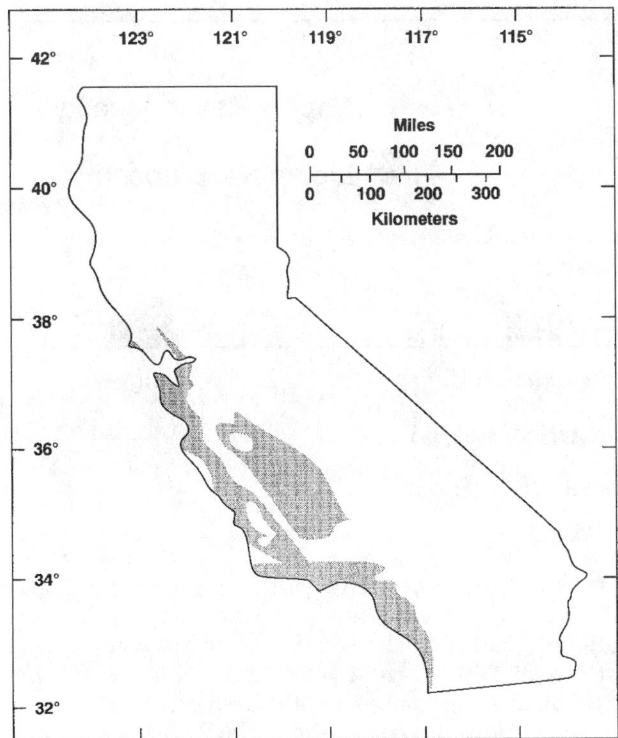

Figure 1. Onshore distribution of the Monterey Formation (after Obradovich and Naeser, 1981).

Figure 2. Lithologic variability of the Monterey Formation at the Naples Beach sampling locality. Note differences in bedding character and coloring.

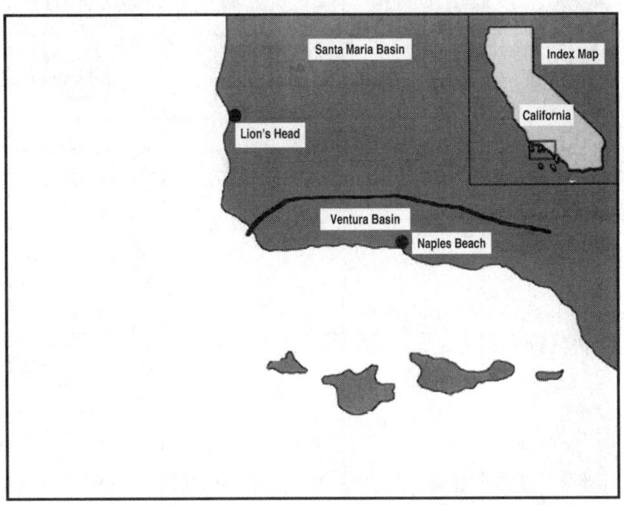

Figure 3. Location index map for the Monterey Formation sampling localities at Naples Beach and Lion's Head, California. Solid line represents boundary between the Ventura and Santa Maria basins.

than anoxic conditions within the water column are indicated by the presence of limited populations of benthic foraminifera (Govean and Garrison, 1981).

However, depositional conditions were not constant either areally or stratigraphically during Monterey time. Stratigraphic variability in the depositional system is clearly indicated by the lithologic variability of the unit (Figure 2). Lithologic changes occur at scales ranging from a few to several tens of centimeters. It is, in part, this high frequency variation in rock character which has hampered the use of conventional well logs for the identification of possibly productive intervals within the Monterey Formation (Cannon, 1981).

Isaacs (1987) has suggested that the Monterey Formation can be subdivided into four informal members based on major lithologic characteristics. These members are differentiated by inorganic chemistry and mineralogy (i.e., proportions of carbonate, silica, detrital material, and phosphate abundance), as well as by the physical attributes of the rock (i.e., degree and type of laminations). Because these lithologic characteristics are a response to both the physico-chemical conditions at the time of deposition and the nature of the input, systematic differences in organic geochemical characteristics (e.g., richness, organic quality) would also be expected throughout the unit. Isaacs (1987) suggested that the abundance of organic matter is related to bulk sedimentary composition. For example, the quantity of organic matter decreases as the abundance of silica increases.

Detailed sampling at two outcrop localities (Naples Beach, Ventura basin, and Lion's Head, onshore Santa Maria basin; Figure 3) provides information on the nature of the organic geochemical heterogeneity of the Monterey Formation. This heterogeneity can be observed in even the most basic geochemical data. Organic carbon content varied from 0.45 to 17.76 wt.% (Figure 4). At the Naples Beach locality, the most organically enriched material was found within the carbonaceous marl member (Figure 5). This member is phosphatic in nature and displays the highest total generation potentials (S_1+S_2 = free + generatable hydrocarbons). The hydrogen

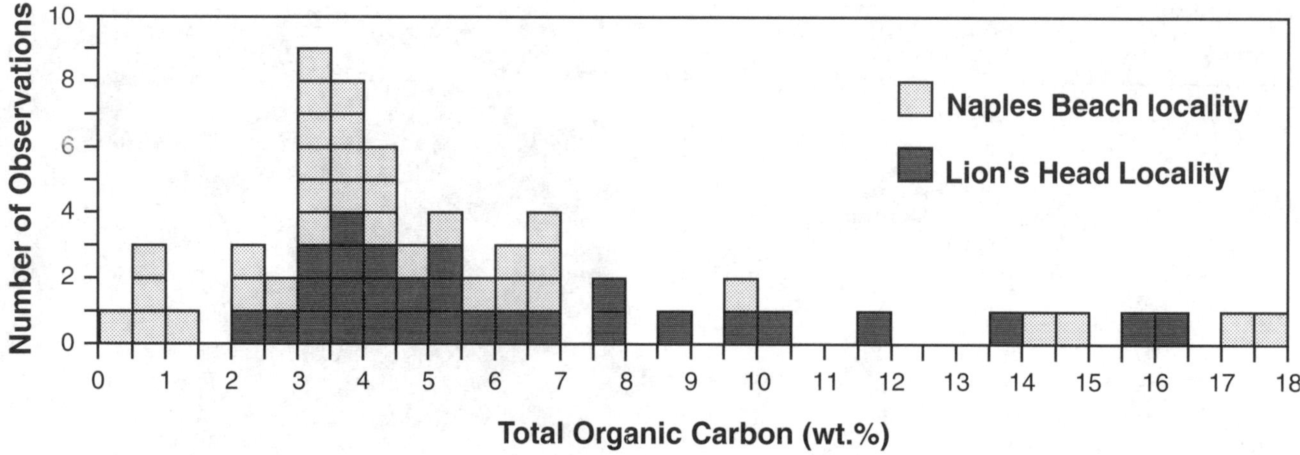

Figure 4. Histogram of total organic carbon determinations, Monterey Formation.

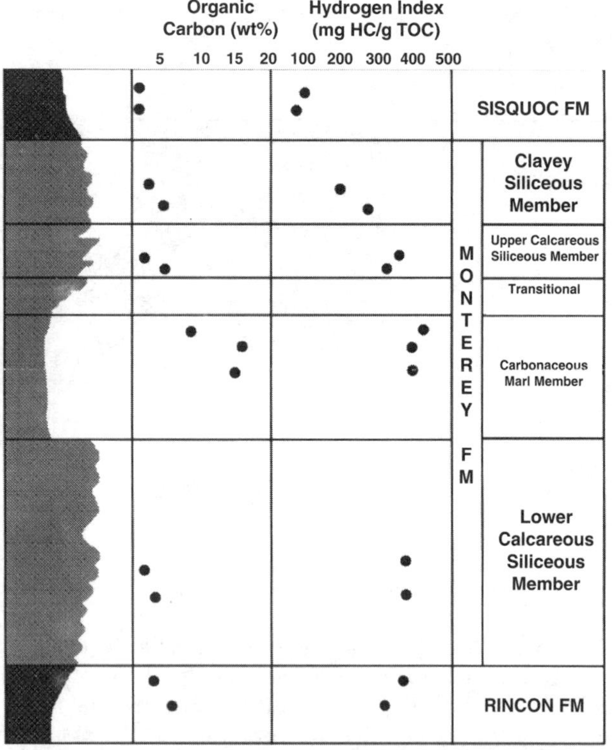

Figure 5. Geochemical log of the Naples Beach locality. (Sample positions from Isaacs et al., 1992.) Shaded area represents a weathering profile. Light shading highlights the Monterey Formation.

indices within this member were also slightly elevated compared with the other three members at Naples Beach.

As a consequence of the level of organic enrichment, the observed stratigraphic trend in the hydrogen index appears to be related to the nature of the organic matter rather than analytical interferences (Katz, 1983). The observed differences in apparent hydrogen enrichment, as manifested by the hydrogen indices, could reflect either differences in the degree of organic preservation of autochthonous marine organic matter or changes in the relative abundance of marine and terrestrial organic matter. Higher hydrogen index values would be indicative of better preservation and/or higher relative abundance of marine organic matter. Pyrolysis-gas chromatography results favor differences in preservation rather than the nature of the organic input. The pyrolysis-gas chromatograms do not display any increase in the relative abundance of either longer chain alkane–alkene doublets or aromatic compounds with decreasing hydrogen index values as would be expected if the lower hydrogen index values were a reflection of an increase in higher land plant input (Figure 6; Horsfield, 1989).

The Lion's Head locality displays similar geochemical variability in organic richness, generation potential, and hydrogen enrichment (Figure 7). When the datasets from the two sampling localities are compared, the Lion's Head locality displays elevated generation potentials and higher hydrogen index values versus the Naples Beach locality. The similarity in the pyrolysis-gas chromatographic fingerprints suggests that these differences are a reflection of organic preservation state rather than the abundance of terrestrial input. Conditions appear to have been more favorable for organic preservation in the Santa Maria basin than in the Ventura basin. This could be a reflection of free oxygen content in the water column, the depth in the sedimentary column at which sulfate reduction occurs, sedimentation rate, as well as water depth, all of which could influence settling and exposure time.

Differences in preservation state between the two basins are also suggested by the observed relationships between organic carbon and total sulfur (Figure 8). In general, the carbon/sulfur ratios at the Lion's Head locality are lower than those at Naples Beach. Lower C/S ratios typically are associated with conditions more favorable for organic preservation (Leventhal, 1983).

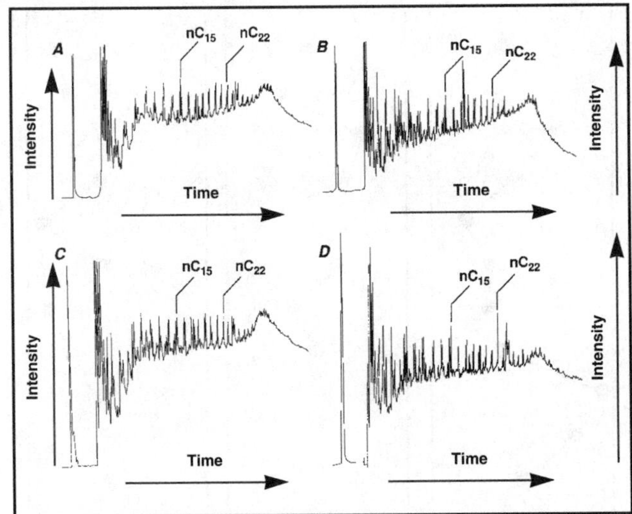

Figure 6. Representative pyrolysis-gas chromatograms of the four different Monterey Formation members. A—Lower calcareous-siliceous member; B—Carbonaceous marl member; C—Upper calcareous-siliceous member; and D—Clayey-siliceous member.

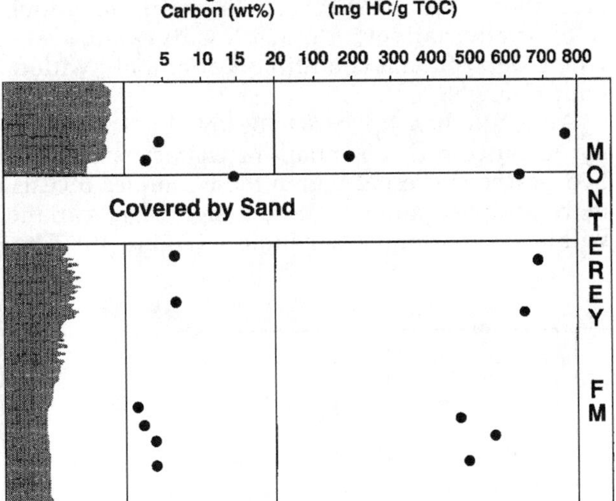

Figure 7. Geochemical log of the Lion's Head locality. (Sample positions from Isaacs et al., 1992.) Shaded area represents a weathering profile.

The intrabasinal variations in organic carbon and hydrogen enrichment could have significant impact on the assessment of potential resources. If the Naples Beach section is considered representative of the Ventura basin as a whole, the observed levels of organic enrichment could result in estimates of available oil-in-place which differ by a factor of nine if individual samples are considered representative of the locality and if all other factors (i.e., kerogen type,

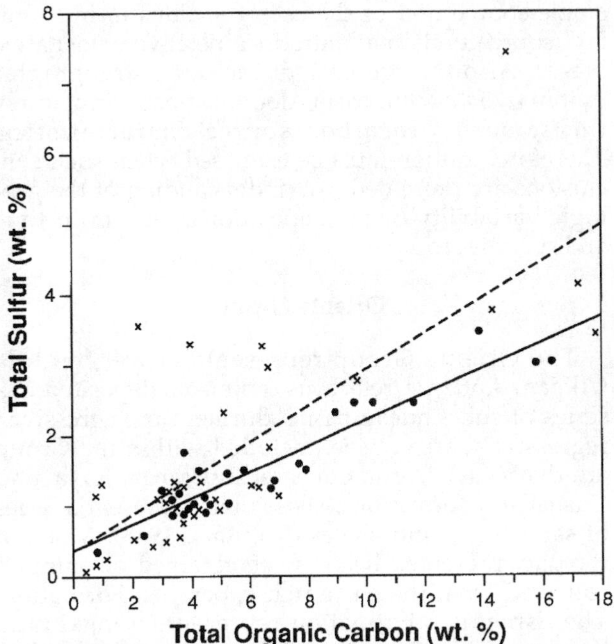

Figure 8. Relationship between organic carbon and total sulfur in the Ventura (•) and the Santa Maria (×) basins. The solid line represents a least-squares fit of the data from the Ventura basin. The dashed line represents a least-squares fit of the data from the Santa Maria basin.

thermal maturity, expulsion and migration efficiency, etc.) are constant. These differences in potential yield would be even greater if the differences in kerogen, suggested by the hydrogen indices, are incorporated into calculations. Assuming that the stratigraphic section would have obtained maturities equivalent to peak oil generation, the hydrocarbon yields would range from 11 to 161 MMBO/km^3, a factor of ~15.

If interbasinal differences are taken into consideration, the potential differences are even further magnified, with hydrocarbon yields in the Santa Maria basin ranging from 4 to 244 MMBO/km^3.

The suggested ranges in oil yields have significant economic implications. The values obtained at the low end of the spectrum suggest only marginal economic potential, something comparable to the Pannonian or Cooper basins. The values calculated for the high end of the range indicate a super-charged system comparable to the West Siberian or Sirte basins. While sub-charged systems are of marginal economic importance, particularly in a frontier area, super-charged systems offer significant exploration potential (Demaison and Huizinga, 1991).

Clearly, a reliable estimate of hydrocarbon yield would have to take into consideration the variations in organic richness and type. Such calculations would have to integrate these parameters over stratigraphic intervals which, based on lithologic characteristics, could be grouped.

Furthermore, these data indicate that the simple extension of a dataset from adjacent basins for even a single source unit or the use of analogs from "similar" source units may introduce positive or negative biases. Also the presence of the same stratigraphic unit in two adjacent basins does not necessarily imply the same hydrocarbon source characteristics. Therefore, caution must be exercised when such conclusions are presented. An understanding of the geologic variability of the depositional system in time and space is required.

Oriente Group

The Oriente Group represents Lower/middle (Albian/Aptian) Cretaceous sediments deposited in a series of sub-Andean basins during a transgressive/regressive marine cycle. Included within the Group are the Huaya, Agua Caliente, Esperanza, Raya, and Cushabatay formations. These units represent a series of sandstones and shales (Kummel, 1948). Much of the data presented below were obtained on samples collected from the shale-rich Esperanza Formation. The Esperanza Formation represents a maximum flooding event (Macellari, 1988). Although the data presented below clearly indicate that the unit contains intervals of source quality, no hydrocarbon reserves have been attributed to this unit. The Esperanza Formation is, moreover, time equivalent to the Napo Formation of Ecuador, which is believed to have been the source for ~6.8 billion barrels of recoverable oil (Feininger, 1975; Canfield et al., 1982).

The oil-prone source rock potential of the unit is manifested by the above-average levels of total organic carbon (TOC > 1.0 wt.%; Bissada, 1982), above-average generation potentials (S_1+S_2 > 2.5 mg HC/g rock; Bissada, 1982) and the elevated hydrogen index values throughout much of the sampled interval (Figure 9). The variability of source rock potential shown by this sample suite appears to be significantly less than previously discussed in the Monterey Formation. This could reflect the size of the population, or alternatively a much less dynamic system where depositional conditions varied only within a limited range.

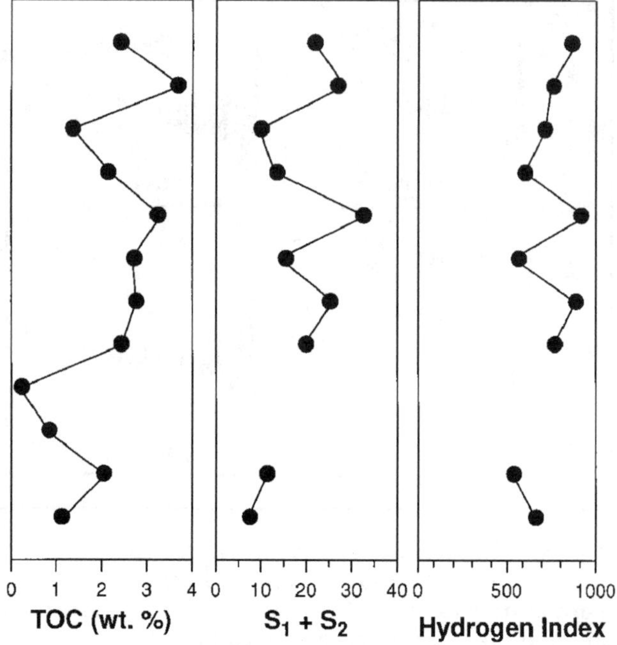

Figure 9. Geochemical log Oriente Group, eastern Peru. (No vertical scale implied.)

Although there appears to be only limited variability in source rock potential, the character of the free hydrocarbons present within these samples reveals a more complex pattern. There is significant variation within several biomarker indices (Figure 10). These

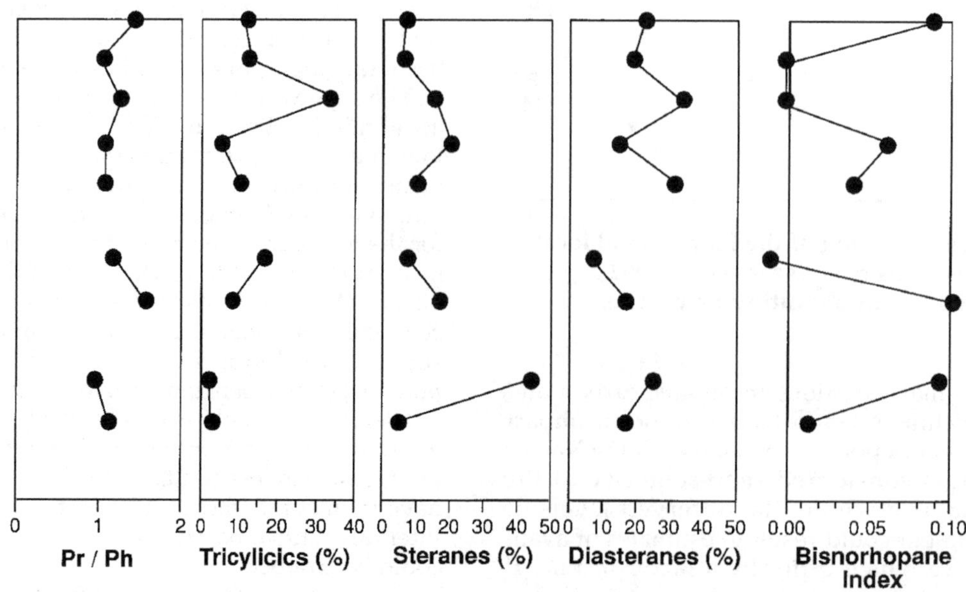

Figure 10. Biomarker log Oriente Group, eastern Peru. (No vertical scale implied.)

differences are observed in the relative abundance of gammacerane, bisnorhopane, steranes, diasteranes, and tricyclic compounds. In addition, there are differences in the distributions of normal steranes (Figure 11) and tricyclics (Figure 12). Considering that these samples were obtained over a discrete interval from a single outcrop locality, the observed differences reflect changes in the nature of the depositional system (e.g., salinity, oxygen level, detrital influx), rather than differences in maturation.

The specific causes for the observed variations are beyond the scope of this chapter. However, these geochemical indices are commonly utilized during oil to source rock correlations. The variability between samples indicates that a positive correlation based on these indices to a discrete rock sample may be fortuitous. A reservoired oil provides an integrated sampling of the geochemical character of the entire source unit rather than information on any discrete horizon. Alternatively, explanations would be necessary to explain why only discrete beds would actively contribute to a hydrocarbon accumulation, when an entire unit displays source rock potential.

Pematang Brown Shale

The Pematang Brown Shale is a lacustrine unit that was deposited in a series of isolated subbasins in central Sumatra, Indonesia (Williams et al., 1985). The unit has generated a minimum of 10 billion barrels of recoverable oil. The amount of oil attributed to this unit has recently increased as a result of the identification of mature Brown Shale or its equivalent in additional subbasins (Longley et al., 1990). The Brown Shale is generally considered Eocene to Oligocene in age, but this age assignment is problematic because of a lack of age-diagnostic fossils.

An examination of the bulk geochemical character of the entire Brown Shale interval reveals significant

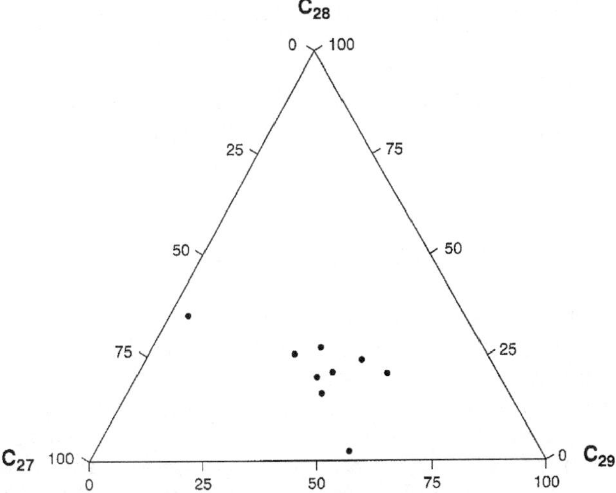

Figure 11. Normal sterane distribution, Oriente Group, eastern Peru.

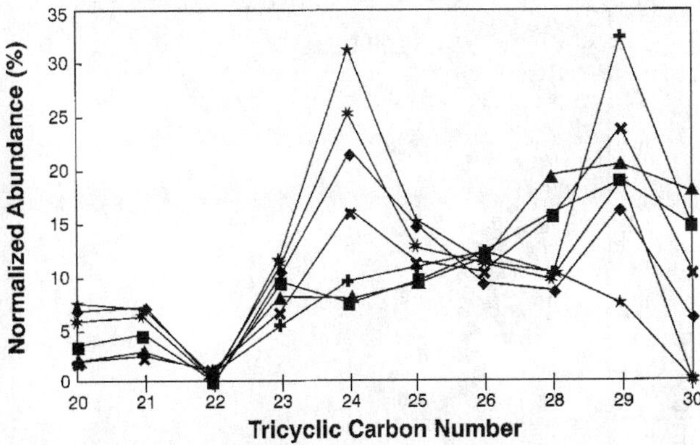

Figure 12. Relative abundance of tricyclic compounds, Oriente Group, eastern Peru.

differences between the upper and lower portions of the Brown Shale. The lower Brown Shale kerogens are gas-prone. In contrast, the upper Brown Shale kerogens are oil-prone. These differences can be seen in the atomic H/C and O/C ratios of the isolated kerogens (Figure 13). The upper Brown Shale kerogens have atomic H/C ratios typically > 1.4, with the lower Brown Shale kerogen atomic H/C ratios being ~1.0.

The observed stratigraphic distribution of the oil-prone facies within the Brown Shale may be reflecting a combination of several factors including the transition from a fluvial to lacustrine depositional setting, the development of the nutrient pool within the lakes, and changes in climatic conditions. As noted by Lambiase (1990) and Schlische and Olsen (1990), within many rift basins there appears to be a tectonostratigraphic evolution from a fluvial to a deep lacustrine and finally a shallow lacustrine sequence. This sequence develops as a result of changes in the relative subsidence and sedimentation rates. Initially, subsidence rates are low compared to sediment influx. Organic matter available to this system would typically be poorly preserved and, even if algal in origin, would commonly appear gas-prone. The initial phase of basin development is followed by a period where subsidence typically exceeds sediment influx. During this second phase of basin development, deep lakes form. It is during this period that, if appropriate conditions exist (i.e., stratification and elevated productivity levels), well-preserved, oil-prone organic matter accumulates (Watson et al., 1987). During the more mature phases of basin development, subsidence rates decrease and the lake basin is filled. Material deposited during this depositional episode is commonly coal-rich and gas-prone. This is the general sedimentary sequence observed in the central Sumatra subbasins (Lambiase, 1990).

Overprinting the response to subsidence and sediment influx are other factors which may impact either organic productivity or preservation within lacustrine sequences. For example, in immature lacustrine systems the nutrient pool is typically insufficient to

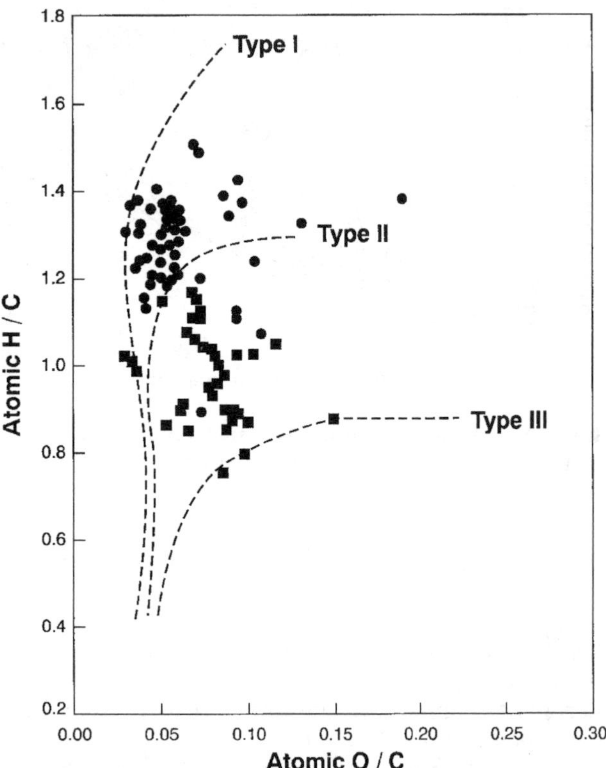

Figure 13. Van Krevelen diagram for the upper (●) and lower (■) Brown Shale, Sumatra. Reference curves for the three primary kerogen types were defined by Tissot et al. (1974).

establish elevated levels of productivity (Bloesch et al., 1977). Elevated levels of productivity are typically assumed to be necessary for the development of organically enriched, oil-prone units. In immature lake systems, external sources of nutrients are required to maintain high trophic levels (Dean, 1981). Therefore, it can be assumed that initial productivity levels within the basin were insufficient for source quality material to develop. Productivity levels within the lakes would have increased as the lake system matured and the nutrient pool became sufficient to maintain the elevated levels necessary for source rock development. This changing nutrient pool results in an increase in the relative importance of autochthonous oil-prone material with increasing lake maturity prior to the lake fill stage.

A further control on the stratigraphic distribution of the Pematang source appears to be climatic. Although central Sumatra was in a tropical setting during the period of interest there is both isotopic evidence (Williams et al., 1985) and modeling results (Katz, 1991) that support temporal changes in climatic conditions. The most significant of these changes appear to be associated with the hydrologic cycle. These changes have an impact on both the size and the depth of the lake, as well as on the nature of the terrestrial input into the lake.

Model results indicate a strongly positive water-balance in the Sumatra region during the early and middle Eocene. Such conditions would favor the development and maintenance of large perennial lakes, which lack large seasonal lake level fluctuations. By early Oligocene time, there is evidence for strong seasonality in the relationship between precipitation and evaporation, with the development of well-defined wet and dry seasons. The rainy season occurred during the austral winter, with the dry season occurring during the austral summer. This seasonality could have resulted in significant changes in lake level. Such changes in lake level would spatially restrict the perennial lake and decrease the system's ability to preserve organic matter by increasing the potential for oxygen renewal.

Although sample spacing in this dataset is insufficient to establish high frequency changes in the Brown Shale, evidence presented by Longley et al. (1990) also infers the presence of high frequency climatic oscillations probably as a result of Milankovitch forcing factors. Within the tropics, these high frequency climatic changes would also have the greatest impact on the relative humidity and aridity. Although geologically short-lived, these high frequency climatic cycles may have substantial impact on lake level even in tropical settings. Scholz and Rosendahl (1990) have noted that two east African tropical rift lakes have experienced lake level fluctuations in excess of 350 meters. Such high frequency climate changes could have as dramatic impact on the net source rock footage as the previously noted long-term climate trends.

In addition to the observed stratigraphic variability in the Brown Shale, there is also indirect evidence of lateral (i.e., interbasinal) variability. Although the Brown Shale is the only thermally mature oil-prone source rock in central Sumatra (Katz and Kahle, 1988; and Katz and Mertani, 1989), there is clear evidence that multiple oil families exist within the basin (Seifert and Moldowan, 1980).

The different oil families appear to be geographically restricted (Figure 14) and in most cases can be related to the different subbasins. It, therefore, can be assumed that these individual subbasins probably contained differing algal populations. These different populations could be the result of differences in water chemistry, turbidity, and circulation patterns.

SAMPLING SCHEMES

Petroleum exploration utilizes geochemistry to estimate volumes and product types and to establish migration patterns. Accurate assessments of these factors can only be obtained if the proper samples are available. It is important to note that the quality of the derived data can be no better than the quality of the samples collected (Matthews and Leimer, 1985). The acquisition of representative samples can only occur if the purpose of study, the potential variability, and

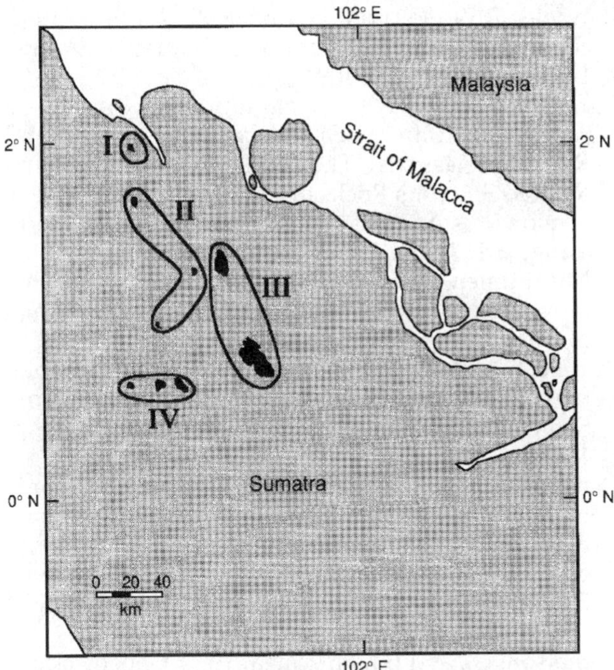

Figure 14. Distribution of the four primary central Sumatran oil families derived from the Brown Shale (after Seifert and Moldowan, 1980).

the level of detail required are known before the sampling scheme is designed.

Because of the stratigraphic and lateral variability observed in source rocks, sampling programs need to incorporate both randomness (to avoid sampling bias) and an understanding of the geologic character of the system. Geologic oddities should be lightly sampled because they are not representative of the unit under examination. Although there may be an apparent contradiction between random sampling and the incorporation of geologic information into sampling program design, the reality is that certain lithofacies have been universally shown not to be hydrocarbon sources. Sampling for source rock analysis of these lithofacies would be nonproductive. Therefore, although a purely random sampling would suggest that sandstones and other high energy lithofacies should be sampled, such analyses would not provide useful information regarding the source rock generation potential of a basin.

When attempting to establish the stratigraphic distribution and variability of hydrocarbon source rocks in either the subsurface or in outcrop, there are several approaches to sampling which may be utilized. One of these is sampling by geologic unit or formation, where a single sample is obtained and is considered representative of the entire unit under study. As noted previously, this sampling approach does not provide an adequate assessment of either source potential or organic facies variability. It is quite possible that if such a sampling program is utilized, the significance of the source may be overestimated or underestimated.

A composite or channel sampling approach may be utilized, either in the subsurface or in outcrop. This provides a more representative overview of the source potential of a given unit or interval. However, because of the nature of the sample, the better source intervals may be effectively diluted, and stratigraphically limited sources capable of generating significant quantities of hydrocarbons can be overlooked. The composite sampling approach is most appropriate when one is attempting to correlate an oil to a specific source because oils represent a stratigraphically and areally integrated product.

Two other discrete sampling approaches are: (1) random sampling with the distance between samples being established following an examination of bed thickness, and (2) a random sampling scheme supplemented by additional sampling based on observed geologic/lithologic variability. These approaches appear to be better suited to establish the distribution and variability of organic matter within a source rock sequence. The data obtained from such a sampling program, when integrated over the stratigraphic thickness of a unit, provide a means of obtaining a reliable estimate of a unit's hydrocarbon generation potential.

However, obtaining a reliable estimate of the mean effective organic richness or generative potential and net thickness of a source is not sufficient to accurately describe the effectiveness of a source. The actual stratigraphic distribution of organic matter with respect to potential carrier beds is also critical because of its relationship to hydrocarbon expulsion efficiency. Thick source units, independent of richness, tend to have lower effective expulsion efficiencies than do thinner source horizons interbedded with carrier beds (Leythaeuser et al., 1984; 1987).

Consequently, the distribution of organic matter plays a key role in establishing effective source thicknesses. This is in contrast to the recently suggested basin classification scheme of Demaison and Huizinga (1991) which is based on an average generation potential over the entire source unit thickness.

CONCLUSIONS

Source rock variability can be observed at several different scales ranging from millimeters to tens or hundreds of meters. The observed variability occurs in both the abundance and type of organic matter present. This variability in the nature and distribution of organic matter can impact the conclusions of various types of geochemical investigations including estimates of available hydrocarbons, the relative abundance of oil and gas, and the waxiness of crude oils, as well as oil-to-source rock correlations. These variations can be related to a number of factors including sea level, climate, basin circulation, and proximity to source.

As a consequence of this variability, sampling programs need to be designed based on program goals. Discrete sampling is probably appropriate for source rock assessment. Such a sampling program should be random, but sampling frequency should be established based on observed lithologic variability. The integration of these samples into a lithostratigraphic framework also permits an understanding of the potential effectiveness of hydrocarbon expulsion. Composite sampling appears to be more appropriate for oil-source rock correlations. This type of sampling allows for the integration of hydrocarbon products in a manner similar to that which occurs within a basin during generation, migration, and accumulation.

ACKNOWLEDGMENTS

This chapter was originally presented as part of the AAPG/SEPM Symposium on Source Rocks in a Sequence Stratigraphic Framework, held at the 1991 Annual AAPG Convention. Samples from the Monterey Formation were obtained in conjunction with CMOGS (Cooperative Monterey Organic Geochemistry Study). Participants in CMOGS include J. Rullkötter, B. J. Katz, R. A. Royle, N. Telnaes, T. Hanesand, J. W. de Leeuw, S. Schouten, M. D. Lewan, J. D. King, P. G. Lillis, R. E. Summons, D. M. Jarvie, E. Michael, D. K. Baskin, M. Schoell, A. Y. Huc, J. Martigny, S. Berlin, P. Sundararaman, G. D. Abbott, G. A. D. Law, B. Bennett, S. G. Talukdar, A. Schimmelmann, L. ten Haven, J. Curiale, B. Bromley, W. A. Michaelis, H. H. Richnow, W. L. Orr, A. Waseda, M. R. B. Loureiro, L. L. Lundell, T. P. Goldstein, C. E. Reimers, and D. Zaback. CMOGS also benefited from geologic data and discussions provided by C. M. Isaacs, J. H. Tomson, D. Z. Piper, R. M. Pollastro, J. A. Barron, R. G. Arends, M. L. Cotton, M. V. Filewicz, T. R. Baumgartner, B. P. Flower, and J. P. Kennett. Organizational leadership for CMOGS was provided by C. M. Isaacs and J. Rullkötter. The authors would like to thank Texaco Inc. for permission to publish. Assistance with the preparation of this manuscript by Mary Hill is appreciated. Dr. Roger Sassen and Dr. George Grabowski, Jr. provided useful critiques of an earlier version of this paper.

REFERENCES CITED

Bissada, K.K., 1982, Geochemical constraints on petroleum generation and migration—a review: Proceedings ASCOPE '81, p. 69–87.

Bloesch, J., P. Stadelman, and H. Buhrer, 1977, Primary production, mineralization, and sedimentation in the euphotic zone of two Swiss lakes: Limnology and Oceanography, 22, p. 511–526.

Brassell, S.C., A.M.K. Wardroper, I.D. Thompson, J.R. Maxwell, and G. Eglinton, 1981, Specific acyclic isoprenoids as biological markers of methanogenic bacteria in marine sediments: Nature, 290 , p. 683–696.

Canfield, R.W., G. Bonilla, and R.K. Robbins, 1982, Sacha oil field of Ecuadorian Oriente: AAPG Bulletin, 66, p. 1076–1090.

Cannon, D.E., 1981, Log evaluation of a fractured reservoir Monterey Shale, *in* R.E. Garrison and R.G. Douglas, eds., The Monterey Formation and Related Siliceous Rocks of California: SEPM Pacific Section (Los Angeles), p. 249–255.

Colling, E.L., B.H. Burda, and P.A. Kelley, 1986, Multidimensional pyrolysis-gas chromatography: Applications in petroleum geochemistry: Journal of Chromatographic Science, 24, p. 7–12.

Curiale, J. A., and J.R. Odermatt, 1989, Short-term biomarker variability in the Monterey Formation, Santa Maria basin: Organic Geochemistry, 14, p. 1–13.

Dean, W.E., 1981, Carbonate minerals and organic matter in sediments of modern north temperate hard-water lakes *in* F.G. Ethride, and R.M. Flores, eds., Recent and Ancient Nonmarine Depositional Environments—Models for Exploration: SEPM (Tulsa), Special Publication 31, p. 213–231.

Demaison, G., and B.J. Huizinga, 1991, Genetic classification of petroleum systems: AAPG Bulletin, 75, p. 1626–1643.

Dolton, G.L., K.H. Carlson, R.R. Charpentier, A.B. Coury, R.A. Crovelli, S.E. Frezon, A.S. Khan, J.H. Lister, R.H. McMullin, R.S. Pike, R.B. Powers, E.W. Scott, and K.L. Varnes, 1981, Estimates of undiscovered recoverable conventional resources of oil and gas in the United States: U. S. Geological Survey Circular 860, 87pp.

Espitalié, J., J.L. Laporte, M. Madec, F. Marquis, P. Leplat, J. Poulet, and A. Boutefeu, 1977, Méthode rapide de caractérisation des roches mères de leur potential pétrolier et de leur degré d'évolution. Revue de l'Institute Francais du Pétrole, 32 p. 23–42.

Feininger, T., 1975, Origin of petroleum in the Oriente of Ecuador: AAPG Bulletin, 59, p. 1166–1175.

Govean, F.M., and R.E. Garrison, 1981, Significance of laminated and massive diatomites in the upper part of the Monterey Formation, California, *in* R.E. Garrison, R.G. Douglas, K.E. Pisciotto, C.M. Isaacs, and J.C. Ingle, eds., The Monterey Formation and Related Siliceous Rocks of California: Pacific Section Society of Economic Paleontologists and Mineralogists (Los Angeles), p. 181–198.

Hayes, J.M., K.H. Freeman, B.N. Popp, C.H. Hoham, 1990, Compound-specific isotopic analyses: A novel tool for reconstruction of ancient biogeochemical processes: Organic Geochemistry, 16, p. 1115–1128.

Horsfield, B., 1989, Practical criteria for classifying kerogens: Some observations from pyrolysis of kerogen. Geochimica et Cosmochimica Acta, 44, p. 1119–1131.

Isaacs, C.M., 1987, Sources and deposition of organic matter in the Monterey Formation, south-central coastal basins of California, *in* R.F. Meyer, ed., Exploration for Heavy Crude Oil and Natural Bitumen: AAPG Studies in Geology 25, p. 193–205.

Isaacs, C.M., J.H. Tomson, M.D. Lewan, R.G. Arends, M.L. Cotton, and M.V. Filewicz, 1992, Preliminary correlation and age of rock samples in the Cooperative Monterey Organic Geochemistry Study, Santa Maria and Santa Barbara-Ventura basins, California: USGS Open File Report. 92-539-D, 29 pp.

Katz, B.J., 1983, Limitations of "Rock-Eval" pyrolysis for typing organic matter: Organic Geochemistry, 4, p. 195–199.

Katz, B.J., 1991, Controls on lacustrine source rock development: A model for Indonesia: Proceedings 20th Indonesian Petroleum Association Convention 1, p. 587–619.

Katz, B.J., and L.W. Elrod, 1983, Organic geochemistry of DSDP Site 467, offshore California, Middle Miocene to Lower Pliocene strata: Geochimica et Cosmochimica Acta, 47, p. 389–386.

Katz, B.J., and G.M. Kahle, 1988, Basin evaluation: A supply-side approach to resource assessment: Proceedings 17th Indonesian Petroleum Association Convention, 1, p. 135–168.

Katz, B.J., and B. Mertani, 1989, Central Sumatra—A geochemical paradox. Proceedings 18th Indonesian Petroleum Association Convention, 1, p. 403–425.

Kummel, B., Jr., 1948. Geological reconnaissance of the Contamana region, Peru. Bulletin of the Geological Society of America, 59, p. 1217-1266.

Lambiase, J.J., 1990, A model for tectonic control of lacustrine stratigraphic sequences in continental rift basins, *in* B.J. Katz, ed., Lacustrine Basin Exploration—Case Studies and Modern Analogs: American Association of Petroleum Geologists (Tulsa), Memoir 50, p. 265–286.

Larter, S.R., and A.G. Douglas, 1980, A pyrolysis-gas chromatographic method for kerogen typing, *in* A.G. Douglas and J.R. Maxwell, eds., Advances in Organic Geochemistry, 1979: Pergamon Press (Oxford), p. 579–584.

Leventhal, J.S., 1983, An interpretation of carbon and sulfur relationships in Black Sea sediments as indicators of environments of deposition: Geochimica et Cosmochimica Acta, 47, p. 133–137.

Leythaeuser, D., A. Mackenzie, R.G. Schaefer, and M. Bjorøy, 1984, A novel approach for recognition and quantification of hydrocarbon migration effects in shale-sandstone sequences: AAPG Bulletin, 68, p. 196–219.

Leythaeuser, D., R.G. Schaefer, and M. Radke, 1987, On the primary migration of petroleum: Proceedings 12th World Petroleum Congress, 2, p. 227–236.

Longley, I.M., R. Barraclough, M.A. Bridden, and S. Brown, 1990, Pematang lacustrine petroleum source rocks from the Malacca Strait PSC, central Sumatra, Indonesia: Proceedings 19th Indonesian Petroleum Association Convention, 1, p. 279–297.

Macellari, C.E., 1988, Cretaceous paleogeography and depositional cycles of western South America: Journal of South American Earth Sciences, 1, p. 373–418.

Macko, S.A., and R.S. Quick, 1986, A geochemical study of oil migration at source rock reservoir contacts: Stable isotopes: Organic Geochemistry, 10, p. 199–205.

Matthews, R.D., and H.W. Leimer, 1985, A review of geological sampling methods: Proceedings 1985 Eastern Oil Shale Symposium, p. 283–290.

Moldowan, J.M., 1984, C_{30}-steranes, novel markers for marine petroleums and sedimentary rocks: Geochimica et Cosmochimica Acta, 48, p. 2767–2768.

Nolte, D.G., 1991, Separation of a mixture of normal paraffins, branched chain paraffins, and cyclic paraffins. United States Patent 4,982,052.

Nolte, D.G., and E.L. Colling, Jr., 1989, Separation of oil into fractions of asphaltenes, resins, aromatics, and saturated hydrocarbons. United States Patent 4,865,741.

Obradovich, J.D., and C.W. Naeser, 1981, Geochronology bearing on the age of the Monterey Formation and siliceous rocks in California, *in* R.E. Garrison, R.G. Douglas, K.A. Pisciotto, C.A. Issacs, and J.C. Ingle, eds., The Monterey Formation and Related Siliceous Rocks of California: Pacific Section SEPM, p. 87–95.

Pisciotto, K.A., and R.E. Garrison, 1981, Lithofacies and depositional environments of the Monterey Formation, *in* R.E. Garrison, R.G. Douglas, K.A. Pisciotto, C.A. Issacs, and J.C. Ingle, eds., The Monterey Formation and Related Siliceous Rocks of California, Pacific Section SEPM, p. 97–122.

Schlische, R.W., and P.E. Olsen, 1990, Quantitative filling models for continental extensional basins with applications to early Mesozoic rifts of eastern North America: Journal of Geology, 98, p. 135–155.

Schoell, M., M.A. McCaffrey, F.J. Fago, and J.M. Moldowan, 1992, Carbon isotopic compositions of 28,30-bisnorhopanes and other biological markers in a Monterey crude oil: Geochimica et Cosmochimica Acta, 56, p. 1391–1399.

Scholz, C.A., and B.R. Rosendahl, 1990, Coarse-clastic facies and stratigraphic sequence models from Lakes Malawi and Tanganyika, east Africa: *in* B.J. Katz, ed., Lacustrine Basin Exploration—Case Studies and Modern Analogs: AAPG (Tulsa), Memoir 50, p. 151–168.

Seifert, W.K., and J.M. Moldowan, 1980, Paleoreconstruction by biological markers: Proceedings 9th Indonesian Petroleum Association Convention, p. 189–212.

Seifert, W.K., J.M. Moldowan, G.W. Smith, and E.V. Whitehead, 1978, First proof of structure of a C_{28}-pentacyclic triterpane in petroleum: Nature 271, p. 456–457.

Simoneit, B.R.T., J.O. Grimalt, T.G. Wang, R.E. Cox, P.G. Hatcher, and A. Nissenbaum, 1986, Cyclic terpenoids of contemporary resinous plant detritus and fossil woods, ambers and coals: Organic Geochemistry, 10, p. 877–889.

Tissot, B., B. Durand, J. Espitalié, and A. Combaz, 1974, Influence of nature and diagenesis of organic matter in formation of petroleum. AAPG Bulletin, 58, p. 499-506.

Tornabene, T.G., T.A. Langworthy, G. Holzer, and J. Oró, 1979, Squalenes, phytanes and other isoprenoids as major neutral lipids of methanogenic and thermoacidic "archaebacteria": Journal of Molecular Evolution, 13, p. 73–83.

Watson, M.P., A.B. Hayward, D.N. Parkinson, and Z.M. Zhang, 1987, Plate tectonic history, basin development and petroleum source rock deposition onshore China: Marine and Petroleum Geology, 4, p. 205–225.

Wenger, L.M., and D.R. Baker, 1986, Variations in organic geochemistry of anoxic-oxic black shale-carbonate sequences in the Pennsylvanian of the midcontinent, U. S. A.: Organic Geochemistry 10, p. 85–92.

Williams, H.H., P.A. Kelley, J.S. Janks, and R.M. Christensen, 1985, The Paleogene rift basin source rocks of central Sumatra: Proceedings 14th Indonesian Petroleum Association Convention, 2, p. 57–90.

Chapter 3

Geochemical and Micropaleontological Characterization of Lacustrine and Marine Hypersaline Environments from Brazilian Sedimentary Basins

Márcio R. Mello
Eduardo A. M. Koutsoukos
Eugênio V. Santos Neto
Augusto C. Silva Telles, Jr.
Petrobrás-Cenpes/Divex
Rio de Janeiro, Brazil

ABSTRACT

Geological studies integrated with paleogeographical, geochemical, and micropaleontological evidence, suggest that part of the organic-rich Permian and Lower Cretaceous to middle Albian successions in the Brazilian sedimentary basins were deposited in marine and lacustrine to marine hypersaline environments. Biological marker and micropaleontological data can be used to characterize four distinct hypersaline systems:

1. a Late Permian epicontinental hypersaline system (Irati Formation, Paraná basin), composed of calcareous black shales that contain abundant lipid-rich organic matter (total organic carbon [TOC] up to 24%);
2. an Early Cretaceous, shallow lacustrine, hypersaline system composed of thick, organic-rich, calcareous black shales (TOC up to 9%) containing lipid-rich organic matter;
3. an Aptian, shallow marine evaporitic system (proto-oceanic gulf) mainly composed of marlstones, carbonates, and calcareous black shales containing abundant lipid-rich amorphous organic matter (TOC up to 14%); and
4. a late Aptian-middle Albian, marine semi-restricted, dominantly neritic, hypersaline system, composed of marlstones and calcareous black shales containing abundant lipid-rich algal organic matter (TOC up to 12%).

The main molecular features diagnostic of these environments are: phytane more abundant than pristane, low abundances of diasteranes and tricyclic terpanes, high abundances of gammacerane, $\alpha\beta$-hopanes, β-carotane, regular C_{27} to C_{29} steranes, regular C_{25} isoprenoid and squalane, Ts<Tm, and C_{35} $\alpha\beta$-hopanes more abundant than C_{34} $\alpha\beta$-hopanes.

The micropaleontological features that characterize these systems are the occurrence of abundant low-diversity communities representing opportunistic species, and common occurrence of millimeter-thick strata with census populations, representing mass mortality events during short-term environmental turnovers.

INTRODUCTION

High organic productivity associated with excellent preservational conditions make hypersaline environments of great importance for the petroleum industry. Prolific source rocks derived from this type of environment have been recorded from upper Precambrian to upper Miocene strata (Peterson and Hite, 1969; Friedman, 1980; Kirkland and Evans, 1980; ten Haven et al., 1985, 1988; Fu Jiamo et al., 1986; Mello et al., 1988a, b; Philp et al., 1989; Fan Pu et al., 1990).

In the last few years, many authors have shown that the organic matter in hypersaline sedimentary rocks shows very similar geochemical, micropaleontological, and biological marker characteristics regardless of whether it is marine or lacustrine in origin (e.g., Mello et al., 1988a, b; Thomas, 1990; de Leeuw and Sinninghe Damsté, 1989; Fan Pu et al., 1990). These characteristics reflect the chemical and physical conditions of these environments. High salinity restricts the range of composition of the biomass (generally halophilic organisms), and hence, the occurrence of specific compounds, complicating the molecular differentiation of hypersaline systems (Kirkland and Evans, 1980; Albaigés et al., 1986; Fu Jiamo et al., 1986; ten Haven et al., 1985, 1988; Connan and Dessort, 1987; Mello et al., 1988a, b; Philp et al., 1989; Fan Pu et al., 1990; de Leeuw and Sinninghe Damsté, 1989; Koutsoukos et al., 1991a, b).

Recently, the availability of new analytical instrumentation (e.g., GC-MS/MS), affording more specific identification of individual hydrocarbon components, has allowed the assessment and differentiation of organic-rich sedimentary rocks and oils derived from lacustrine and marine hypersaline depositional settings. This study summarizes a multidisciplinary geochemical and micropaleontological approach used to assess and characterize hypersaline depositional environments containing organic-rich rocks in Paleozoic and Mesozoic Brazilian sedimentary basins.

In the present study, rock samples recovered from sedimentary successions from the Paraná, Sergipe, Ceará, and Potiguar basins (Figure 1), ranging from Permian to middle Albian in age, were analyzed using geochemical and micropaleontological techniques. As a result, four distinct organic-rich hypersaline depositional systems were identified (Table 1 and Figure 1).

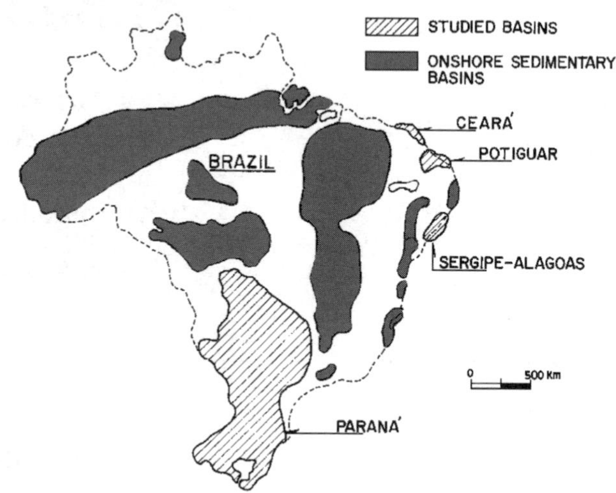

Figure 1. Location map of the four studied Brazilian sedimentary basins.

ANALYTICAL PROCEDURES

The analytical procedures for the micropaleontological analyses are described elsewhere (e.g., Koutsoukos et al., 1991a; Silva-Telles and Viana, 1990).

The oil and rock samples were submitted to elemental, carbon isotope and liquid chromatography analyses according to routine methods (Mello, 1988).

The GC analysis of alkanes was carried out using a Hewlett Packard 5890A equipped with a splitless injector and fitted with a 30m DB-5 column. Hydrogen was employed as carrier gas with a temperature program of 40 to 80°C at 8°C/min and 80 to 320°C at 4°C/min. The GC-MS analyses were performed using an HP-5890 mass spectrometer coupled to a Hewlett Packard 5890A GC equipped with oncolumn injector, and fitted with a 25m SE-54 column. Helium was employed as a carrier gas with temperature program of 70-190°C at 30°C/min and 190 to 290°C, at 2°C/min. The spectrometer was operated in multiple ions detection (MID). The GC-MS/MS technique was performed using a triple-quadrupole Finnigan TSQ-70 instrument coupled to HP 5890A splitless gas chromatograph fitted with a DB-5 fused

Table 1. Geological and geochemical data for the selected hypersaline samples.

Sample	1	2	3	4
Type	Core	Cuttings	Cuttings	Core
Depth (m)	150	2346	1920	1138
Age	Late Permian	Early Cretaceous	Aptian	m. Albian l. Aptian
Lithology	Calcareous Black Shale	Calcareous Black Shale	Calcareous Black Shale	Marl
Depositional Environment	Marine Hypersaline	Lacustrine Hypersaline	Marine Evaporitic	Marine Carbonate
$CaCO_3$ (%)	27	23	20	73
TOC (%)	4.9	1.3	3.9	13.6
S_2 (Kg/ton rock)	28	3	16	106
HI (mg HC/g TOC)	579	168	409	780
T_{max} (°C)	420	438	433	427
R_o (%)	0.50	0.64	0.52	0.55
$\delta^{13}C$ (‰)	-24.30	-30.13	-26.00	-25.30
HOP/STE index*	7.0	2.2	1.7	0.6
Dinosteranes	absent	absent	present	present
C_{30} Steranes	present	absent	present	present

*Peak area of C_{30} 17α, 21β(H)-hopane (35) in m/z 191 chromatogram over sum of peak areas of C_{27} 20R and 20S 5α, 14α, 17α(H)-cholestane (8+10) in m/z 217 chromatogram.

silica column. Helium was the carrier gas and the temperature program used was 60 to 200°C at 15°C/min and 200 to 300°C/min at 2°C/min.

RESULTS AND DISCUSSION

A distinct advantage in the examination of the geochemical, biological marker, and micropaleontological characteristics of hypersaline petroleum source rocks is the availability of samples from a variety of hypersaline depositional environments whose geological features are well documented and described. The Brazilian Paleozoic and Mesozoic sedimentary basins provide an ideal opportunity for such investigations, since they contain a succession of rocks deposited in different hypersaline environments, all represented in this study (Figures 1, 2, and Table 1). Also, where assumptions have to be made from geological and paleontological studies, previous geochemical and micropaleontological data for samples from well-defined depositional environments provide a background for the present investigation.

Although some of the molecular parameters discussed herein are maturity-dependent (e.g., biological marker concentrations; Mello, 1988a, b), the samples chosen for this study cover a relatively narrow maturity range (% R_o from 0.50 to 0.64%; Table 1), allowing the various features to be ascribed exclusively to source input.

The integration of geological, geochemical, and micropaleontological data allowed the identification and characterization of four distinct organic-rich hypersaline depositional systems, which are discussed separately in the following sections.

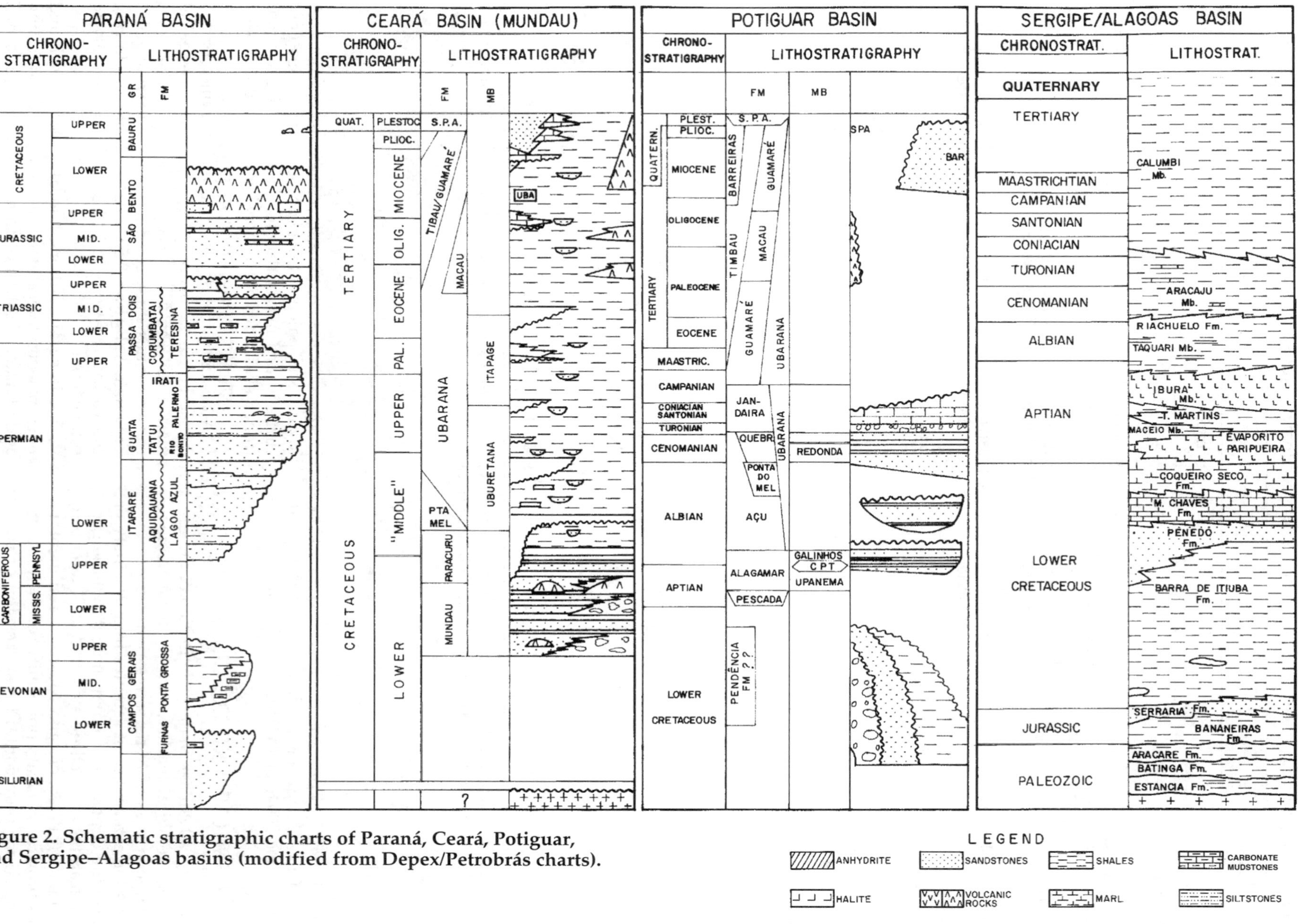

Figure 2. Schematic stratigraphic charts of Paraná, Ceará, Potiguar, and Sergipe–Alagoas basins (modified from Depex/Petrobrás charts).

Permian Epicontinental Hypersaline System

The deposits belonging to this group comprise interbedded terrigenous and carbonate rocks from the Irati Formation, Paraná basin, Brazil (Figures 1, 2). The Paraná basin is a large intracratonic Paleozoic basin, filled with up to 6000 m of Silurian to Cretaceous rocks and Upper Jurassic to Lower Cretaceous igneous rocks (Zalán et al., 1990). The Upper Permian bituminous Irati shales are the best hydrocarbon source rock in the basin (Figures 2, 3). This source rock generated subcommercial oil accumulations and seeps in the Paraná basin. Geochemical evidence suggests that the Irati shales overreached the oil window by the influence of Jurassic-Cretaceous igneous intrusives. The oil generation, although small, is always associated with igneous intrusives (Cerqueira and Santos Neto, 1986; Santos Neto and Cerqueira, 1990; Mello et al., 1991).

The Irati calcareous black shales are organic rich (average of 6 wt.% TOC) and show excellent hydrocarbon yield (around 30 kg HC/ton rock, from Rock-Eval pyrolysis; Figure 3 and Table 1). These features occur mostly in the eastern part of the basin, where these beds are up to 30 m thick (Cerqueira and Santos Neto, 1986; Santos Neto and Cerqueira, 1990).

Organic petrography shows that organic matter is composed of a mixture of amorphous and liptinitic kerogen showing an intense fluorescence under ultraviolet light (Trigüis, 1991). The high hydrogen index values (average 600 mg HC/g TOC; Figure 3 and Table 1) confirm the very good hydrocarbon yield characteristic of type I/II kerogen.

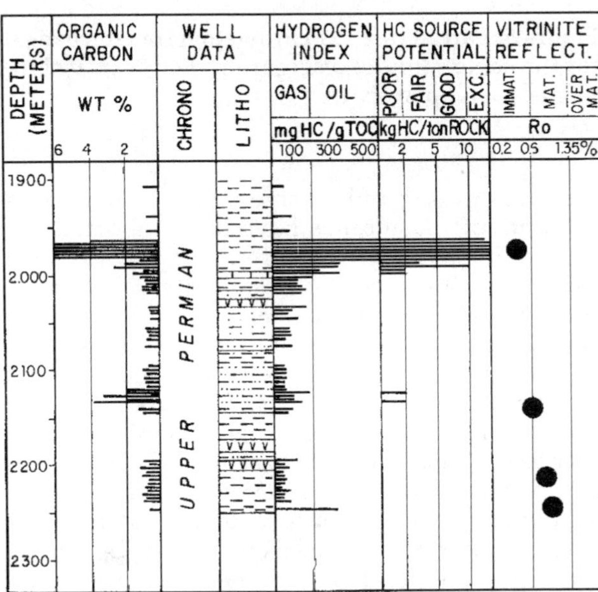

Figure 3. Geochemical log of the well 1-ES-2-RS showing the stratigraphic position of the Permian shallow marine organic-rich hypersaline deposits of the Irati Formation, Paraná basin (see Figure 2 for lithological interpretation).

The geochemical and molecular data suggest a good correlation between reservoired oils and organic extracts from the Irati calcareous black shales (cf. Cerqueira and Santos Neto, 1986; Santos Neto, 1990; Mello et al., 1991). It is worthy to mention that the organic extracts studied from the Irati Formation presented similar carbon isotope data (around –24‰; Table 1). Sample 1 in Figures 4 to 7 and Table 1 illustrates the biological markers from a typical rock extract of the organic-rich facies from the Irati Formation (see Appendix 1 for peak identification). The gas and mass chromatograms (CG, CG-MS, and GC-MS/MS) reveal features diagnostic of a dysoxic–anoxic hypersaline environment (e.g., Fu Jiamo et al., 1986; ten Haven et al., 1985, 1988; Mello, 1988; Mello et al., 1988a, b; Philp et al., 1989; de Leeuw and Sinninghe Damsté, 1989; Thomas, 1990). The most important features are the dominance of phytane over pristane (Pr<<Ph), high abundance of acyclic isoprenoids (mainly regular i-C_{25} and i-C_{30}), β–carotane, gammacerane (peak 40), regular C_{27} to C_{29} steranes (peaks 8–16), C_{30} αβ-hopane (peak 35), low relative abundance of diasteranes (see peaks 6 and 7), Ts lower than Tm, and C_{35} αβ hopanes in higher abundance than their C_{34} counterparts (peaks 45 and 44, respectively, in Figure 5). Such molecular features were found in organic extracts and oils derived from hypersaline environments described previously in Brazilian sedimentary basins (Mello, 1988; Mello et al., 1988a, b). Also some of these features have been reported for hypersaline environments in other basins around the world (cf. Shi Jiyang et al., 1982; Jiang and Fowler, 1986; Fu Jiamo et al., 1986; ten Haven et al., 1985, 1988; Wang Tieguan et al., 1988). It is interesting to note the high relative abundance of phytane, regular i-C_{25} and i-C_{30} long chain isoprenoids and gammacerane (sometimes the major peaks in the GC and m/z 191 fingerprints, respectively; e.g., Figures 4, 5). Such compounds have been reported to arise from an archaebacterial population, mainly halophilic, and therefore, their abundances might be expected to increase with an increase in salinity (Waples et al., 1974; ten Haven et al., 1987, 1988; Mello et al., 1988a). Extremely high abundances of these compounds have been reported for samples deposited in hypersaline conditions (Fu Jiamo et al., 1986; ten Haven et al., 1988; Wang Tieguan et al., 1988; Mello et al., 1988a, b; de Leeuw and Sinninghe Damsté, 1989). Also, low relative abundance of diasteranes (peaks 6 and 7 in Figure 6) is in keeping with the idea of hypersaline conditions during the deposition of the bituminous Irati black shales. Since the reported condition for diasteranes formation is acid catalysis from clay minerals (e.g., Rubinstein et al., 1975), the lack of or low concentrations of such compounds, in the samples analyzed, suggest either the lack of or a low input of clay minerals in the original environment of deposition. Hence, hypersaline-derived organic extracts might be expected to contain lower amounts of diasteranes than those derived from clay-rich depositional environments (e.g., lacus-

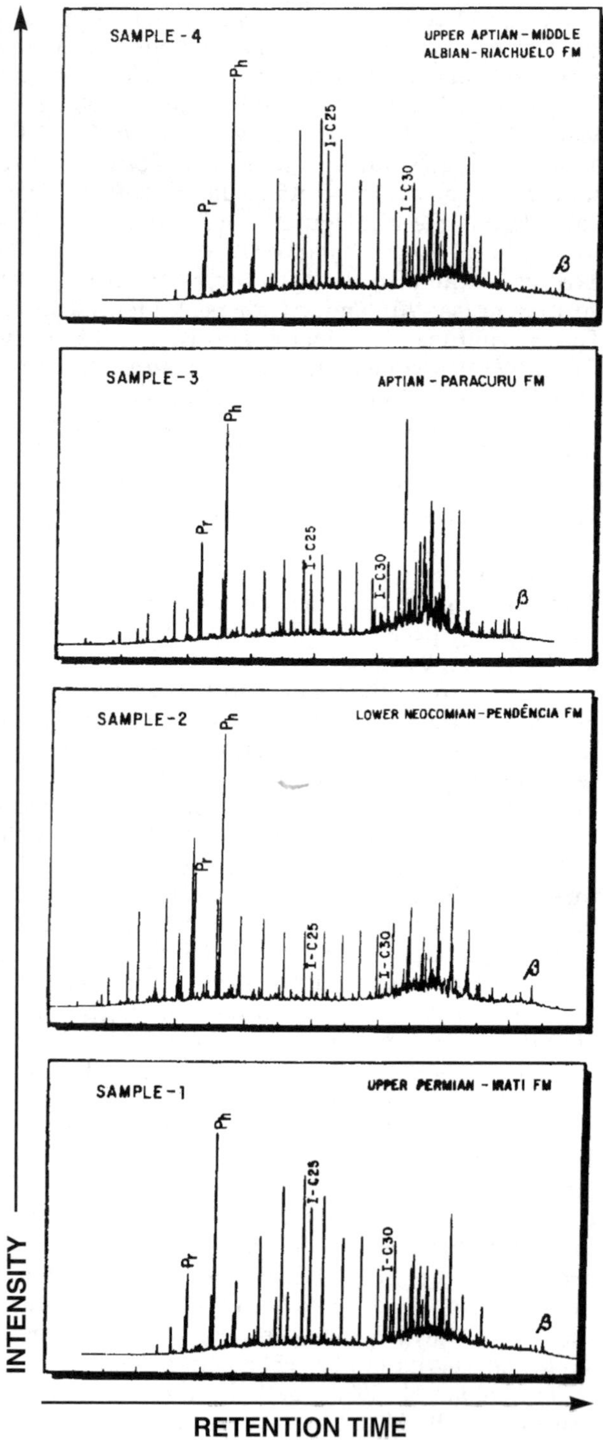

Figure 4. Gas chromatograms of total alkanes for typical hypersaline organic extract samples (1) shallow epicontinental marine, Irati Fm.; (2) shallow lacustrine, part of the Pendência Fm.; (3) shallow marine evaporitic, part of the Paracuru Fm.; and (4) neritic marine carbonate, Riachuelo Fm. (For peak assignments see Appendix 1.)

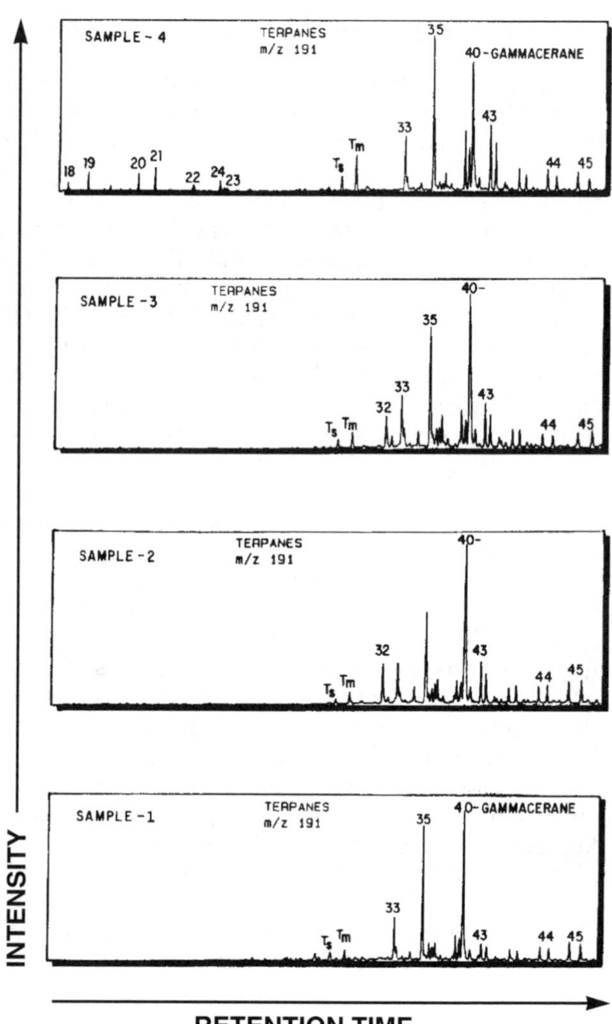

Figure 5. Partial m/z 191 chromatograms (terpanes) for the same samples as Figure 4. (For peak assignments see Appendix 1.)

trine freshwater and marine deltaic). Indeed, low amounts of diasteranes have been reported from hypersaline environments around the world (cf. Fu Jiamo et al., 1986; ten Haven et al., 1988; Connan and Dessort, 1987; Mello et al., 1988a, b; Philp et al., 1989). Such evidence, together with the occurrence of biological markers diagnostic of hypersaline water conditions (see above), suggest for part of the Irati Formation an environment that generally occurs in restricted basins with arid to semi-arid climates (low precipitation/intense evaporation). Other characteristics of the Irati bituminous section are the dominance of C_{29} steranes relative to their C_{27} and C_{28} counterparts (peaks 14–16, 8–10, and 11–13, respectively, in Figure 6); high hopane/sterane ratio (ranging from 4 to 10), low relative abundance of tricyclic terpanes (peaks 18–23), pregnanes and homopregnanes (e.g., peaks 1–5), the absence of 28,30-bisnorhopane and Ts<Tm (Figures 4–6). The values of the hopane/sterane ratio (ranging from 4 to 10; Table 1) in the sam-

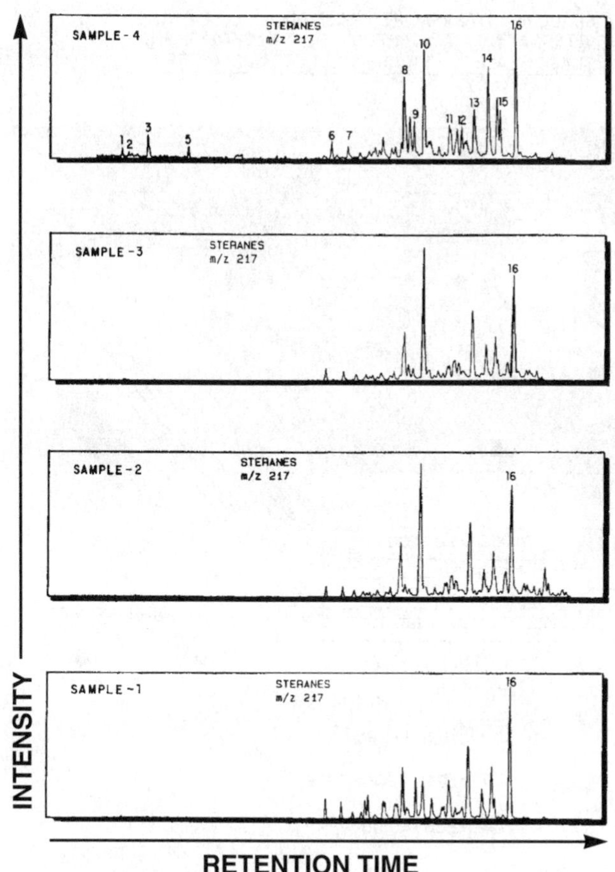

Figure 6. Partial m/z 217 chromatograms (steranes) for the same samples as Figure 4. (For peak assignments see Appendix 1.)

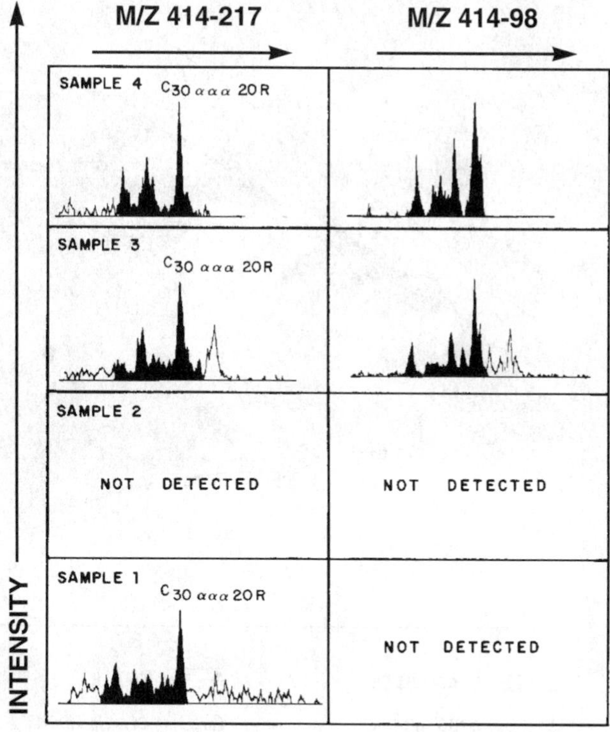

Figure 7. GC-MS/MS data showing the C_{30} steranes (24-*n*-propylcholestanes; m/z 414-217) and dinosteranes (m/z 414-98) distributions for the same samples as Figure 4.

ples of this group are not typical of the reported marine-derived samples (usually <4; Mello et al., 1988a, b). An explanation could be very high concentrations of bacterially derived hopanoids in the Irati kerogen compared to medium to high populations of halotolerant algal species using sterols as rigidifiers and protectors of cell wall materials (cf. Mello et al., 1988a). The presence of C_{35} αβ hopanes in higher abundance than their C_{34} counterparts has been observed in oils and rock extracts derived from hypersaline paleoenvironments (cf. Fu Jiamo et al., 1986; ten Haven et al., 1985, 1988; Mello et al., 1988a, b; de Leeuw and Sinninghe Damsté, 1989). The C_{35}/C_{34} αβ hopanes >1 (peaks 45 and 44, respectively, in Figure 5) in the samples studied support the idea of hypersaline conditions in the environment of deposition of the Irati Formation. Also, the Ts/Tm ratios, typically <1 (Figure 5) may reflect a specific hypersaline source input or mineral matrix (cf. Mello, 1988; Mello et al., 1988a, b; Philp et al., 1989). The low relative abundance of C_{20} to C_{29} tricyclic terpanes (peaks 18–23 and 25–26) appears also to be an indicator of hypersaline conditions, since it has been reported that their precursors are suppressed by hypersaline water conditions (Mello, 1988; De Grande, 1991). It is also important to note the presence, although in low abundance, of the C_{30} steranes (24-*n*-propylcholestanes obtained using GC-MS/MS; Figure 7 and Table 1). Such compounds, believed to be marine source indicators, are derived from *Chrysophyta* marine algae (e.g., Moldowan et al., 1985; Summons et al., 1987; Goodwin et al., 1988; Moldowan et al., 1990). The low abundance of these biological markers in the Irati black shales perhaps reflects a restricted depositional environment as opposed to open marine conditions (e.g., Wolf and Silva, 1974; Oelofsen and Araújo, 1983; Thomas, 1990).

Dinosteranes, considered to be diagnostic of marine environments (e.g., Volkman et al., 1990), are absent in all the analyzed samples (Upper Permian) (Table 1). This appears to be a result of the appearance of dinosterane precursors (marine dinoflagellates), only after Rhaetian times (Upper Triassic) (Summons et al., 1987; Thomas, 1990).

The fossil biota in the Irati black shales is characterized by some organisms considered to be diagnostic of a semi-restricted, shallow marine environment. The most important are: low diversity foraminiferal assemblages composed of agglutinated species (foraminifera that, for shell construction, agglutinate sediment grains to compensate for poor conditions of calcium carbonate bioprecipitation, e.g., *Ammodiscus*

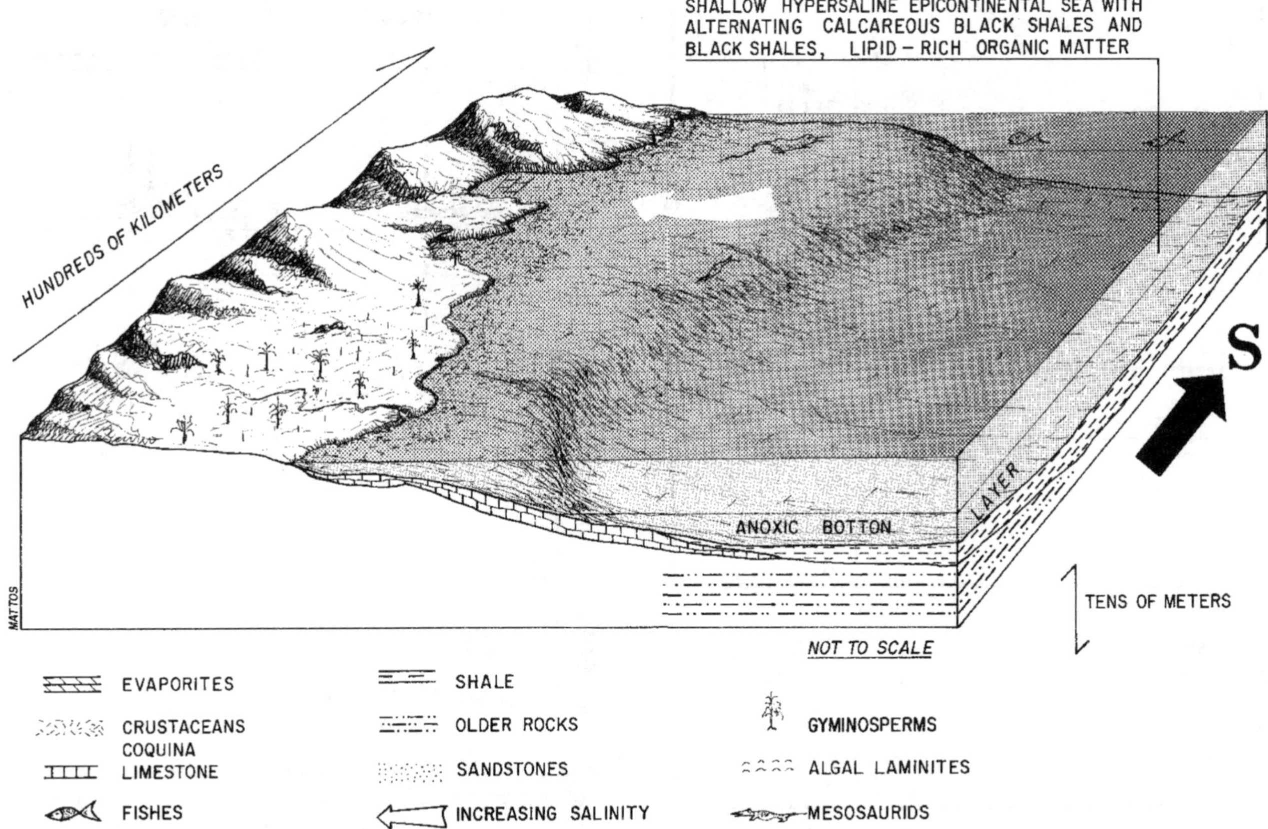

Figure 8. Schematic block diagram showing the sedimentary facies in a shallow epicontinental hypersaline sea setting envisaged for the Upper Permian system (Irati Fm.) of the Paraná basin (see Figure 2 for lithological interpretation).

spp. and *Sorosphaera* spp.) and calcareous specimens (*Fusulinina* spp.), brachiopods (*Lingula* spp.), sponge spicules, crinoids, and ostracods of the genus *Bairdia* (Amaral, 1971; Campanha, 1985; Campanha and Zaine, 1989).

The block diagram in Figure 8 depicts the interpreted environment in which the calcareous black shales and carbonate rocks of the Irati Formation were deposited. The environment appears to have been dominated by arid to semi-arid conditions with deposition of organic-poor, carbonate rocks, in relatively shallow and calm conditions (margins of an epicontinental sea). The organic-rich calcareous black shales were deposited in the deeper parts of the basin, under anoxic, hypersaline conditions, in density stratified water (e.g., below the halocline/pycnocline; Demaison and Moore, 1980). Such interpretation is supported by the presence of biomarkers diagnostic of hypersaline water conditions, deposition of nodular anhydrite, absence of benthic fossils, and lack of bioturbation. Seasonal laminations of shale/limestone with persistent lateral continuity are present. Such features indicate little water movement and lethal conditions (H_2S) for the benthonic fauna in the bottom water. Catastrophic events (storms) cause water overturns that promote and enhance the dysoxic–anoxic conditions of the environment, with mass mortality even in the reptile and fish communities that thrived in the oxygenated surface waters of the Irati sea. Evidence of such phenomena is reported in the literature (Della Fávera, 1987; Lavina et al., 1989).

The geochemical, paleontological, and sedimentological data suggest that the Irati Formation was deposited in a very large epicontinental sea with some continental influence along the margins and normal saline to hypersaline conditions within isolated gulfs. The lack of modern analogs complicates the depiction of such a geological scenario.

Lower Cretaceous Shallow Lacustrine Hypersaline System

The Lower Cretaceous organic-rich sedimentary rocks investigated in this system are confined to the Sergipe and Potiguar basins in the northern area of the Brazilian continental margin (Figures 1, 2, and Table 1).

The Lower Cretaceous, shallow lacustrine, hypersaline system identified in the Sergipe (parts of the Coqueiro Sêco Formation) and Potiguar basins (parts of the Pendência Formation; Figure 2) is composed of thick beds of organic-rich, calcareous black shales (TOC up to 6%), with moderate sulfur content (up to 0.6%). The hydrogen indices (up to 700 mg HC/g organic carbon) and organic petrography data identify the organic matter as being almost entirely composed of lipid-rich material (type I kerogen composed of amorphous and herbaceous material). The excellent hydrocarbon source potential (S_2 from Rock-Eval pyrolysis up to 50 kg HC/ton of rock), indicates that they can be good source rocks under ideal thermal maturity conditions (Table 1 and Figure 9).

The compositional and biological marker data of a typical organic extract from the Pendência hypersaline facies are illustrated by sample 2 (Table 1 and Figures 4–7). As can be observed, excepted by the dominance of C_{27} steranes over their C_{28} and C_{29} counterparts (peaks 8 to 16 in m/z 217), presence of 28, 30-bisnorhopane (peak 32 in m/z 191), and absence of C_{30} steranes (24-*n*-propylcholestanes obtained using GC-MS/MS; Figure 7 and Table 1), the sample from the Pendência Formation shows molecular features identical to those of samples of the Irati calcareous black shales. It is interesting to note also, the absence of dinosteranes compounds (4, 23, 24-trimethyl cholestanes; Figure 7 and Table 1) ascribed to be diagnostic of marine environment (e.g., Summons et al., 1987; Thomas, 1990). Such data are the only molecular evidence that suggest a lacustrine origin for the samples studied. The $\delta^{13}C$ values around –30‰ (Table 1) also suggest a lacustrine origin for the samples, since similar data have been extensively reported for lacustrine samples from

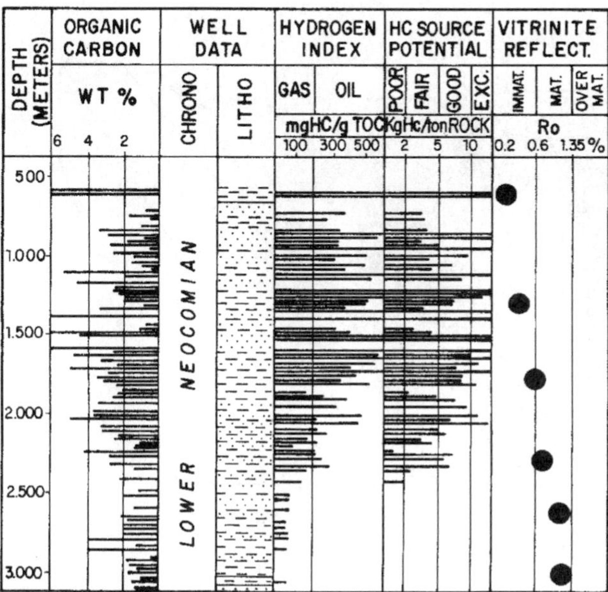

Figure 9. Geochemical log of the well 1-UPN-1-RN showing the stratigraphic position of the Lower Cretaceous shallow lacustrine organic-rich hypersaline deposits of the Pendência Formation, Potiguar basin (see Figure 2 for lithological interpretation).

Brazilian sedimentary basins (Mello, 1988; Mello et al., 1988a, b; Mello and Maxwell, 1991).

As observed for the Irati calcareous black shales, the molecular data presented by sample 2 in Figures 4 to 6, suggest also the presence of arid to semi-arid climates during the deposition of the calcareous black shale from the Pendência Formation (e.g., paucity of diasteranes and very high abundances of phytane, gammacerane and acyclic isoprenoids (regular *i*-C_{25} and *i*-C_{30}; Mello, 1988). The low values of the hopane/sterane ratio (<3; Table 1) in the hypersaline lacustrine samples of this group differ with the reported paucity of steranes in lacustrine environments (generally hopane/sterane ratio > 4; Mello et al., 1988a, b; Mello and Maxwell, 1991). One explanation could be that hypersaline water conditions increase the development of halophilic algal species that use sterols other than lipids as rigidifiers and protectors of cell membranes (cf. Mello et al., 1988a).

The ostracofauna of the Coqueiro Sêco Formation is very similar to that of the coquina sequence in the Lagoa Feia Formation (saline lacustrine system, Campos basin, southeastern Brazil; Bertani and Carozzi, 1985). In both formations, the microfaunal diversity is high, decreasing from the bottom to the top of the unit. *Cypridea* species are common (typical of low salinity paleoenvironments ranging from freshwater to oligohaline; Anderson et al., 1967). The occurrence of low diversity to monospecific communities, alternating with well diversified communities associated with *Cypridea* species, throughout most of the section could be related to alternating low and high salinity cycles, as found by Anderson et al. (1967) in the Purbeck/Wealden Formation of southern England. It is worthy to mention the good correlation between high salinity cycles and the very high abundances of phytane, gammacerane and regular *i*-C_{25} and *i*-C_{30} isoprenoids (biomarker compounds considered to be diagnostic of hypersaline conditions; e.g., Mello, 1988). During low salinity periods the abundance of these molecular compounds drops considerably, suggesting brackish to freshwater conditions (e.g., Mello and Maxwell, 1991). In spite of the high diversity assemblages and the presence of low-salinity species in the ostracofauna of the studied section, qualitative and quantitative fluctuations in the microfaunal parameters are observed, mainly an upward decreasing of diversity.

The Pendência Formation, Potiguar basin (Figures 1, 2) contains a low diversity, low abundant, and poorly preserved ostracofauna. Thick barren intervals preclude detailed studies of microfaunal evolution or population density. The Pendência Formation shows lower microfaunal species diversity than the Candeias Formation, Recôncavo basin, considered to be deposited in a typical lacustrine freshwater environment (e.g., Mello and Maxwell, 1991).

Palynological data of the Pendência Formation show high abundance of spores, suggesting dominance of a humid paleoclimate during its deposition. Perhaps the apparent conflict between geochemical and palynological data could be reconciled if there had been periodic alternations from humid to arid or semi-arid climates.

Because lakes are extremely dependent on climatic changes, significant expansions and contractions of these water bodies during short periods of the geological time are common. Such variations can provide diagnostic biological marker features that characterize not only hypersaline end members, but also saline to brackish end members (cf. Mello and Maxwell, 1991).

In summary, the most important biological marker features shown by sample 2 (Figures 4–7) that can be ascribed to deposition under lacustrine hypersaline conditions are: low pristane/phytane ratio, medium to high sulfur content, $\delta^{13}C$ values around –30‰, high abundances of β-carotane, regular i-C_{25} and i-C_{30} long chain isoprenoids and gammacerane, Ts/Tm<1, low hopane/sterane ratio (<3), low relative abundance of tricyclic terpanes, predominance of C_{35} hopanes over the C_{34} counterparts, and the absence of C_{30} steranes and dinosteranes (Figures 4–7, Table 1).

The block diagram in Figure 10 is an idealized setting for the Lower Cretaceous shallow lacustrine hypersaline system proposed for the Sergipe and Potiguar basins. The model assumes semi-arid to arid climates, with high evaporation, leading to water density stratification and permanent bottom-water dysoxia–anoxia. Such conditions, associated with high primary productivity in the well oxygenated surface waters, enhanced the degree of organic-matter production, accumulation, and preservation, resulting in deposition of laminated organic-rich rocks (Kelts, 1988; Talbot, 1988; De Dekker, 1988; Katz, 1991; Mello and Maxwell, 1991).

Few analogous examples of ancient shallow lacustrine hypersaline systems have been reported in the literature. The best comparisons to the Brazilian examples appear to be the well studied Eocene Green River Formation in the Uinta basin, USA (Reed, 1977; Tissot et al., 1978; Demaison and Moore, 1980; Dean and Fouch, 1983), and the Chaidamu and Jianghan basins in China (Chen Changming et al., 1984; Powell, 1986; Fu Jiamo et al., 1986).

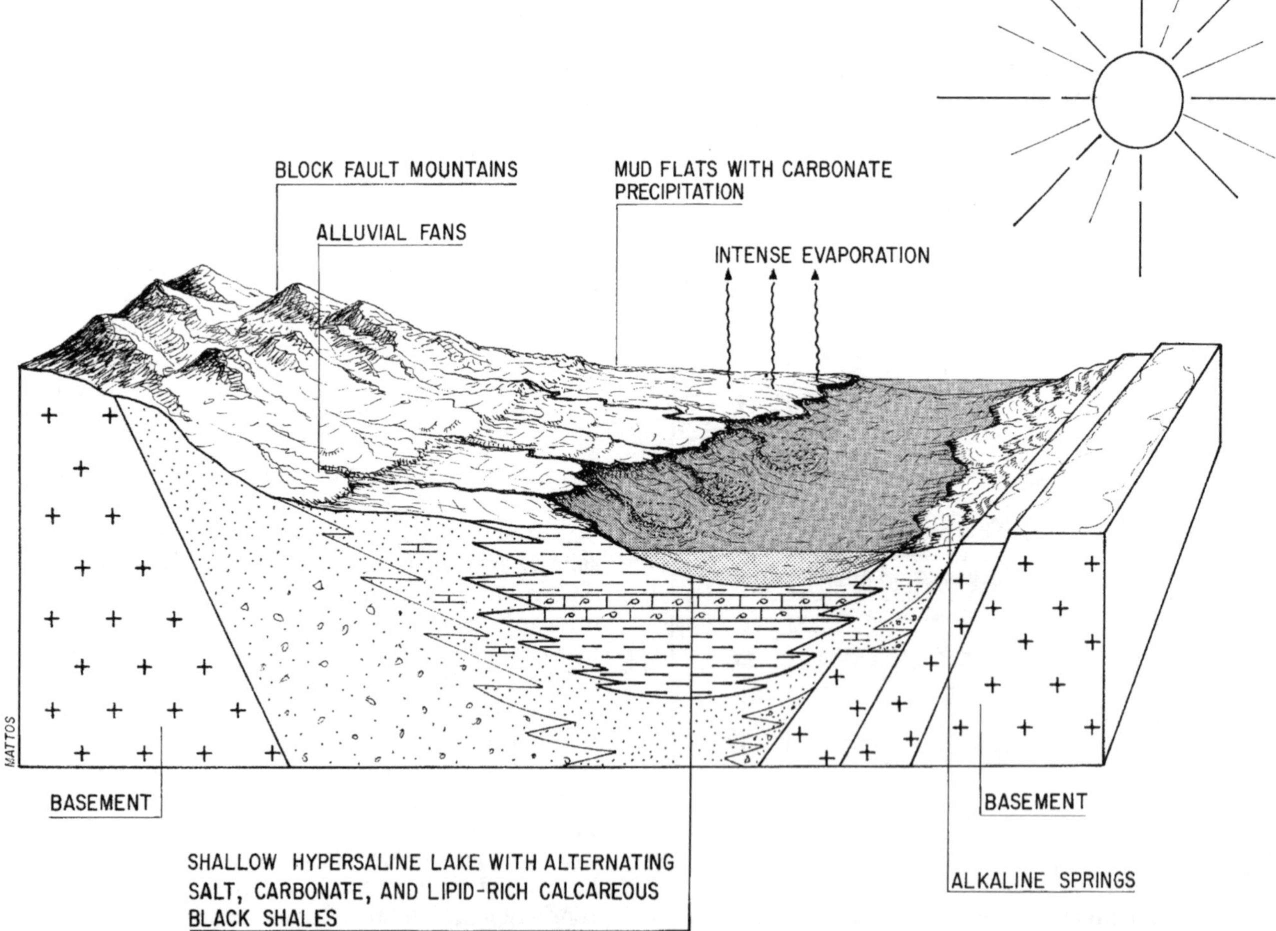

Figure 10. Schematic block diagram showing the sedimentary facies in a shallow hypersaline lake of the Rift phase envisaged for the Lower Cretaceous system (part of the Coqueiro Sêco and Pendência formations) from Sergipe and Potiguar basins, respectively (see Figure 2 for lithological interpretation).

Aptian Shallow Marine Evaporitic System

Organic-rich sedimentary rocks of this system were studied in the Potiguar (Alagamar Formation) and Ceará (Paracuru Formation) basins (Figures 1, 2, and Table 1). These rocks were deposited in the Aptian and are characterized by unusual geochemical, mineralogical, and micropaleontological data. They are composed mainly of calcareous black shales and marls, associated with evaporites. Generally, the lack of invertebrates in these depositional environments is due to the extreme high salinity of evaporitic brines. Such environments are so harsh that no normal marine microfauna (e.g., dinoflagellates, calcareous nannoplankton, or foraminifera) survive (e.g., Kendall, 1978; Friedman, 1980).

The sedimentary rocks studied are mainly composed of organic-rich (TOC up to 13%) calcareous black shales and marls ($CaCO_3$ up to 45%), usually rich in sulfur content (S ranging from 0.5 to 2.5%). Rock-Eval pyrolysis data and organic petrology indicate a predominance of type II kerogen (hydrogen indices up to 750 mg of HC/g TOC; e.g., Figure 11 and Table 1), mainly made up of a mixture of amorphous organic matter (45–60%) with herbaceous (15–25%) and woody plus coaly material (10–25%).

Sample 3 in Figures 4–7 shows a gas chromatogram and molecular data for a typical organic-rich rock extract from the Paracuru Formation, Ceará basin (see Figures 1, 2). The molecular features of this sample are identical to those observed in the sample of the Pendência Formation (except GC-MS/MS data in Figure 7), and very similar to the one from the Irati

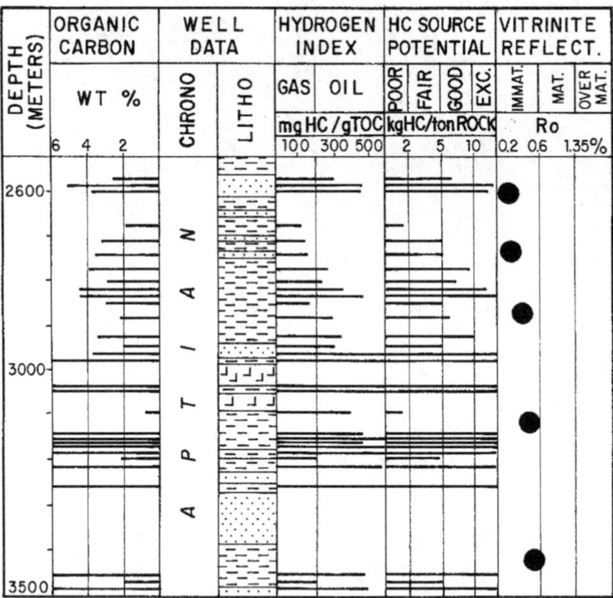

Figure 11. Geochemical log of the well 1-CES-42A showing the stratigraphic position of the Aptian proto-oceanic gulf-type shallow evaporitic organic-rich deposits of the Paracuru Formation, Ceará basin (see Figure 2 for lithological interpretation).

Formation (lacustrine hypersaline system and epicontinental marine hypersaline system, respectively). As observed in the other hypersaline samples (Figures 4–6), the very high abundances of phytane, gammacerane, and regular i-C_{25} and i-C_{30} isoprenoids, together with low abundance of diasteranes (cf. peaks 6 and 7 in Figure 6), suggest an arid climate leading to a low clastic influx (lack of clay minerals in the depositional environment). Therefore, such specific molecular characteristics reflect not the marine or lacustrine character, but the hypersaline condition of deposition in which specific groups of precursors have thrived. These "end member" features provide perhaps the most straightforward of the classifications into groups when compared with other samples from different depositional settings (e.g., freshwater lacustrine and normal marine environments; c.f. Mello et al., 1988a, b). Useful features for characterizing this environmental system are the same described for the hypersaline systems shown above (samples 1 and 2 in Figures 4–7 and Table 1), except for the presence of the C_{30} steranes and dinosteranes (Figure 7 and Table 1). Such compounds, believed to indicate a marine source (Moldowan et al., 1985; Summons et al., 1987; Goodwin et al., 1988), are absent in the hypersaline lacustrine samples (e.g., sample 2 in Figure 7 and Table 1).

The bulk, elemental, and molecular features for these organic-rich rocks are in agreement with the micropaleontological data (see below), and compare favorably with oils and source rocks from marine hypersaline environments in Messinian basins (northern Apennines), Italy (ten Haven et al., 1985, 1988); Tarragona basin, Spain (Albaiges et al., 1986); Camargue basin, southern France (Connan and Dessort, 1987); Williston basin, Canada; Cuanza basin, Angola; and onshore basins of Dubai and Oman (Thomas, 1990).

The ostracofauna recorded in the organic-rich facies from the Alagamar Formation (Figure 2) is very similar to that in the Santana Formation, Araripe basin, deposited during the Aptian in marine conditions (Silva-Telles and Viana, 1990). Two distinct ostracod assemblages can be differentiated in the Alagamar succession: limnetic to oligohaline lacustrine biotopes with species of *Cypridea* and abundant specimens of the ostracod "gen. et sp. indet. 207" nomem nudun in the lower part, and restricted hypersaline paralic/lagoonal biotopes characterized by abundant specimens of *Pattersoncypris angulata* spp. in the upper part of the section.

As the microfauna in the samples from Alagamar and Paracuru formations are very similar, a single paleoecological investigation approach was undertaken. The lower microfaunal association, equivalent to the Upanema member, in the Potiguar basin is characterized by a lacustrine limnetic to oligohaline fauna with strong variations of abiotic factors (e.g., salinity, oxygen, and lake area). This is marked by cyclic mass mortality events characterized by the occurrence of millimeter-thick strata with census populations (populations of organisms all living at the same time which may occasionally occur when some catastrophic event causes the sudden killing of the entire living population and allows its burial separate from preexisting specimens), and also by community replacement of the ostracods, with faunal recurrence (the return of a faunal association to a certain place after its disappearance due to a catastrophic event) without phylogenetic extinctions. The upper Aptian lacustrine ostracofaunas of the Ceará and Potiguar basins have similar faunal density, preservation, and diversity assemblages containing about 5 and 7 species, respectively.

In the Potiguar and Ceará basins (Figures 1 and 2) the Ponta do Tubarão and Trairi beds represent a transgressive event with restricted marine (paralic to lagoonal) character.

The upper Aptian marine ostracofauna is composed mainly of subspecies of *Pattersoncypris angulata* (Krommelbein and Weber, 1971). This species is commonly associated with dinoflagellates (Arai and Coimbra, 1990). Both lacustrine and marine ostracofaunas are characterized by census populations with abundant juvenile forms, indicating mass mortality, which can be related to climate-induced paleosalinity variations.

The block diagram in Figure 12 shows the envisaged evaporitic proto-oceanic gulf-type paleoenvironment that dominated the Potiguar and Ceará basins during the Aptian. This model, based on the integration of geological, geochemical, and paleontological data, assumes that intermittent incursions of seawater account for the filling of preexisting, deep topographic depressions (rift basins). These periodic marine incursions, combined with tectonic quiescence, isolation by topographical barriers, and an arid climate, led to a low clastic influx and sluggish circulation conditions. Such a setting is appropriate for high evaporation, with subsequent cyclic deposition of evaporites (e.g., halite, anhydrite, and dolomite), mixed carbonates, and siliciclastics. The organic-rich rocks are associated with periods after the marine transgressions, in which less saline conditions were established. The tremendous biomass productivity of such environments is due to the extensive supply of nutrients provided to a selective number of well-adapted species that, without competition, are able to thrive, providing a high input of organic matter (cf. Kirkland and Evans, 1980). Furthermore, the high density of hypersaline waters resulted in water column stability, causing water density stratification and permanent bottom-water dysoxia-anoxia. Such condi-

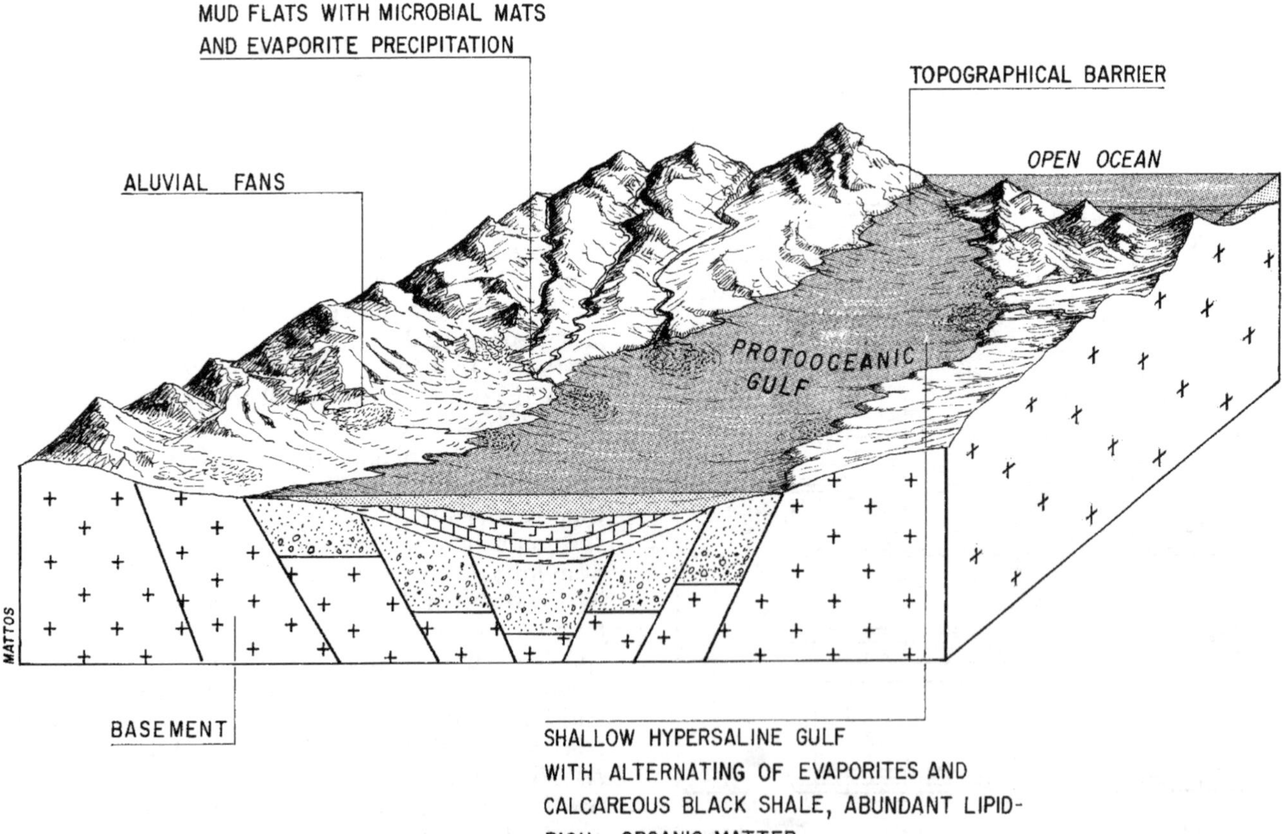

Figure 12. Schematic block diagram showing the sedimentary facies in a proto-oceanic gulf-type evaporitic environment envisaged for the Aptian (Paracuru Formation) of the Ceará basin (see Figure 2 for lithological interpretation).

tions, lethal for macrofauna and benthonic organisms, dramatically enhance the preservation potential of the organic matter. The result is the deposition of laminated, organic-rich, calcareous black shales and marls (e.g., Demaison and Moore, 1980; Kirkland and Evans, 1980; Mello, 1988).

Some examples of analogous ancient evaporitic environments have been reported in the literature: middle Miocene evaporites of the western Mediterranean (Hsu, 1972); the Tyro (eastern Mediterranean) and Messinian basins (northern Apennines), Italy (ten Haven, 1985, 1988); Tarragona basin, Spain (Albaigés et al., 1986); Camargue basin, southern France (Connan and Dessort, 1987); Marl Slate member of the Zechstein, England (Gibbons, 1987); and the Gulf of Suez, Egypt and El Lajjun, Jordan (Barwise and Roberts, 1984; Barwise, 1987).

Upper Aptian–Middle Albian Marine Hypersaline System

The studied organic-rich rocks of this hypersaline system are confined to the Sergipe basin, in the northern area of the Brazilian continental margin, and range in age from late Aptian to middle Albian (Riachuelo Formation, Figures 1, 2, and Table 1). Figure 12 shows a typical geochemical well log, along with the stratigraphic position of the organic-rich upper Aptian–middle Albian carbonates. The sequence is characterized by organic-rich (TOC up to 12%), dark gray marlstones ($CaCO_3$ up to 30%), and calcareous black shales. The Rock-Eval pyrolysis data indicate a predominance of good to excellent hydrocarbon source potential (S_2 up to 46 kg HC/ton rock), largely arising from the presence of type II kerogen (hydrogen indices up to 700 mg HC/g TOC; Figure 13 and Table 1). Microscopic examination of this organic-rich sequence shows generally 40 to 60% of amorphous organic matter, around 20 to 30% herbaceous material, comprising mainly pollen and spores, and 20 to 30% woody and coaly organic matter. The presence of organic and hydrogen-rich layers throughout the upper Aptian–middle Albian in the basin indicates that most of the succession was deposited under dysoxic-anoxic hypersaline bottom conditions (cf. Koutsoukos et al., 1991a, b).

The biological marker distributions (sample 4 in Figures 4–7 and Table 1) show a number of features similar to those observed in hypersaline environments described above and in rocks deposited in marine carbonate hypersaline environments (cf. McKirdy et al., 1984; Palacas et al., 1984; Zumberge, 1984; Mello et al., 1988a, b). These features include a dominance of phytane over pristane, low abundances of diasteranes relative to steranes (e.g., peaks 6 and 7 vs. peaks 8 and 10 for C_{27} steranes; Appendix 1), low hopane/sterane ratios (peak 35/peaks 8 + 10; Table 1), high abundances of gammacerane (peak 40), β-carotane, regular *i*-C_{25} isoprenoid alkanes and squalane (*i*-C_{30}), high abundances of vanadyl relative to nickel porphyrins, and the occurrence of C_{35} αβ hopanes in higher or

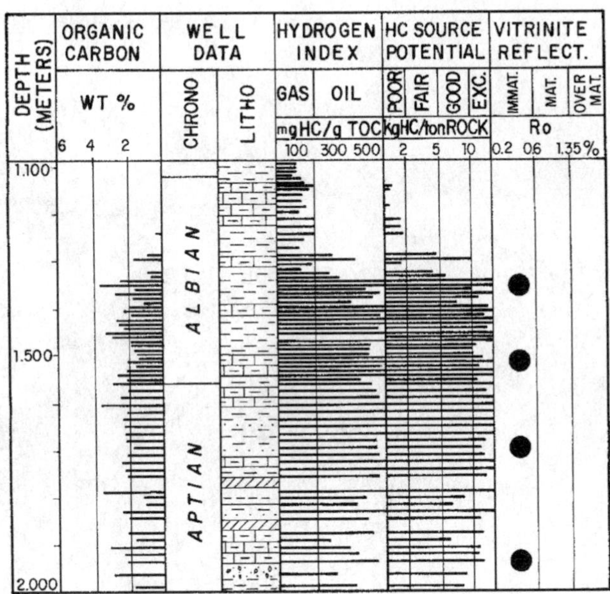

Figure 13. Geochemical log of the well 1-CR-2-SE showing the stratigraphic position of the upper Aptian–lower Albian dominantly neritic marine carbonate organic-rich hypersaline deposits of the Riachuelo Formation, Sergipe basin (see Figure 2 for lithological interpretation).

similar abundance than the C_{34} counterparts (peaks 45 vs. 44; cf. Mello et al., 1988a, b). Such features have been widely reported in deposits associated with dysoxic-anoxic hypersaline water columns (e.g., Waples et al., 1974; Jiang and Fowler, 1986; Fu Jiamo et al., 1986; ten Haven et al., 1987; Mello, 1988; Mello et al., 1988a, b; Koutsoukos et al., 1991a, b; Peters and Moldowan, 1991). The diagnostic molecular features that allowed the differentiation of this marine hypersaline carbonate environment, when compared with the others shown in Figures 4–7 and Table 1, are the presence of high abundances of dinosteranes, C_{30} regular steranes, pregnanes and homopregnanes (peaks 1–5), and tricyclic terpanes up to C_{29} (peaks 18–23 and 25–26).

Figure 14 shows the paleogeographical reconstruction for the late Aptian-middle Albian of the Sergipe basin, in the southwestern area. High source area relief supplied voluminous siliciclastic sediments, consisting of conglomerates and graded sandstones deposited by gravity flows, with subordinate siltstones and shales. The main carbonate packages consist of oolitic/oncolitic/bioclastic packstones/grainstones, and red algal (*Solenoporacea*) patch reefs, deposited in shallow water banks. The organic-rich rocks were mostly accumulated in structural lows, from middle neritic to upper bathyal settings. The deposits consist of fine-grained limestones, marlstones, and shales of the Taquari Member (Figure 2; Koutsoukos et al., 1991a).

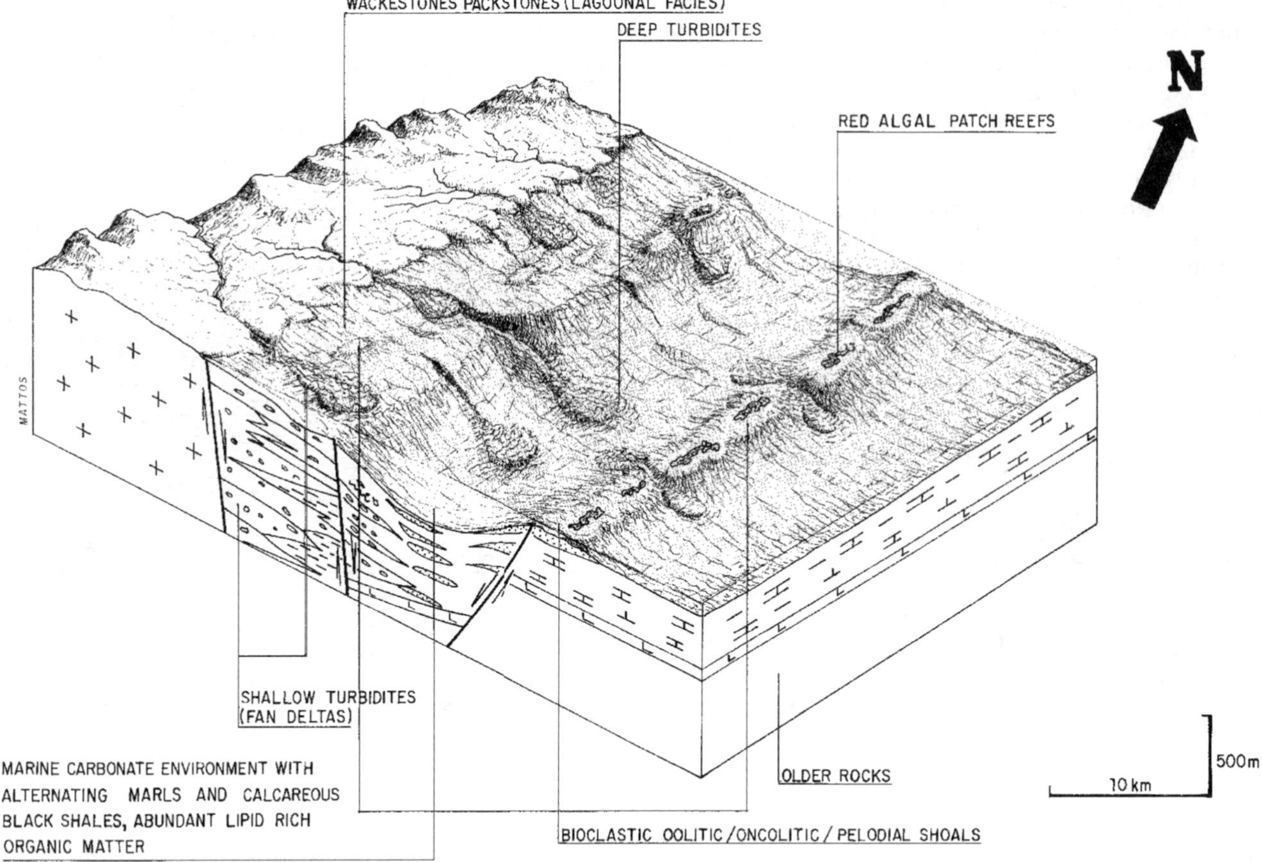

Figure 14. Schematic block diagram showing the sedimentary facies in a dominantly neritic marine carbonate hypersaline environment envisaged for the upper Aptian–middle Albian (Riachuelo Formation) of the Sergipe basin (see Figure 2 for lithological interpretation).

Possible microfaunal responses to hypersaline bottom waters (Figure 15) are: (1) abundant and well diversified microbivalves in marginal marine, paralic (tidal flats) to middle neritic biotypes (associated with the microbivalves are the microgastropods which probably fed on extensive hypersaline shallow water algal mats); (2) the predominance of large populations of favusellids, middle Cretaceous benthonic foraminifera (hedbergellid ecophenotypes) adapted to shallow, warm, hypersaline, carbonate-saturated environments in nearshore biotopes (Koutsoukos et al., 1989, 1991a); and (3) organic-rich layers deposited in middle neritic to upper bathyal dysaerobic to quasi-anaerobic settings, containing either a very impoverished low-diversity benthonic assemblage of calcareous and agglutinated benthonic foraminiferids or only a low-diversity agglutinated assemblage (Koutsoukos et al., 1991a, b).

CONCLUSIONS

Biological marker and micropaleontological data allowed the characterization and differentiation of four hypersaline systems in the Brazilian sedimentary basins:

1. an Upper Permian epicontinental shallow marine hypersaline system;
2. a Lower Cretaceous shallow lacustrine hypersaline system;
3. an Aptian gulf-type shallow evaporitic system; and
4. an upper Aptian–middle Albian semi-restricted dominantly neritic hypersaline system.

The main diagnostic molecular features that characterize the high water salinity of these environments are: phytane more abundant than pristane; Tm higher than Ts; low relative abundances of tricyclic terpanes, pregnanes, homopregnanes, and diasteranes; high abundances of gammacerane (sometimes the major peak in m/z 191); β-carotane, regular i-C_{25} isoprenoid, squalane, and C_{35} hopanes predominating over C_{34} counterparts.

The diagnostic molecular features that allow some differentiation between lacustrine and marine hypersaline systems are the presence and abundance of

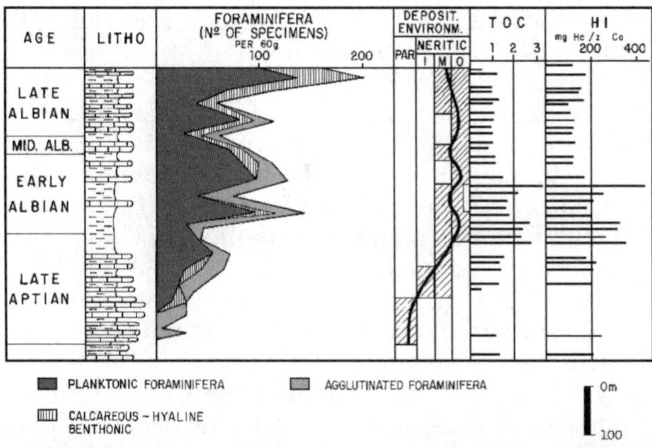

Figure 15. Stratigraphic distribution, relative abundance, and depositional paleoenvironment of the major foraminiferal groups in the upper Aptian to Albian succession, with geochemical data, of a typical well-section of the Sergipe basin (see Figure 2 for lithological interpretation).

dinosteranes (limited to Upper Triassic and younger strata) and C_{30} steranes (24-*n*-propylcholestanes) detected using GC-MS/MS technique.

The main micropaleontological features that characterize hypersaline environments are: low diversity to monospecific assemblages; high density of specimens; common occurrence of millimeter-thick strata with census populations; abundant microgastropods and microbivalves in marginal marine, paralic (tidal flats) to middle neritic biotypes; lack of bioturbation during maximum hypersalinity episodes; large populations of favusellids in middle Cretaceous shallow, warm, hypersaline, carbonate-saturated environments; organic-rich layers deposited in middle neritic to upper bathyal dysaerobic to quasi-anaerobic settings, containing either a very impoverished low-diversity benthonic assemblage of calcareous and agglutinated benthonic foraminiferids or only a low-diversity agglutinated assemblage.

The specific molecular features that characterize the hypersaline environments, discussed in this study, must reflect, not the marine or lacustrine character, but the hypersaline conditions of deposition in which specific groups of precursors thrived. These "end member" features provide perhaps the most straightforward means of classification into groups when compared with other samples from different depositional settings (e.g., freshwater lacustrine and normal marine environments). However, the presence of typical marine biota and biological markers (e.g., foraminifera, crinoids, and dinoflagellates, and dinosteranes and C_{30} regular steranes) can be used to distinguish lacustrine and marine hypersaline environments.

ACKNOWLEDGMENTS

The authors are indebted to the Geochemistry Group of Petrobrás for all the geochemical data provided. The authors are also grateful to K.E. Peters for his helpful comments and suggestions during the review of this chapter. We also thank Petrobrás for permission to publish.

REFERENCES CITED

Amaral, S.E., 1971, Contribuição ao conhecimento geológico, petrográfico e sedimentológico da Formação Irati no Estado de São Paulo: Boletim IGA, Instituto de Geociências e Astronomia, Universidade de São Paulo, n. 2, 81 p. São Paulo.

Albaigés, J., J. Algaba, E. Glavell, J. Grimalt, 1986, Petroleum geochemistry of the Tarragona Basin (Spanish Mediterranean off-shore), *in* D. Leythauser and J. Rullkötter, eds., Advances in Organic Geochemistry 1985: Elmsford, N.Y., Pergamon Journals Ltd, p. 441-450.

Anderson, F.W., R.A.B. Bazley, and E.R. Shephard-Thorn, 1967, The sedimentary and faunal sequence of the Wadhurst Clay (Wealden) in boreholes at Wadhurst Park, Sussex: Bulletin of the Geological Survey of Great Britain, v. 27, p. 171-235.

Arai, M., and J.C. Coimbra, 1990, Análise paleoecológica do registro das primeiras ingressões marinhas na Formação Santana (Cretáceo Inferior da Chapada do Araripe), *in* Simpósio sobre a Bacia do Araripe e bacias interiores do Nordeste, Crato, Departamento Nacional de Produção Mineral, v. 1, p. 225-239.

Barwise, A.J.G., and I. Roberts, 1984, Diagenetic and catagenetic pathways for porphyrins in sediments: Organic Geochemistry, 6:167-176.

Barwise, A.J.G., 1987, Mechanisms involved in altering deoxophylloerythroetioporphyrin-etioporphyrin ratios in sediments and oils, *in* R.H. Filby and G.J. Van Berkel, eds., Geochemistry, Characterization and Processing Symposium Series: American Chemical Society, n. 344, p. 01-109.

Bertani, R.T., and A.V. Carozzi, 1985, Lagoa Feia Formation (Lower Cretaceous) Campos Basin offshore Brazil: Rift Valley stage carbonate reservoirs I & II: Journal of Petroleum Geology 8, 37-58, 199-220.

Campanha, V.A., 1985, Ocorréncia de braquiópodes inarticulados na Formação Irati, no Estado de São Paulo. An. Acad. brasil. Ciênc. 57 (1):115.

Campanha, V.A., and M.F. Zaine, 1989, Foraminíferos da Formação Irati no Estado de São Paulo: XI Congresso Brasileiro de Paleontologia, Abstract, p. 38.

Chen Changming, Huang Jiakuan, Chen Jingshan, Tian Xingyou, 1984, Depositional models of Tertiary rift basins, eastern China, and their application to petroleum prediction: Sedimentary Geology, 40:73-88.

Cerqueira, J.R., and E.V. Santos Neto, 1986, Papel das intrusões de diabásio no processo de geração de hidrocarbonetos na Bacia do Paraná: Anais do 3° Congresso Brasileiro de Petróleo, Rio de Janeiro, TT-73, 15 p.

Connan, J., and D. Dessort, 1987, Novel family of hexacyclic hopanoid alkanes (C_{32}–C_{35}) occurring in sediments and oils from anoxic paleoenvironments: Organic Geochemistry, 11:103-113.

Dean, W., and T.D. Fouch, 1983, Lacustrine environment, *in* P.A. Scholle et al., eds., Carbonate Depositional Environments, AAPG Memoir 33, p. 97-130.

De Dekker, P., 1988, Large Australian lakes during the last 20 million years—Sites for petroleum source rocks or metal ore deposition or both, *in* A.J. Fleet, K. Kelts, and M.R. Talbot, eds., Lacustrine Petroleum Source Rocks: Geological Society Special Publication 40, p. 45-59.

De Grande, S.M.B., 1991, Extended tricyclic triterpanes in sediments and petroleums, *in* D. Manning, coordinating ed., Organic Geochemistry, Advances and Applications in Energy and the Natural Environment, 15th Meeting of the European Association of Organic Geochemists, Poster Abstracts, pt. 1 (Petroleum geochemistry: molecular characterization), p. 181-183.

Della Fávera, J.C., 1987, Tempestades como agentes de poluição ambiental e mortandade em massa no passado geológico: caso das formações Santana (Bacia do Araripe) e Irati (Bacia do Paraná). B. Geoci. Petrobrás, 1 (2):239.

Demaison, G.J., and G.T. Moore, 1980, Anoxic environments and oil source bed genesis: AAPG Bulletin, n. 64, p. 1179-1209.

Fan Pu, R.P. Philp, Li Zhenxi, and Ying Guangguo, 1990, Geochemical characteristics of aromatic hydrocarbons of crude oils and source rocks from different sedimentary environments: Organic Geochemistry 16:(1–3), p. 427-435.

Friedman, G.M., 1980, Review of depositional environments in evaporite deposits and the role of evaporites in hydrocarbon accumulation: AAPG Bulletin, n. 56, p. 1072-1086.

Fu Jiamo, S. Guoying, P. Pingan, S.C. Brassell, G. Eglinton, and J. Jigang, 1986, Peculiarities of salt lake sediments as potential source rocks in China, *in* D. Leythauser and J. Rullkötter, eds., Advances in Organic Geochemistry 1985: Elmsford, N.Y., Pergamon Journals Ltd., p. 119-127.

Gibbons, M.J., 1987, The depositional environment and petroleum geochemistry of the Marl Slate-Kupferschiefer, *in* J. Brooks and A.J. Fleet, eds., Geol. Soc. Spec. Marine Petroleum Source Rocks Publ. 26, 249.

Goodwin, N.S., A.L. Mann, R.L. Patience, 1988, Structure and significance of C_{30} 4-methylsteranes in lacustrine shales and oils: Organic Geochemistry, v. 12, p. 495-506.

ten Haven, H.L., J.W. De Leeuw, and P.A. Schenk, 1985, Organic geochemical studies of a Messinian evaporitic basin, northern Apennines (Italy), part I—Hydrocarbon biological markers for a hypersaline environment. Geochemica et Cosmochimica Acta, 49:2181-2191.

ten Haven, H.L., J.W. De Leeuw, J.W. Rullkötter, and J.S. Sinninghe Damsté, 1987, Can the pristane/phytane ratio be used as a palaeoenvironmental indicator?: Nature 330, p. 641-643.

ten Haven, H.L., J.W. De Leeuw, J.S. Sinninghe Damsté, P. A. Schenck, S.E. Palmer, and J.E. Zumberge, 1988, Application of biological markers in the recognition of paleo hypersaline environments, *in* A.J. Fleet, K. Kelts, and M.R. Talbot, eds., Lacustrine Petroleum Source Rocks: Geological Society Special Publication 40, p. 123-130.

Hsu, K., 1972, Origin of saline giants: a critical review after the discovery of Mediterranean evaporite: Earth Science Review 8, p. 371-396.

Jiang, Z.S., and M.G. Fowler, 1986, Carotenoid-derived alkanes in oils from northwestern China, *in* D. Leythauser and J. Rullkötter, eds. Advances in Organic Geochemistry 1985: Elmsford, N.Y., Pergamon Journals Ltd, p. 831-839.

Katz, B.J., 1991, Controls on distribution of lacustrine source rocks through time and space, *in* B.J. Katz, ed., Lacustrine Basin Exploration—Case Studies and Modern Analogs, AAPG Memoir 50, chapter 4, p. 61-76.

Kelts, K., 1988, Environments of deposition of lacustrine petroleum source rocks—An introduction, *in* A.J. Fleet, K. Kelts, and M.R. Talbot, eds., Lacustrine Petroleum Source Rocks: Geological Society Special Publication 40, p. 3-26.

Kendall, A.C., 1978, Facies models 12. Subaqueous evaporites. Geosci. Can., 5:124-139.

Kirkland, D.W., and R. Evans, 1980, Source-rock potential of evaporitic environment. AAPG Bulletin 69:181-190.

Koutsoukos, E.A.M., P. N. Leary, and M.B. Hart, 1989, *Favusella* Michael (1972): evidence of ecophenotypic adaptation of a planktonic foraminifer to shallow-water carbonate environments during the mid-Cretaceous: Journal of Foraminiferal Research, 19 (4): 324-336.

Koutsoukos, E.A.M., M.R. Mello, N.C. Azambuja Filho, M.B. Hart, and J.R. Maxwell, 1991a, The upper Aptian–Albian succession of the Sergipe Basin, Brazil: an integrated paleoenvironmental assessment: AAPG Bulletin 75 (3): 479-498.

Koutsoukos, E.A.M., M.R. Mello, and N.C. Azambuja Filho, 1991b, Micropalaeontological and geochemical evidence of mid-Cretaceous dysoxic-anoxic palaeoenvironments in the Sergipe Basin, northeastern Brazil, *in*: R.V. Tyson, and T.H. Pearson, (eds), Modern and Ancient Continental Shelf Anoxia: Geological Society Special Publication, 58: 427-447.

Kroemmelbein, K., and R. Weber, 1971, Ostracoden des "Nordost-Brasilianischen Wealden": Geologischen Jahrbuch 115, 1-93. Hannover.

Lavina, E.L., D.C.A. Barberena, and U.F. Faccini, 1989, Os Mesossaurídeos da Formação Irati na localidade de Passo de São Borja (RS): Evidência de uma paleocatástrofe ecológica. Abstracts of the XI Congresso Brasileiro de Paleontologia, p. 35-36.

de Leeuw, J., and J.S. Sinninghe Damsté, 1989, Organic sulfur compounds and other biomarkers as indicators of palaeosalinity, *in* Geochemistry of

Sulfur in Fossil Fuels: W.L. Orr and C.M. White, eds., ACS Symposium Series; American Chemical Society: Washington, DC, p. 417-443.

McKirdy, D.M., A.J. Kantsler, J.K. Emmett, and A.K. Aldridge, 1984, Hydrocarbon genesis and organic facies in Cambrian carbonates of the eastern Officer Basin South Australia, *in* J.G. Palacas, ed., Petroleum Geochemistry and Source Rock Potential of Carbonate Rocks: AAPG Studies in Geology, 18, p. 13-32.

Mello, M.R., 1988, Geochemical and molecular studies of the depositional environments of source rocks and their derived oils from the Brazilian marginal basins, PhD. Thesis, University of Bristol, 240 p.

Mello, M.R., N. Telnaes, P.C. Gaglianone, M.I. Chicarelli, S.C. Brassel, and J.R. Maxwell, 1988a, Organic geochemical characterization of depositional palaeoenvironment of source rocks and oils in Brazilian marginal basins, *in* L. Mattavelli and L. Novelli, eds., Advances of Organic Geochemistry, v. 13, n. 1-3, p. 31-45.

Mello, M.R., P.C. Gaglianone, S.C. Brassell, and J.R. Maxwell, J.R., 1988b, Geochemical and biological marker assessment of depositional environment using Brazilian offshore oils: Marine and Petroleum Geology, 5:205-223.

Mello, M.R., E.V. Santos Neto, J.R. Cerqueira, R. Rodrigues, and F.T.T. Gonçalves, 1991, *in* D. Manning, coordinating editor, Organic Geochemistry, Advances and Applications in Energy and the Natural Environment, 15th Meeting of the European Association of Organic Geochemists, Poster Abstracts, part 1 (Petroleum geochemistry: case histories), p. 76-78.

Mello, M.R., and J.R. Maxwell, 1991, Organic geochemical and biological marker characterization of source rocks and oils derived from lacustrine environments in the Brazilian continental margin, *in* B.J. Katz, ed., Lacustrine Basin Exploration—Case Studies and Modern Analogs, AAPG Memoir 50, chapter 5, p. 77-99.

Moldowan, J.M., S.K. Seifert, and E.J. Gallegos, 1985, Relationship between petroleum composition and depositional environment of petroleum source rocks. AAPG Bulletin 69:1255-1268.

Moldowan, J.M., F.J. Fago, C.Y. Lee, S.R. Jacobson, D.S. Watt, N.E. Slougui, A. Jeganathan, and D.C. Young, 1990, Sedimentary 24-*n*-Propylcholestanes, molecular fossils diagnostic of marine algae: Science, v.247, p. 309-312.

Oelofsen, B., and D.C. Araújo, 1983, Palaeoecological implications of the distribution of mesosaurid reptiles in the Permian Irati sea (Paraná Basin), South America, Revista Brasileira de Geociências, 13(1):1-6.

Palacas, J.G., D.E. Anders, and J.D. King, 1984, South Florida Basin—A prime example of carbonate source rocks of petroleum, *in* J.G. Palacas, ed., Petroleum Geochemistry and Source Rock Potential of Carbonate Rocks, AAPG Studies in Geology, 18, p.71-96.

Peterson, J.A., and R.J. Hite, 1969, Pennsylvanian evaporite-carbonate cycles and their relation to petroleum occurrence, southern Rocky Mountains. AAPG Bulletin 53:884-908.

Peters, K.E., and J.M. Moldowan, 1991, Effects of source, thermal maturity, and biodegradation on the distribution and isomerization of homohopanes in petroleum. Organic Geochemistry, v. 17, n. 1, p. 47-61.

Philp, P.R., Li Jinggui, C.A. Lewis, 1989, An organic geochemical investigation of crude oils from Shanganning, Jianghan, Chaidamu and Zhungeer Basins, People's Republic of China. Organic Geochemistry, v.14, 4:447-460.

Powell, T.G., 1986, Petroleum geochemistry and depositional settings of lacustrine source rocks. Marine and Petroleum Geology, 3:200-219.

Reed, W.E., 1977, Molecular compositions of weathered petroleum and comparison with its possible source. Geochimica et Cosmochimica Acta. 41:237-247.

Rubinstein, I., O. Sieskind, and P. Albrecht, 1975, Rearranged sterenes in a shale: occurrence and simulated formation. J. Chem. Soc., Perkin Trans. 1, p. 1833-1836.

Santos Neto, E.V., and J.R. Cerqueira, 1990, Avaliação geoquímica da Bacia do Paraná. 4° Congresso Brasileiro de Petróleo, Rio de Janeiro, TT-108, 7p.

Shi Jiyang, A.S. Mackenzie, R. Alexander, G. Eglinton, A.P. Gowar, G.A. Wolff, and J.R. Maxwell, 1982, A biological marker investigation of petroleums and shales from the Shangli oilfield, People's Republic of China. Chem. Geol., 35:1-31.

Silva-Telles, Jr., A. and M. S. S. Viana, 1990, Paleoecologia dos ostracodes d Formação Santana (Bacia do Araripe): um estudo ontogenético de populações, *in* Simpósio sobre a Bacia do Araripe e bacias interiores do Nordeste, Crato, Departamento Nacional de Produção Mineral, v. 1, p. 309-327.

Summons, R.E., J.K. Volkman, and C.J. Boreham, 1987, Dinosterane and other steroidal hydrocarbons of dinoflagellate origin in sediments and petroleum. Geochim. Cosmochim. Acta 51, 3075-3082.

Talbot, M.R., 1988, Non-marine oil-source rock accumulation: evidence from the lakes of tropical Africa, *in* A.J. Fleet, K. Kelts, and M.R. Talbot, eds., Lacustrine Petroleum Source Rocks: Geological Society Special Publication 40, p. 29-45.

Thomas, J., 1990, Biological markers in sediments with respect to geological time. PhD. Thesis, University of Bristol, 123 p.

Tissot, B., G. Derro, and A. Hood, 1978, Geochemical study of the Uinta Basin—Formation of petroleum from the Green River Formation. Geochimica et Cosmochimica Acta, 42:1459-1465.

Tissot, B.P., and D.H. Welte, 1984, Petroleum Formation and Occurrences. 2nd Edition: Springer-Verlag, Berlin, 699 p.

Trigüis, J.A., 1991, Organic facies of Irati Formation, Paraná Basin, Brazil. An organic geochemical overview. XII International Congress on Carboniferous and Permian Geology and Stratigraphy. Abstracts, Buenos Aires, p. 81-82.

Volkman, J.K., P.S. Kearney, and S.W. Jeffrey, 1990, A new source of 4-methyl sterols and 5α(H)-stanols in sediments: prymnesiophyte microalgae of the genus *Pavlova*. Org. Geochem. v.15, 489-497.

Wang Tieguan, Fan Pu, and F.M. Swain, 1988, Geochemical characteristics of crude oils and source beds in different continental facies of four oil bearing basins, China, *in* A.J. Fleet, K. Kelts, and M.R. Talbot, eds., Lacustrine Petroleum Source Rocks: Geological Society Special Publication 40, p.

Waples, D.W., P. Haug, and D.H. Welte, 1974, Occurrence of a regular C_{25} isoprenoid hydrocarbon in Tertiary sediments representing a lagoonal-type, saline environment. Geochimica et Cosmochimica Acta, v. 38, p. 381-387.

Wolf, M., and Z.C.C. Silva, 1974, Petrographic description and facies analysis of some samples from the oil shale of the Irati Formation (Permian). Anais do XXVIII Congresso Brasileiro de Geologia, p. 159-170.

Zalán, P.V., S. Wolff, M.A.M. Astolfi, I.S. Vieira, J.C.J. Conceição, V.T. Appi, E.V. Santos Neto, J.R. Cerqueira, A. Marques, 1990, The Paraná Basin, Brazil, *in* M.W. Leignton, et al., Interior Cratonic Basins. Tulsa, AAPG Memoir 51, p. 681-708.

Zumberge, J.E., 1984, Source rocks of the La Luna Formation (Upper Cretaceous) in the Middle Magdalena Valley, Colombia, *in* J.G. Palacas, ed., Petroleum Geochemistry and Source Rock Potential of Carbonate Rocks, AAPG Studies in Geology, 18:127-134.

APPENDIX 1

Pr	2,6,10,14-tetramethylpentadecane (pristane)
Ph	2,6,10,14-tetramethylhexadecane (phytane)
i-C_{25}	2,6,10,14,18-pentamethyleicosane (regular)
i-C_{25}	2,6,10,15,19-pentamethyleicosane (irregular)
i-C_{30}	squalane
β-	β-carotane
1	13β(H), 17α(H)-diapregnane(C_{21})
2	5α(H), 14β(H), 17α(H)-pregnane (C_{21})
3	5α(H), 14β(H), 17β(H) + 5α(H), 14α(H), 17α(H)-pregnane (C_{21})
4	4α-methyl-5α(H),14β(H),17β(H) + 4α-methyl-5α(H),14α(H),17α(H) homopregnane (C_{22})
5	5α(H), 14β(H), 17β(H) + 5α(H),14α(H),17α(H)-homopregnane (C_{22})
6	13β(H),17α(H)-diacholestane, 20S (C_{27}-diasterane)
7	13β(H),17α(H)-diacholestane, 20R (C_{27}-diasterane)
8	5α(H), 14α(H), 17α(H), 20S (C_{27}-cholestane)
9	5α(H), 14β(H), 17β(H), 20R + 20S (C_{27}-cholestane)
10	5α(H), 14β(H), 17β(H), 20R (C_{27}-cholestane)
11	5α(H), 14α(H), 17α(H), 20S (C_{28}-methyl-cholestane)
12	5α(H), 14β(H), 17β(H), 20R + 20S (C_{28}-methyl-cholestane)
13	5α(H), 14α(H), 17α(H), 20R (C_{28}-methyl-cholestane)
14	5α(H), 14α(H), 17α(H), 20S (C_{29}-ethyl-cholestane)
15	5α(H), 14β(H), 17β(H), 20R + 20S (C_{29}-ethyl-cholestane)
16	5α(H), 14α(H), 17α(H), 20R (C_{29}-ethyl-cholestane)
17	C_{19} tricyclic terpane
18	C_{20} tricyclic terpane
19	C_{21} tricyclic terpane
20	C_{23} tricyclic terpane
21	C_{24} tricyclic terpane
22	C_{25} tricyclic terpane
23	C_{26} tricyclic terpanes
24	C_{24} tetracyclic (Des-E)
25	C_{28} tricyclic terpanes
26	C_{29} tricyclic terpanes
27	C_{25} tetracyclic
28	C_{27} 18α(H)-trisnorneohopane(Ts)
29	C_{30} tricyclic terpanes
30	C_{27} 17α(H)-trisnorhopane(Tm)
31	C_{31} tricyclic terpanes
32	17α(H),18α(H),21β(H)-28,30-bisnorhopane(C_{28})
33	C_{29} 17α(H),21β(H)-norhopane
34	C_{29} 17β(H),21α(H)-norhopane
35	C_{30} 17α(H),21β(H)-hopane
36	C_{33} tricyclic terpanes
37	C_{30} 17β(H),21α(H)-hopane
38	C_{34} tricyclic terpanes
39	C_{31} 17α(H),21β(H)-homohopane (22S + 22R)
40	C_{30} gammacerane
41	C_{32} 17α(H),21β(H)-bishomopane (22S + 22R)
42	C_{35} tricyclic terpanes
43	C_{33} 17α(H),21β(H)-trishomohopane (22S + 22R)
44	C_{34} 17α(H), 21β(H)-tetrakishomohopane (22S + 22R)
45	C_{35} 17α(H),21β(H)-pentakishomohopane (22S + 22R)

Chapter 4

The Sequence Stratigraphy of Transgressive Black Shales

Paul B. Wignall
Department of Earth Sciences
University of Leeds
Leeds, United Kingdom

James R. Maynard
Mobil North Sea Ltd.
London, United Kingdom

ABSTRACT

Black shales are relatively deep-water facies that form in basin center locations in epicontinental settings. At times of rapid relative sea level rise, conditions in epicontinental basins can become sufficiently deep (even in marginal settings) for the accumulation of thin black shales in close association with shallow-water facies. Using examples from the British Lower Jurassic (the Jet Rock) and Upper Carboniferous (marine bands) these thin transgressive black shales are classified into two categories: (1) Maximum flooding black shales—which are usually condensed, and pass into considerably thicker nearshore facies. Condensed sections are separated from sequence boundaries by sediments of the transgressive systems tract. (2) Basal transgressive black shales—which are associated with the initial flooding surface across sequence boundaries. They show no lateral facies variation, and do not connect laterally with contemporary nearshore or shoreline facies.

INTRODUCTION

It is well established that, in many epicontinental seas, distal sediment-starved basins accumulate black shales (Demaison and Moore, 1980). In such deep-water settings the benthic environment is poorly oxygenated (anoxic or dysoxic) due to isolation of the bottom waters from effects such as the wind-driven vertical advection of oxygen-rich surface waters. The establishment of a pycnocline can further isolate bottom waters (e.g., Wignall, 1989). The damping of tidal circulation in broad epicontinental seas can also reduce bottom water mixing (Hallam and Bradshaw, 1979). In contrast, some deep-water black-shale models invoke bottom-water circulation (Heckel, 1977; Oschmann, 1988; Miller, 1990), but they share a crucial factor in common with stagnant models, i.e., that bottom waters are effectively isolated from more oxygen-rich surface waters by a density interface such as a thermocline.

The occurrence of black shales within the sequence stratigraphic models of the Exxon Production Research group (EPR) has not been fully evaluated. Condensed sections (downlap surfaces in their seismic expression) of the EPR model form in sediment-starved, deep-water conditions. Black shales are therefore good candidates for at least some condensed sections. For example, Loutit et al. (1988) consider the black shales associated with the celebrated Cenomanian/Turonian anoxic event to be an organic-rich condensed section. Condensed sections can also extend into more marginal settings at times of rapid sea level rise where they form on the maximum flooding surface separating the transgressive from the highstand systems tracts. This is an important mechanism for introducing black shales into areas normally characterized by shallower depositional conditions. Deep-water deposition also persists into the early stages of highstand systems tracts, particularly if sediment supply is low (Loutit et al., 1988). For example, Savrda (1991) documented oxygen-restricted deep-water facies from a distal highstand systems tract in the Paleocene of Alabama.

Hallam and Bradshaw (1979) have noted that many black shales are associated with sites of greatest subsidence and this led them to propose a model of irregular bottom topography (abbreviated to the "puddle model" in Wignall, 1991). In this model the deep-water conditions essential for black shales are achieved by rapid subsidence only and the potential exists for black shale formation throughout the sea level cycle. However, black shales will be most likely to form in sites of rapid subsidence at the inflexion point of relative sea level rise.

Black shales are classic basinal facies and they therefore tend to occur as the central "bull's-eye" of paleogeographic facies maps (e.g., Calver, 1968; Hallam and Bradshaw, 1979) where they are surrounded by shallower water facies. However, thin black shales are also associated with shallower water facies and this has led numerous authors to infer a distinct shallow water origin for some black shales (e.g., Zangerl and Richardson, 1963; Hallam, 1967b; Hallam and Bradshaw, 1979; Coveney and Shaffer, 1988). In many cases these occurrences can probably be attributed to deposition on maximum flooding surfaces separated by thin transgressive systems tracts from sequence boundaries, although they require detailed analysis in the light of EPR's model. However, instances of black shales directly overlying very shallow water facies are common. Leckie et al. (1990) record a late Albian black shale resting on a thin bone bed which in turn lies on estuarine facies. Coveney et al. (1991) record Pennsylvanian black shales directly overlying coals (see also Zangerl and Richardson, 1963). An even more spectacular juxtaposition of facies occurs in the Late Permian of northeast England where the Marl Slate (an organic-rich, thinly laminated limestone rather than a true black shale) rests directly on eolian dunes and breccias containing dreikanters (Bell et al., 1979). In no instances do the black-shale facies interdigitate with the marginal facies, for they are separated from them by erosion surfaces that probably mark sequence boundaries. In these examples the black shales appear to occur within the lower part of the transgressive systems tracts which is normally characterized by shallow water deposition (e.g., Nummedal and Swift, 1988). In order to investigate whether there are indeed two distinct depositional scenarios for black shales (basin margin and basin center) a sequence-stratigraphic analysis of the early Toarcian (Early Jurassic) of central and northern England was undertaken (Wignall, 1991; Wignall and Hallam, 1991). Further insights into the sequence stratigraphic occurrence of black shales have been obtained by a detailed study of the Late Carboniferous of northern England.

THE JET ROCK

Organic-rich facies are widespread in the lower Toarcian of both Europe and elsewhere around the world (Jenkyns, 1988). The Jet Rock Member, a black-shale facies, is the local representative of this anoxic event in northern England.

Early Jurassic marine deposition in England occurred in a series of small basins in an epicontinental setting (Figure 1). The main positive structural feature was the London Platform, which was a site of minimal subsidence associated with marginal facies (Donovan et al., 1979). The principal depocenters to the north of the platform were the East Midlands Shelf and the Cleveland basin, which were separated by the Market Weighton High.

During the Spinatum Zone, in the latest Pliensbachian, carbonate deposition was established across most of the region and water depths appear to have been generally shallow (Hallam, 1967a). The broad extent of shallow-marine facies suggests highstand deposits. These are capped by a sequence boundary in the earliest Tenuicostatum Zone and the corresponding hiatus spans much of this zone on the margins of the London Platform where a weathered (karstic?) surface caps the limestone (Figure 2A). In the Cleveland basin a sharp junction separates limestones from black and dark-gray shales but no biostratigraphic gap is detectable (Figure 2B). A minor phase of onlap occurred during the Tenuicostatum Zone to be followed by a sequence boundary at the Tenuicostatum–Falciferum boundary. This basal Toarcian sequence contains anaerobic and dysaerobic facies in basin-center locations only, while skeletal lag of minimal thickness accumulated around the margins (Figure 2A). A minor sequence boundary that appears to occur at the base of the Exaratum Subzone for the *Eleganticeras elegantulum* ammonite Chron at the base of the subzone is only present in the basinal Port Mulgrave section (Howarth, 1962).

The succeeding interval of onlap, in the mid-Exaratum Subzone, witnessed the rapid expansion of the area of black-shale deposition, hitherto restricted to the basin centers (Figure 2). Thus, on the margins of the London Platform black shales rest directly on shallow-marine carbonates. At some localities a thin veneer of reworked and abraded phosphatic material overlies the sequence boundary and immediately

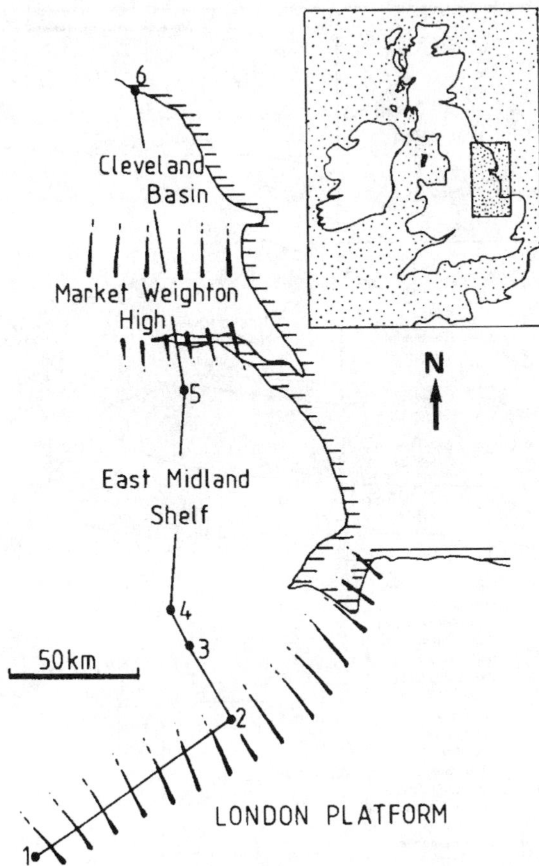

Figure 1. Toarcian paleogeography of northern and central England showing location of cross sections in Figure 2.

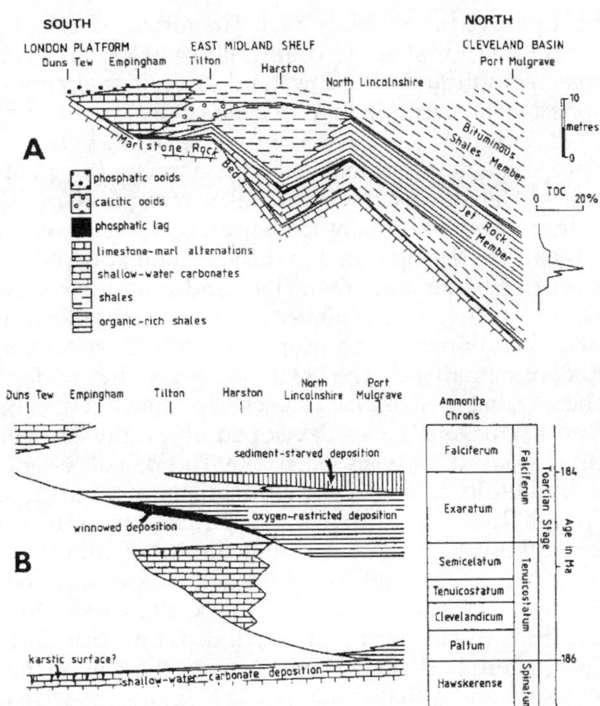

Figure 2. A—Lithostratigraphic correlation of the late Pliensbachian and early Toarcian sediments of England, based upon references in Wignall (1991). Total organic carbon (TOC) values for the Port Mulgrave section taken from Myers and Wignall (1987). B—Chronostratigraphic interpretation showing development of two sequences. Ages from Haq et al. (1988).

underlies the shale suggesting a period of sediment starvation immediately prior to black-shale deposition. The upper limit of the transgressive systems tract appears to be marked by a thin, laminated limestone with numerous belemnites in the topmost Exaratum Subzone at Port Mulgrave. This marks the level of minimum terrigenous sediment input in the Cleveland basin, a characteristic of maximum flooding surfaces. At a similar level on the margins of the London Platform the maximum flooding surface appears to be marked by calcitic ooids scattered in mudstone—an enigmatic facies. The highest values of total organic carbon occur toward the base of the Jet Rock (Figure 2A).

The Jet Rock and the laterally equivalent black shales are classic early-transgressive black shale and as Grabowski and Glaser (1990) have noted, "No time-equivalent shoreline facies are present." Thus the black shales have developed in the deeper parts of a transgressing sea which was accumulating virtually no sediment around its margins. The organic-matter type (algal-amorphous kerogen) remains constant throughout the black shale outcrops (Grabowski and Glaser, 1990).

The absolute and relative depths of black-shale deposition have been the subject of controversy. Hallam (1967b), Morris (1979), and O'Brien (1990) all suggest relatively shallow deposition distinct from the basinal deep-water black shales, whereas Wignall (1991) has suggested that water depths during transgression were relatively deep. The argument for shallow water hinges on the limited extent of marine deposition at the time of black shale formation while the greatest water depth is considered to have occurred in the Bifrons Zone when marine deposition was at its maximum extent (and no black shales were forming in the region). The model for shallow water deposition relies on restricted circulation in semi-isolated epicontinental basins to induce oxygen-restricted benthic conditions (Hallam and Bradshaw, 1979).

None of these arguments for shallow-water black shales is entirely convincing. There is no a priori reason why the extent of marine deposition should directly correlate with water depth, particularly during early transgression when a large area of the shelf is not accumulating sediment (for example the patchy development of a phosphatic lag beneath the black shale at some locations points to a period of sediment starvation prior to black shale formation). In the EPR model, the greatest water depths are predicted to occur around the base of the highstand systems tract while the peak extent of deposition occurs in the topmost part of this systems tract. The restricted-circulation model for shallow water black shales has even greater applicability to lowstand depositional conditions when epicontinental marine basins are at their most isolated, yet there is no correlation between black shale development and lowstand conditions (as indeed Hallam and Bradshaw, 1979, recognized).

In a reassessment of transgressive black shales, Wignall (1991) suggested that the combination of sea level rise, sediment starvation, and ongoing, background subsidence regimes created relatively deep water conditions over a proportionately large area of epicontinental seas, the "expanding-puddle model." Thus, transgressive black shales are considered to be typically basinal facies developed atypically in marginal areas. This is supported by the paleoecologic, sedimentologic, and geochemical characteristics of the Jet Rock black shales, which remain uniform throughout their outcrop from the basin centers to more marginal locations. To investigate the sequence stratigraphic occurrence of black shales further, another example is presented from the Late Carboniferous (Pennsylvanian) of northern England, of a succession that contains black shales associated with both early transgression (as in the Jet Rock example) and also maximum flooding.

UPPER CARBONIFEROUS BLACK SHALES OF NORTHERN ENGLAND

Late Namurian (Upper Carboniferous) sedimentation in northern England occurred in a complex series of fault-bounded basins undergoing a thermal phase of subsidence following active rifting in the early Namurian (Leeder and McMahon, 1988). The Rossendale and Huddersfield subbasins were the two principal depocenters in the central part of northern England (Figure 3) where they contain 2750 m of Upper Carboniferous sediments. The basin infill primarily consists of deltaic sandstones and their associated shales (Collinson, 1988). The latter are commonly ascribed a nonmarine epithet for they contain no fauna other than occasional bands of "nonmarine" bivalves. Deposition is considered to have occurred in fresh or brackish water (Spears and Amin, 1981). The only marine fauna is restricted to thin bands, commonly associated with black, organic-rich, paper-shale lithologies that covered extensive areas of northern England on more than 60 occasions in the Namurian (Ramsbottom, 1977). Virtually every marine band contains a diagnostic fauna of goniatites, enabling the detailed subdivision of the basin infill. In sequence-stratigraphy terminology, the marine bands represent maximum flooding surfaces (Read, 1991; Maynard et al., 1991) when terrigenous input to the basin center was at a minimum. Deep-water deposition beneath thermoclines seems a likely scenario, as envisaged by Heckel (1991) for similar, nearly contemporary facies in North America.

Despite more than 60 years of intensive research since Bisat's (1924) pioneering work on goniatites, the

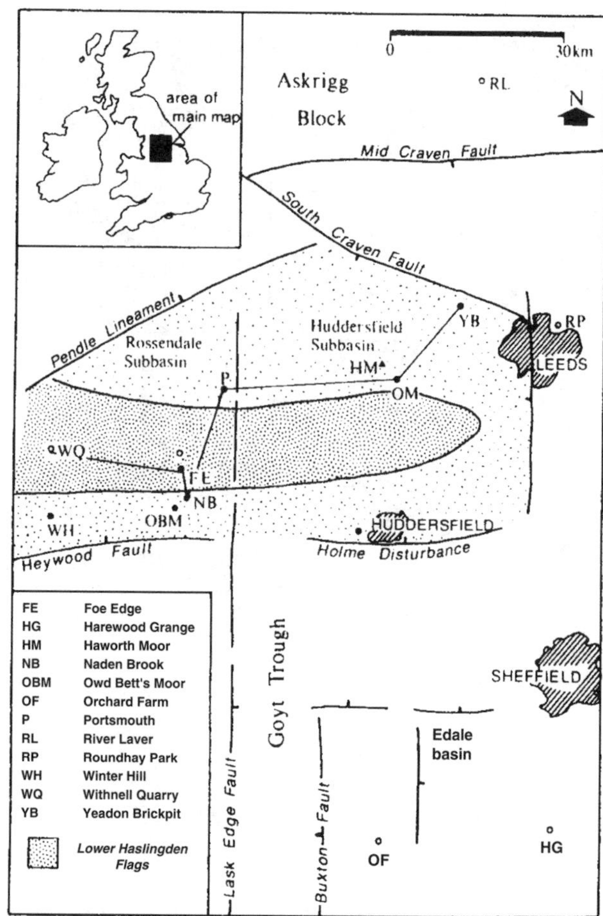

Figure 3. Upper Carboniferous basins of northern England (from Lee, 1988) showing the extent of lower Haslingden Flags deposition within the Rossendale and Huddersfield subbasins. Study localities indicated: ○ locality at which Owd Bett's Horizon is not present, ● locality at which Owd Bett's is present.

stratigraphy and sedimentology of the "nonmarine" mudrock facies is still only vaguely understood. However, a preliminary analysis of late Yeadonian (late Namurian) mudrocks by Maynard et al. (1991) revealed a previously unrecognized basal transgressive marine black shale.

Yeadonian Marine Bands

Gastrioceras cumbriense Marine Band

The fine-grained facies associated with the Gastrioceras cumbriense Marine Band have been logged at ten localities within the Rossendale and Huddersfield subbasins (Figures 3, 4) and at four more localities to the north, east, and south of this major depocenter (Figure 5). Note was made of grain size, sediment color, and fissility (the basic divisions being wavy, platey, and papery, the last characteristic commonly occurring in black shales) and the pres-

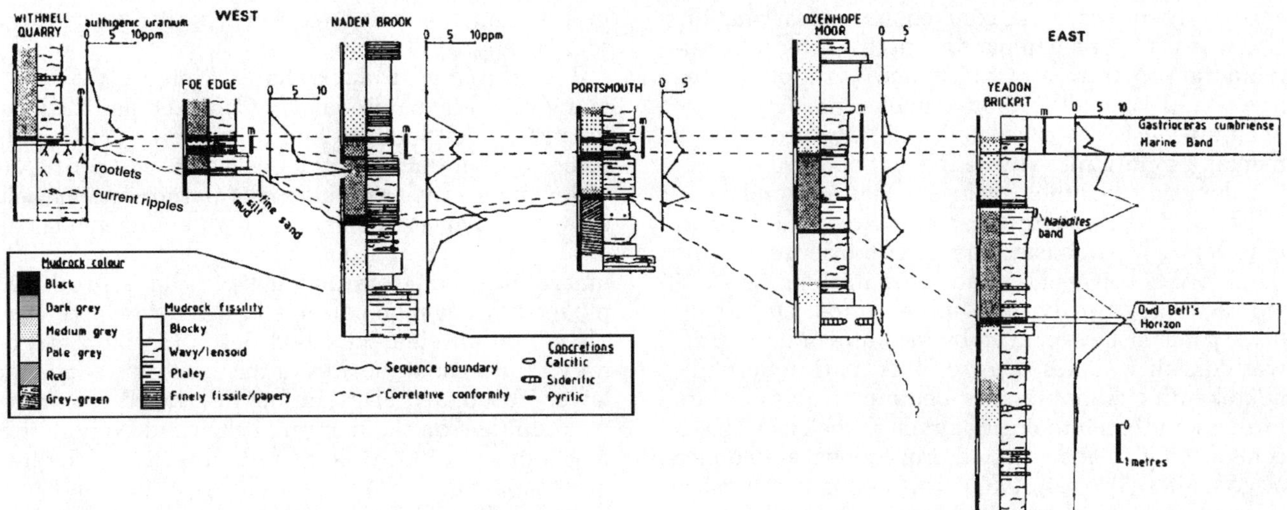

Figure 4. Correlation of the shale successions associated with the G. cumbriense Marine Band from the westernmost Rossendale subbasin where the Marine Band directly overlies the lower Haslingden Flags (Withnell Quarry) to the eastern margin of the Huddersfield subbasin where a more complete shale succession is developed. Scale bars marked in meters. Left-hand column indicates mudrock colors, right-hand column depicts fissility and grain size. Solid vertical lines, labeled m, indicate the occurrence of a shelly marine fauna. Plots of authigenic uranium are calculated from data collected using a portable gamma ray spectrometer.

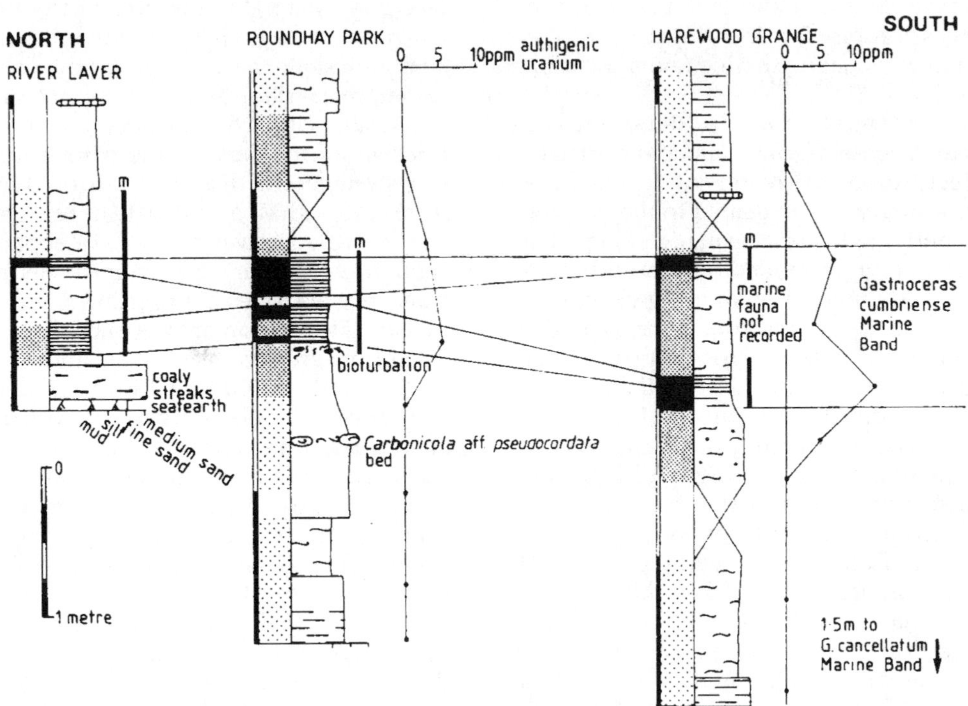

Figure 5. Correlation of G. cumbriense Marine Band localities in north and central England. Localities shown in Figure 3, key as for Figure 4. Occurrence of *Carbonicola* aff. *pseudocordata* from R. M. C. Eager (personal communication, 1991).

ence or absence of fauna, concretions, and lamination. Use was made of a gamma-ray spectrometer to determine the proportions of K, Th, and U. Applying the Myers and Wignall (1987) technique, the abundance of authigenic U was calculated, and this paleoenvironmentally significant value (Leeder et al., 1990) is plotted alongside the lithologic logs in Figures 4 and 5.

Beyond the margins of the Rossendale and Huddersfield subbasins the G. cumbriense Marine Band ranges between 0.6 and 1.0 m thick in the studied sections. It is divisible into an upper and lower black paper shale separated by medium- or dark-gray, wavy-fissility shales (Figure 5). A marine fauna of nektobenthic goniatites and benthic bivalves occurs throughout these lithologies although the black shales contain a higher absolute and proportional abundance of goniatites (Wignall, 1987). This points to a harsher benthic environment in the black shales probably in the lower reaches of the dysoxic zone (also indicated by the elevated authigenic U values; Figure 5). The presence of some authigenic U and only a few benthic species in the wavy-fissility shales indicates that they also belong to dysoxic facies. At the northernmost locality, in the banks of the River Laver, the lowest black shale is only 0.3 m above a seat earth (a sandy paleosol with abundant rootlets), but it is separated by a fine sandstone containing abundant coaly material. The Marine Band therefore appears not to sit directly on the potential sequence boundary at the top of the seat earth but is separated from it by a very thin transgressive systems tract. Twenty kilometers to the north of the River Laver locality, black shales are not developed and the Marine Band is developed in an expanded succession of coarser-grained facies containing a diverse, stenohaline marine fauna including brachiopods (Wignall, 1987).

Within the Rossendale and Huddersfield subbasins the G. cumbriense Marine Band and associated strata shows considerably greater lateral-thickness variations. These appear to be related to the development of the underlying lower Haslingden Flags. This deltaic-sandstone body prograded into the subbasins from west to east in the form of an elongate, bar-finger sand, according to Collinson and Banks (1975), although see Bristow (1988). Deposition of the succeeding shales was clearly influenced by the presence of this sandstone body for the thinnest and most incomplete successions directly overlie the thickest development of the lower Haslingden Flags. The thickest shale sections occur at the margins of the subbasins at Yeadon Brickpit (type locality for the Yeadonian) and at Winter Hill (Maynard et al., 1991). At both of these localities, a total of four thin, black or dark-gray paper shales are developed (Figure 4). The upper two at Yeadon contain a marine fauna like that found in the G. cumbriense Marine Band beyond the subbasin. The third-lowest black shale is unfossiliferous and directly overlies a coarsening-up sequence which contains in its topmost part a fauna of *Naiadites* associated with *Spirorbis*. At Winter Hill a marine fauna is found within all three of the uppermost black shales and in the intervening sediments, indicating a greater marine influence in the southwest of the region (Wignall, 1987).

It is the lowest black shale, named Owd Bett's Horizon in Maynard et al. (1991), which has the most interesting development and distribution. It is readily recognized at outcrop by its resistant weathering, its yellow patina (due to oxidation of pyrite), and its well-developed papery fissility and high authigenic U values. Although it does not contain any visible macrofauna, microfaunal samples have produced moderately common paleoniscid fish debris and a single conodont element. Owd Bett's Horizon can be traced throughout the area of the subbasins where the lower Haslingden Flags are not developed (Figure 4). At Foe Edge, on the margin of the sand body, a thin development of Owd Bett's rests directly on the topmost sandstone. On the central axis of the sandstone (Withnell Quarry, Figure 3) only one of the two uppermost leaves of the G. cumbriense Marine Band overlies a seat earth while the other black shales are not present.

The sequence-stratigraphic interpretation of this shale succession within the subbasins is best visualized in a chronostratigraphic chart (Figure 6). A sequence boundary caps the lower Haslingden Flags, and subaerial exposure is indicated by rootlets in the most westerly outcrop. The presence of reddened mudstones at Portsmouth, immediately below Owd Bett's Horizon, may record subaerial oxidation even in the center of the Rossendale subbasin. The position of the correlative conformity in the expanded Yeadon succession is more difficult to locate. The succeeding Owd Bett's Horizon lies on a basal transgressive surface and onlaps the margins of the preceding lower Haslingden Flags delta. It is overlain by more extensive gray shale facies. Importantly, Owd Bett's shows no lateral variability and does not grade into more marginal facies. The three black shales of the G. cumbriense Marine Band fit the criteria for marine flooding surfaces in the EPR model. They grade into dark-gray shales and ultimately into expanded nearshore successions and are separated from the sequence boundary by a thin development of shaly transgressive facies in the basins.

This late Yeadonian sequence thus contains both a basal transgressive black shale and black shales associated with maximum flooding within the same sequence. The G. cumbriense Marine Band black shales pass into expanded marginal sediments while the Owd Bett's Horizon has no lateral facies equivalents. It is possible that two minor cycles are represented in this succession with the base of the second cycle occurring immediately above the *Naiadites* band seen at Yeadon (Figure 4). However, the occurrence of the black shales in this alternative interpretation is not altered. In order to investigate whether different depositional conditions produced these distinct sequence stratigraphic occurrences a geochemical study was undertaken.

Geochemistry

Authigenic uranium. Values of authigenic U were determined by gamma-ray spectrometry. Both the G.

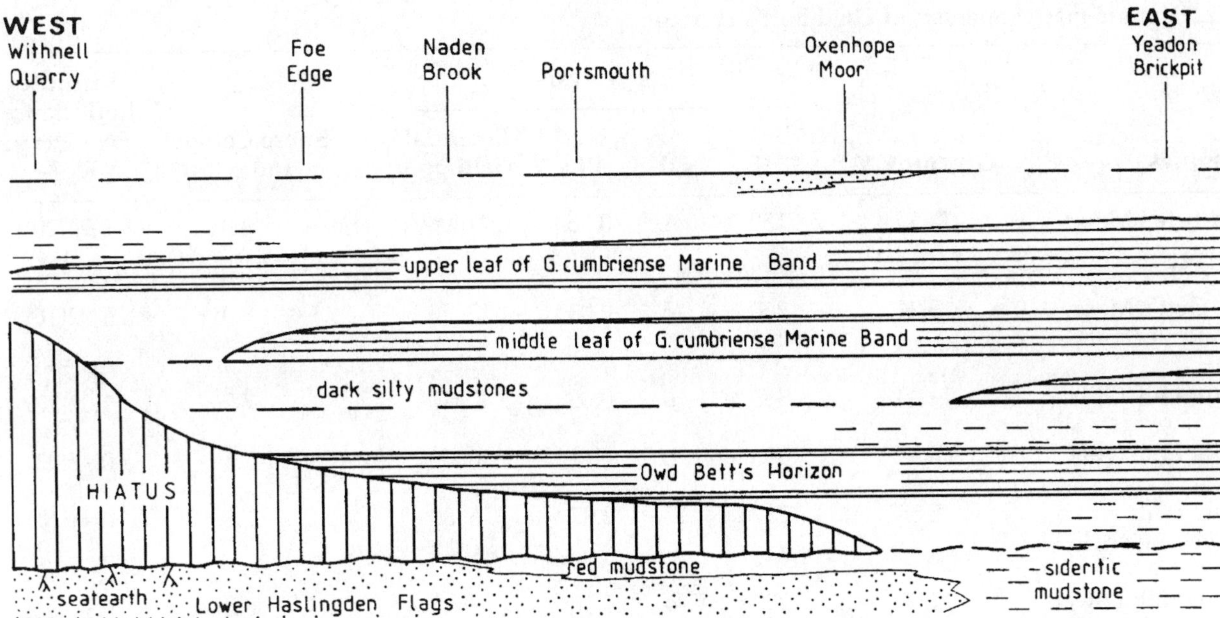

Figure 6. Chronostratigraphic chart for the upper Yeadonian succession from the west of the Rossendale subbasin (Withnell Quarry) to the east of the Huddersfield subbasin (Yeadon Brickpit). See Figure 3 for location.

Table 1. Geochemical characteristics of Owd Bett's Horizon (OBH) vs. the black shales of the G. cumbriense Marine Band (GCMB).

	GCMB	OBH
Organic C %	4.5 (n = 12)	12.5 (n = 6)
C/S_0	2.3 (n = 11)	2.08 (n = 5)
DOP	0.3–0.5	0.9
Authigenic U ppm	6.1 (n = 10)	14.2 (n = 8)

cumbriense Marine Band black shales and Owd Bett's Horizon show enrichment of authigenic U (Table 1), testifying to oxygen-restricted depositional conditions (Wignall and Myers, 1988). The average Owd Bett's value is considerably higher than that for Marine Band due mainly to the high value of 35 ppm authigenic U at the former's type locality at Owd Bett's Moor (Maynard et al., 1991). There appears to be no consistent variation of values for Owd Bett's Horizon, whereas the Marine Band black shales show higher values around the margins of the subbasins (Figure 7). This may be due to a greater abundance of humic organic matter around the margins closer to the terrigenous sources. This type of organic matter is known to have a strong affinity for uranium (Leventhal, 1981).

Degree of pyritization (DOP). The DOP index has proved of utility in assessing the degree of anoxicity of black shales and is defined as the proportion of total reactive iron incorporated in pyrite (Raiswell et

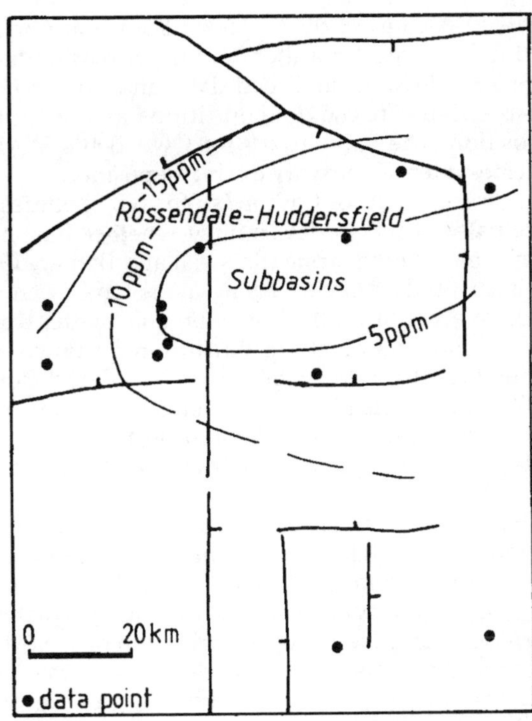

Figure 7. Contoured abundance of maximum measured authigenic uranium values in the G. cumbriense Marine Band in northern and central England showing lowest values in the center of the Rossendale and Huddersfield subbasins. Data in Figures 4 and 5 and in Maynard et al. (1991). See Figure 3 for location of map.

Table 2. Organic-matter analysis of Owd Bett's Horizon.

Samples	Organic C %	Pyrolysis				Spore Colour Index	Vitrinite Reflectance
		HI	OI	PI	Potential Yield ppm		R_o %
1. Owd Bett's Moor: OBH	14.30	150	1	0.05	21490	8.0	0.99(20)
2. Owd Bett's Moor: OBH upper 5 cm	5.24	260	5	0.03	13620	8.0	1.02(16)
3. Winter Hill: OBH	11.00	369	6	0.06	40540	7.5	1.02(22)
4. Winter Hill: OBH	6.82	306	6	0.11	20850	7.0–7.5	0.97(23)
5. Winter Hill: intervening shale	2.85	217	12	0.07	6180	7.5	0.94(14)
6. Winter Hill: Lower leaf GCMB	13.60	293	4	0.10	39810	8.0	0.96(22)
7. Winter Hill: Upper leaf GCMB	10.70	132	7	0.08	14140	7.5	1.01(27)
8. Naden Brook: OBH	12.40	194	3	0.03	24000	7.5–8.0	1.03(31)

al., 1988). Values of 0.3 to 0.5 for the G. cumbriense Marine Band are typical of Namurian marine bands in general (JRM unpublished data) and point to oxygen-restricted (dysoxic) conditions at the time of deposition. The value of 0.9 for Owd Bett's Horizon indicates intense anoxicity during deposition.

Carbon/sulfur plots. Carbon/sulfur (C/S) values are principally of use in determining whether the shales are marine or nonmarine (Raiswell and Berner, 1985). The measured organic carbon values have been corrected to account for the loss of organic matter due to thermal maturation using vitrinite-reflectance data (Table 2) and the technique of Raiswell and Berner (1987). The results give closely similar C/S values for Owd Bett's Horizon and the Marine Band and both are typical normal-marine values (Table 1; Berner and Raiswell, 1983).

Organic-matter type. Samples of Owd Bett's Horizon and black shales from the G. cumbriense Marine Band from Owd Bett's Moor, Naden Brook, and Winter Hill were analyzed by Mobil North Sea Ltd. to ascertain the maturity and origin of the organic matter. Pyrolysis, spore, and vitrinite analyses were undertaken on eight samples while two samples were further subjected to gas chromatography and mass spectrometry in order to determine pristane (Pr/C-17) and phytane (Ph/C-18) ratios. Results are summarized in Table 2.

The potential yield and hydrogen index (HI) of both the Marine Band black shales and Owd Bett's Horizon are high, suggesting that both are good oil source rocks. Vitrinite reflectance for all analyzed samples is in the range of 0.94 to 1.03%, indicating that the shales are mature for oil generation. Both the Marine Band and Owd Bett's contain type 1 kerogen (Figure 8) although the original organic-matter type has probably been altered due to bitumen loss during thermal maturation. The similarity of the organic-matter types in the two black shales is emphasized in the pristane–phytane plot (Figure 9) where they both have the same ratio, indicating a combined terrestrial and marine source for the organic matter. Biodegradation also influences the values of Pr/C-17 against Ph/C-18 by decreasing the C-17 and C-18 relative to pristane and phytane. The enhanced biodegradation of Owd Bett's relative to the Marine Band may point to greater weathering of the sample.

In summary, the black shales of Owd Bett's Horizon and the Marine Band are geochemically similar. Both appear to have been deposited under oxygen-deficient marine conditions. The intensity of anoxia during Owd Bett's deposition appears to have been somewhat greater, giving higher values of authigenic U, DOP, and total organic carbon. Some lateral variation in the organic-matter type in the Marine Band black shales is suggested by variations in authigenic U, whereas the basal transgressive Owd Bett's appears uniform throughout its outcrop. Deposition of the Marine Band probably occurred under dysaerobic, rather than truly anaerobic, deposition as testified by the moderate values of DOP and the presence of some benthic species (Wignall, 1987).

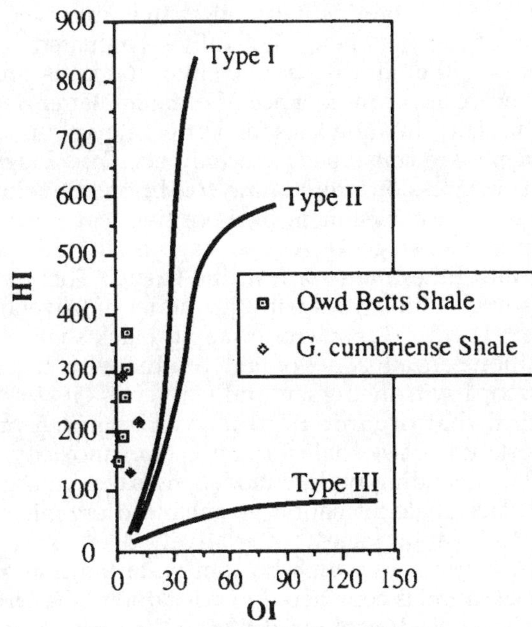

Figure 8. Modified van Krevelen plot for black shales from the G. cumbriense Marine Band and Owd Bett's Horizon. HI—hydrogen index, OI—oxygen index, kerogen Types I–III shown.

Further examples of previously undocumented Owd Bett's type black shales have been located in the Rossendale–Huddersfield subbasins; for example, a similar authigenic U-enriched black shale closely underlies the Gastrioceras cancellatum Marine Band.

DISCUSSION

The occurrence of both basal transgressive and maximum-flooding black shales within the British Namurian appears to parallel the Pennsylvanian successions of North America. Moore (1964) was among the first to recognize two distinct depositional sites for black shales—basinal and nearshore. In some instances the latter rest directly on paralic facies such as coals. Previously Zangerl and Richardson (1963) proposed rapid deposition in shallow water for a nearshore example, the Mecca Quarry Shale. But, Heckel (1977, p. 1059) noted, "Although coals are non-marine to shoreline deposits, the overlying (black shale) beds do not necessarily have to be very shallow-water deposits . . . the coal swamp could have foundered with no significant deposition."

Zangerl and Richardson's shallow-water model has been recently revived by Coveney and Shaffer (1988) and Coveney et al. (1991). They noted that nearshore shales are enriched in certain metals, particularly molybdenum, in distinction to offshore shales which are rich in phosphate. The nearshore shales are also characterized by variable sulfur-isotopic ratios, claimed to be indicative of deposition in shallow brackish water (Coveney and Shaffer, 1988). The nearshore shales also contain abundant, well-preserved fish which suggest rapid burial (Allison, 1988) and "a prolific fish population" (Coveney and Shaffer, 1988, p. 20). The main evidence for shallow-water deposition lies in the observation that the black shales overlie coals (but see Heckel's [1977] pertinent comments quoted above). No stratigraphic data was presented to show what the correlative facies of the nearshore black shales are in distal settings but the model of Coveney et al. (1991) predicts that the black shales should become thinner offshore and pass into more normal-marine facies. In fact Heckel and Hatch (1992) showed that the nearshore black shales are commonly extensive and do not grade laterally into shoreline facies. At least some of the Pennsylvanian black shales are condensed sections associated with maximum flooding surfaces (e.g., Heckel, 1977, Figure 7) like the G. cumbriense Marine Band.

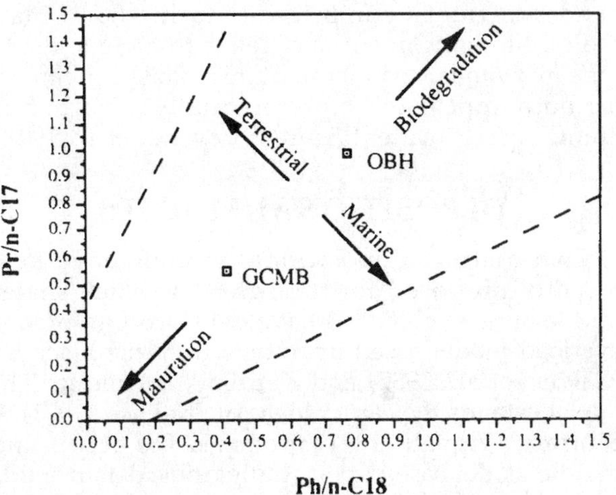

Figure 9. Pristane–phytane plot for Owd Bett's Horizon (OBH) and G. cumbriense Marine Band (GCMB) organic matter. The upper line represents a ratio for terrestrial organic matter (e.g., coals) and the lower line represents a ratio for marine organic matter (cf, Dydik et al., 1978).

Owd Bett's Horizon shows many of the features of the nearshore black shales described by Coveney et al. (1991), as it contains fish debris and overlies shallow-water (deltaic) facies. The presence of fish does not necessarily indicate high sedimentation rates; indeed, concentrations of such pelagic fauna are usually indicative of very slow sedimentation rates (Loutit et al., 1988). Well-preserved fish may indicate high episodic sedimentation rates but not high bulk sedimentation rates. Alternatively, fish may be preserved by overgrowth of bacterial mats in very low oxygen settings with no implications for sedimentation rates. The sparsity of conodonts in Owd Bett's Horizon may be indicative of high sedimentation rates although this could equally be caused by intense anoxicity in the water column.

The possibility of Mo enrichment in Owd Bett's Horizon has yet to be evaluated although other geochemical factors show little variation between this band and the G. cumbriense Marine Band black shales. Mo enrichment may relate to organic-matter type in a similar manner to authigenic U enrichment, for both appear to be preferentially adsorbed by humic organic matter (Figure 7; Coveney et al., 1991).

DEPOSITIONAL MODELS

Two distinct depositional models have been recently proposed for transgressive black shales. Middelburg et al. (1991) have developed an organic overload model based upon data from the Black Sea (Calvert et al., 1987) and Kau Bay, Indonesia. This model extends the earlier ideas of Strakhov (1971). In contrast, Wignall (1991) proposed the expanding-puddle model based upon stratigraphic data from the geological record, particularly the Jet Rock.

Middelburg et al. (1991) have traced the development of barred basins during the Holocene transgression (Figure 10A). Initially the basins are isolated freshwater lakes, but with a rise in sea level a saline underflow develops (Figure 10A, stage 2). Eventually the freshwater wedge is displaced from much of the basin and the surface waters become saline. The result is "higher plankton production brought about by displacement of nutrients from the deep waters into the euphotic zone" (Middelburg et al., 1991, p. 680). The high productivity causes the rapid accumulation of sapropel beneath a fully oxygenated water column, although the benthic environment is probably anoxic as a consequence of organic matter degradation. This situation lasts for a brief interval until the nutrients are consumed, and only at a later stage of the transgression are euxinic conditions developed due to the establishment of a positive water balance (Figure 10A, stage 4). At this stage finely laminated coccolith limestones form in the basins. This model has several similarities with the model of Coveney et al. (1991) who also suggested rapid black shale accumulation beneath zones of high productivity. It is also in accord with Pedersen and Calvert's (1990) contention that organic productivity is much more important in black shale formation than anoxicity.

The expanding-puddle model envisages transgressive black shale formation by enhanced organic matter preservation beneath a relatively deep, stratified water column in which bottom waters are anoxic. This situation is considered to arise due to the virtual cessation of sediment input during the early stages of transgression (Figure 10B, stage 2). Productivity may be quite low during this interval due to the paucity of land-derived nutrients.

The two models are irreconcilable, for Coveney et al. (1991) predict very rapid black shale accumulation

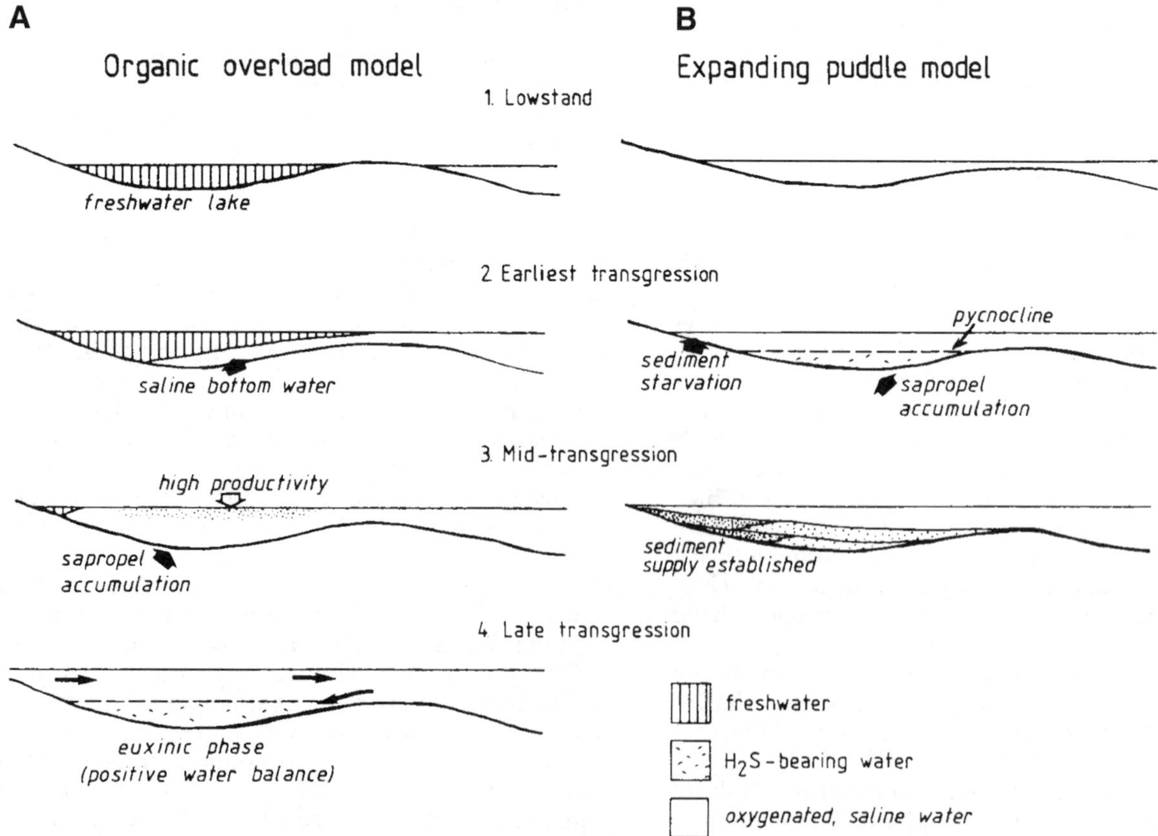

Figure 10. Competing models for the development of transgressive black shales. A—Organic overload model of Middelburg et al. (1991). B—Expanding-puddle model of Wignall (1991). Stage 1—Lowstand; stage 2—Earliest transgression; stage 3—Mid-transgression; stage 4—Late transgression.

beneath an oxygenated water column, whereas Wignall (1991) predicts very slow accumulation, in association with sediment starvation, beneath anoxic bottom waters. It would be convenient if transgressive black shales could be distinguished from other black shales by some geochemical criteria; Coveney et al. (1991) have suggested Mo enrichment of the former, but Heckel and Hatch (1992) have shown this to be erroneous. In the present study, the transgressive Owd Bett's black shales is geochemically indistinguishable from the G. cumbriense black shales. The stratigraphic occurrence of Owd Bett's Horizon at the transition from nonmarine to marine facies is in close agreement with the Middelburg et al. (1991) model and the rarity of conodonts may indicate rapid sedimentation during organic overloading. The Jet Rock black shales cannot be explained by this model because their development is not preceded by freshwater lacustrine conditions. If two different models are invoked for the Owd Bett's and Jet Rock black shales then it is curiously coincidental that they have identical sequence stratigraphic occurrences.

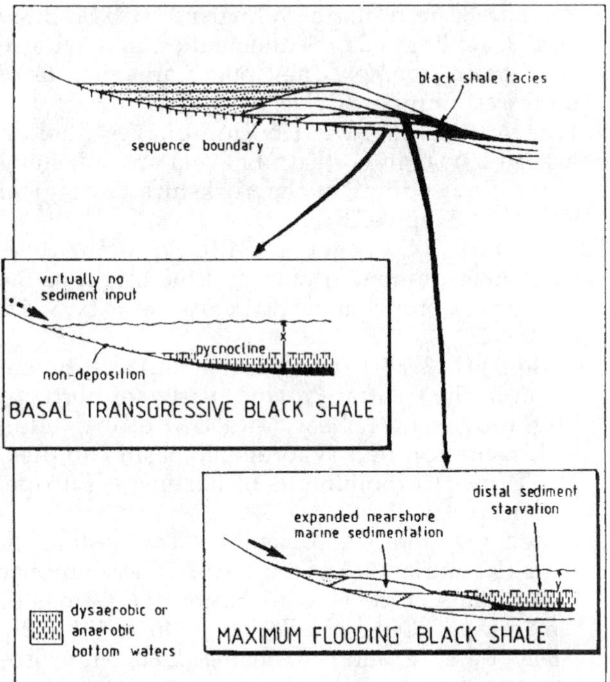

Figure 11. Sequence stratigraphic occurrences of basal transgressive and maximum flooding black shales illustrating their depositional environments based upon the expanding-puddle model. It is unknown if water depths *x* and *y* are comparable, although both are considered to be relatively deep. In both models, black shale accumulation occurs beneath a density interface.

CONCLUSIONS

Thick black shales may form all or part of a sequence in distal marine settings where the water is deep because of rapid sea level rise and/or because subsidence is great. These occurrences are adequately explained by the puddle model (Hallam and Bradshaw, 1979; Wignall, 1991). Thin black shales extend from basinal into marginal settings where they occur in close stratigraphic association with potential reservoir rocks. These thin black shales fall into two categories (Figure 11).

Basal transgressive black shales are distinguished by the following features:

1. They rest directly on sequence boundaries and their correlative conformities.
2. They pass proximally into sediment-starved marginal facies such as phosphatic lags (Jet Rock example) or surfaces of nondeposition (Owd Bett's example).
3. They exhibit no obvious lateral facies variation.

Maximum flooding surface black shales (the G. cumbriense Marine Band examples) are distinguished by the following:

1. They rest conformably on sediments of the transgressive systems tract.
2. They pass proximally into thicker sections of shallower marine facies.
3. They exhibit the most oxygen-restricted facies in the basin center.

Both black-shale types can be closely associated with shallow-water facies around the basin margin although the basal transgressive black shales rest directly on such facies, whereas the maximum flooding black shales are usually separated from them by transgressive system tracts, although these may only be very thin (e.g., River Laver locality, Figure 5).

ACKNOWLEDGMENTS

We thank George Grabowski (Exxon), Phil Heckel (University of Iowa), Richard Hodgekinson (Mobil), Michael Eager (Knaresborough), and Nick Riley (British Geological Survey), and at Leeds University we thank John Varker, Rob Raiswell, Simon Dean, and Mike Leeder for their help.

REFERENCES CITED

Allison, P.A., 1988, The role of anoxia in the decay and mineralization of proteinaceous macro-fossils: Paleobiology, v. 14, p. 139-154.

Bell, J., J. Holden, T.H. Pettigrew, and K.W. Sedman, 1979, The Marl Slate and Basal Permian breccia at Middridge, Co. Durham: Proceedings of the Yorkshire Geological Society, v. 42, p. 439-460.

Berner, R.A., and R. Raiswell, 1983, Burial of organic carbon and pyrite sulfur in sediments over Phanerozoic time: a new theory: Geochimica et Cosmochimica Acta, v. 47, p. 855-862.

Bisat, W.S., 1924, The Carboniferous goniatites of the north of England and their zones: Proceedings of the Yorkshire Geological Society, v. 20, p. 40-124.

Bristow, C.S., 1988, Controls on the sedimentation of the Rough Rock Group (Namurian) from the

Pennine Basin of northern England, *in* B.M. Besly and G. Kelling, eds., Sedimentation in a synorogenic basin complex. The Upper Carboniferous of northwest Europe, Blackie, p. 114-131.

Calver, M.A., 1968, Distribution of Westphalian marine faunas in northern England and adjoining areas. Proceedings of the Yorkshire Geological Society, v. 37, p. 1-72.

Calvert, S.E., J.C. Vogel, and J.R. Southon, 1987, Carbon accumulation rates and the origin of the Holocene sapropel in the Black Sea. Geology, v. 15, p. 918-922.

Collinson, J.D., 1988, Controls on Namurian sedimentation in the Central Province basins of northern England, *in* B.M. Besly and G. Kelling, eds., Sedimentation in a synorogenic basin complex. The Upper Carboniferous of northwest Europe, Blackie, p. 85-101.

Collinson, J.D., and N.L. Banks, 1975, The Haslingden Flags (Namurian G_1) of south-east Lancashire: Bar finger sands in the Pennine Basin: Proceedings of the Yorkshire Geological Society, v. 40, p. 431-458.

Coveney, R.M., Jr., and N.R. Shaffer, 1988, Sulfur-isotope variations in Pennsylvanian shales of the midwestern United States: Geology, v. 16, p. 18-21.

Coveney, R.M., Jr., W.L. Watney, and C.G. Maples, 1991, Contrasting depositional models for Pennsylvanian black shale discerned from molybdenum abundances: Geology, v. 19, p. 147-150.

Demaison, G.J., and G.T. Moore, 1980, Anoxic environments and oil source bed genesis: Bulletin of the American Association of Petroleum Geologists, v. 64, p. 1179-1209.

Donovan, D.T., A. Morten, and H.C. Ivimey-Cook, 1979, The transgression of the lower Lias over the northern flank of the London platform: Journal of the Geological Society of London, v. 136, p. 165-173.

Dydik, B.M., B.R.T. Simoneit, S.C. Brassell, and G. Eglinton, 1978, Organic geochemical indicators of palaeoenvironmental conditions of sedimentation: Nature, v. 272, p. 216-222.

Grabowski, G.J., Jr., and K.S. Glaser, 1990, Depositional model for transgressive marine organic-rich rocks formed in platformal settings: Middle Callovian and Lower Toarcian examples from onshore England: Abstracts volume, 13th International Sedimentological Congress, Nottingham, U.K., p. 196.

Hallam, A., 1967a, An environmental study of the Upper Devonian and Lower Toarcian in Great Britain: Philosophical Transactions of the Royal Society, v. 252, p. 393-445.

Hallam, A., 1967b, The depth significance of shales with bituminous laminae: Marine Geology, v. 5, p. 481-493.

Hallam, A., and M.J. Bradshaw, 1979, Bituminous shales and oolitic ironstones as indicators of transgressions and regressions: Journal of the Geological Society of London, v. 136, p. 157-164.

Haq, B.U., J. Hardenbol, and P.R. Vail, 1988, Mesozoic and Cenozoic chronostratigraphy and cycles of sea-level change, *in* C.K. Wilgus et al., eds., Sea-level changes an integrated approach, SEPM Special Publication, v. 42, p. 71-108.

Heckel, P.H., 1977, Origin of phosphatic black shale facies in Pennsylvanian cyclothems of Mid-Continent North America: American Association of Petroleum Geologists Bulletin, v. 61, p. 1045-1068.

Heckel, P.H., 1991, Thin widespread Pennsylvanian black shales of midcontinent North America, a record of a cyclic succession of widespread pycnoclines in a fluctuating epeiric sea *in* R.V. Tyson and T.H. Pearson, eds., Modern and ancient continental shelf anoxia: Geological Society of London Special Publication, v. 58, p. 259-273.

Heckel, P.H., and J.R. Hatch, 1992, Comment on: Coveney, J.R. Jr, W.L. Watney, C.G. Maples, 1991, Contrasting depositional models for Pennsylvanian black shale discerned from molybdenum abundances: Geology, v. 20, p. 88-90.

Howarth, M.K., 1962, The Jet Rock Series and Alum Shale Series of the Yorkshire Coast. Proceedings of the Yorkshire Geological Society, v. 33, p. 381-422.

Jenkyns, H.C., 1988, The early Toarcian (Jurassic) anoxic event: Stratigraphic, sedimentary, and geochemical evidence: American Journal of Science, v. 288, p. 101-151.

Leckie, D.A., C. Singh, F. Goodarzi, and J.H. Wall, 1990, Organic-rich, radioactive marine shale: A case study of a shallow-water condensed section, Cretaceous Shaftesbury Formation, Alberta, Canada: Journal of Sedimentary Petrology, v. 60, p. 101-117.

Lee, A.G., 1988, Carboniferous basin configuration of Central and Northern England modelled using gravity data, *in* B.M. Besly and G. Kelling, eds., Sedimentation in a synorogenic basin complex. The Upper Carboniferous of northwest Europe: Blackie, p. 69-84.

Leeder, M.R., and A. McMahon, 1988, Upper Carboniferous (Silesian) basin subsidence in northern Britain, *in* B.M. Besly and G. Kelling, eds., Sedimentation in a synorogenic basin complex. The Upper Carboniferous of northwest Europe: Blackie, p. 43-52.

Leeder, M.R., R. Raiswell, H. Al-Biatty, A. McMahon, and M. Hardman, 1990, Carboniferous stratigraphy, sedimentation and correlation of well 48/3-3 in the Southern North Sea Basin: Integrated use of palynology, natural gamma/sonic logs and carbon/sulphur geochemistry: Journal of the Geological Society of London, v. 147, p. 287-300.

Leventhal, J.S., 1981, Pyrolysis gas chromatography-mass spectrometry to characterise organic matter and its relationship to uranium content of Appalachian Devonian black shales: Geochimica et Cosmochimica Acta, v. 45, p. 883-889.

Loutit, T.S., J. Hardenbol, P.R. Vail, and G.R. Baum, 1988, Condensed sections: The key to age dating and correlation of continental margin sequences, *in*

C.K. Wilgus et al., eds., Sea-level changes: an integrated approach: SEPM Special Publication, v. 42, p. 183-213.

Maynard, J.R., P.B. Wignall, and W.J. Varker, 1991, A "hot" new shale facies from the Upper Carboniferous of northern England: Journal of the Geological Society of London, v. 148, in press.

Middelburg, J.J, S.E. Calvert, and R. Karlin, 1991, Organic-rich transitional facies in silled basins: response to sea-level change. Geology, v. 19, p. 679-682.

Miller, A.G., 1990, A paleoceanographic approach to the Kimmeridge Clay Formation, *in* A.Y. Huc, ed., Deposition of organic facies: American Association of Petroleum Geologists, Studies in Geology, No. 30, p. 13-26.

Moore, R.C., 1964, Paleoecological aspects of Kansas Pennsylvanian and Permian cyclothems. Kansas Geological Survey Bulletin, v. 169, p. 287-380.

Morris, K.A., 1979, A model for the deposition of bituminous shales in the Lower Toarcian *in* Association Sedimentologistes Français, Symposium Sedimentation jurassique W-Européen, Publication Spéciale, v. 1, p. 397-406.

Myers, K.J., and P.B. Wignall, 1987, Understanding Jurassic organic-rich mudrocks—new concepts using gamma-ray spectrometry and palaeocology: Example from the Kimmeridge Clay of Dorset and the Jet Rock of Yorkshire, *in* J.K. Legget and G.G. Zuffa, eds., Marine clastic environments: Concepts and case studies: Graham and Trotman Ltd., London, p. 175-192.

Nummedal, D., and D.J.P. Swift, 1988, Transgressive stratigraphy at sequence-bounding unconformities: Some principles derived from Holocene and Cretaceous examples, *in* D. Nummedal, O.H. Pilkey, and J.D. Howard, eds., Sea-level fluctuations and coastal evolution. SEPM Special Publication. No. 41, p. 241-260.

O'Brien, N.R., 1990, Significance of lamination in Toarcian (Lower Jurassic) shales from Yorkshire, Great Britain: Sedimentary Geology, v. 67, p. 25-34.

Oschmann, W., 1988, Kimmeridge Clay sedimentation—a new cyclic model: Palaeogeography, Palaeoclimatology and Palaeoecology, v. 65, p. 217-251.

Pedersen, T.F., and S.E. Calvert, 1990, Anoxia vs. productivity: what controls the formation of organic-carbon-rich sediments and sedimentary rocks: American Association of Petroleum Geologists Bulletin, v. 74, p. 454-466.

Raiswell, R., and R. A. Berner, 1985, Pyrite formation in euxinic and semi-euxinic sediments: American Journal of Science, v. 285, p. 710-724.

Raiswell, R., and R.A. Berner, 1987, Organic carbon losses during burial and thermal maturation of normal marine shales: Geology, v. 15, p. 853-856.

Raiswell, R., F. Buckley, R.A. Berner, and T.F. Anderson, 1988, Degree of pyritisation of iron as a paleoenvironmental indicator of bottom-water oxygenation: Journal of Sedimentary Petrology: v. 58, p. 812-819.

Ramsbottom, W.H.C., 1977, Major cycles of transgression and regression (mesothems) in the Namurian: Proceedings of the Yorkshire Geological Society, v. 41, p. 261-291.

Read, W.A., 1991, The Millstone Grit (Namurian) of the southern Pennines viewed in the light of eustatically controlled sequence stratigraphy: Geological Journal, v. 26, p. 157-166.

Savrda, C.E., 1991, Ichnology in sequence stratigraphic studies: An example from the Lower Paleocene of Alabama: Palaios, v. 6, p. 39-53.

Spears, D.A., and M.A. Amin, 1981, Geochemistry and mineralogy of marine and non-marine Namurian black shales from the Tansley Borehole, Derbyshire: Sedimentology, v. 28, p. 407-417.

Strakhov, N.M., 1971, Geochemical evolution of the Black Sea in the Holocene. Lithology and Mineral Resources, no. 3, p. 263-274.

Wignall, P.B., 1987, A biofacies analysis of the Gastrioceras cumbriense Marine Band (Namurian) of the central Pennines: Proceeedings of the Yorkshire Geological Society, v. 46, p. 111-121.

Wignall, P.B., 1989, Sedimentary dynamics of the Kimmeridge Clay: Tempests and earthquakes: Journal of the Geological Society of London, v. 146, p. 273-284.

Wignall, P.B., 1991, Model for transgressive black shales?: Geology, v. 19, p. 167-170.

Wignall, P.B., and A. Hallam, 1991, Biofacies, stratigraphic distribution and depositional models of British onshore Jurassic black shales, *in* R.V. Tyson and T.H. Pearson, eds., Modern and ancient continental shelf anoxia: Geological Society of London Special Paper 58, p. 291-309.

Wignall, P.B., and K.J. Myers, 1988, Interpreting benthic oxygen levels in mudrocks: A new approach: Geology, v. 16, p. 452-455.

Zangerl, R., and E.S. Richardson, 1963, The paleoecological history of two Pennsylvanian black shales: Fieldiana Geological Memoir, v. 4, 352 p.

Chapter 5

Sedimentology of Organic Matter in Upper Tithonian–Berriasian Deep-Sea Carbonates of Southeast France: Evidence of Eustatic Control

D. Steffen
G. E. Gorin
University of Geneva
Department of Geology & Paleontology
Geneva, Switzerland

ABSTRACT

A quantitative palynofacies study was performed on upper Tithonian–Berriasian, open-marine, fine-grained carbonates outcropping at Berrias (Berriasian stage stratotype), Broyon, and Angles in the Vocontian basin of southeast France. Organic constituents are subdivided into a relatively autochthonous marine fraction (mainly dinocysts and secondarily foraminiferal linings) and an allochthonous land-derived fraction (higher plant debris and sporomorphs). Fluorescent amorphous organic matter is absent throughout these organic-lean carbonates.

Within a precise biostratigraphic framework based on ammonites and calpionellids, the three studied sections are interpreted in terms of sequence stratigraphy, using field observations, micro- and macrofacies analysis, and palynofacies data. Within this sequence stratigraphy framework, trends in the distribution of organic matter bear the imprint of sea level fluctuations. Moreover, a model of an idealized palynofacies sequence in open marine carbonates is proposed for each systems tract:

- The lowstand systems tract (LST) is dominated by large-scale, angular, and/or degraded terrigenous fragments;
- The transgressive systems tract (TST) is marked by the upward decrease in abundance, size, and angularity of terrigenous fragments, and by the upward increase in blade-shaped black humic fragments (approx. inertinite) and in dinocysts diversity and abundance, which peak at the maximum flooding surface (mfs);
- The highstand systems tract (HST) shows reversed trends, both dinocysts and blade-shaped black humic fragments decreasing, and terrigenous fragments increasing.

The outlined methodology has a definite application in petroleum geology. It can be used in different depositional environments, notably in more organic-rich sediments, where the potential presence of fluorescent amorphous organic matter, particularly abundant in transgressive episodes, provides an extra palynofacies parameter.

INTRODUCTION

The upper Tithonian–Berriasian open-marine carbonates in the Vocontian basin of southeast France are of high stratigraphic interest because they are well dated by ammonites and calpionellids. Furthermore, the Berrias section represents the Berriasian stage stratotype. Detailed analysis of the particulate organic matter in the Berrias stratotype has been previously described by Gorin and Steffen (1991). Two other sections are studied in this chapter, i.e., at Broyon and Angles (Figure 1), to provide lateral correlation across the basin.

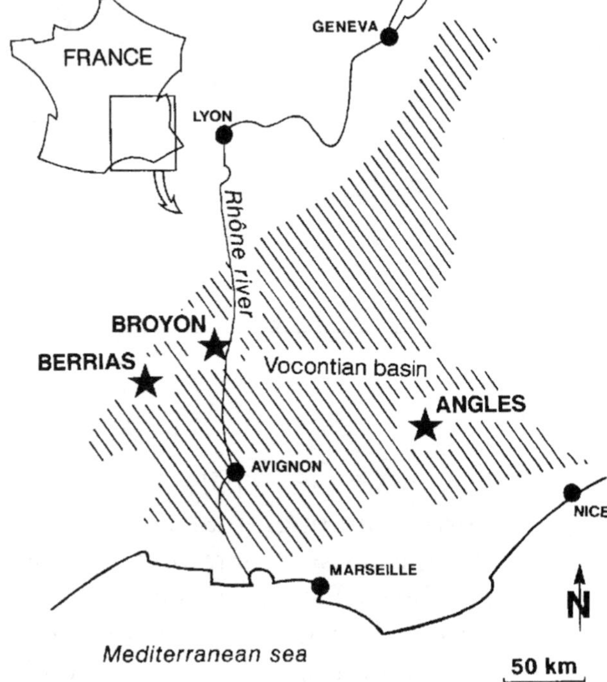

Figure 1. Location map.

In this study we aim to provide a clear picture of organic matter distribution in an open marine setting. This approach is not based on the total amount of organic matter, but only on the relative proportions of the different organic constituents and their variations in the studied sections. In a second stage, these variations are integrated into a sequence stratigraphy framework derived from the synthesis of field observations, standard sedimentology, and palynofacies analysis. Correlation between palynofacies parameters and this sequence stratigraphy framework may reveal the possible influence of eustacy on organic matter distribution across the three sections.

The upper Tithonian–Berriasian deposits of southeast France studied in this chapter do not display any source rock characteristics: their total organic carbon (TOC) content is too low (below 0.2%) and their palynofacies does not display any evidence of anoxia. However, the precision of the biostratigraphic framework allows us to precisely position palynofacies variations in time. If it can be demonstrated that these variations are associated with sea level changes, they may provide a good reference for studying Berriasian deposits in other basins. The same methodology can subsequently be applied to other more organic-rich open marine deposits (see discussion below).

METHODS

Fieldwork

This study relies on detailed lithologic and sedimentologic field observations. Special attention was given to the identification of key surfaces (e.g., erosion or nondeposition) and bed stacking patterns (e.g., evolution of bed thickness, cyclic features). Sampling was concentrated near biozone limits and important lithologic changes or any horizons showing indications of a sedimentary break.

A preliminary sequence stratigraphy interpretation was established on the basis of outcrop information. It then was compared to palynofacies indications. When some contradiction appeared between palynofacies and sequence stratigraphy, we went back to the field to collect supplementary samples in the problematic zone to improve systems tracts breakdown. To confirm or precisely define some limits, the final interpretation was compared with data obtained from geochemistry (Emmanuel and Renard, 1992) and clay mineralogy (Deconinck, 1992).

Preparation of Samples

Some 150 samples collected in the three studied sections have been analyzed. Because of very low organic content in these carbonates, large amounts of rock (usually 0.5 kg; exceptionally up to 2 kg per sample) were processed in order to get residues rich enough in particulate organic matter. These quantities are very high; palynologists working with shale samples usually process quantities around 5 to 20 g.

Organic residues were obtained using a standard palynologic treatment with nonoxidizing acids to remove the mineral fraction, followed by heavy liquid separation; no oxidation by nitric acid was used. The samples were subsequently sieved (15 µm) to eliminate the finely disseminated organic matter. All samples were processed by the same procedure to provide a valid comparison between different horizons, and to preclude changes in organic matter content that could result from preparation artifacts.

Observation Techniques and Procedure

Dispersed organic matter (sensu Burgess, 1974), or particulate organic matter (Hart, 1986; Tyson, 1987), was observed both in natural transmitted light and reflected UV-fluorescent light microscopy. Observations were made according to the following procedure (steps 1 to 3 in transmitted light):

1. Identification of organic constituents.
2. Establishment of the relative proportion of each organic constituent by counting at least 200 particles. (We made some tests by counting up to 500 particles, but it appears that, after a count of 200, the relative proportions expressed as percentages show only minor changes [plus or minus 1%]. However, in the analyzed sections, constituents variations of less than 5% will be considered insignificant to reduce analytical artifacts.)
3. Determination of other palynofacies parameters, such as: diversity and type of plankton, measurement of size, and visual assessment of sorting, shape, and degree of biodegradation of some constituents.
4. Observation under incident UV-fluorescent light.

USE OF PALYNOFACIES

Choice of Classification

Combaz (1964) introduced the term *palynofacies* for the global microscopic image of organic constituents in a rock, which gives information on stratigraphy, paleogeography, paleoecology, and diagenesis. In the present study, the contribution of palynofacies is essentially paleogeographic and paleoecologic. Additional methods, mainly biostratigraphy and sedimentology, were required to complement palynofacies data and help provide a reliable sequence stratigraphy interpretation.

There are many classifications of sedimentary organic particles (e.g., Combaz, 1964, 1980; Staplin, 1969; Burgess, 1974; Bujak et al., 1977; Parry et al., 1981; Whitaker, 1984; Boulter and Riddick, 1986; Powell et al., 1990; Davies et al., 1991; and see Tyson, 1987, for the correspondence between the main organic groups). This diversity in classification schemes is a result of different analysis techniques and/or the specific scope of each study (transmitted vs. reflected light, geochemistry, etc., see Tissot and Welte, 1984, p. 541).

There is presently considerable debate among palynologists and organic petrologists about defining a complete working classification (e.g., Open Workshop on Organic Matter Classification, Amsterdam, June 27–28, 1991). Proposing a new terminology would simply add more confusion to this discussion. Consequently, an existing classification has been used in the present work, i.e., the so-called "Shell palynomaceral classification" proposed by Whitaker (1984).

Description of Organic Constituents

Only the organic particles encountered in the Berriasian carbonates of southeast France (Figure 2) will be discussed below:

- Palynomaceral 1 (PM1)—orange to brown translucent fragments, visual "gelatinized aspect," generally no visible original structure; commonly designated as vitrinite in reference to its approximate coal maceral equivalent; originates from partially oxidized higher plant debris.

	ORIGIN	GROUP	CONSTITUENT		SYMBOLS
CONTINENTAL (ALLOCHTHONOUS)	higher plant debris (macrophytes tissues)	palynomacerals (PM)	PM1		
			PM2		
			PM3		
			PM4	equidimensional	
				blade-shaped (T)	
	pollen & spores	sporomorphs (1)	bisaccates		
			non-saccates		
	highly degraded macrophytes tissues	amorphous organic matter (AOM) (2)	non-fluorescent AOM		
MARINE (REL. AUTOCHTHONOUS)	mainly degraded phytoplankton		fluorescent AOM		
	marine phytoplankton (3)		dinoflagellate cysts (dinocysts)	proximate	
				chorate	
			acritarchs		
			tasmanitids		
	foraminifera		foraminifera test linings		

Figure 2. Classification of organic constituents valid for Lower Cretaceous carbonates of southeast France (adapted from Whitaker, 1984). (1) Sporomorphs—being rare in the studied sections, no distinction was made in this chapter between bisaccates and non-saccates. (2) Amorphous organic matter is not present in Berriasian carbonates. (3) Dinoflagellate cysts and acritarchs are not all strictly marine; although a few lacustrine species are known (Traverse, 1988), all species encountered in this study are marine (E. Monteil, personal communication, 1992).

- Palynomaceral 2 (PM2)—orange to brown translucent fragments, visual "gelatinized aspect" not so pronounced as in PM1, cellular structure visible; intermediate stage between PM1 and PM3; originates from poorly oxidized higher plant debris.
- Palynomaceral 3 (PM3)—yellow fragments, often clear, cellular structure clearly visible; commonly designated as cutinite in reference to its approximate coal maceral equivalent; originates from cuticles of higher plant leaves.
- Palynomaceral 4 (PM4)—black opaque fragments; commonly designated as inertinite in reference to its approximate coal maceral equivalent; originates from highly oxidized (carbonized) higher plant debris.
- Sporomorphs—land plant spores and pollen; because of the scarcity of this group in the studied samples, no distinction was made between saccate and non-saccate sporomorphs.
- Dinocysts—cysts of dinoflagellates; marine microplankton.
- Foraminiferal linings—organic (tectin) linings of some foraminifera; considered as a marine indicator (Stancliffe, 1989).

Whitaker (1984) proposed suffixes to qualify palynofacies types (B—degraded; L—bleached; R—markedly rounded; S—notably small; T—blade-shaped). They are not used here exactly in the same sense. A suffix was added to groups of organic particles and not to the global palynofacies, e.g., PM4-T for blade-shaped palynomaceral 4 or PM3-L for bleached palynomaceral 3.

Fluorescent (hydrogen-rich) amorphous organic matter of algal/bacterial origin characterizes low-energy, stagnant, oxygen-depleted paleoenvironments (Staplin, 1969; Bujak et al., 1977; Tyson, 1987). The lack of amorphous organic matter throughout the upper Tithonian–Berriasian sediments analyzed here indicates relatively oxidizing depositional conditions.

Principles of Organic Sedimentology

Particulate organic matter is composed of physical particles that behave as any other sedimentary grains with their own physical characteristics (Huc, 1988). Most organic particles are hydrodynamically similar to clay and silt (Tyson, 1987). Consequently, one can apply parameters traditionally used by "mineral" sedimentologists, such as size, sorting, or rounding of fragments. These criteria may give an estimation of the energy conditions at the time of deposition (Denison and Fowler, 1980; Whitaker, 1984). However, caution must be taken when using this approach because organic residues are composed of two fractions: allochthonous (land-derived) and relatively autochthonous (marine) (Whitaker, 1984). The distribution of the allochthonous fraction in a marine environment (mainly land-derived organic fragments and sporomorphs) is essentially controlled by physical processes. By contrast, the relatively autochthonous fraction (mainly marine algae and foraminiferal linings) can be significantly affected by biological and/or ecological processes, although physical processes may remain important (e.g., redeposited sediments).

Relative buoyancy seems to be an essential factor controlling the distribution of the allochthonous organic fraction. It is a function of specific gravity, size, and shape of a given particle, as explained below:

Specific gravity—the following scale of increasing buoyancy has been proposed (Fisher, 1980; Van der Zwan 1990) for equidimensional palynomacerals: PM4, PM1, PM2, and PM3. However, this scale does not take into consideration the size and shape of palynomaceral particles. Hence it is a specific gravity scale rather than a buoyancy scale;

Size and shape—rounding, sorting, and size of allochthonous fragments are a function of the depositional energy (Parry et al., 1981; Bryant et al., 1988; Van der Zwan, 1990). In recent sediments a seaward decrease in phytoclasts size is also reported (e.g., Cross et al., 1966; Caratini et al., 1983). The blade-shaped outline of some PM4 fragments (PM4-T) strongly influences their hydrodynamic behavior: because of their flat shape, their settling to the sea floor is delayed in the same way as that of mica flakes (Stanley, 1986). Consequently they are more buoyant than equidimensional PM4 fragments. PM4-T is even qualified by Van der Zwan (1990, p. 175) as "extremely" buoyant, what may seem contradictory to the buoyancy scale proposed above, where PM4 is considered as the least buoyant palynomaceral. This ambiguity only reflects that the shape of a given fragment may become a predominant parameter in defining its buoyancy.

In addition to buoyancy, distribution of the allochthonous fraction also can be affected by its resistance to degradation. PM4 is considered a highly stable palynomaceral and can be transported far out to sea before being degraded. This may locally emphasize the relative proportion of PM4 fragments, irrespective of their hydrodynamic properties.

With regard to phytoplankton, it has been observed that dinocyst abundance and diversity increase with major transgressions (Wall, 1965; Tissot, 1979; Smelror and Leereveld, 1989; Habib and Miller, 1989). This may indicate a trend toward more open marine conditions (Van der Zwan, 1990). Van Pelt and Habib (1988) observed a marked increase in dinocyst abundance and diversity during transgressions and, inversely, abundant assemblages of tracheids, cuticles, and sporomorphs in regressive intervals.

Palynofacies can be affected additionally by local ecological or chemical processes. Sedimentation of organic particles can be strongly influenced when they are incorporated within zooplankton fecal pellets (Honjo and Roman, 1978; Turner and Ferrante, 1979) and such pellets are known to play an important role in organic matter fluxes in the marine environment. Assessing the provenance of fecal pellets, or any aggregate, is a major problem: if they are transported, they will be subjected to the same hydrodynamic processes as other particles, but if they are produced in situ their size and shape will not have a significant effect. Bioturbation related to macrobenthic fauna activity is another cause of perturbation, as it reworks sediments and renews the exposure of organic fragments to degradation at the sediments/water interface (Müller and Suess, 1979; Tyson 1987).

Finally, conditions of preservation (i.e., oxidation level of the depositional environment) largely influence the type of palynofacies observed in fossil sediments (Hart, 1986; Tyson, 1987; Tribovillard and Gorin, 1991).

LOCATION AND GEOLOGIC FRAMEWORK OF THE STUDIED SECTIONS

Berrias

The Berrias section is located in the southern part of the Ardèche area, southeast France (Figure 1). It consists of 35 m of well-bedded, fine-grained limestones that contain more than 90% $CaCO_3$ (Gorin and Steffen, 1991) but becomes slightly more marly at the top (Figure 3). Breccia horizons occur episodically in the lowermost part (top of *Jacobi-Grandis* ammonite zone and base of *Subalpina* ammonite subzone); a major breccia event marks the middle part (base of *Privasensis* ammonite zone).

The section is precisely dated by ammonites and calpionellids (Busnardo et al., 1965; Le Hégarat and Remane, 1968; Le Hégarat, 1973, 1980) and has been calibrated with magnetostratigraphy (Galbrun et al., 1986). Although the Tithonian–Berriasian boundary has not been precisely located, calpionellid evidence indicates that the section begins in the very early Berriasian (base of B calpionellid zone). At the top are lower Valanginian marls (*Otopeta* ammonite zone). Calpionellid zonation shows that the base of the C zone is missing; this small hiatus coincides with the major breccia event. The biostratigraphic subdivision presented here is the latest update by R. Busnardo, B. Clavel, and G. Le Hégarat (personal communication, 1992).

The fine-grained carbonates of the Berrias section are considered to be hemipelagic outer platform deposits (Gorin and Steffen, 1991; Strohmenger and Strasser, 1993).

Broyon

The Broyon quarry is located in the Rhône valley, in the northern part of the Ardèche area, southeast France (Figure 1). It consists of a 25-m-high cliff made of white fine-grained limestones, overlain by 15 m of marl (Figure 4). The section starts in the early Tithonian (*Richteri* ammonite zone) and ends in the middle part of the Berriasian (the lower part of *Privasensis* ammonite zone; see Cecca et al., 1989, for biostratigraphy).

The lower Tithonian (*Richteri* ammonite zone) deposits consist of a limestone/marl alternation interpreted as calciturbidites (G. Dromart, personal communication, 1992). The upper Tithonian to lower Berriasian interval (*Microcanthum* and *Durangites* to base of *Jacobi–Grandis* ammonite zones) corresponds to a massive, poorly bedded, fine-grained limestone, with sporadic breccia events. It has been described by Cecca et al. (1989) under the name "Formation des Calcaires blancs." An important biostratigraphic hiatus reportedly occurs between the early and late Tithonian (Le Hégarat, 1973; Cecca et al., 1989). The overlying lower Berriasian interval (upper part of *Jacobi–Grandis* ammonite zone) is dominated by marl deposits with some intercalated marly limestone beds. The uppermost part of the section is poorly exposed (*Subalpina* ammonite zone). The last bed, identified as a fluxoturbidite (Strohmenger and Strasser, 1993), belongs to the *Privasensis* ammonite zone (C calpionellid zone; Le Hégarat, 1973).

Lithologies observed in Broyon are interpreted as hemipelagic slope deposits (G. Dromart, personal communication, 1992; Strohmenger and Strasser, 1993).

Angles

The Angles section is located in the Haute-Provence Alps area, southeast France (Figure 1). It shows a 75-m-thick limestone/marl alternation; limestones dominate at the base and marls become more abundant towards the top (Figure 5). The section is dated by calpionellids as late Tithonian to early Valanginian (Le Hégarat and Ferry, 1990). The Berriasian–Valanginian boundary has been precisely located by ammonites (R. Busnardo, personal communication, 1992). Calpionellid zonation shows two hiatuses in the section: the first one coincides with the slumped horizon between B and D1 zones, where the top B and the C calpionellid zones are missing, and the second one is located in the channelized beds between D1 and D2 zones (top of D1 and base of D2 missing). The Angles limestone–marl alternations are interpreted as pelagic basin deposits (Le Hégarat and Ferry, 1990; Strohmenger and Strasser, 1993).

RESULTS OF PALYNOFACIES STUDY

Palynofacies analysis, including relative abundances (% relative numeric particle frequency) of organic constituents and comments concerning specific horizons are presented for Berrias, Broyon, and Angles (Figures 3–5). The relationship between these palynofacies and sequence stratigraphy also is addressed.

Berrias

At the Berrias locality, some 60 samples were analyzed. The palynofacies is dominated by allochthonous fragments (mainly PM4 and PM1/2-B groups)

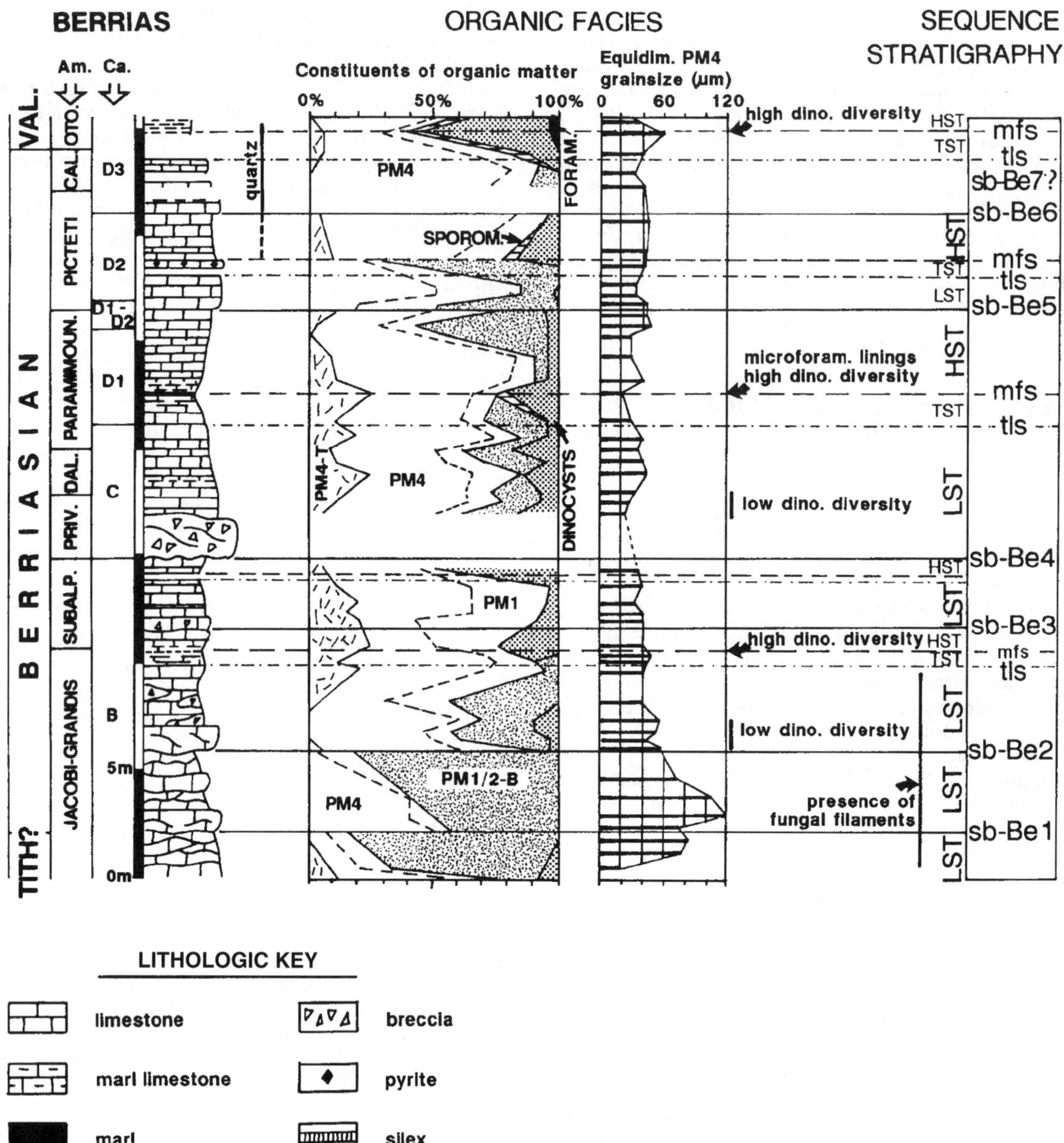

Figure 3. Berrias section—biostratigraphic (Am = ammonite subzones; Ca = calpionellid zones), lithologic, and palynofacies data, and derived sequence stratigraphy interpretation. The nomenclature used for sequence boundaries is the latest regional update by T. Jacquin and P. R. Vail (personal communication, 1992). For the sake of clarity and to bring out the main trends, the palynofacies diagram has been smoothed over densely sampled intervals. Sequence stratigraphy abbreviations: sb = sequence boundary, tls = top lowstand surface, mfs = maximum flooding surface, LST = lowstand systems tract, TST = transgressive systems tract, HST = highstand systems tract.

in the basal part. In the upper part, dinocysts become relatively abundant. Fluorescent amorphous organic matter (AOM) is absent throughout the section. Sporomorphs are rare and only appear in the upper half of the section where they never exceed 5% of the total assemblage (Figure 3).

The base of the section (*Jacobi–Grandis* ammonite zone) is dominated by an assemblage of PM1-B and

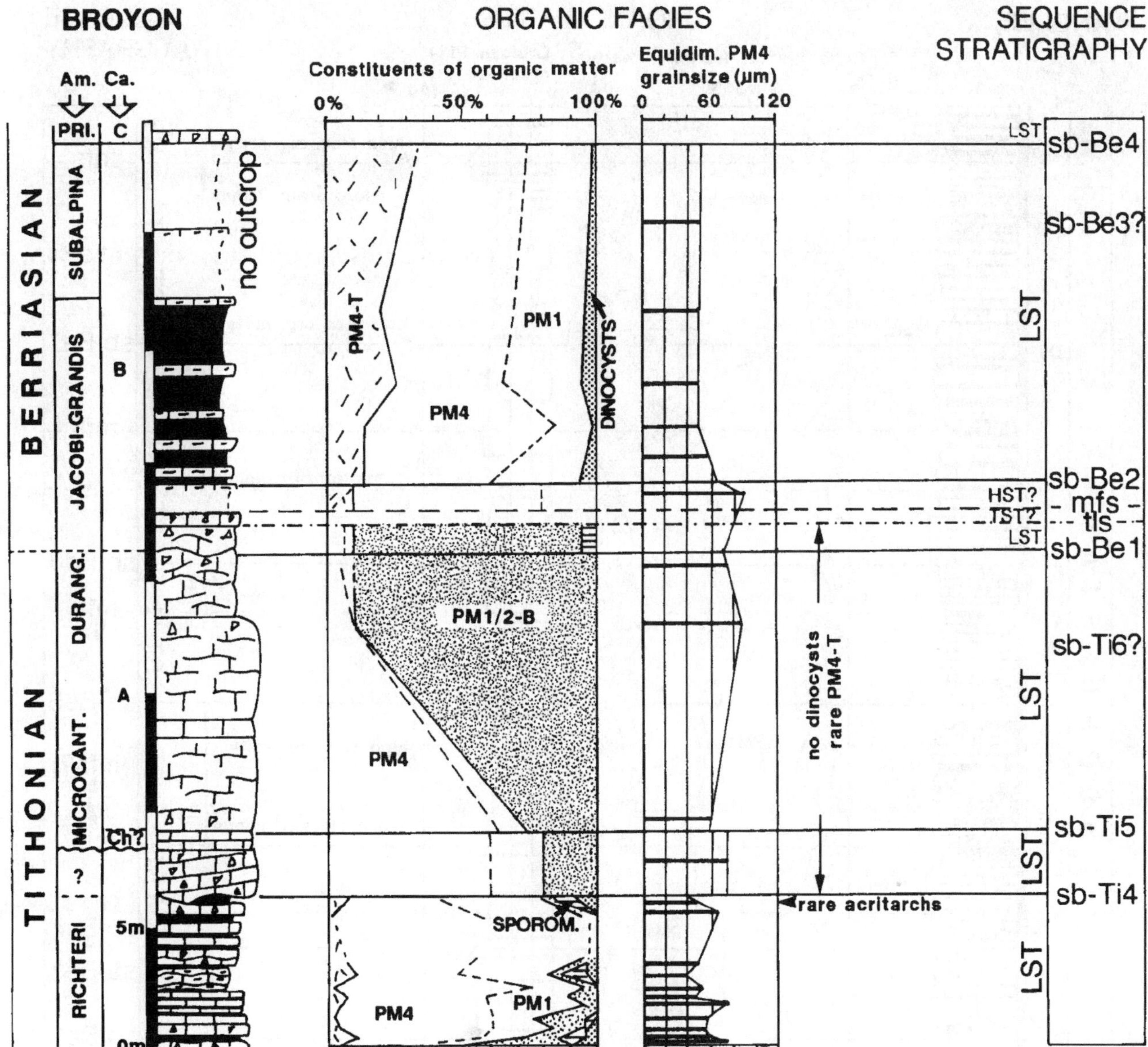

Figure 4. Broyon section—biostratigraphic (Am = ammonite subzones; Ca = calpionellid zones), lithologic and palynofacies data, and derived sequence stratigraphy interpretation. The sequence boundaries nomenclature is based on the latest regional update by T. Jacquin and P. R. Vail (personal communication, 1992). For the sake of clarity and to bring out the main trends, the palynofacies diagram has been smoothed over densely sampled intervals. For other abbreviations, see caption for Figure 3. For lithologic key, see Figure 3.

PM2-B (in this study, degraded phytoclast category, noted PM1/2-B) and by filaments (of probable fungal origin). The poorly defined shape of PM1/2-B evokes fecal pellets in some cases (Gorin and Steffen, 1991). Their common occurrence as nonfluorescent aggregates is indicative of hydrogen-poor, terrigenous, allochthonous organic matter. Equidimensional PM4 grains are the largest for the whole section (up to 120 µm in diameter), poorly sorted, and quite angular. Dinocysts are rare and their species diversity is very low (two to five species).

The *Subalpina* ammonite zone is marked by the disappearance of the PM1/2-B assemblage. Dinocysts become more abundant (10% in average; up to 50% at the top of the zone) and are associated with large amounts of PM4-T. Equidimensional PM4 grains are smaller and better sorted.

The *Privasensis* to *Paramimounum* ammonite zone displays a complete assemblage of PM4-T, PM4, PM1, PM1/2-B, and dinocysts in variable amounts. The proportion of dinocysts varies similarly to that of PM4-T, but inversely from that of PM1/2-B.

The *Picteticeras* ammonite zone reflects two different intervals. The lower interval is characterized by the near absence of dinocysts and large amounts of completely degraded fragments, mainly PM1/2-B

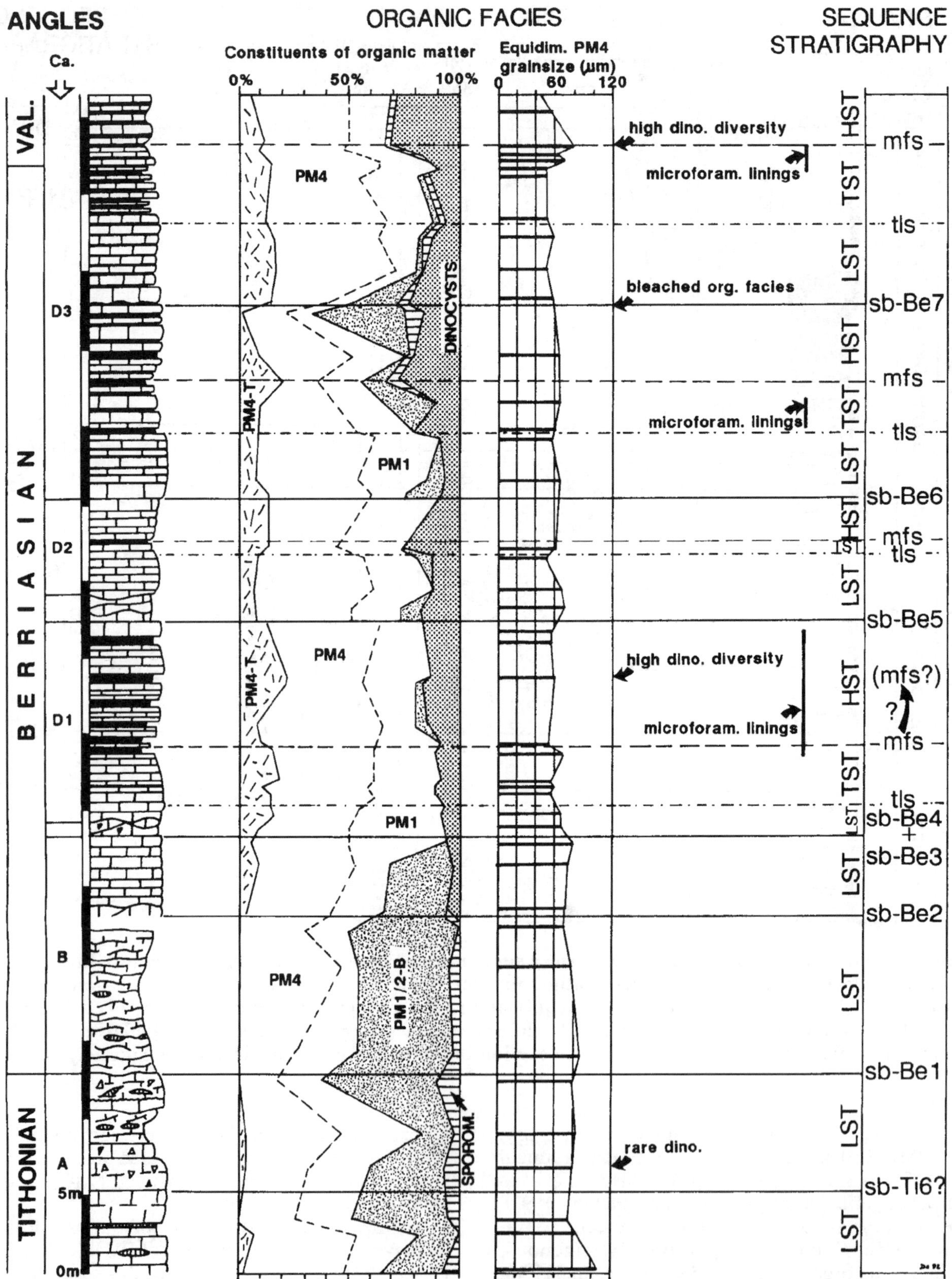

Figure 5. Angles section—biostratigraphic (Am = ammonite subzones; Ca = calpionellid zones), lithologic and palynofacies data, and derived sequence stratigraphy interpretation. The nomenclature used for sequence boundaries is the latest regional update by T. Jacquin and P. R. Vail (personal communication, 1992). For the sake of clarity and to bring out the main trends, the palynofacies diagram has been smoothed over densely sampled intervals. For other abbreviations, see caption for Figure 3. For lithologic key, see Figure 3.

and probably unrecognizable dinocysts. Well-preserved dinocysts reappear in the upper interval, coinciding, as in the *Privasensis* and *Paramimounum* ammonite zones, with an increase in PM4-T and the disappearance of PM1/2-B.

Although the upper part of the section (*Callisto* and *Otopeta* ammonite zones) is poorly exposed, it was possible to collect some samples. They yielded an exceptionally rich and well preserved dinocysts assemblage accompanied by some sporomorphs and foraminiferal linings.

Broyon

At the Broyon locality some 30 samples were analyzed. The section consists of three different lithological parts, each characterized by a typical palynofacies (Figure 4). At the base of the section (early Tithonian, *Richteri* ammonite zone), the palynofacies shows, in various proportions, all the major organic constituents observed in Berrias: i.e., all types of palynomacerals (PM1 to PM4), some sporomorphs, and rare dinocysts. Their distribution does not appear to reflect any specific trend. The size, rounding, and sorting of PM4 grains is also quite variable from one sample to the next.

In the middle part of the section ("Formation des Calcaires blancs," late Tithonian–early Berriasian, *Microcanthum* to base of *Jacobi–Grandis* ammonite zones), the palynofacies is limited to an assemblage of PM4 and highly degraded PM1/2-B. The PM1/2-B particles can be as large as 600 µm. A small percentage show cell structures but most are so degraded as to have become structureless. Some are mixed with fungal filaments and others display a very rounded shape reminiscent of fecal pellets.

The palynofacies changes sharply in the upper part of the section (middle of *Jacobi–Grandis* to *Subalpina* ammonite zones). The PM1/2-B assemblages observed in the underlying interval disappears completely. Dinocysts and PM4-T are present. Distribution and characteristics of each organic group are quite constant and contrast markedly with the fluctuating palynofacies described at the base of the section.

As in the Berrias section, fluorescent amorphous organic matter is absent. UV-fluorescence permits accurate determination of hydrogen-poor, nonfluorescent, PM1/2-B fragments.

Angles

Some 60 samples were analyzed at the Angles locality. All the organic constituents observed in the Berrias section have been found in this section and show some similarities in their distribution: PM1/2-B assemblages dominate in the lower part and dinocysts become more abundant toward the top (Figure 5).

At the base of the section (A and base of B calpionellid zones) PM1/2-B assemblages are abundant, whereas dinocysts (less than 1%), PM4-T, and sporomorphs are rare. Equidimensional PM4 grains are large, subangular, and poorly sorted.

The top of the B calpionellid zone corresponds to a transition interval characterized by an increasing proportion of PM4-T and dinocysts, whereas PM1/2-B decrease.

In the upper part of the section (from D1 calpionellid zone to the top) PM1/2-B assemblages reoccur episodically. Dinocysts are present throughout in variable proportions. Peaks in their distribution correspond to maxima of PM4-T and of small, rounded, well sorted, equidimensional PM4 grains (see Figure 5, e.g., the marly bed in the middle of D2 calpionellid zone and that located 10 m above in D3 calpionellid zone). As in the Berrias section, sporomorphs reappear in the upper part, and dinocysts are richest in the lowermost Valanginian beds.

RELATIONSHIP BETWEEN PALYNOFACIES AND SEQUENCE STRATIGRAPHY SYSTEMS TRACTS

Before interpreting the palynofacies results in the Berriasian sections, some basic concepts in sequence stratigraphy interpretation will be reviewed. These concepts directly control our interpretation of depositional patterns (for sequence stratigraphy tenets, see Posamentier and Vail, 1988; Posamentier et al., 1988; Van Wagoner et al., 1988; Vail et al., 1991), and subsequently the distribution of organic matter in sediments, according to the principles of organic sedimentology presented above. Figure 6 summarizes the relationships between the type of palynofacies and the different systems tracts and surfaces in open marine sediments deposited in a relatively oxidizing environment (i.e., no fluorescent amorphous organic matter). It provides a means to translate palynofacies observations into sequence stratigraphy interpretation. Some of these principles have been previously tested in the Berrias section by Gorin and Steffen (1991).

Lowstand Systems Tract (LST)

Relative sea level fall leads to a total or partial shelf emersion (type I and type II unconformities are not distinguished here), thereby creating intense erosional activity on land and infill of the basin through sediment bypass. This implies reworking of shallow sediments and reoxygenation of bottom conditions. Turbidites, channels, and mass-slides, associated with basin floor fan and slope fan deposits, will bring proximal organic fragments into a relatively deeper environment. The lowstand prograding complex is formed by a vertical succession of deposits becoming more proximal toward the top (shallowing-up). This results in an increase, both in relative proportion and size, of allochthonous particles (PM1 to PM4 and sporomorphs) (Pasley et al., 1991; Gregory and Hart, 1992) and in a decrease of the relatively autochtho-

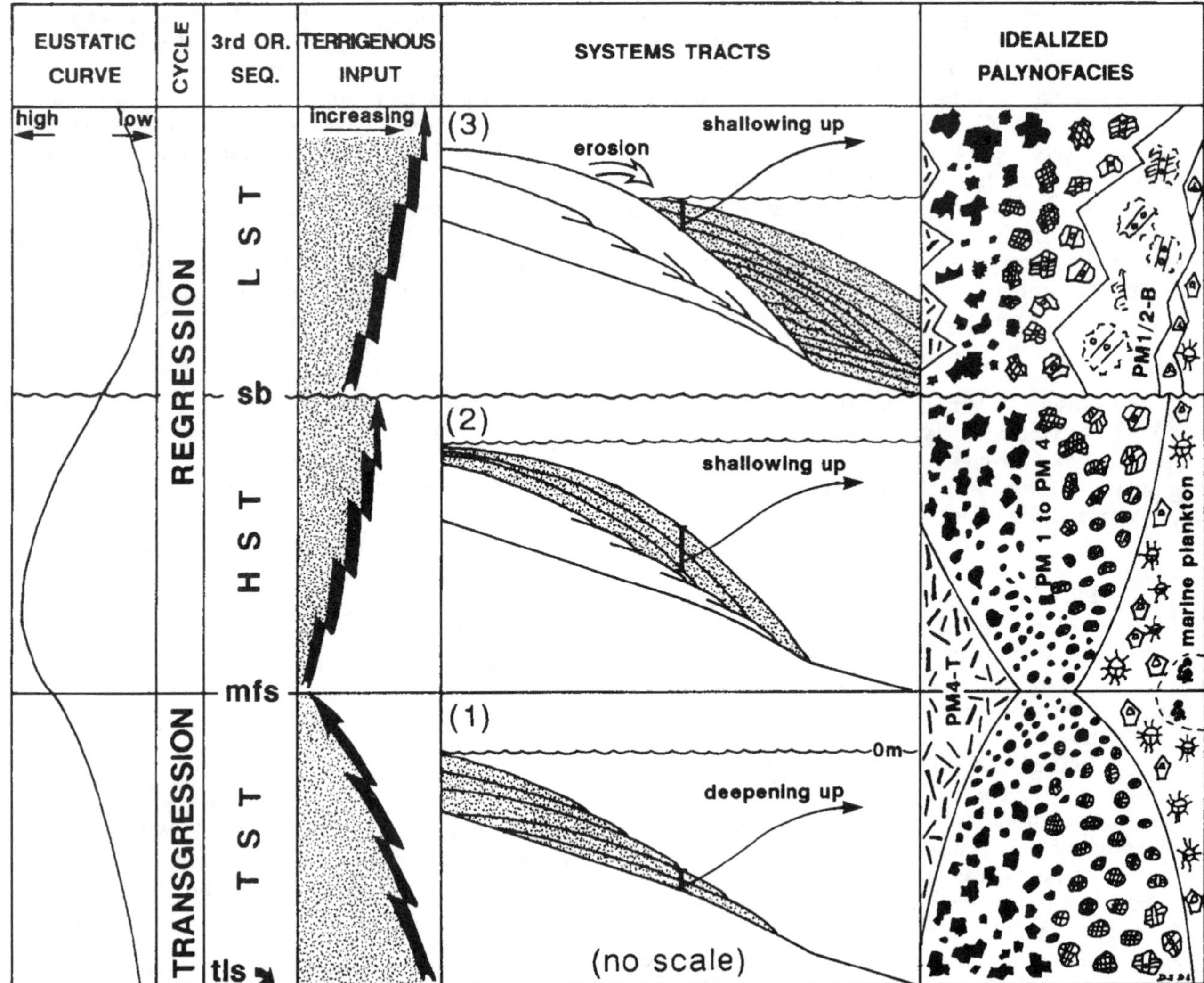

Figure 6. Relationship between sequence stratigraphy and palynofacies. This model is basically applicable to open marine sediments deposited in a relatively oxidizing environment (absence of fluorescent amorphous organic matter). This corresponds to an ideal organic matter distribution: in fact, detailed studies show that there is always a "background noise" and that sufficient, properly located samples are needed to provide a reliable interpretation. Sequence stratigraphy abbreviations, see caption for Figure 3. Symbols of organic constituents, see Figure 2.

nous components (mainly dinocysts). The top lowstand surface (tls) marks the end of the lowstand systems tract.

Transgressive Systems Tract (TST)

The transgressive systems tract corresponds to a retrograding unit marked by the landward shift of sediment accommodation. It results in a vertical succession of deposits becoming more distal up section (deepening-up). Terrigenous fragments become smaller and more rounded toward the top (abundance of PM4-SR). Flooding of the shelf moves erosional activity landward, carbonate productivity progressively starves, resulting in bed thinning and enrichment in fossils (Vail et al., 1991). The organic content is marked by an increase of relatively autochthonous organisms (dinocysts, foraminiferal linings), both in abundance and diversity. Because of their hydrodynamic properties, the most buoyant fragments (PM4-T and PM4-S) become dominant in the allochthonous fraction. The transgressive systems tract culminates with the maximum flooding surface (mfs), which corresponds basinward to a condensed section. The palynofacies associated with the maximum flooding surface is characterized by peaks in the trends developed during the transgressive phase, i.e., it normally contains the highest relatively autochthonous fraction, the maximum planktonic diversity, and the most buoyant allochthonous fraction.

Highstand Systems Tract (HST)

Trends observed in the highstand systems tract are inverted with respect to the transgressive systems tract. The highstand systems tract is a prograding unit showing shallowing-up patterns analogous to those described in the lowstand prograding complex. The major difference is that the shelf is not exposed during early highstand deposits, hence large degraded terrigenous fragments (PM1/2-B) observed in lowstand deposits are absent. They can reappear towards the top when sediments become more proximal and the terrigenous input associated with the late highstand deposits on the shelf becomes more abundant. A sequence boundary (sb) indicates the top of the highstand systems tract.

With regard to stacking patterns and lithologies, Vail et al. (1991) propose the following schematic model for deep-water carbonate sequences: the lowstand systems tract is characterized by thick, cyclic deposits and is followed by thinning up in the transgressive systems tract due to starvation in the basin. This starvation culminates at the maximum flooding surface, which generally corresponds to the thinnest beds in a sequence, and is the result of depletion in the $CaCO_3$ budget, caused by the development of platform carbonates (Berger and Winterer, 1974, cited in Vail et al., 1991, p. 657). Decrease in starvation induces thickening up in the highstand systems tract.

SEQUENCE STRATIGRAPHY INTERPRETATION

The interpretation proposed below is based on field observation, palynofacies analysis (this study), and standard sedimentological facies analysis (Strohmenger and Strasser, 1993). The nomenclature used for sequence boundaries on Figures 3–5 and 7 corresponds to the latest regional update by T. Jacquin and P. R. Vail (personal communication, 1992).

Tithonian and Lower Part of B Calpionellid Zone

Dinocysts are virtually absent in this interval, except at the base of the Broyon section (*Richteri* ammonite zone) where they appear sporadically, accompanied by low amounts of PM4-T. Spores and pollen are present. Equidimensional PM4 is relatively larger than in the overlying parts. No specific trend appears in the distribution of organic matter constituents. This dominance by terrigenous (allochthonous) fragments suggests lowstand systems tract deposits. It is confirmed by field observations: calciturbidites at the base of the Broyon section (*Richteri* ammonite zone) followed by massive, poorly bedded limestones in the three sections, with channels features. This interval is interpreted as a succession of lowstand systems tract deposits (stacked lowstands). Palynofacies show no limits between these stacked lowstands, and the location of sequence boundaries summarized in Figure 7 is mainly based on field observations. The transgressive systems tract and the highstand systems tract below sequence boundary Be2 in the Broyon section are not clearly established because of poor outcrop availability.

Upper Part of B Calpionellid Zone

Dinocysts appear in small quantities in the three sections. Two sequences are present in the Berrias section. In the first one, PM1/2-B is abundant in the lowstand systems tract above sequence boundary Be2 and disappears in the transgressive systems tract, where dinocysts and PM4-T increase towards the maximum flooding surface. The second sequence is not so clearly identifiable, except for the peak of dinocysts indicating the maximum flooding surface. In the other two sections, these two sequences are poorly documented because only the base of the lower one can be sampled (lack of outcrop in Broyon and stratigraphic gap in Angles). In both sections studied material continuously shows small amounts of plankton, suggesting lowstand systems tract deposits. At Broyon, the sharp palynofacies change across sequence boundary Be2 (complete disappearance of PM1/2-B) reveals a difference in the type of the two lowstand deposits. The channels and breccias in the massive limestones below sequence boundary Be2 suggest reworked carbonates from the slope (slope fan as defined by Vail et al., 1991), whereas the superimposed marlstone–limestone alternations (*Jacobi–Grandis* and *Subalpina* ammonite zone) are interpreted as a lowstand prograding wedge (Strohmenger and Strasser, 1993).

C and D Calpionellid Zones

Sequence boundary Be4 is well marked in the field by a breccia horizon in the Berrias and Broyon sections. It is missing in the Angles section where a stratigraphic gap coincides with slumped beds. These slumps are not thought to be directly associated with eustatic variations, but rather related to slope instability, which is well-documented in the Tithonian–Berriasian interval of the Vocontian basin (Beaudoin, 1980).

Based on palynofacies analysis, the upper part of the Berrias and Angles sections can be interpreted as a succession of sequences that follow the general trends of organic matter distribution proposed in Figure 6. Maximum flooding surfaces are the clearest horizons, due to maxima in both abundance and diversity of dinocysts, as well as abundance in PM4-T. PM1/2-B are mainly found in the lowstand systems tract, but sometimes reoccur near the top of the highstand systems tract, where they may indicate the lateral equivalent of the shelf late highstand deposits.

Systems tracts around sequence boundary Be7 have not been identified in Berrias because of lack of outcrop (no sampling possible). The transgressive systems tract of the uppermost sequence (Berriasian–Valanginian boundary) is clearly marked in both sec-

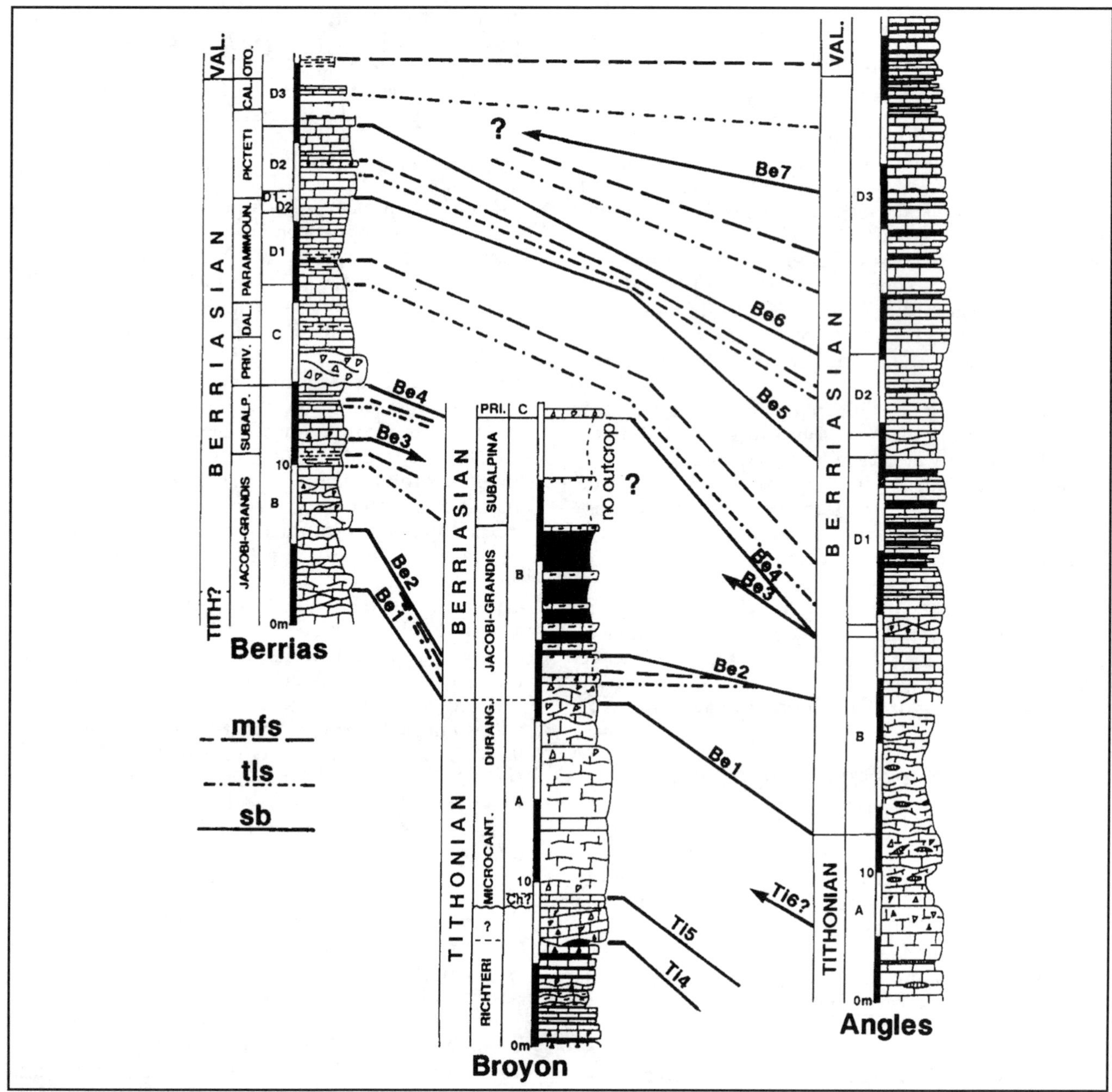

Figure 7. Correlation of the three studied sections summarizing the sequence stratigraphy interpretation proposed in Figures 3–5.

tions by a significant increase in abundance and diversity of dinocysts. The overall palynofacies of this interval in the Berrias and Angles sections is strikingly similar.

PERTURBATIONS IN THE THIRD-ORDER SEQUENCES RECORD

There are local discrepancies between the "idealized palynofacies" proposed in Figure 6 and the organic matter distribution observed across the studied sections. They are due to the fact that the "idealized palynofacies" is related to a systems tract breakdown, therefore taking into account only the third-order sea level changes. Actually, high-order oscillations (fourth- to sixth-order, sensu Vail et al., 1991) or low-order oscillations (first- and second-order, sensu Vail et al., 1991) do certainly influence the distribution of organic matter.

Effect of High-order Sea Level Changes (Fourth to Sixth)

The effect of high frequency oscillations are observed at the scale of a bed or a bedset (Vail et al.,

1991). In the studied sections, palynofacies parameters from closely spaced samples, collected within one systems tract, show only small differences, clearly weaker than the general trend observed throughout. This probably indicates that the effect of high-order eustatic variations is very attenuated in the basinal setting where the studied hemipelagic to pelagic carbonates were deposited. This is certainly different for shallow water sediments, where even a small eustatic variation drastically changes the depositional environment. However, the erratic distribution of organic matter in some lowstand deposits (e.g., *Richteri* ammonite subzone in Broyon; Figure 4) could be related to high-order eustatic changes.

Effect of Low-order Sea Level Changes (First and Second)

Low frequency oscillations are characterized by durations of over 3 m.y. (Vail et al., 1991). Their effect could thus be observable at the scale of the whole section (length of the Berriasian = 5 to 5.5 m.y., Haq et al., 1987; Harland et al., 1989). The Haq et al. (1987) long-term eustatic curve indicates an important regression throughout the Tithonian, followed by a transgression during the early Berriasian (Figure 7). In the three sections studied here, the latter coincides with the disappearance of PM1/2-B and the appearance of dinocysts in significant proportions (Figures 3–5). Although this very general trend could create some distortions in the palynofacies, it happens at such a different scale that it cannot be confused with the third-order sequences signals. Trends in the organic matter distribution may eventually be enhanced or attenuated, if second- and third-order eustatic variations are in phase or not.

PALYNOFACIES AND SEQUENCE STRATIGRAPHY: APPLICATIONS IN THE PETROLEUM INDUSTRY

Identification and delineation of reservoir and source rocks are of prime importance for petroleum geology. The recent development of sequence stratigraphy coupled with standard sedimentology is proving a powerful tool for predicting the vertical and lateral distribution of these rocks. The methodology outlined in this chapter tends to demonstrate the value of palynofacies studies as a complementary method to standard sedimentology in establishing a sequence stratigraphy framework.

The approach presented here is still preliminary and relatively little documented in the literature. Upper Tithonian–Berriasian carbonates of southeast France were originally chosen because they offer a high biostratigraphical resolution (a prerequisite to testing a sequence stratigraphy model) and not for their reservoir or source rock potential. Once the validity of the model is tested, the same approach can subsequently be applied to rock intervals more attractive for the petroleum industry. The upper Tithonian–Berriasian carbonates of the Vocontian basin, deposited in fully oxic conditions, are devoid of fluorescent amorphous organic matter. In the same area, the Valanginian to Barremian interval is exposed as an alternation of open marine carbonates and marls locally rich enough in fluorescent amorphous organic matter to become potential source rocks. In the Angles area, amorphous organic matter appears in the early Valanginian and its overall content generally increases in the late Hauterivian to peak in the early Barremian (J.-F. Raynaud, personal communication, 1992; De Renéville and Raynaud, 1981). This increase indicates a trend towards dysoxic conditions and correlates with a second-order transgression (Figure 8). Total organic carbon content may reach up to 4% in the Barremian.

The value of palynofacies as a predictive tool in sequence stratigraphy has been recently demonstrated in a few other different depositional systems offering potential interest to the petroleum industry. In a shelf setting, lower Toarcian source rocks containing abundant fluorescent amorphous organic matter were deposited during a third-order transgressive systems tract both in England (Grabowski and Glaser, 1990) and on the northern Tethyan margin (Gorin and Feist-Burkhardt, 1990).

In fine-grained siliciclastic systems, the relationship between palynofacies and sequence stratigraphy systems tracts has been demonstrated, from outcrop and core samples, in Upper Cretaceous deposits of New Mexico (Pasley et al., 1991): palynofacies allow these authors to distinguish between transgressive and regressive shales, the former having the highest total organic carbon associated with fluorescent amorphous organic matter. Subsurface lower Tertiary deltaic sediments of Louisiana show a correspondence between palynofloral composition and systems tracts (Gregory and Hart, 1992). In the latter two siliciclastic systems, the content in terrigenous palynomorphs is the lowest in the late transgressive and early highstand systems tracts.

In mixed carbonate–siliciclastic coastal sediments of Miocene age outcropping in central Tunisia, Blondel et al. (1993) demonstrate the close association between palynofacies and sea level changes. Finally, palynofacies can also be used as a sequence stratigraphy predictive tool in subsurface European Permian Zechstein carbonates, which form a significant gas reservoir (personal observations). In this case, most of the constituents are terrestrial and variations in their level of degradation are closely associated with systems tracts.

Therefore, these examples taken from various locations show that the distribution of organic constituents is clearly associated with sequence stratigraphy. However, the value of palynofacies parameters as indicators of sea level changes may locally be enhanced or attenuated, depending on the depositional environment.

The foregoing studies have been performed both on outcrop and well data. Except for rocks particular-

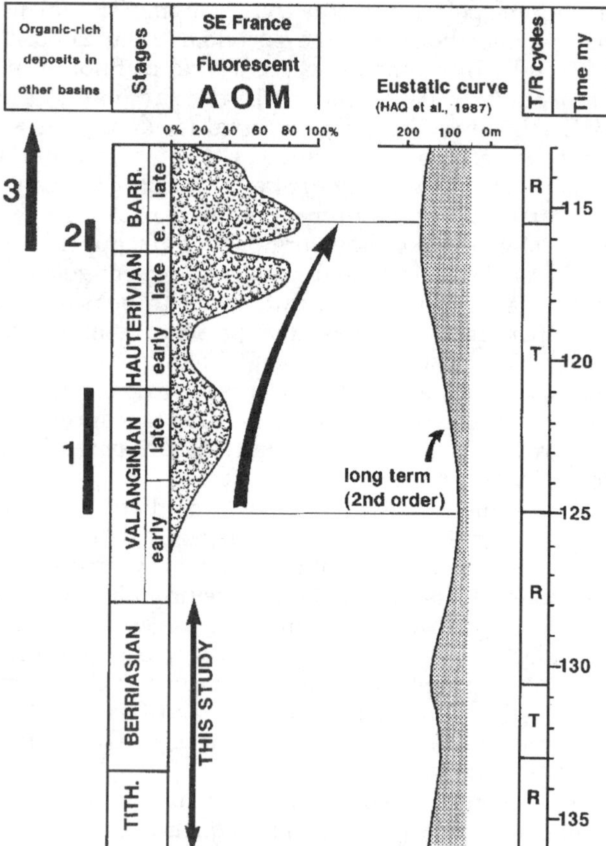

Figure 8. Distribution of fluorescent amorphous organic matter (AOM) in the upper Tithonian–Barremian deposits in the Angles area. Amorphous organic matter appears in the early Valanginian and reaches a maximum in the early Barremian. This increase coincides with a 2nd order transgression. A gradual enrichment in TOC across the Early Cretaceous has also been reported in the North Atlantic by Summerhayes and Masran (1982). Source of data: this paper, Tithonian–Berriasian, J. F. Raynaud (personal communication, 1992), Valanginian–Hauterivian, and De Renéville and Raynaud (1981), Barremian. T = transgression, R = regression. (1) Yorkshire (D5 beds, Speeton Clay) and Lower Saxony (Tyson and Funnell, 1987). (2) Lower Saxony (Blattertonsteine), Yorkshire (Speeton Clay, LB1-LB2 beds), southern Sweden (black clay shales) and southern North Sea (dark brown-black clays) (Tyson and Funnell, 1987). (3) North Atlantic (Hatteras Formation, unit 4d) (Summerhayes and Masran, 1982).

ly lean in organic matter, where sufficient core material would be required, palynofacies studies can be carried out on the data set available from an industry well. If cores are not available, enough sidewall sample material is needed over critical intervals; in this case, electrical log stacking patterns (Gregory and Hart, 1992) are decisive in properly positioning the samples. They replace the bed stacking patterns that can be observed in outcrops.

CONCLUSIONS

This study demonstrates that significant vertical variations in the distribution of organic constituents can be detected in open marine upper Tithonian–Berriasian carbonates of southeast France. These sediments were originally considered as unfavorable for the sedimentation and preservation of organic matter. Fluorescent amorphous organic matter, indicative of low-energy, oxygen-depleted depositional conditions is absent throughout the studied interval.

The combination of precise biostratigraphic data, field observations (e.g., bed-stacking patterns and key sedimentary surfaces), sedimentologic data (micro- and macrofacies analysis), and palynofacies parameters allow us to propose a reliable sequence stratigraphy interpretation for the analyzed sections. The correlation of palynofacies and sequence stratigraphy shows the close relationship between eustatic variations and the distribution of particulate organic matter. Based on these observations, the following model is proposed:

- lowstand systems tract (LST)—predominance of PM1/2-B and large-scale, angular PM4, dinocysts being rare;
- transgressive systems tract (TST)—upward decrease in abundance, size, and angularity of allochthonous fragments; upward increase in PM4-T and in dinocysts diversity and abundance, peaking at the maximum flooding surface (mfs);
- highstand systems tract (HST)—upward increase in abundance, size, and angularity of allochthonous fragments; upward decrease in PM4-T and in dinocysts abundance and diversity.

The method used in this study relies on relative variations of organic constituents and not on the organic richness of a sediment. Consequently, this approach uses general trends that can only be derived from properly sampled sections. Isolated samples cannot provide any reliable information on eustatic variations.

This methodology has definite applications in petroleum geology by offering an additional tool in sequence stratigraphy interpretation geared to reservoir and source rock delineation. The precise tie between palynofacies and short-term eustatic oscillations (third-order) can be applied to studies of the late Tithonian–Berriasian deposits in other sedimentary basins. Such applications can be done either regionally or globally, depending on the nature and extent of the identified eustatic variations.

The same approach can be similarly applied to other time intervals, e.g., the more organic-rich Valanginian–Barremian sediments in the Vocontian basin. There, in addition to the organic constituents observed in the late Tithonian–Berriasian interval, fluorescent amorphous organic matter will be an important component reflecting eustatic variations.

ACKNOWLEDGMENTS

This work was supported by the Swiss National Science Foundation (grant no. 20-30276.90). The authors are part of a working group studying the Lower Cretaceous of the Vocontian basin, coordinated by R. Jan du Chêne and J.-F. Raynaud. The proposed sequence stratigraphy interpretation was discussed in the field with P. R. Vail. We are indebted to R. Busnardo, B. Clavel, and G. Le Hégarat for kindly providing us with their latest updates of the Berriasian stratotype and the Angles section; to G. Dromart for his information on the Broyon section; to T. Jacquin and P. R. Vail for allowing us to use their latest regional update of sequence boundary nomenclature; to E. Monteil for letting us benefit from his vast knowledge of dinocysts, and to J.-F. Raynaud for communicating to us unpublished data on amorphous organic matter in the Angles area. Our thanks also go to M. Floquet for preparing the numerous palynological slides and to P. Kindler for his shrewd comments on the manuscript. Finally, we are grateful to R. V. Tyson and R. J. Witmer for their critical review.

REFERENCES CITED

Beaudoin, B., 1980, Le Bassin subalpin in A. Autran and J. Dercourt, coord., Colloque C7, Géologie de la France: Publications du 26ème Congrès Géologique International, Mémoire BRGM 107, p. 284-291.

Berger, W. H., and E. L. Winterer, 1974, Plate stratigraphy and fluctuating carbonate line, in K. J. Hsü and H. C. Jenkyns, eds., Pelagic sediments on land and under the sea: International Association of Sedimentologists, Special Publication 1, p. 11-48.

Blondel, T., G. E. Gorin, and R. Jan du Chêne, 1993, Sequence stratigraphy in coastal environment: sedimentology and palynofacies of Miocene in Central Tunisia, in H. W. Posamentier, C. P. Summerhayes, B. U. Haq, and G. P. Allen, eds., Sequence stratigraphy and facies association: Spec. Publs. Int. Ass. Sediment. 18, p. 161-179.

Boulter, M. C., and A. Riddick, 1986, Classification and analysis of palynodebris from the Palaeocene sediments of the Forties Field: Sedimentology, v. 33, p. 871-886.

Bryant, I. D., J. D. Kantorowicz, and C. F. Love, 1988, The origin and recognition of laterally continuous carbonate-cemented horizons in the Upper Lias Sands of southern England: Marine Petroleum Geology, v. 5, p. 108-133.

Bujak J. P., M. S. Barss, and G. L. Williams, 1977, Offshore eastern Canada–Part I: offshore east Canada's organic type and hydrocarbon potential: Oil and Gas Journal, v. 75, p. 198-201.

Burgess, J. D., 1974, Microscopic examination of kerogen (dispersed organic matter) in petroleum exploration: Geological Society of America, Special Paper 153, p. 19-30.

Busnardo, R., G. Le Hégarat, and J. Magne, 1965, Le stratotype du Berriasien: Mémoire BRGM 34, p. 5-33.

Caratini, C., J. Bellet, and C. Tissot, 1983, Les palynofaciès: représentation graphique, intérêt de leur étude pour les reconstitutions paléogéographiques, in Géochimie organique des sédiments marins. D'Orgon à Misedor: éditions CNRS, Paris, p. 327-352.

Cecca, F., R. Enay, and G. Le Hégarat, 1989, L'Ardescien (Tithonique supérieur) de la région stratotypique: séries de références et faunes (ammonites, calpionelles) de la bordure ardéchoise: Documents des Laboratoires de Géologie, Lyon, 107, 115 p.

Combaz, A., 1964, Les palynofaciès: Revue de Micropaléontologie, v. 7, p. 205-218.

Combaz, A., 1980, Les kérogènes vus au microscope, in B. Durand, ed., Kerogen: Insoluble Organic Matter in Sedimentary Rocks: Editions Technip, Paris, p. 55-112.

Cross, A. T., G. G. Thompson, and J. B. Zaitzeff, 1966, Source and distribution of palynomorphs in bottom sediments, southern part of Gulf of California: Marine Geology, v. 4, p. 467-524.

Davies, J. R., A. McNestry, and R. A. Waters, 1991, Palaeoenvironments and palynofacies of a pulsed transgression: the late Devonian and early Dinantian (Lower Carboniferous) rocks of southeast Wales: Geological Magazine, v. 28, p. 355-380.

Deconinck, J.-F., 1992, Clay mineralogy of Early Cretaceous sediments of south-east France: Berriasian stratotype, Berriasian and Barremian of Angles: Abstracts, Sequence stratigraphy of European basins, Dijon, p. 366-367.

Denison, C. N., and R. M. Fowler, 1980, Palynological identification of facies in a deltaic environment, in The sedimentation of the North Sea reservoir rocks: Norwegian Petroleum Society, v. 12, p. 1-22.

De Renéville, P., and J.-F. Raynaud, 1981, Palynologie du stratotype du Barrémien: Bulletin des Centres de Recherches Exploration-Production, Elf-Aquitaine, v. 5, p. 1-29.

Emmanuel, L., and M. Renard, 1992, Carbonate geochemistry (Mn, d13C, d18O) of Berriasian pelagic limestones. Relationship with sequence stratigraphy. (Angles section, Vocontian Trough, S-E France): Abstracts, Sequence stratigraphy of European basins, Dijon, p. 370-371.

Fisher, M. J., 1980, Kerogen distribution and depositional environments in the Middle Jurassic of Yorkshire, UK: Proceedings 4th International Palynological Conference, Lucknow, 1976-1977, v. 2, p. 574-580.

Galbrun, B., L. Rasplus, and G. Le Hégarat, 1986, Données nouvelles sur le stratotype du Berriasien: corrélations entre magnétostratigraphie et biostratigraphie: Bulletin de la Société Géologique de France, v. 8, t. II/4, p. 575-584.

Gorin, G. E., and S. Feist-Burkhardt, 1990, Organic facies of Lower to Middle Jurassic sediments in the Jura Mountains, Switzerland: Review of Palaeobotany and Palynology, v. 65, p. 349-355.

Gorin, G. E., and D. Steffen, 1991, Organic facies as a tool for recording eustatic variations in marine fine-grained carbonates—example of the Berriasian stratotype at Berrias (Ardèche, SE France): Palaeogeography, Palaeoclimatology, Palaeoecology, v. 85, p. 303-320.

Grabowski G. J., Jr., and K. S. Glaser, 1990, Depositional model for transgressive marine organic-rich rocks formed in platformal settings: Middle Callovian and Lower Toarcian examples from onshore England: Abstracts, 13th International Sedimentological Congress, Nottingham, p. 196.

Gregory, W. A., and G. F. Hart, 1992, Towards a predictive model for the palynologic response to sea-level changes: Palaios, v. 7, p. 3-33.

Habib, D., and J. A. Miller, 1989, Dinoflagellate species and organic facies evidence of marine transgression and regression in the Atlantic Coastal Plain: Palaeogeography, Palaeoclimatology, Palaeoecology, v. 74, p. 23-47.

Haq, B. U., J. Hardenbol, and P. R. Vail, 1987, Chronology of fluctuating sea levels since the Triassic: Science, 235, p. 1156-1166.

Harland, W. B., R. L. Armstrong, A. V. Cox, L. E. Craig, A. G. Smith, and D. G. Smith, 1989, A geologic time scale: Cambridge University Press, Cambridge, 263 p.

Hart, G. F., 1986, Origin and classification of organic matter in clastic systems: Palynology, v. 10, p. 1-23.

Honjo, S., and M. R. Roman, 1978, Marine copepod fecal pellets: production, sedimentation and preservation: Journal of Marine Research, v. 36, p. 45-57.

Huc, A. Y., 1988, Sedimentology of organic matter, in F. H. Frimmel and R. F. Christman, eds., Humic Substances and their Role in the Environment: Wiley, New York, p. 215-243.

Le Hégarat, G., 1973, Le Berriasien du Sud-Est de la France: Documents du Laboratoire de Géologie de la Faculté des Sciences de Lyon, v. 43, 575 p.

Le Hégarat, G., 1980, Le Berriasien, in C. Cavelier and J. Roger, coord., Les étages français et leur stratotypes: Mémoire BRGM 34, p. 9-16.

Le Hégarat, G., and S. Ferry, 1990, Le Berriasien d'Angles (Alpes-de-Haute-Provence, France): Géobios, 23, p. 369-373.

Le Hégarat, G., and J. Remane, 1968, Tithonique supérieur et Berriasien de la bordure cévenole. Corrélation des Ammonites et des Calpionelles: Géobios, v. 1, p. 7-70.

Müller, P. J., and E. Suess, 1979, Productivity, sedimentation rate and sedimentary organic matter in the oceans - 1. Organic carbon preservation: Deep Sea Research, v. 26A, p. 1347-1362.

Parry, C. C., P. J. K. Whitley, and R. D. H. Simpson, 1981, Integration of palynological and sedimentological methods in facies analysis of the Brent Formation, in L. V. Illing and G. B. Hobson, eds., Petroleum Geology of the Continental Shelf of North-West Europe: Heyden, London, p. 205-215.

Pasley, M. A., W. A. Gregory, and G. F. Hart, 1991, Organic matter variations in transgressive and regressive shales: Organic Geochemistry, v. 17, p. 483-509.

Posamentier, H. W., M. T. Jervey, and P. R. Vail, 1988, Eustatic controls on clastic deposition I—conceptual framework, in C. K. Wilgus, B. S. Hastings, C. G. St. C. Kendall, H. W. Posamentier, C. A. Ross, and J. C. Van Wagoner, eds., Sea-level changes—an integrated approach: SEPM Special Publication 42, p. 109-124.

Posamentier, H. W., and P. R. Vail, 1988, Eustatic controls on clastic deposition II - sequence and systems tract models, in C. K. Wilgus, B. S. Hastings, C. G. St. C. Kendall, H. W. Posamentier, C. A. Ross, and J. C. Van Wagoner, eds., Sea-level changes—an integrated approach: SEPM Special Publication 42, p. 125-154.

Powell, A. J., J. D. Dodge, and J. Lewis, 1990, Late Neogene to Pleistocene palynological facies of the Peruvian continental margin upwelling, leg 112, in E. Suess, R. Von Huene et al.: Proceedings of the Ocean Drilling Program, Scientific Results, v. 112, p. 297-321.

Smelror, M., and H. Leereveld, 1989, Dinoflagellate and acritarch assemblages from the late Bathonian to Early Oxfordian of Montagne de Crussol, Rhône Valley, southern France: Palynology, v. 13, p. 121-141.

Stancliffe, R. P. W., 1989, Microforaminiferal linings: their classification, biostratigraphy and palaeoecology, with special reference to specimens from British Oxfordian sediments: Micropaleontology, v. 35, p. 337-352.

Stanley, D. J., 1986, Turbidity current transport of organic-rich sediments: Alpine and Mediterranean examples, in P. A. Meyers and R. M. Mitterer, eds., Deep ocean black shales: organic geochemistry and paleoceanography setting: Marine Geology, v. 70, p. 85-101.

Staplin, F. L., 1969, Sedimentary organic matter, organic metamorphism, and oil and gas occurence: Bulletin of Canadian Petroleum Geology, v. 17, p. 47-66.

Strohmenger, C., and A. Strasser, 1993, Eustatic controls on the depositional evolution of Upper Tithonian and Berriasian deep-water carbonates (Vocontian Trough, SE France): Bulletin des Centres de Recherches Exploration-Production, Elf-Aquitaine, in press.

Summerhayes, C. P., and T. C. Masran, 1982, Organic facies of Cretaceous and Jurassic sediments from Deep Sea Drilling Project site 534 in the Blake–Bahama basin, western North Atlantic, in R. E. Sheridan, F. M. Gradstein et al.: Initial Reports of the Deep Sea Drilling Project, v. 76, p. 469-480.

Tissot, B., 1979, Effects on prolific petroleum source rocks and major coal deposits caused by sea-level changes: Nature, v. 277, p. 463-465.

Tissot, B. P., and D. H. Welte, 1984, Petroleum Formation and Occurence, 2nd edition: Springer-

Verlag, Berlin, 699 p.

Traverse, A., 1988, Paleopalynology: Unwin Hyman, Boston, 600 p.

Tribovillard, N.-P., and G. E. Gorin, 1991, Organic facies of the Early Albian Niveau Paquier, a key black shales horizon of the Marnes Bleues Formation in the Vocontian trough (subalpine ranges, SE France): Palaeogeography, Palaeoclimatology, Palaeoecology, v. 85, p. 227-237.

Turner, J. T., and J. G. Ferrante, 1979, Zooplankton fecal pellets in aquatic ecosystems: Bioscience, v. 29, p. 670-676.

Tyson, R. V., 1987, The genesis and palynofacies characteristics of marine petroleum source rocks, in J. Brooks and A. J. Fleet, eds., Marine Petroleum Source Rocks: Geological Society of London Special Publication 26, p. 47-67.

Tyson, R. V., and B. M. Funnell, 1987, European Cretaceous shorelines, stage by stage: Palaeogeography, Palaeoclimatology, Palaeoecology, v. 59, p. 69-91.

Vail, P. R., F. Audemard, S. A. Bowman, P. N. Eisner, and C. Perez-Cruz, 1991, The stratigraphic signatures of tectonics eustasy and sedimentology—an overview, in G. Einsele, W. Ricken and A. Seilacher, eds., Cycles and Events in Stratigraphy: Springer-Verlag, Berlin, p. 617-659.

Van der Zwan, C. J., 1990, Palynostratigraphy and palynofacies reconstruction of the Upper Jurassic to lowermost Cretaceous of the Draugen Field, offshore Mid Norway: Review of Palaeobotany and Palynology, v. 62, p. 157-186.

Van Pelt, R. S., and D. Habib, 1988, Dinoflagellate species abundance and organic facies in Jurassic Twin Creek Limestone signal episodes of transgression and regression: Abstracts, 7th International Palynological Congress, Brisbane, p. 168.

Van Wagoner, J. C., H. W. Posamentier, R. M. Mitchum, Jr., P. R. Vail, J. F. Sarg, T. S. Loutit, and J. Hardenbol, 1988, An overview of the fundamentals of sequence stratigraphy and key definitions, in C. K. Wilgus, B. S. Hastings, C. G. St. C. Kendall, H. W. Posamentier, C. A. Ross, and J. C. Van Wagoner, eds., Sea-level changes—an integrated approach: SEPM Special Publication 42, p. 39-46.

Wall, D., 1965, Microplankton, pollen and spores from the Lower Jurassic in Britain: Micropaleontology, v. 11, p. 151-190.

Whitaker, M. F., 1984, The usage of palynostratigraphy and palynofacies in definition of Troll Field geology, in Offshore Northern Seas—reduction of uncertainities by innovative reservoir geomodelling: Norskpetroleum forening, article G6.

Chapter 6

Variation of the Distribution of Organic Matter Within a Transgressive System Tract: Kimmeridge Clay (Jurassic), England

Jean Paul Herbin
Carla Müller
Institut Français du Pétrole, France

Jeannine R. Geyssant
Frédéric Mélières
Université Pierre et Marie Curie, France

Ian E. Penn
British Geological Survey, England

Yorkim Group
Institut Français du Pétrole, France

ABSTRACT

The Kimmeridge Clay of northern England is part of a highstand system tract, rich in organic material, whose offshore correlatives sourced much of the North Sea oil province. Spatial and temporal variation over a 35-km-long transect (extended to 100 km by correlation offshore), some 200 m thick ranging from the Cymodoce to the Pallassioides zones, and representing some 6.5 m.y. of marine sedimentation, was evaluated by study of four continuously cored boreholes sited to sample both basin and shelf facies.

Thin-bed stratigraphy, established by geophysical log signatures calibrated by the ammonite succession, enables organic-rich beds to be traced throughout the transect and into other English basins. These results show that total organic carbon may be computed reliably from a combination of resistivity, density, and sonic logs and increases by over 50% as each level is traced from shelf to basin where deeper waters are thought to have favored its entrapment and preservation in an environment more depleted in dissolved oxygen and with a more rapid burial. Associated, coccolith-rich, marker bands of the shelf pass into dolomites in the basin owing to the precipitation of early dolomite through bacterial decay of the organic-rich material.

Vertical distribution of the organic content of each section identifies a hierarchy of sedimentary cycles with periodicities of about 25,000 and 280,000 years. The short period cycles, less than 1 m thick, comprise alternations of more or less organic-rich beds. The analysis of the extracts suggests unity of origin of the organic matter of type II origin (zoo- and phytoplankton) throughout the period of sedimentation. The differences observed in the kerogens can be interpreted in terms of a fluctuation of oxygen deficiency whose variations (dysaerobic, anaerobic, or anoxic) are recorded in time and in space. Some of the variation, near coccolith-rich beds, can be related to a very high organic productivity (up to 40 wt.% total organic carbon). The second order cycles show maximum kerogen enrichment in the middle of the transgressive tract intervals, or at the base of high level, or platform edge prisms. These relationships allow the distribution of the organic matter to be deduced from sequence stratigraphic studies.

INTRODUCTION

Because of their significant contribution to the genesis of the North Sea hydrocarbons, the transgressive deposits of the Upper Jurassic Series were selected in order to study source rock variation, specifically by modeling the variation of the quantity of organic carbon, and the type of organic matter, in time and space.

Of these deposits, the Kimmeridgian clays are the most significant. Hitherto, the most thorough observations of their organic matter were obtained from the type area of the Kimmeridge Clay (Cox and Gallois, 1981) in both outcrop and in the subsurface around Kimmeridge Bay, and from the cored borehole exploration conducted by the British Geological Survey on behalf of the United Kingdom Department of Energy, intended to estimate the petroleum potential of the UK's "onshore" oil-shale levels. These studies led to the decision that comparison with the "offshore" source rock facies could best be made by studying the Kimmeridge Clay of northern England (Yorkshire), the nearest land area to the North Sea reservoirs boasting a complete sequence. The space–time system selected was that lying between the Mutabilis and Pectinatus zones, since Gallois (1979) demonstrated that the oil-shale levels were more frequently found over that interval. The present study therefore illustrates the variation in distribution of organic matter in an onshore block about 35 km long by 200 m thick, representing nearly 6.5 m.y. of sedimentary history. Extrapolation to the adjacent offshore area increases the lateral distance to about 100 km.

GEOLOGICAL SETTING

On the time scale of Haq et al. (1987), the Kimmeridgian Stage coincides with a period of maximum transgression and, although its causes are still not clearly identified, a relationship appears to exist between the accumulation of planktonic organic matter and the high sea levels (Tissot, 1979). Moreover, source rocks of exceptional quality existed during other major transgressive periods such as during the Cenomanian/Turonian stages (Schlanger and Jenkyns, 1976; Herbin et al., 1986).

In addition, the richness of organic matter deposited during the Kimmeridgian time is not merely a local phenomenon. High-grade potential or effective source rocks exist in northern latitudes, both on the Canadian margin, Egret Member (Grant et al., 1988), in the North Sea, Borglum Formation and clays of the Kimmeridge Formation (Cornford, 1984), in the Aquitaine Basin, Lons Formation (Claret et al., 1981), and in western Siberia, Bazhenov Formation (Grace and Hart, 1986). These are all in very different structural settings, ranging from continental margin to rift and shelf.

The Kimmeridgian Stage is well represented in England (Figure 1), and borders the Mesozoic depositional high of the London Platform or Brabant Massif for a distance of more than 400 km. It crosses the Eastern England Shelf (Whittaker, 1985) and is present in the Cleveland basin, north of the depositional high of the Market Weighton Block. Here, its outcrop is sometimes extremely thin due to faulting and Cretaceous concealment, but the strata are better preserved on the northern limb of the syncline which now marks the southern margin of the Cleveland basin. The Vale of Pickering, the scene of this study, is partly a Quaternary-covered, topographical depression sited over this synclinal outcrop of Upper Jurassic clays. Here, four fully cored boreholes were drilled to test the organic variation in time and space of the Kimmeridge Clay.

The regional structural context of this onshore zone is known from the studies of the Deep Geology

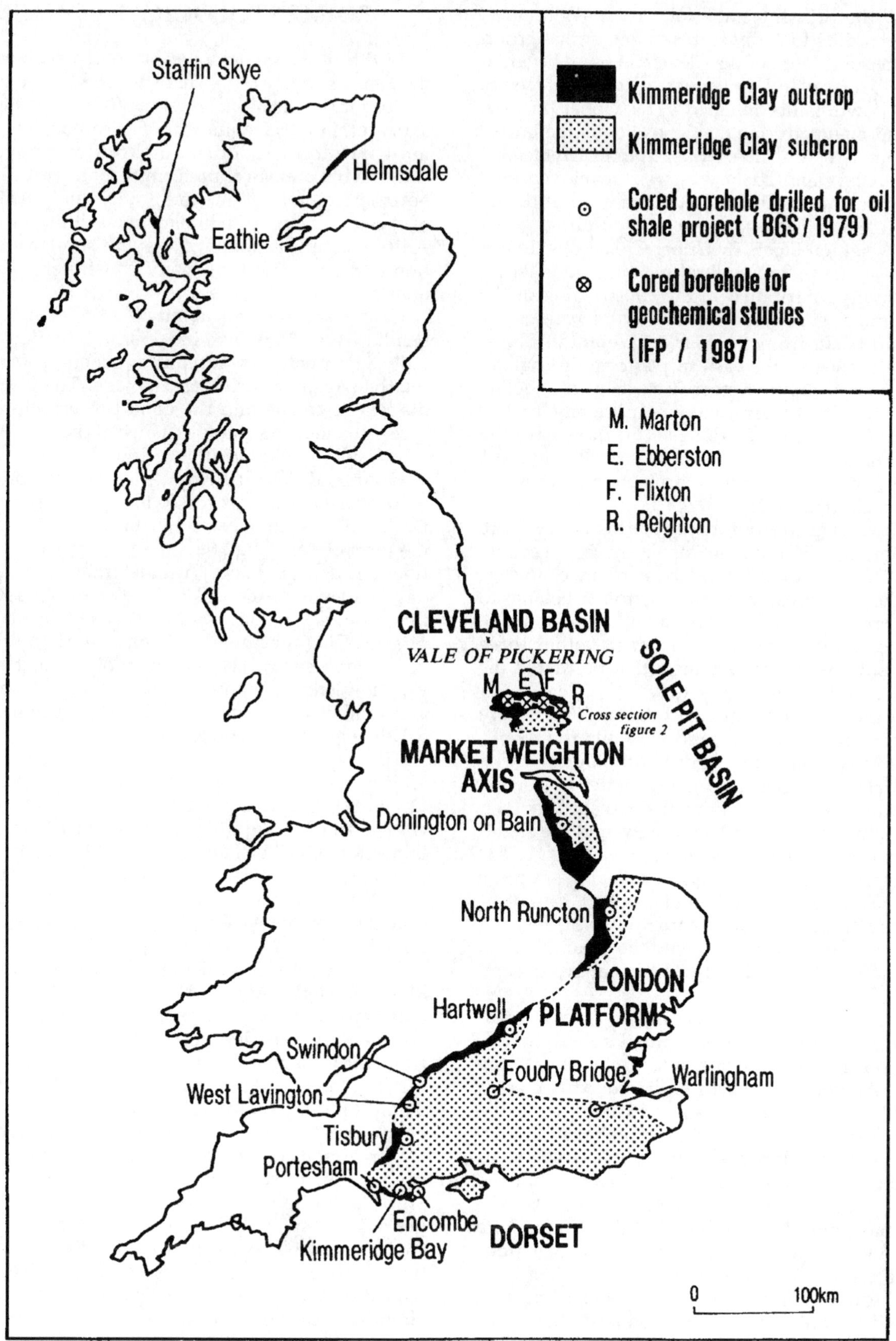

Figure 1. Location of the outcrop and subcrop of the Kimmeridge Clay in Great Britain, and the four IFP boreholes in the Yorkshire area (Marton 87, Ebberston 87/87 bis, Flixton 87, Reigthon 87), and the BGS boreholes (Gallois 1979).

Group of the British Geological Survey (Whittaker, 1985). According to Kirby and Swallow of that group (1987), the subsidence of the Cleveland basin began in the Late Triassic and continued to the Cretaceous period. Growth faults, initiated in the Upper Jurassic, developed mainly during the deposition of the Lower Cretaceous Speeton Clay. Thus, by mid-Cretaceous time, the Cleveland basin was well developed as a result of northern subsidence, contrasted with the area of lesser sedimentation, corresponding to the Market Weighton Block. By Upper Cretaceous time, a reversal of the movements along the pre-existing growth faults contributed to the construction of the fault-bounded synclinal structure that separates the Cleveland basin from the Market Weighton Block. Similarly seawards, the eastern platform of England occurs as an N-dipping, post-Cretaceous monocline interrupted by the fault corresponding to the Dowsing Fault zone. To the east, in the basin, the Upper Jurassic Series is more strongly folded and the Cretaceous deposits lie unconformably on a range of Jurassic rocks (Figure 2).

The investigatory boreholes of this study were aligned along a 35-km-long east–west cross section. Marton 87, Ebberston 87, and Flixton 87 are characteristic of the basin area, whereas Reighton 87 belongs to the Eastern England Shelf (Penn and Abbott, 1989). The major fault separating it from the other three boreholes derives from tectonism that postdated the deposition of the Kimmeridge Clay.

From the four boreholes, a total of 770 m were cored with more than 97% recovery. On the whole, the cored material displays little lithological variation. The levels richest in organic matter (>10 wt.% total organic carbon, TOC) are distinguished by their foliated appearance, their low density, and the slightly brown color of the layers with >20 wt.% TOC. The sedimentary cycles or rhythms, which appear on the geophysical logs (resistivity and density), cannot be confirmed from the macroscopic descriptions, but were correlated subsequently with the organic carbon distributions obtained directly from analyses of samples at 10-cm intervals.

The recovered cores display many structural features attesting to stresses sustained subsequent to their deposition (slip forms, tension joints). At the Marton 87 borehole, the base of the Eudoxus Zone is affected by a thrust fault for about 15 m of the succession. At Flixton 87, a fault at the top of the Hudlestoni Zone appears to reduce the sequence by about 5 m, as estimated from the thickness anticipated from the known gradual westward thickening of the beds.

The maximum depth of burial of the Kimmeridge Clay, however, was reached at the end of Upper Cretaceous times, and corresponds to a depth of 1 km at Reighton 87 borehole (on the Eastern England Shelf) and 1.4 km at Marton 87 borehole (in the center of the Cleveland basin), corroborating the profiles determined by Williams (1986) (Figure 3). These depths are insufficient to have allowed the genesis of hydrocarbons, as confirmed by the average T_{max} of 428°C obtained on about 6000 samples.

SEDIMENTOLOGICAL CONTEXT

Within the area sampled by the cored boreholes, the Kimmeridge Clay is characterized by the extreme uniformity of the black claystone facies, rich in organic matter, in which a number of calcareous, very commonly dolomitic, horizons are developed. The extreme mineralogical uniformity, both vertically and horizontally, reflects the uniformity and permanence of the conditions of sedimentation within the region. All the samples examined display similar composition and comprise four major constituents: calcite, quartz, clays, and pyrite.

Calcite accounts for about 10 wt.% of the rock. It results from planktonic biogenic production, essentially represented by coccoliths, generally forming a relatively monospecific assemblage. The absence of diagenetic calcite and the good preservation of the coccoliths indicate these organisms did not undergo significant dissolution after their burial.

Quartz, determined by x-ray diffraction (XRD, total quartz), accounts for 10 to 15 wt.% of the rock. Under the optical microscope, this mineral appears in the form of very small (less than 30 µm), angular (terrigenous detrital quartz) hyaline grains, whose abundance does not exceed 3 to 5 wt.% of the sediment; this observation suggests the existence of diagenetic quartz. The very small dimension of the detrital quartz grains suggests deposition of this material in a very low energy environment, sheltered from wave and tidal currents, and implying a substantial water depth of at least several tens of meters.

The clays make up 65 to 75 wt.% of the rock. They are mineralogically very constant and comprise illite (20 wt.%), irregular mixed-layer illite/smectite (50 wt.%), and kaolinite (30 wt.%) for the Marton 87, Ebberston 87/87 bis, and Flixton 87 boreholes. In the Reighton 87 borehole, however, kaolinite accounts for only 20 wt.% of the clay mineral assemblage, which is dominated by illite (20 wt.%) and mixed-layer materials (60 wt.%).

Scanning electron microscope observations by F. Mélières (Herbin et al., 1991) show that the clay mineral matrix of the rock pseudomorphs the original sedimentary particles (coccoliths) or those resulting from early diagenesis (dolomite rhombohedrons) in the finest details. Pseudomorphs of such a degree of detail, to a few hundred angströms, could not have been made by detrital particles, which have much larger dimensions (about 1 µm). This morphology suggests the existence of a diagenetic mineral phase. Since kaolinite is considered here as a terrigenous detrital mineral, only illite and mixed-layer materials could result at least partly from the diagenetic evolution of an originally more siliceous clay phase, accompanied by the liberation of silica. Smectite, due to its chemical composition and structure, could have made up this original clay phase.

Pyrite is omnipresent in the sediments, where it accounts for 3 to 5 wt.%, in its framboidal form characteristic of very early diagenesis (Berner, 1971). Its presence attests to the intense activity of sulfate-

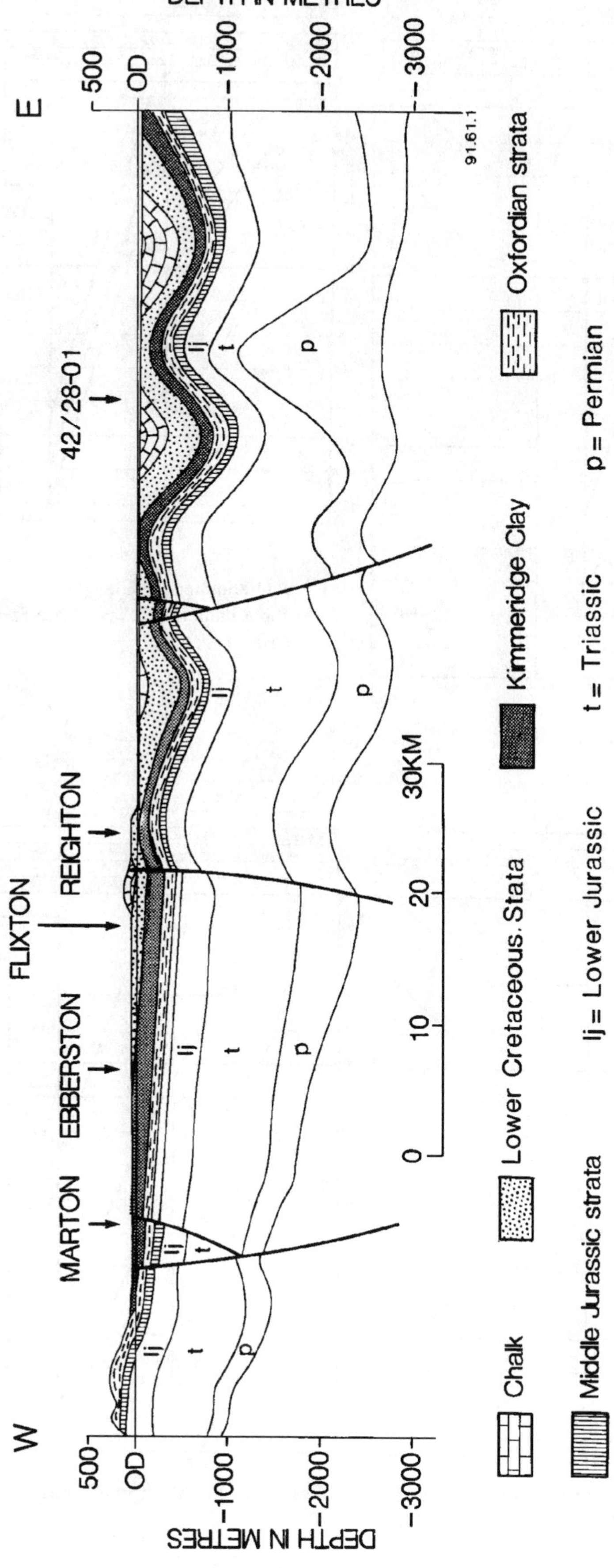

Figure 2. East–west geological cross section from the west of the Cleveland Basin, the Vale of Pickering, to the Sole Pit Basin, southern North Sea (from Penn and Abbott, 1989) See Figure 1 for location of cross section.

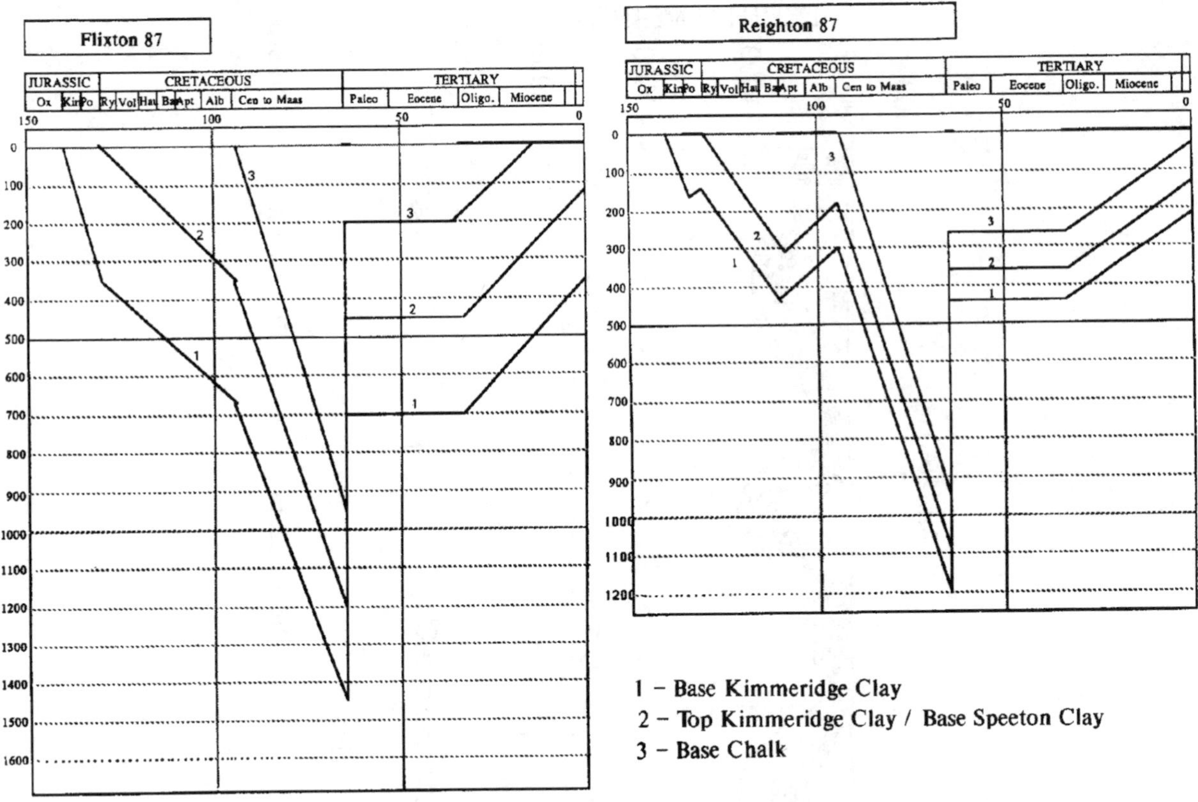

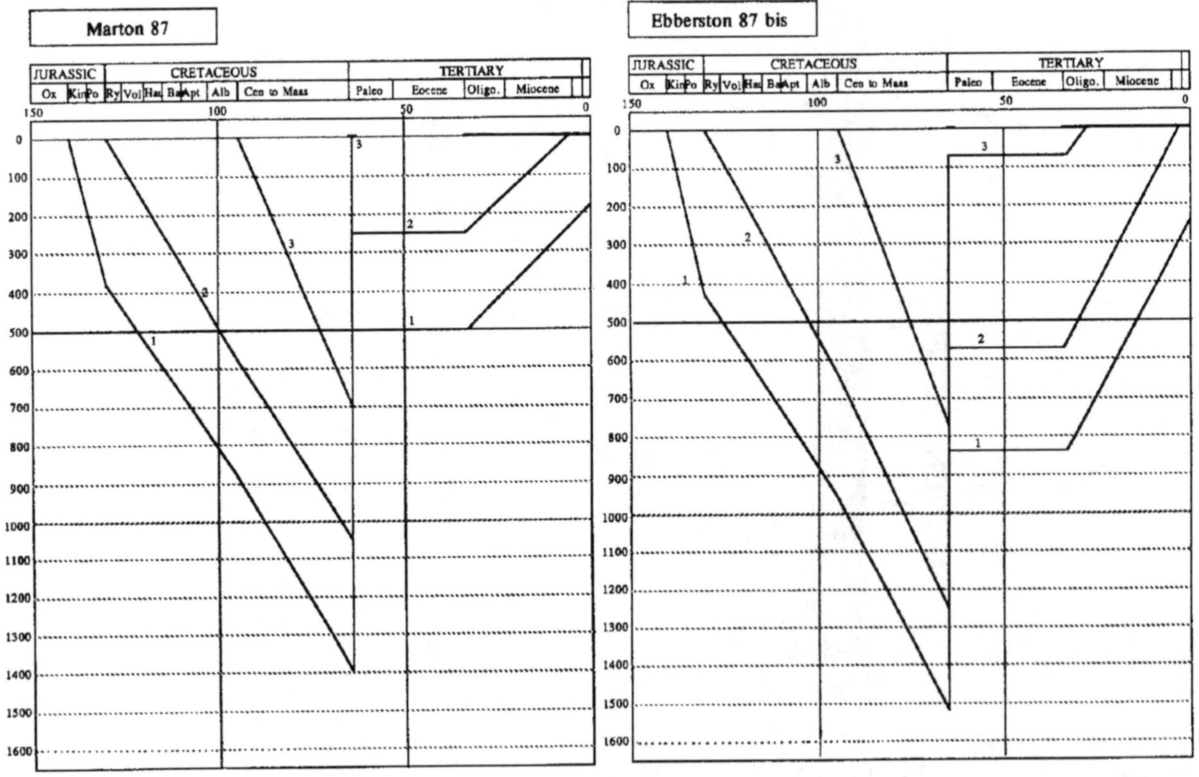

Figure 3. Kimmeridge Clay burial history at Flixton 87, Reighton 87, Marton 87, and Ebberston 87 bis boreholes (from Penn and Abbott, 1989).

reducing bacteria, resulting from the abundance of the organic matter during sedimentation.

Highly indurated beds, less than 1 m thick, appear within this generally uniform sedimentary sequence and can be observed from west to east throughout the study area. These layers consist of dolomite whose abundance decreases slightly from Marton 87 (80 to 90 wt.% of the rock) to Flixton 87 (70 to 80 wt.%). Farther to the east, at the Reighton 87 borehole, however, where the beds are more condensed and appear to have been deposited beneath shallower water, the lateral equivalents of these dolomitic beds are often purely calcitic.

The dolomite was formed relatively early in diagenesis, as shown by isotope geochemistry (Fritz and Smith, 1970). The high negative values of $\delta^{18}O$ (–2 to –8‰) appear to discount the possibility of evaporative confinement, since this would have resulted in an ^{18}O enrichment. Thus, temporary isolation of the basin appears improbable, and permanent communication with the open sea is most likely. The values of $\delta^{13}C$, ranging between –6 and +8‰, indicate that the carbonate of the dolomite derives from the carbon of the organic matter degraded by bacterial processes. The latter occurred either during a slightly oxidizing or in a sulfate-reducing environment in the neighborhood of the water–sediment interface. This is the case for the large majority of the samples analyzed, for which the values of $\delta^{13}C$ are slightly negative. Bacterial degradation also occurred in a methanogenetic hyperconfinement located at greater depth since at 142.82 m depth in Marton 87 borehole $\delta^{13}C$ = +7.8‰ and, in its lateral equivalent at 179.30 m depth at Ebberston 87 bis, $\delta^{13}C$ = +2.8‰.

Within the indurated carbonate beds, the calcite results from coccoliths which make up to 80 wt.% of the rock (Reighton 87, 133.63 m). The assemblages are distinguished by an abundance of extremely fragile forms which are sensitive to dissolution processes. This calcite appears to be a possible source of the calcium involved in the genesis of the dolomite. The juxtaposition between the coccoliths of these layers (with an abundance of fragile forms) and those of the surrounding claystones (assemblages of dissolution-resistant forms) suggests the existence of the most exceptionally good conditions of preservation during the deposition of the calcareous beds. Such excellent preservation may result from an oxygen-depleted episode induced by increased planktonic production (blooms) causing a more intense stratification of the water column (Tyson et al., 1979). Accordingly, very high organic carbon contents are found in the neighborhood of the carbonate layers.

Hence, the dolomite beds appear to result from the diagenetic transformation of coccoliths that accumulated in outstandingly favorable conditions of preservation during depositional burial. This transformation involved dissolution of the calcium of the fragile nannoflora under the action of CO_2 produced by the bacterial transformation of the organic matter. The presence of magnesium in the pore water led to the appearance of a primary dolomite, whose growth continued as burial proceeded, due to the progressive vertical expulsion of the fluids by compaction. A similar origin has already been postulated for the dolomite beds of the Kimmeridge Clay of the Dorset area (Irwin and Curtis, 1977; Irwin, 1980). The dolomitic diagenesis appears to have been more intense in the western part of the basin (Marton 87) than in its eastern part (Flixton 87). Such a gradient may result from the greater accumulation of organic matter in the west than in the east.

STRATIGRAPHIC DETERMINATION AND GEOPHYSICAL LOG CORRELATION

Ten ammonite zones were identified, four belonging to the lower Kimmeridgian (*sensu anglico*) substage, and six to the upper Kimmeridgian (*sensu anglico*) substage confirming the zonal scheme previously established in the Yorkshire area (Cope, 1974a, 1980) more completely. The four lower Kimmeridgian zones (Cymodoce, Mutabilis, Eudoxus, Autissiodorensis) contain a fairly diversified perisphinctid fauna (*Rasenia, Rasenioides, Aulacostephanoides, Aulacostephanus, Tolvericeras, Subdichotomoceras, Sutneria*), Cardioceratinae (*Amoeboceras, Hoplocardioceras, Nannocardioceras*) and Aspidoceratidae. The lowest five upper Kimmeridgian zones (Elegans, Scitulus, Wheatleyensis, Hudlestoni, and Pectinatus) are characterized by the nearly exclusive presence of a single genus of perisphinctid, *Pectinatites*, which replaces *Aulacostephanus* present in the previous Eudoxus and Autissiodorensis zones. A sixth zone (probably the Pallasioides Zone) is represented at Flixton 87 above the Pectinatus Zone. The zones recognized in these four boreholes also serve to establish correlation with the other areas of England (Figure 4): Dorset (Cope, 1967, 1978; Cox and Gallois, 1981), Wiltshire (Birkelund et al., 1983), Warlingham (Callomon and Cope, 1971) and the Wash area (Cope, 1974b; Gallois and Cox, 1974, 1976). The arrival at different levels of the Mutabilis, Eudoxus, Autissiodorensis, and Elegans zones of more southerly elements (*Aspidoceras, Tolvericeras, Sutneria,* and *Gravesia*) may attest to the presence of southerly currents bringing more favorable environmental and climatic conditions northward. It also furnishes interesting detail for correlations with the sub-Mediterranean provinces and the Franco-German region.

The analysis of the geophysical log traces (gamma-ray, resistivity, density) enables very precise stratigraphical correlation between the four boreholes. Apart from the major peaks, which correspond to the dolomite–carbonate levels (with high resistivity and density values) and which nevertheless display a dampening of the signal amplitude relative to the decrease in their thickness from west to east (from the

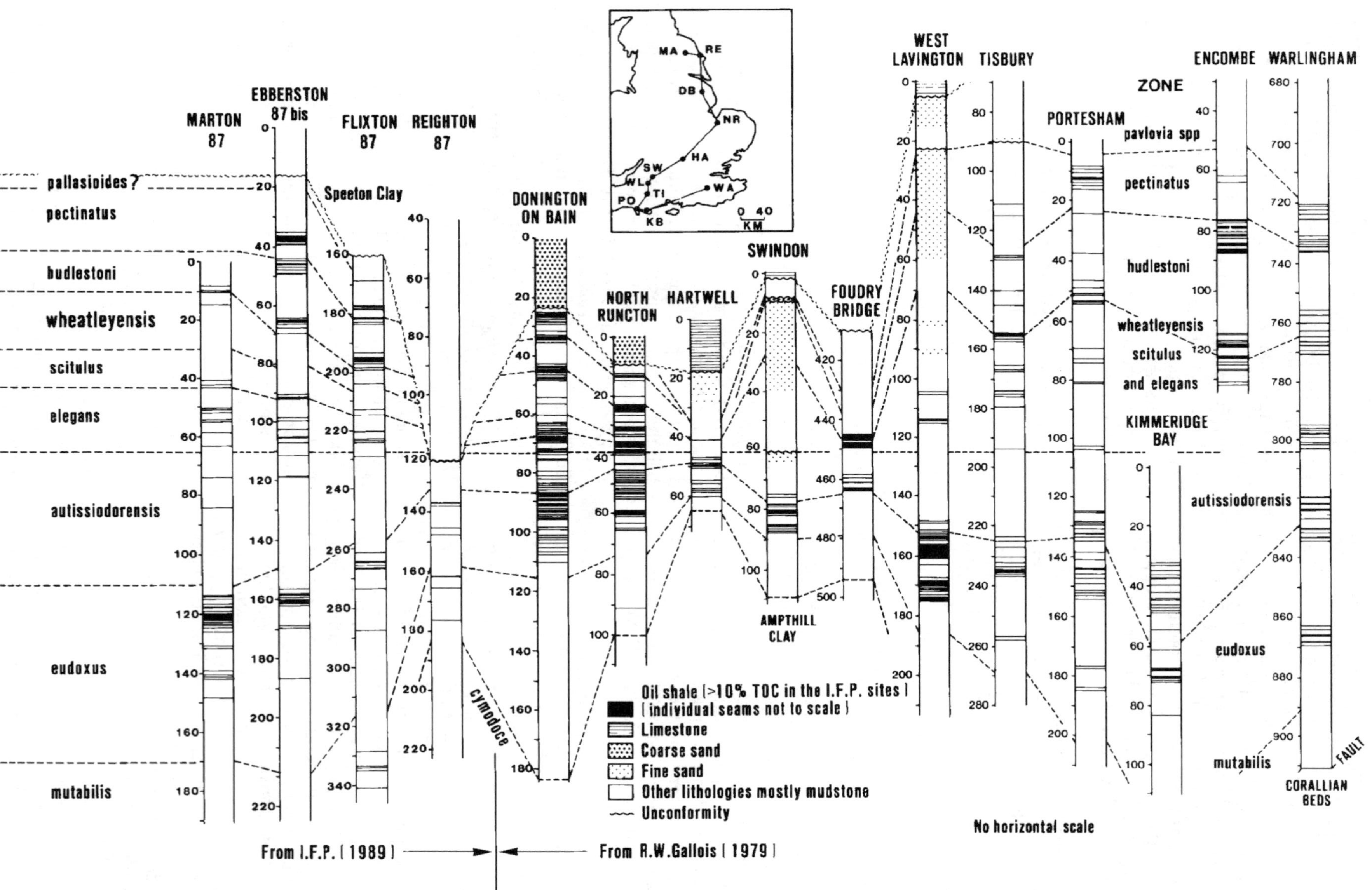

Figure 4. Stratigraphic correlation, based on the ammonite fauna of the oil shale-rich parts of the Kimmeridge Clay, proved in the British Geological Survey and the Institut Français du Pétrole cored boreholes from Figure 1.

basin area to the platform area), the logs reveal good correlation in detail (Figure 5).

One of the dolomite levels, whose lower part is characterized by a unit less than 10 cm thick, consisting of millimeter grains of magnesium calcite in XRD, corresponds to a sharp change in the baseline of the gamma-ray log. These highly characteristic lithological details and corresponding log changes made this bed the base, or "0" level, of correlation of the study. Seventy-five marker levels were identified from the resistivity logs below this reference bed, and 102 above it. Thus, a total of 178 reference marker levels were surveyed and correlated between the boreholes.

The interval of biostratigraphic uncertainty between the definite identification of the Eudoxus and Autissiodorensis zones is less than 1 m, and is located just above the "0" reference level. The comparison of these biostratigraphic and log data with those of the Dorset area (Cox and Gallois, 1981) helps to establish the equivalence of the "0" marker level with the "Flats Stone Band" (Figure 5B). Other lithological reference levels defined in Dorset were identified in Yorkshire by a similar approach (determination of ammonite zones, comparison with log data, and comparison with the Dorset area) and are:

- the "Cementstone Band" (double layer) corresponds to the dolomitic double layer "+36"/"+37" at the top of the Autissiodorensis Zone,
- the "Blackstone Band" is the equivalent of reference mark "+78" at the top of the Wheatleyensis Zone (Figure 5A),
- the "Rope Lake Head Stone Band" corresponds to reference mark "+84" near the base of the Hudlestoni Zone (Figure 5A),
- the "White Stone Band," a layer rich in coccoliths, can be correlated with reference mark "+91" at the base of the Pectinatus Zone.

The probable extension of these lithological reference marks from Dorset to Yorkshire is significant and serves to estimate the scale of the sedimentological processes which occurred in the Kimmeridgian Stage, at least on the Eastern England Platform. Their presence at the boundary of the ammonite zones also gives them some value as isochron. Because the log reference marks are "parallel" to the different stratigraphic boundaries, it can be assumed that they are themselves isochronous.

Thus, the geophysical log analysis of the Kimmeridgian sequence in the Vale of Pickering in the Marton 87, Ebberston 87/87 bis, Flixton 87, and Reighton 87 boreholes serves to subdivide the sedimentary sequence apparently isochronously, and consequently to define relative chronology at any point of the basin from east to west. However, an accurate quantitative estimate of the geological time represented between each such level cannot be given due to the low precision of the known radiometric dating. At best, only the concept of "equal duration" of the zones can be applied to evaluate the lower Kimmeridgian (*s.a.*) zones (Cymodoce, Mutabilis, Eudoxus, and Autissiodorensis) at 1 Ma, and the upper Kimmeridgian (*s.a.*) zones (Elegans, Scitulus, Wheatleyensis, Hudlestoni, and Pectinatus) at 0.5 Ma.

QUANTITATIVE VARIATION OF ORGANIC MATTER

The study of the quantitative variation relies on the Rock-Eval organic carbon distribution (Espitalié et al., 1985/1986) obtained from analyses performed every 50 cm. In the highly variable intervals, i.e., at the top of the Eudoxus Zone, within the Elegans Zone, at the base and at the top of the Hudlestoni Zone, and at the base of the Pectinatus Zone, the analyses were conducted every 10 cm.

The organic carbon logs show good agreement with the resistivity and density logs, which helped to establish the inferred isochronous reference levels (Figure 6). The peaks of high resistivity and lower density are related to the tops of organic sequences (high TOC), whereas the high-density peaks associated with low resistivity coincide with the base of organic sequences (low TOC).

The organic sequences, like the geophysical marker levels, correspond to isochrones throughout the Vale of Pickering, and can be extrapolated to Dorset if, for example, reference is made to certain highly characteristic beds, like the "Blackstone," which are lithologically, stratigraphically, and geochemically distinctive. This particularly applies to the sets of peaks marking the top of the Eudoxus Zone, equivalent to "reference bed 32" of the nomenclature of the British Geological Survey, and which can be identified from Yorkshire to Dorset, passing through the Wash area (Gallois, 1988).

SPATIAL VARIATION

The analysis of the distribution of total organic carbon throughout the Kimmeridge Clay beneath the Vale of Pickering (Figure 7) reveals that the rich zones (over 10 wt.% TOC) are easily correlated between the four boreholes: the top of the Eudoxus Zone, the top of the Wheatleyensis Zone, and the base of the Hudlestoni Zone, or the top of the Hudlestoni Zone with the lower part of the Pectinatus Zone. These units also demonstrate a tendency to enrichment in TOC when traced from east to west. This is especially clear for the bimodal distribution of the top of the Eudoxus Zone, where the maximum TOC contents increase by nearly 50%, going from Reighton 87 (TOC < 20 wt.%) to Marton 87 (TOC > 35 wt.%).

A detailed study of this zone, located below the "Flats Stone Band" corresponding to reference mark "0" (Figure 8), helps to correlate the organic sequences throughout the region (from Reighton 87 to Marton 87), and to quantify the general thickening between marker levels (identified on the resistivity logs) "0" and "–32" from east to west. These are at 13.65 m at Reighton 87, 28.75 m at Flixton 87, 34.25 m at Ebberston 87 bis, and 36.50 m at Marton 87. That is

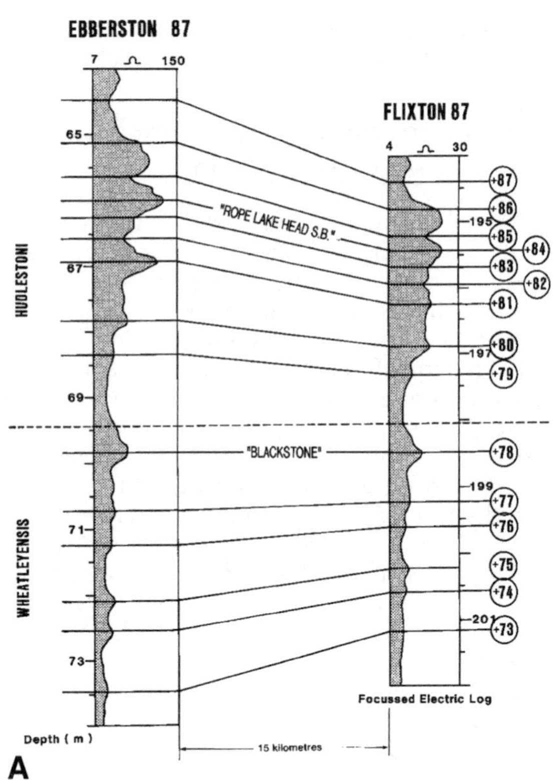

Figure 5. A—Geophysical log correlation based on the resistivity logs of the Ebberston 87 bis, and Flixton 87 boreholes with the reference marks "+73" and "+87" at the boundary of the Wheatleyensis and Hudlestoni zones. Note the reference mark "+84" corresponding to a dolomitic cementstone equivalent of the "Rope Lake Head Stone Band" and the reference mark "+78" corresponding to the so-called "Blackstone" in the type section of Dorset. B—Geophysical log correlation based on the resistivity logs of the Marton 87, Ebberston 87 bis, and Flixton 87 boreholes with the reference marks "+2" and "-6" at the boundary of the Eudoxus and Autissiodorensis zones. Note the reference mark "0" corresponding to a dolomitic cementstone equivalent of the "Flats Stone Band" in the type section of Dorset.

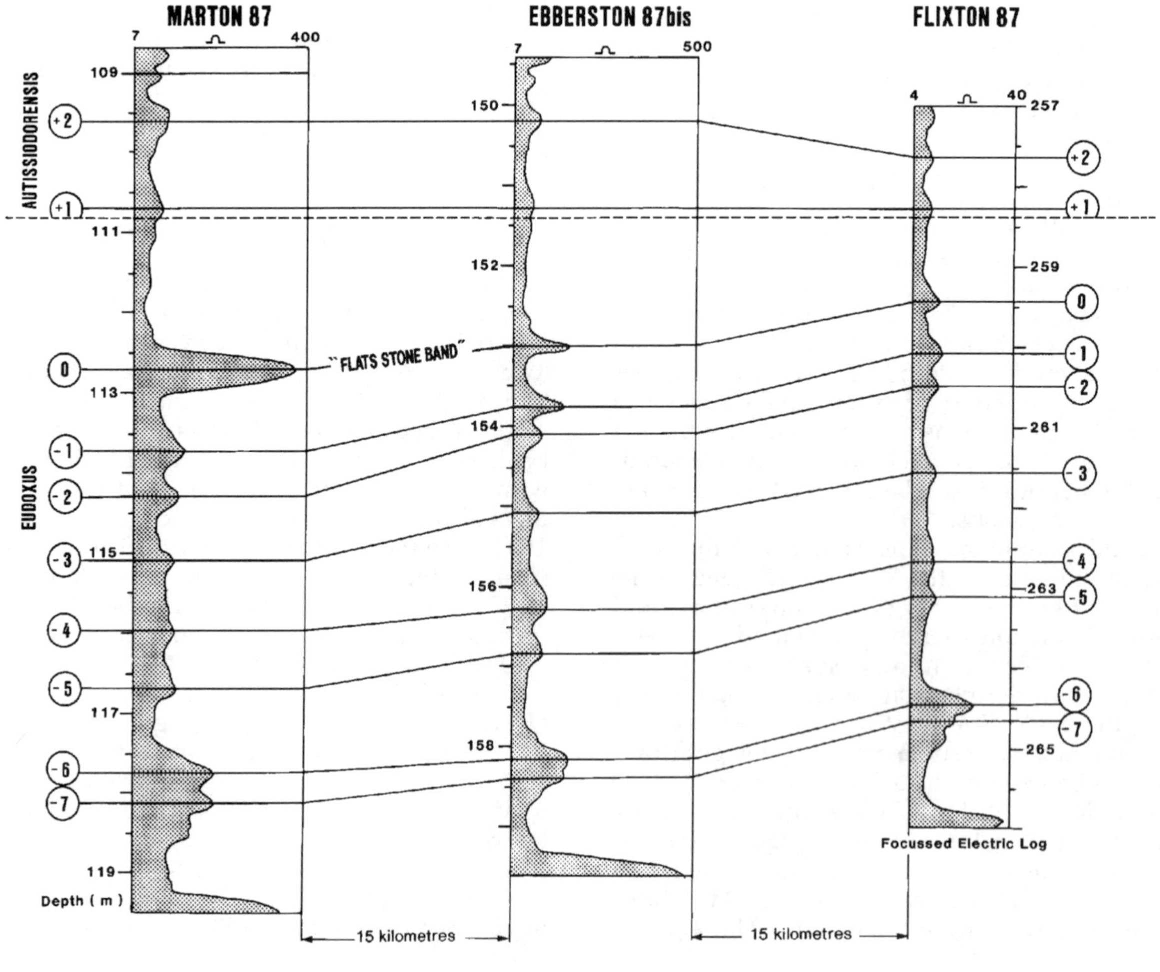

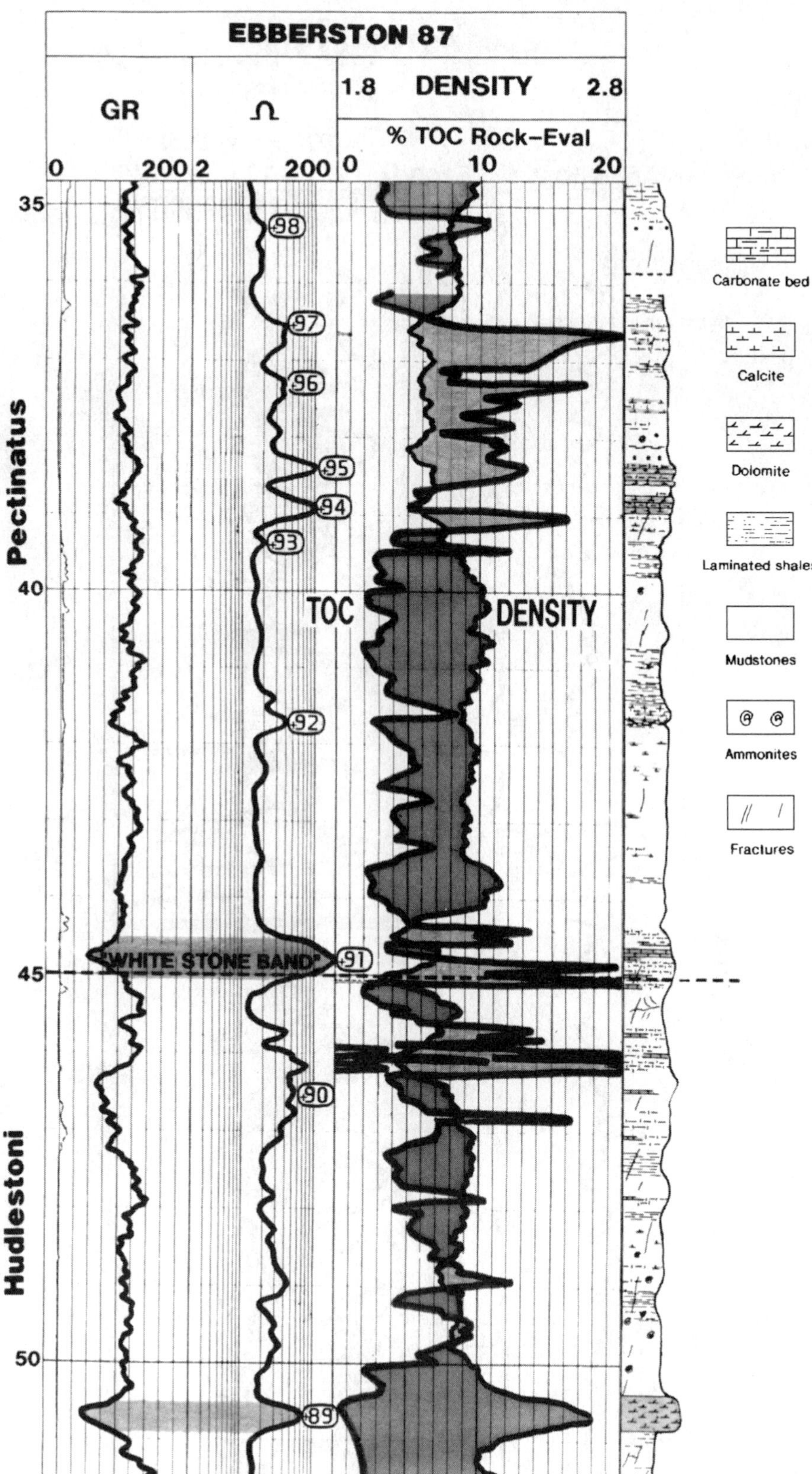

Figure 6. Correlation between the total organic carbon distribution resulting from Rock-Eval analyses and the gamma-ray (GR), resistivity (Ω), and density logs of Ebberston 87 and Ebberston 87 bis. A—Ebberston 87 borehole from 35 to 51 m depth at the boundary between the Hudlestoni and the Pectinatus zones, with the coccolith-rich layer equivalent to the "White Stone Band" (reference mark "+91") and the enrichment in organic matter on both sides. This enrichment corresponds to one of the peaks of accumulation in organic matter. (Figure 6 continues on pp. 78–79.)

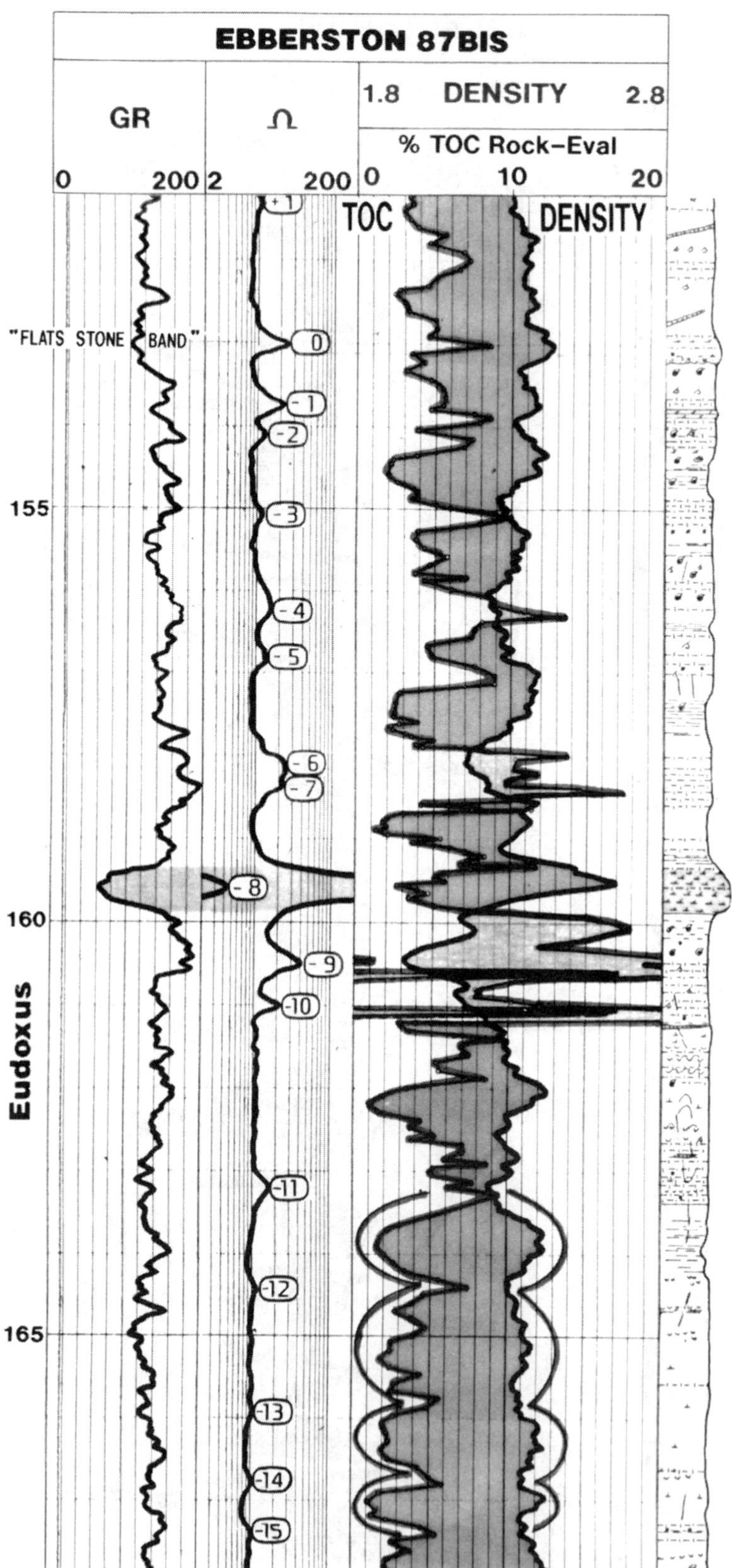

Figure 6 (continued). B—Ebberston 87 bis borehole, from 151 to 168 m (reference marks "+1" to "–15"), with the "0" reference mark corresponding to the "Flats Stone Band" and the enrichment in organic matter characteristic of the top of the Eudoxus Zone on both sides of the carbonate bed (reference mark "–8"). This maximum enrichment in organic matter, corresponding to the "reference bed 32" of the nomenclature of the British Geological Survey, can be identified from Yorkshire to Dorset, passing through the Wash area (Gallois, 1988).

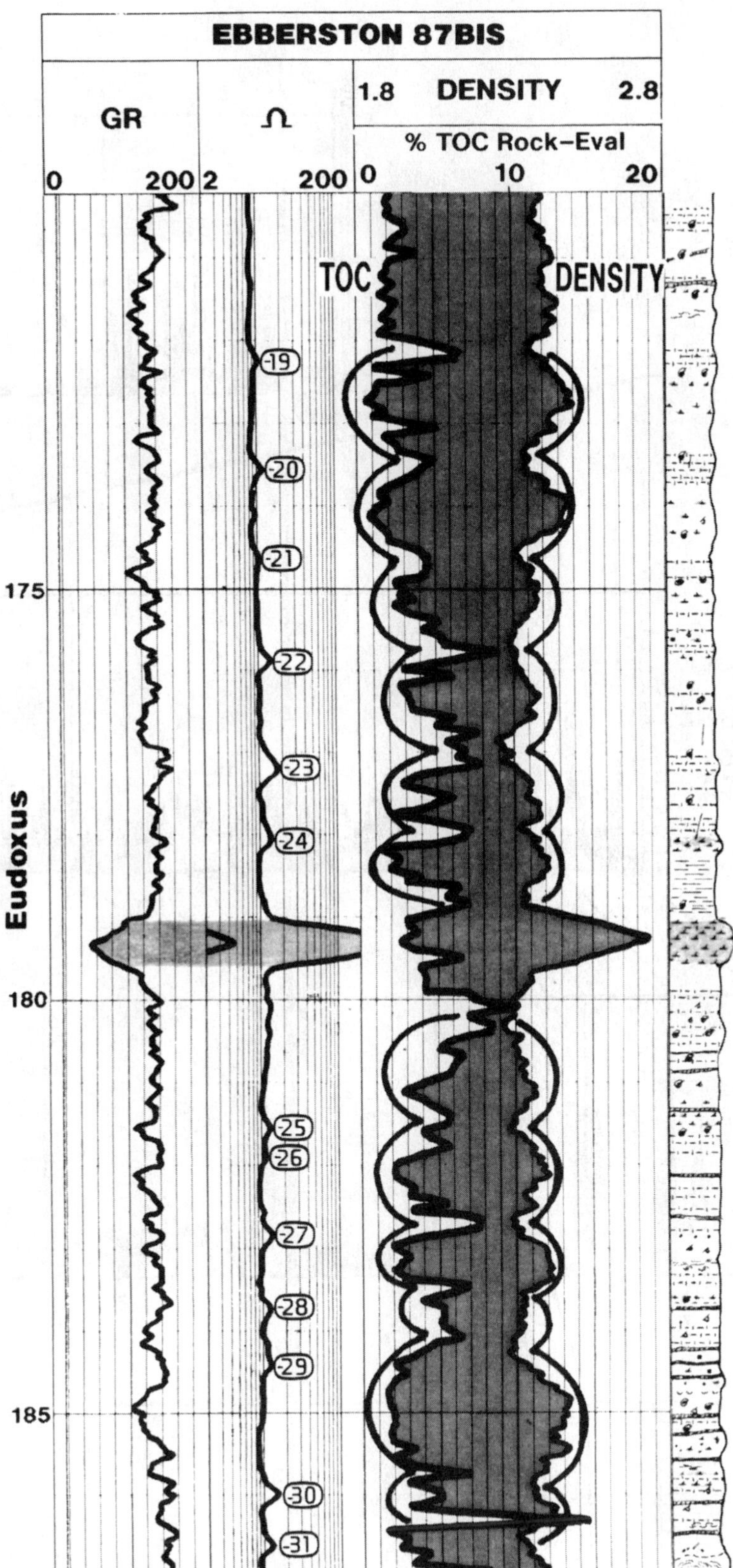

Figure 6 (continued). C—Ebberston 87 bis borehole within the Eudoxus Zone from 170 to 187 m (reference marks "–19" to "–31") showing typical cyclicity of total organic carbon content.

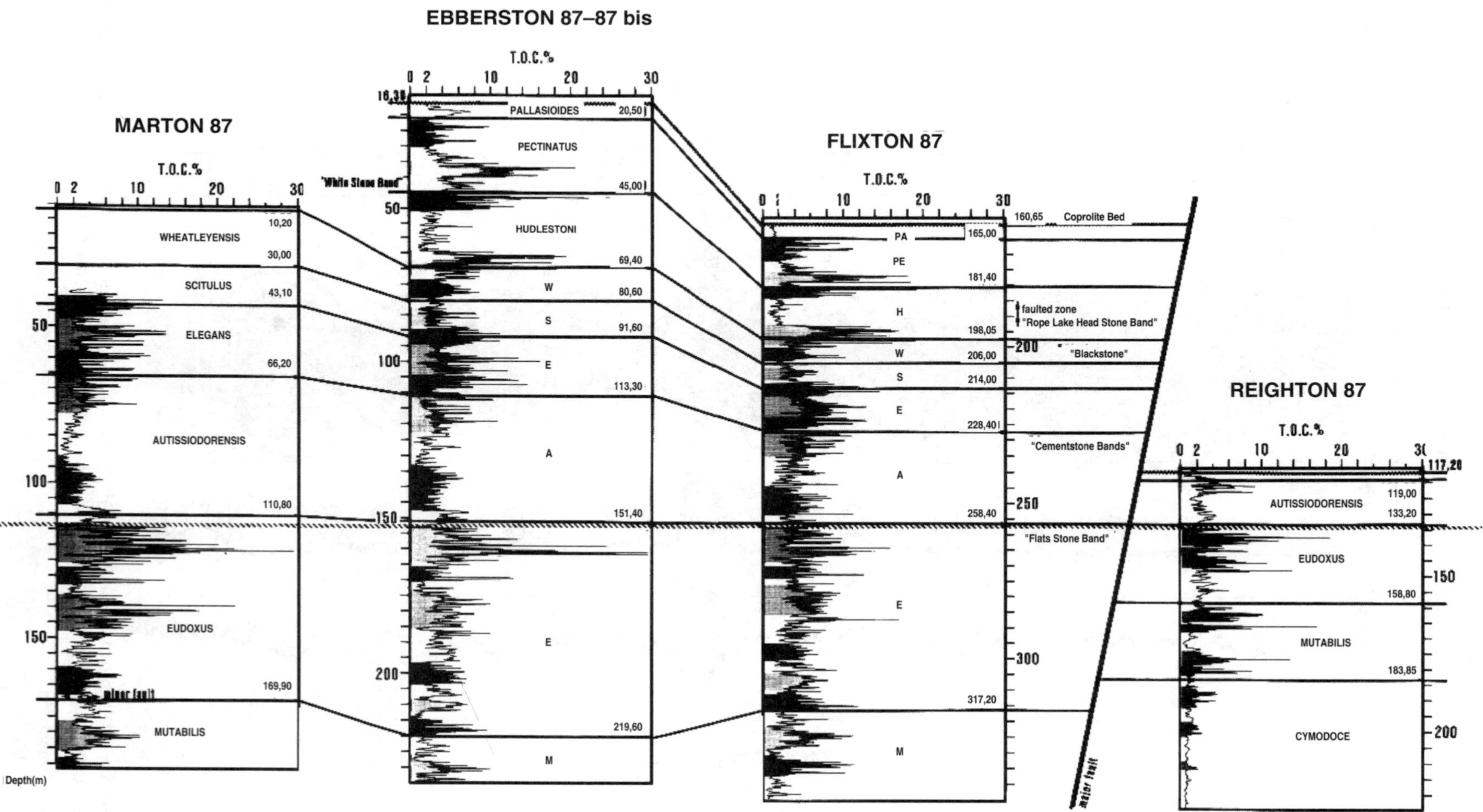

Figure 7. Correlation of the total organic carbon distribution of the four boreholes (issued from Rock-Eval analysis and Carbolog). The frames correspond to the correlation of the organic sequences at 10-m intervals.

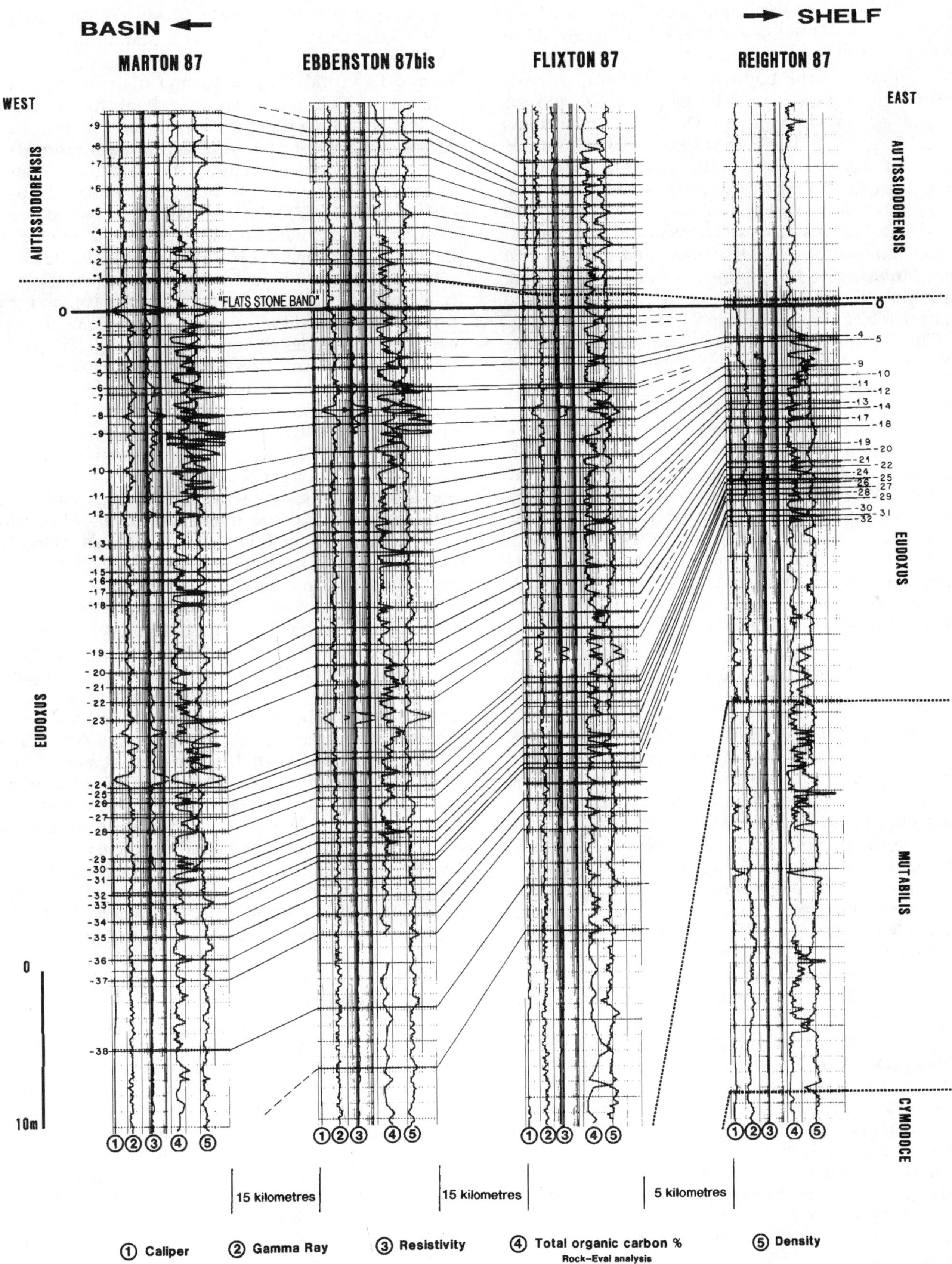

Figure 8. Detailed electrical correlation from the "basin" to the "shelf" of the reference marks "+9" and "–38" and the total organic carbon distribution at the boundary between the Autissiodorensis Zone and the Eudoxus Zone in the four boreholes: Marton 87, Ebberston 87 bis, Flixton 87, and Reighton 87. See Figure 1 for locations.

to say there is a 110% increase in thickness from Reighton 87 to Flixton 87, 19% from Flixton 87 to Ebberston 87, and 7% from Ebberston 87 to Marton 87. A glance at the Eudoxus Zone overall reveals equivalent respective percentages of 130, 16, and 8% over an estimated period of 1 Ma. The distinctiveness of the Reighton 87 site with respect to the other three sites, all located within the basin (Flixton 87, Ebberston 87 bis, and Marton 87), is connected with its position on the Eastern England Shelf. It should be observed that this increase in thickness from east to west (from platform to basin) goes hand-in-hand with the general rise in organic matter content from Reighton 87 (TOC ranging from 1 to 20 wt.%) to Marton 87 (TOC ranging from 2 to 40 wt.%).

The organic geochemical variations reflect the degree of organic matter preservation through the water column and at the sediment–water interface (Pelet, 1984). The existence of organic-rich layers in the zone exhibiting the highest rate of sedimentary accumulation (basinal deposition) can be explained partly by a more favorable environment for the preservation of organic matter in a deeper bathymetric area (and a lower energy environment than the shelf area) and partly by a better preservation associated with more rapid burial, which confirms the model described by Demaison et al. (1983) established for the Kimmeridge Clay Formation in the Central Graben of the North Sea.

VARIATION IN TIME

Whereas the spatial variation in total organic carbon fluctuates by a factor of two (the maximum TOC of about 20 wt.% at Reighton 87 reaching nearly 40 wt.% at Marton 87), the variation in time in the same borehole is about ten times greater (fluctuating in time by a factor of 20), ranging from 2 to 40 wt.% between the poorest and richest layers in Marton 87 or from 1 to 20 wt.% in Reighton 87.

This vertical variation is not a random occurrence, but passes alternately from low TOC values to high values. These cycles were identified on the geophysical logs as resistivity or density variations (Figure 6C). This cannot be identified visually in the cores, but is obvious in weathered outcrops such as at Kimmeridge Bay, Dorset (Figure 9).

To obtain a standardized image of the stratigraphic distribution of the organic matter of the entire transect (each borehole providing a complement either downward, e.g., Cymodoce Zone at Reighton 87, or upward, e.g., Hudlestoni, Pectinatus, Pallassioides zones at Flixton 87 and Ebberston 87), the TOC logs were synthesized to summate and then to average the data, to obtain a composite log (in a manner analogous to the stacking of seismic traces). The reference site selected for the depth correlation was Flixton 87, located near the boundary between the basin and the platform. By using the correlations from the log reference marks within each ammonite zone and their inferred chronological equivalence, this synthetic composite log fixes the highs (layers with maximum TOC) and the lows (layers with minimum TOC) of the sequences in a standard Kimmeridgian time frame. The total organic carbon distribution thus obtained is no longer of local value (at the borehole), but can be extended to the entire transect investigated representing about 6.5 m.y. of sedimentary history.

A quantitative estimate of the duration of these cycles can be suggested from the chronological curve of Haq et al. (1987) which set the upper boundary of the Cymodoce Zone at 143 Ma, and that of the top of the Pectinatus Zone at 137.5 Ma. If the synthetic log is filtered by using a Fourier series analysis program, a first order of cyclicity appears (Figure 10), and, by processing the residual, a second order that comprises at least 23 cycles (or one cycle per 280,000 years) also appears. A third order of cycle corresponds to the cycles (less than 1 m thick) that can be observed directly on the logs (Figure 6C). These number about 200 over all the zones studied (Cymodoce to Pectinatus) and have a periodicity of about 25,000 years. Because of the uncertainty of the absolute datings, the composite log was not related in detail to a precise time scale. The thickening of the Mutabilis and Eudoxus zones (Figure 10) may correspond to twice the rate of accumulation of the Elegans to Pectinatus zone interval.

When the synthetic, composite log is compared with the coastal onlap of the sediments summarized by the work of Haq et al. (1987), the pulses corresponding to the maximum accumulation of organic matter cannot be correlated with the sequence boundaries of these authors, but lie either at the middle of a transgressive interval (top of the Eudoxus Zone) or at the base of the high level prism (base of the Hudlestoni Zone) or at the base of a platform edge prism (base of the Pectinatus Zone). This correlation must accordingly be refined regionally, at the level of the Cleveland Basin, to construct local curves of relative variation of coastal onlap of the sediments and eustatic variation in order to define more precisely the relationships existing between the periods of high sea levels and the accumulation of marine organic matter.

Furthermore the synthetic, composite log (Figure 10) shows a general tendency of the total organic carbon to increase in amount as the sequence passes from lower to upper Kimmeridgian, Cymodoce Zone to Pectinatus Zone. This drift from the base to the top of the overall transgressive sequence is comparable to the one recorded during the Cenomanian/Turonian which is another period of very high sea level. The mid-Cretaceous deposits of DSDP 603 site located east of Cape Hatteras (North Atlantic), for which total organic carbon distribution was plotted virtually continuously (one sample every 2 cm) (Herbin et al., 1987), exhibit a similar type of cyclicity with alternation of different clays of black and green color associated with a general upward increase of the TOC content (Figure 11). The similarity of the two vertical distributions is not fortuitous, but may reflect a par-

Figure 9. Example of cyclicity from the type section in the Dorset area (Kimmeridge Bay), each cycle represents about 80 cm of thickness, alternations are obvious in weathered outcrops.

ticular process of preservation of the organic matter of type II during the period of high sea level stands, as if the aquatic environment was progressively more and more depleted in dissolved oxygen during the transgression.

Within the Kimmeridge Clay the cyclicity cannot be understood by contrasting (as was done for the North Atlantic, mid-Cretaceous sequence) two types of sedimentation, one autochthonous and the other allochthonous (turbiditic). This hypothesis, sometimes volunteered to explain the existence of alternations in the Mesozoic sediments of the sites located on a continental margin setting, has no equivalent in the Yorkshire area. Large-scale dynamic processes (slumps, turbidite flows) from east to west, i.e., from the sedimentary high to the basin did not exist here, the two areas having shallow paleobathymetric levels. In fact, alternating periods of more or less preservation or accumulation of organic matter during time is a common feature within sedimentary sequences. Numerous examples exist in the geological column. Such alternations are frequent in Lower Cretaceous formations of the North Atlantic where alternating Aptian white and dark marly limestones or Albian/Cenomanian green and black clays show regular fluctuations in the quantity and the quality of the organic matter (Herbin et al., 1984). They also occur in the South Atlantic where their autochthonous character is unquestionable (Deroo et al., 1984). The cyclicity consequently appears as a major feature of the sedimentology of the organic matter. However the process of alternation is not invariably subordinated to the richness in organic matter. For example, the Upper Jurassic succession in the Atlantic Basin shows red and green clays corresponding to fluctuation in the state of oxidation without any preservation or accumulation of organic matter (Cat Gap Formation, DSDP Site 367, Dean and Gardner, 1982). This means that at the same period of time the sedimentary content may differ although the process was similar—with alternations.

Hence, these alternations tend to reflect the sedimentological record resulting from processes likely to occur in harmony (e.g., eustatic variation and climatic

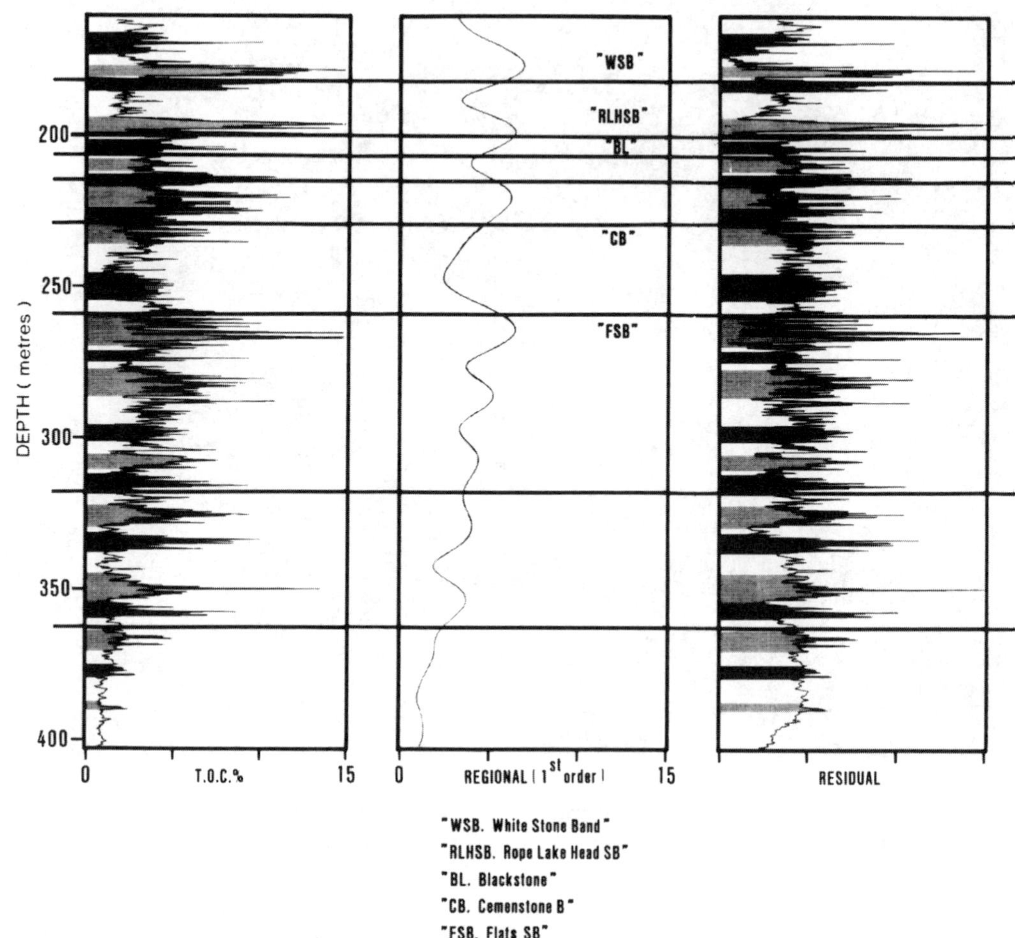

Figure 10. Correlation between the synthetic organic carbon log (resulting from the data available on the four boreholes), the coastal onlap, and the eustatic curves of Haq et al. (1987). The frames correspond to the 10-m interval organic sequences as in Figure 7.

cycles), so creating a characteristic frequency and amplitude variation resulting in a hierarchy of cycles of several scales. This cyclicity was also identified in the Kimmeridge Clay at outcrop in the type section in Dorset (Oschmann, 1988a, b, 1990). The influence of climatic variations between wet and semi-arid periods has been pointed out on the basis of the results of palynological (Waterhouse, 1990), geochemical (Mann and Myers, 1989), and mineralogical (Wignall and Ruffell, 1990) studies. In the present state of knowledge, the assumption of a cyclic climatic origin affecting the thermocline appears most probable to explain the widespread Kimmeridgian cycles. Such cycles were also expressed elsewhere during Jurassic time (House, 1987) and may have a similar explanation.

QUALITATIVE VARIATION

Detailed studies of the elementary analysis of the kerogen and the gas chromatography (GPC) of the saturated hydrocarbons were conducted on selected samples from Ebberston 87/87 bis borehole to characterize the origin of the organic matter. These samples were obtained from two units: (1) at the boundary of the Pectinatus–Hudlestoni zones, on both sides of the "White Stone Band" from 44.30 to 51.15 m, and (2) from 159.65 to 161.17 m and from 176.20 to 185.20 m, characterizing the bimodal distribution located at the top of the Eudoxus Zone. Within these units, the samples were distributed between "sequence top" for the high organic carbon contents (accordingly corre-

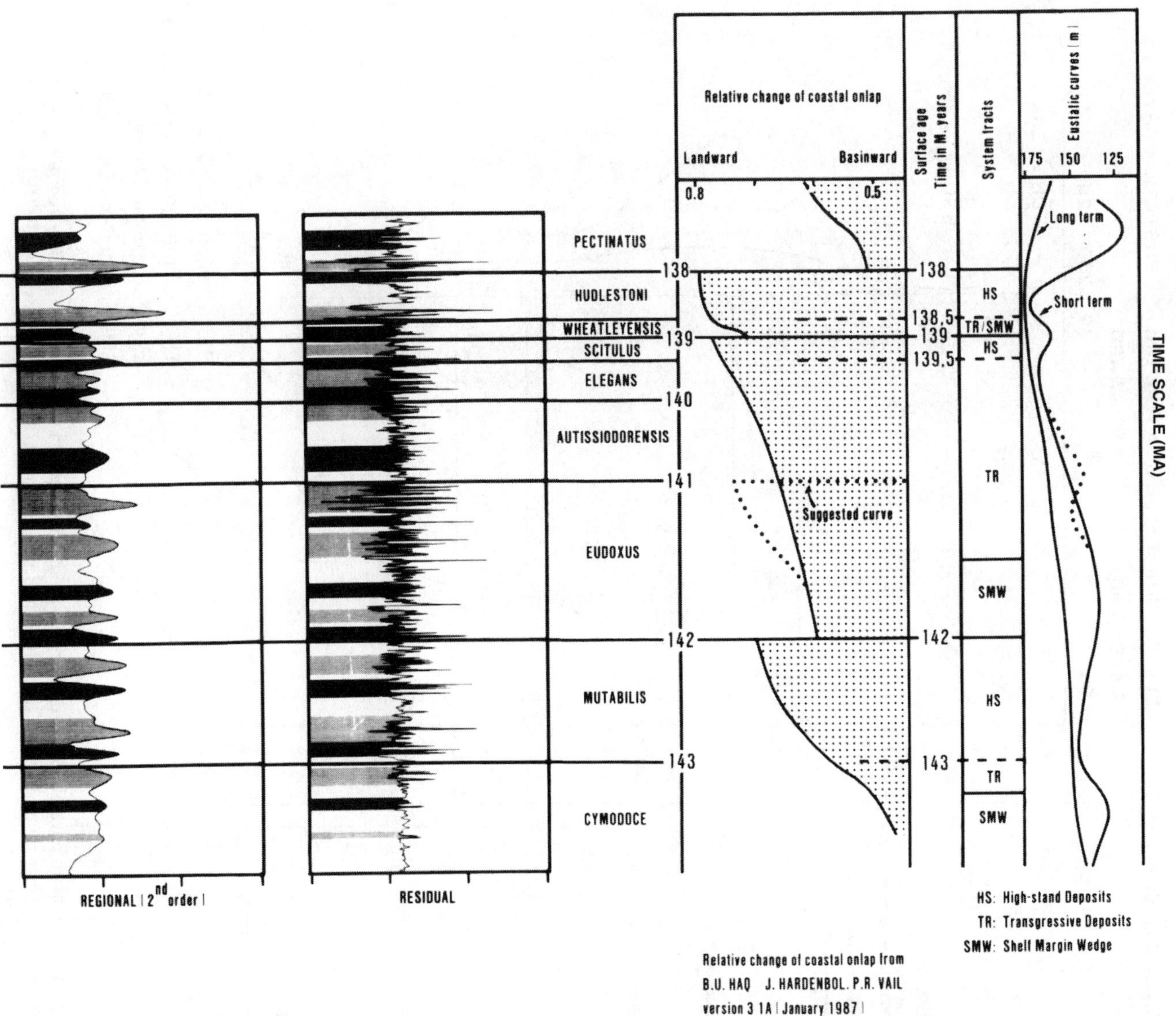

Figure 10 (continued).

sponding to the lowest values of the density log), "sequence bottom" for the low organic carbon contents, and "sequence middle" for samples in intermediate position.

The results of the elementary analysis serve to differentiate the "sequence top" and "sequence bottom" levels by an H/C ratio very close to type II for the high organic carbon contents ("sequence top"), decreasing at the same time as the O/C increases for the low organic carbon contents ("sequence bottom") (Figure 12). This discrimination also appears in the hydrogen index/oxygen index (HI/OI) diagrams whether total rock or kerogen only is considered. The similarity of the HI/OI diagrams between the total rock and kerogen populations suggests that the differentiation existing between high and low organic carbon layers is not related to the rock matrix, but clearly reflects qualitative variations in the organic matter.

The use of the hydrogen index and the oxygen index determined by Rock-Eval pyrolysis on total rock has also been expressed in terms of "sequence top," "sequence middle," and "sequence bottom" on the four boreholes, Marton 87, Ebberston 87 bis, Flixton 87, and Reigthon 87, i.e., incorporating the entire bimodal distribution located at the top of the Eudoxus Zone (Figure 8). In the Marton 87 borehole (Figure 13A), the population of the "sequence bottom" samples is generally characterized by TOC < 2 wt.%, exceptionally as high as 6 wt.%, and HI fluctu-

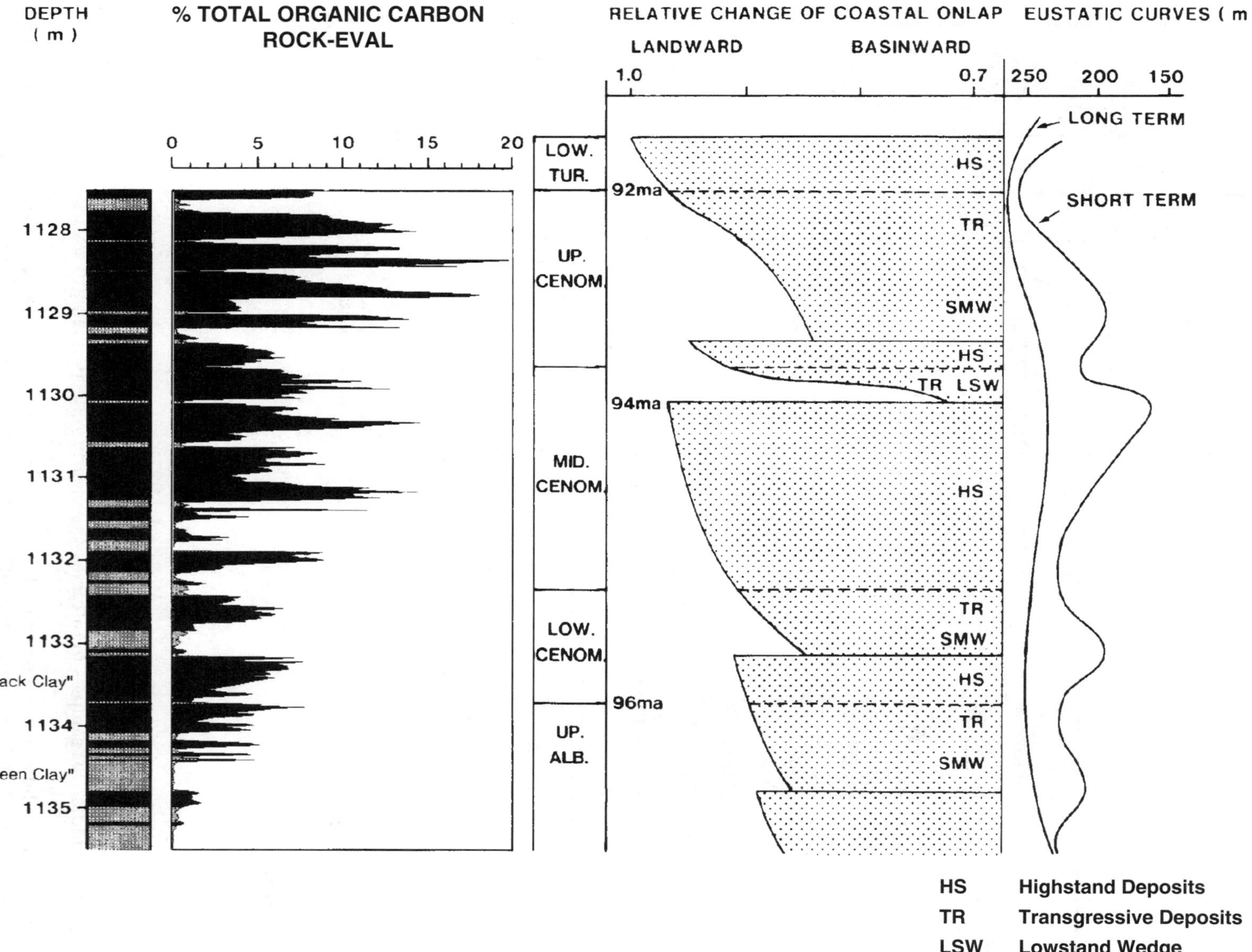

HS	Highstand Deposits
TR	Transgressive Deposits
LSW	Lowstand Wedge
SMW	Shelf-margin Wedge

Figure 11. Vertical distribution of the total organic carbon content at the top of the Hatteras Formation with alternations of green and black clays during another period of high sea level: the Cenomanian–Turonian Boundary Event (CTBE) in DSDP Site 603 located off Cape Hatteras (North Atlantic).

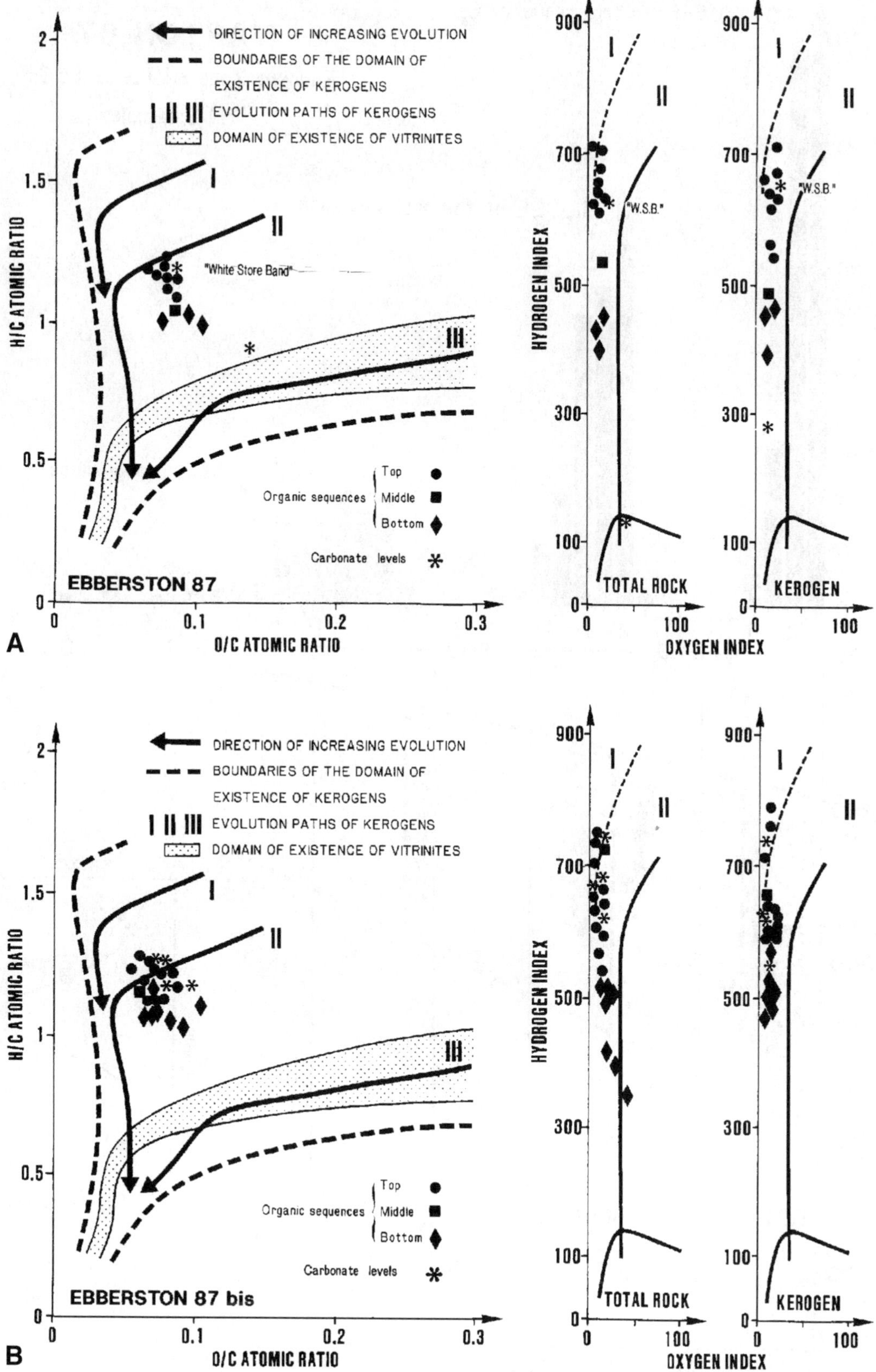

Figure 12. Elemental analyses (van Krevelen diagram) of kerogen and Rock-Eval pyrolysis (HI/OI diagram) of total rock and kerogen. A—Samples located at the boundary between the Hudlestoni and Pectinatus zones, on both sides of the "White Stone Band" (43.83 to 50.63 m depth in the Ebberston 87 borehole). B—Samples located at the top of the Eudoxus Zone (159.55 to 184.83 m depth in the Ebberston 87 bis borehole).

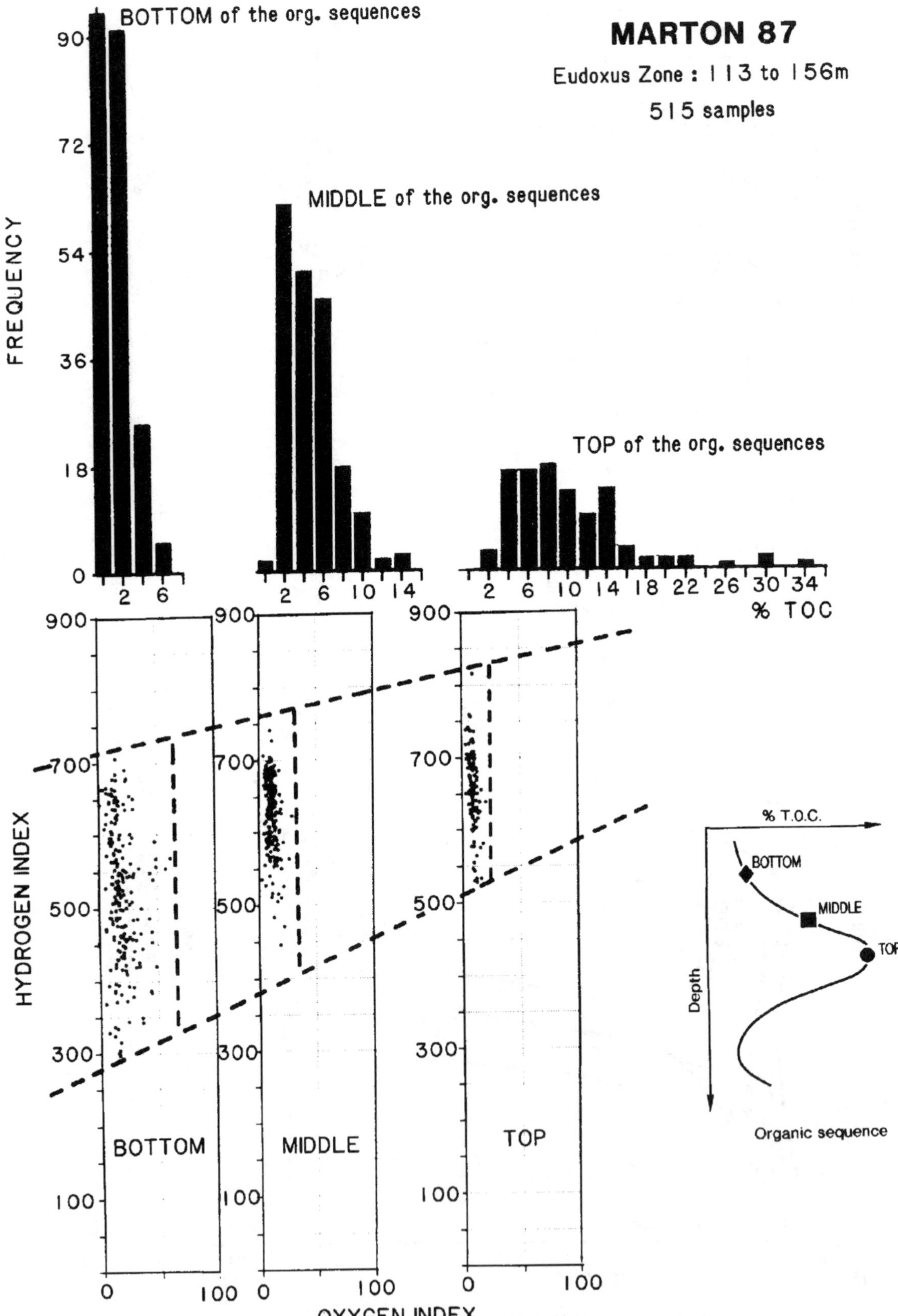

Figure 13. Histograms of total organic carbon content and HI/OI diagrams. A—Populations corresponding to the "top," "middle," and "bottom" of the organic sequences in the Eudoxus Zone of Marton 87 borehole (113 to 156 m). (Figure 13 continues on pp. 89–91.)

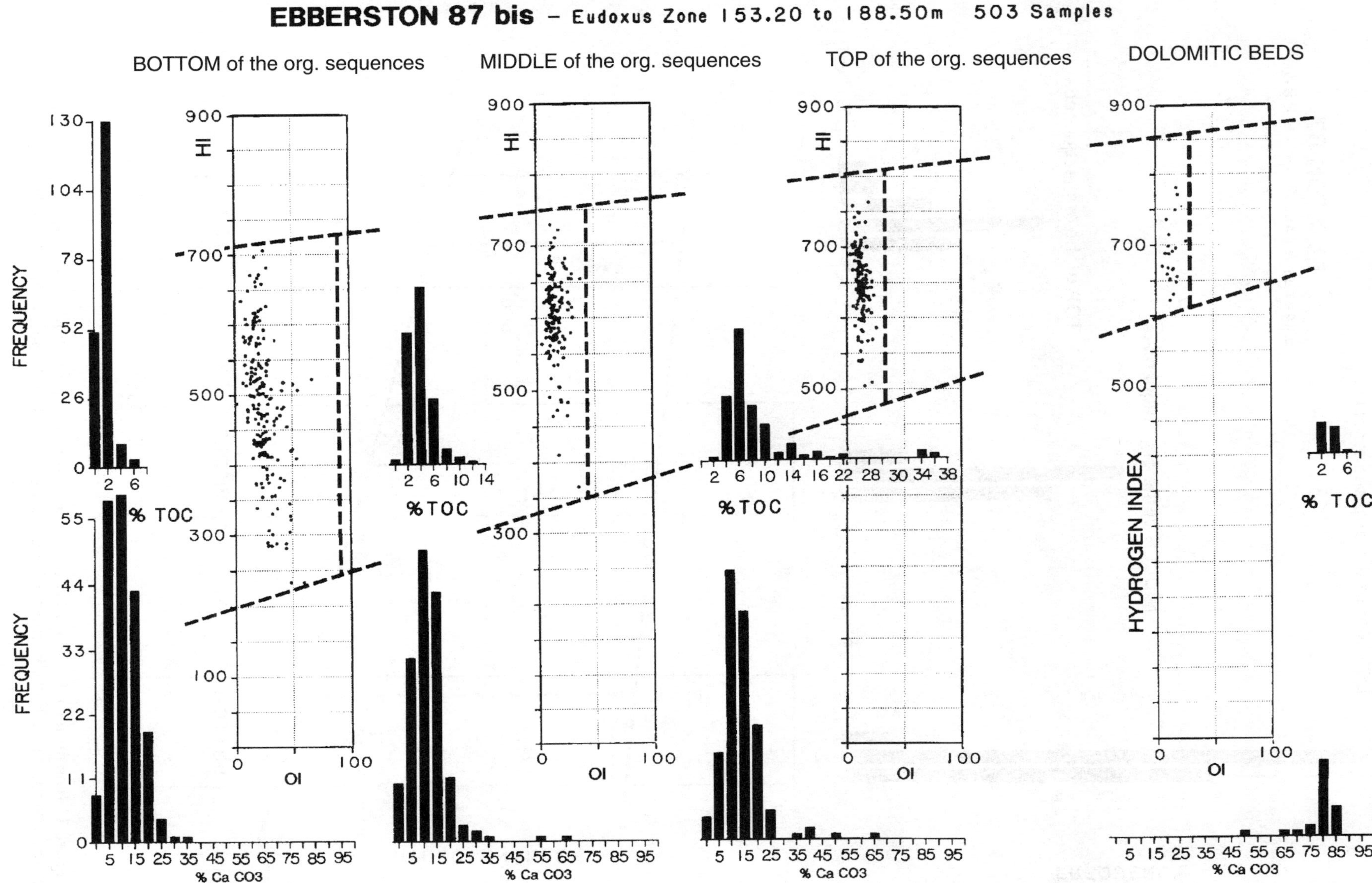

Figure 13 (continued). B—Populations corresponding to the "top," "middle," and "bottom" of the organic sequences and the dolomitic beds in the Eudoxus Zone of Ebberston 87 bis borehole (153.20 to 188.50 m).

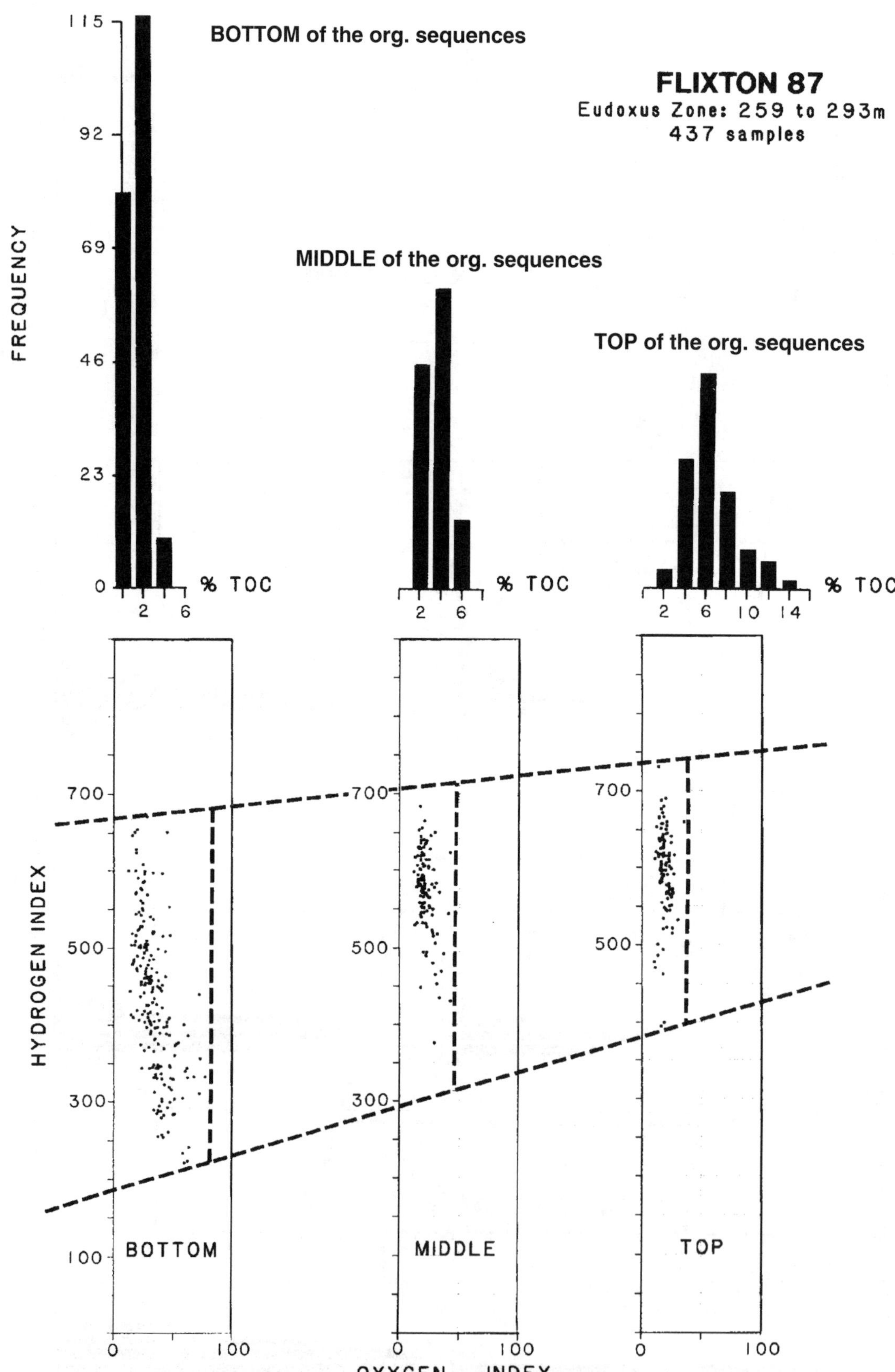

Figure 13 (continued). C—Populations corresponding to the "top," "middle," and "bottom" of the organic sequences in the Eudoxus Zone of Flixton 87 borehole (259 to 293 m).

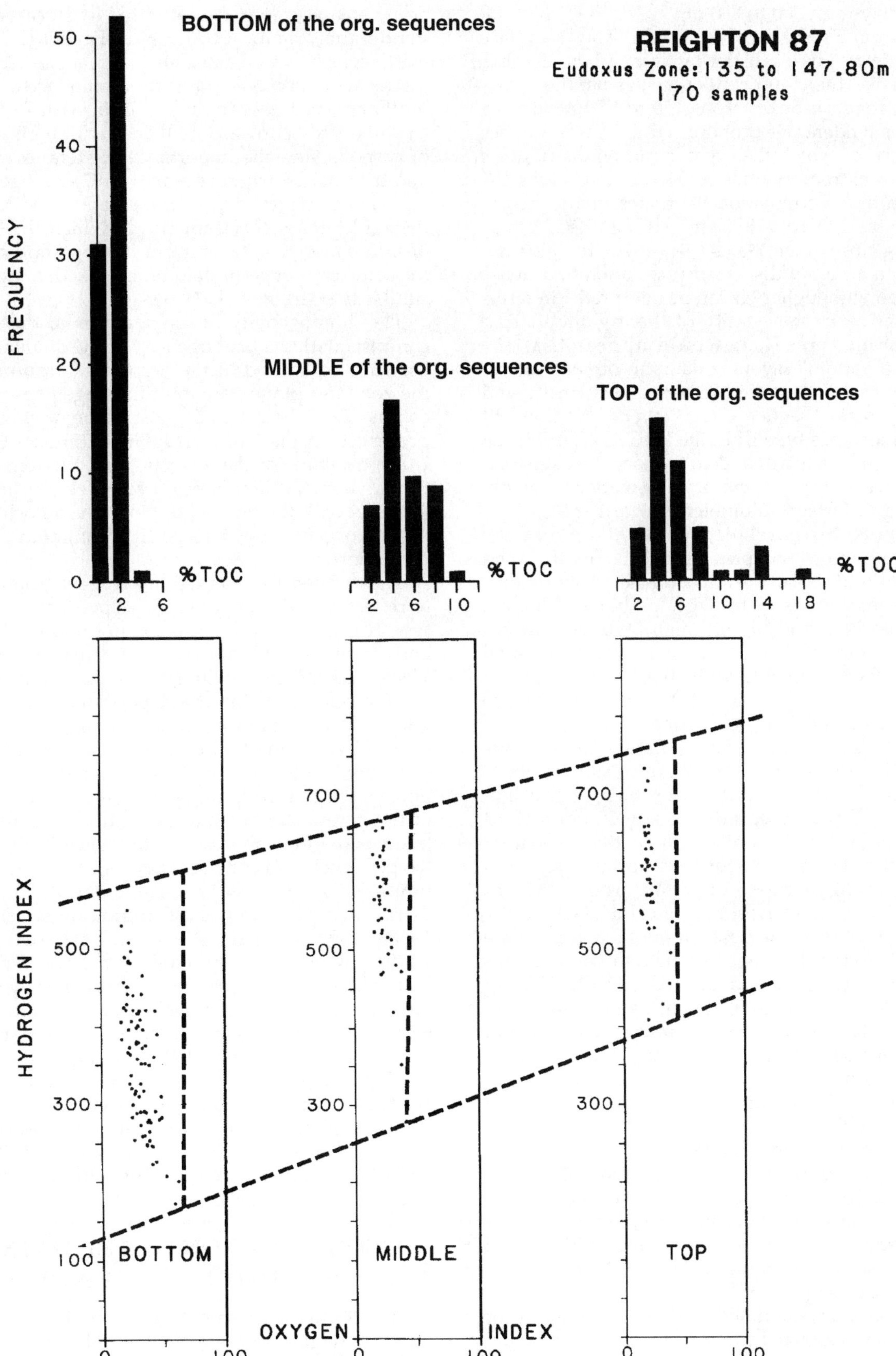

Figure 13 (continued). D—Populations corresponding to the "top," "middle," and "bottom" of the organic sequences in the Eudoxus Zone of Reighton 87 borehole (135 to 147.80 m).

ating widely from 300 to 700 mg HC/g TOC, with OI ranging from 5 to 50 mg CO_2/g TOC. The "sequence middle" figure has a broader organic carbon distribution, whereas the cloud condenses against the Y-axis with hydrogen indexes from 450 to 750 mg HC/g TOC. This tendency is pronounced for the "sequence top" layers, of which the organic carbon distribution spreads to extreme values of 34 wt.%, whereas the hydrogen index population fluctuates in a narrower range from 525 to > 800 mg HC/g TOC, while remaining close to the Y-axis (very low OI). This transition from a highly dispersed population to a dense population with higher HI can be interpreted in terms of a variation in the quality of the organic matter. Although of type II, it would appear that the "sequence bottom" layers contain an organic matter that is less well preserved in terms of quantity and quality than the "sequence top" layers, and the drift in type II towards type III in the H/C, O/C or HI/OI diagrams can be attributed to a process of synsedimentary weathering of the organic matter of planktonic origin. Fifteen kilometers toward the east, in Ebberston 87 bis borehole, the population of the "sequence bottom" samples, "sequence middle" and "sequence top" is very similar to the previous one. The carbonate beds (> 65 wt.% carbonate) have a peculiar position and present high hydrogen indices (600 to 800 mg HC/g TOC) in spite of the low total organic carbon content (2 wt.% TOC) comparative to the clays (Figure 13B). Such data are in accordance with the elemental analysis obtained, for example, on the "White Stone Band" (Figure 12A). These results confirm that during the deposition of some carbonate-rich levels, the organic matter of type II origin could be well preserved also. Again, 15 km toward the east, in Flixton 87 borehole the pattern is similar but the population of "sequence bottom" samples shows a slight decrease of the hydrogen indices (Figure 13C). This trend is emphasized in the Reighton 87 borehole located on the shelf area where the population of the "sequence bottom" samples has hydrogen indices fluctuating from 150 to 550 mg HC/g TOC (Figure 13D), much lower than in the other sites. This means that for an equivalent period of time, represented by the Eudoxus Zone, a relative degradation of the condition of preservation of the organic matter in quantity and quality existed from the "basin" to the "shelf."

Consequently, the qualitative variation can be expressed both in time and in space with differences between the position of the samples in the organic sequences: "sequence bottom," "sequence middle," and "sequence top," and also differences within these organic sequences from the basin area (Marton 87) to the shelf area (Reighton 87).

All the samples examined for elemental analysis of kerogen were extracted to characterize the saturated hydrocarbons by gas chromatography and mass spectrometry. Irrespective of the organic carbon and carbonate contents, these samples (whether belonging to the "sequence top" layers or "sequence bottom" layers) display comparable saturated hydrocarbon chromatograms, identical to those of Dorset (Huc et al., 1991), which can be associated with a type II organic matter with its classic traits: predominance of C_{17}, C_{18} n-alkanes, richness in isoprenoids with Pr/Ph > 1, steady decrease toward n-alkanes with high numbers of carbons, and the importance of steranes/sterenes and triterpanes/triterpenes in the nC_{25+} range (Figure 14). This homogeneity of the n-alkanes in GPC is confirmed by mass spectrometry, in which the samples of different layers, taken from the top and bottom of the sequences or from dolomitic beds, display perfect similarity (Herbin et al., 1991).

This homogeneity in the analyses of the extracts contrasts with the heterogeneity of the qualitative distribution identified in the kerogens. Assuming that the zoo- and phytoplankton (dinoflagellates and coccoliths) lie at the origin of most of the organic matter preserved in the Kimmeridgian sediments (Gallois, 1976), the study of the extracts from different parts of the organic and lithological cycles (clay or dolomite layers) reveals the unity of origin of the autochthonous marine organic matter (type II) throughout the sedimentation.

The differences observed in kerogen variation tend to reflect variations in the state of preservation of this type II organic matter, affected in quality (H/C, HI) and in quantity (TOC) within a sedimentary cycle. These alternations are linked to regular fluctuation in oxygen content within the depositional environment which ranges from dysaerobic to anaerobic, as suggested by Tyson (1987). Such cyclic variations are confirmed, moreover, by the paleoecological observations of the benthic fauna seen on the broken surfaces of the cores. The benthic associations (bivalves and gastropods) depend on their tolerance in life to low oxygen levels and previous studies on the Dorset section have shown that an aerobic population can be distinguished from a dysaerobic one (Oschmann, 1988a, 1990; Wignall, 1990). Such paleoecological studies, which are beginning on the cored material from the Cleveland basin, confirm that even with quite large amounts of type II organic matter (2 to 8 wt.%), benthic life was still active (Oschmann, personal communication, 1992). Consequently, during the deposition of the mudstones the environment was not totally depleted in dissolved oxygen and the true anoxic environment was certainly only restricted to the period of very high TOC accumulation (20 to 40 wt.%) corresponding to the deposition of the very finely laminated shales.

EXTRAPOLATION OF THE DATA IN BLOCKS 42 AND 47 IN THE NORTH SEA

In order to extrapolate the onshore data to the Sole Pit Basin (Figure 15), two offshore oil wells (42/28-01 and 47/13-01) have been studied using the Carbolog software (Carpentier et al., 1989). This methodology allows estimation of the total organic carbon content from a combination of the resistivity and sonic logs.

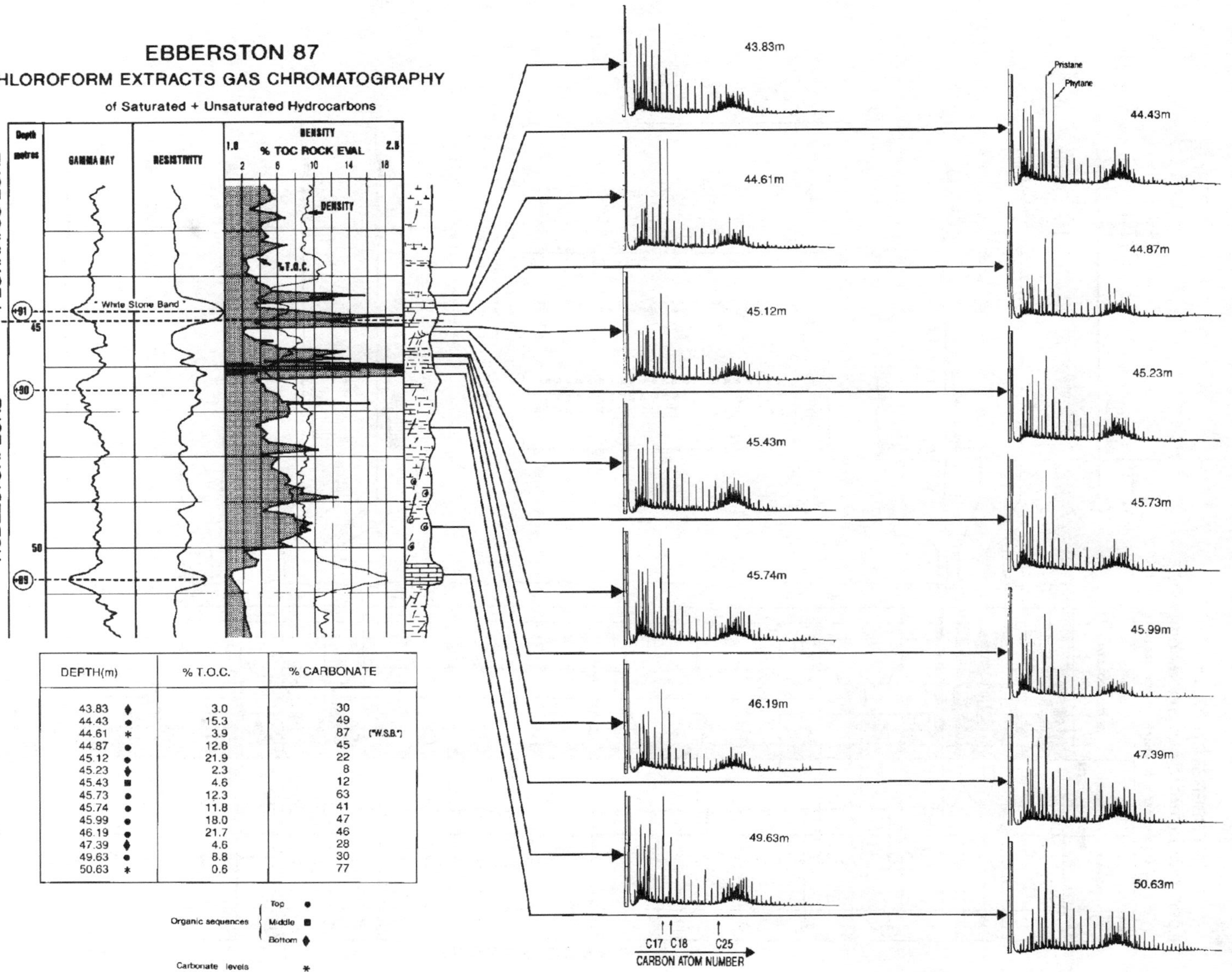

DEPTH(m)	% T.O.C.	% CARBONATE
43.83 ♦	3.0	30
44.43 ●	15.3	49
44.61 *	3.9	87 ("W.S.B.")
44.87 ●	12.8	45
45.12 ●	21.9	22
45.23 ♦	2.3	8
45.43 ■	4.6	12
45.73 ●	12.3	63
45.74 ●	11.8	41
45.99 ●	18.0	47
46.19 ●	21.7	46
47.39 ♦	4.6	28
49.63 ●	8.8	30
50.63 *	0.6	77

Figure 14. Chromatography of the saturated hydrocarbons. A—Samples from the boundary between the Hudlestoni and Pectinatus zones, on both sides of the "White Stone Band" from 43.83 to 50.63 m in the Ebberston 87 borehole (corresponding to the Figure 12A). (Figure 14 continues on p. 94.)

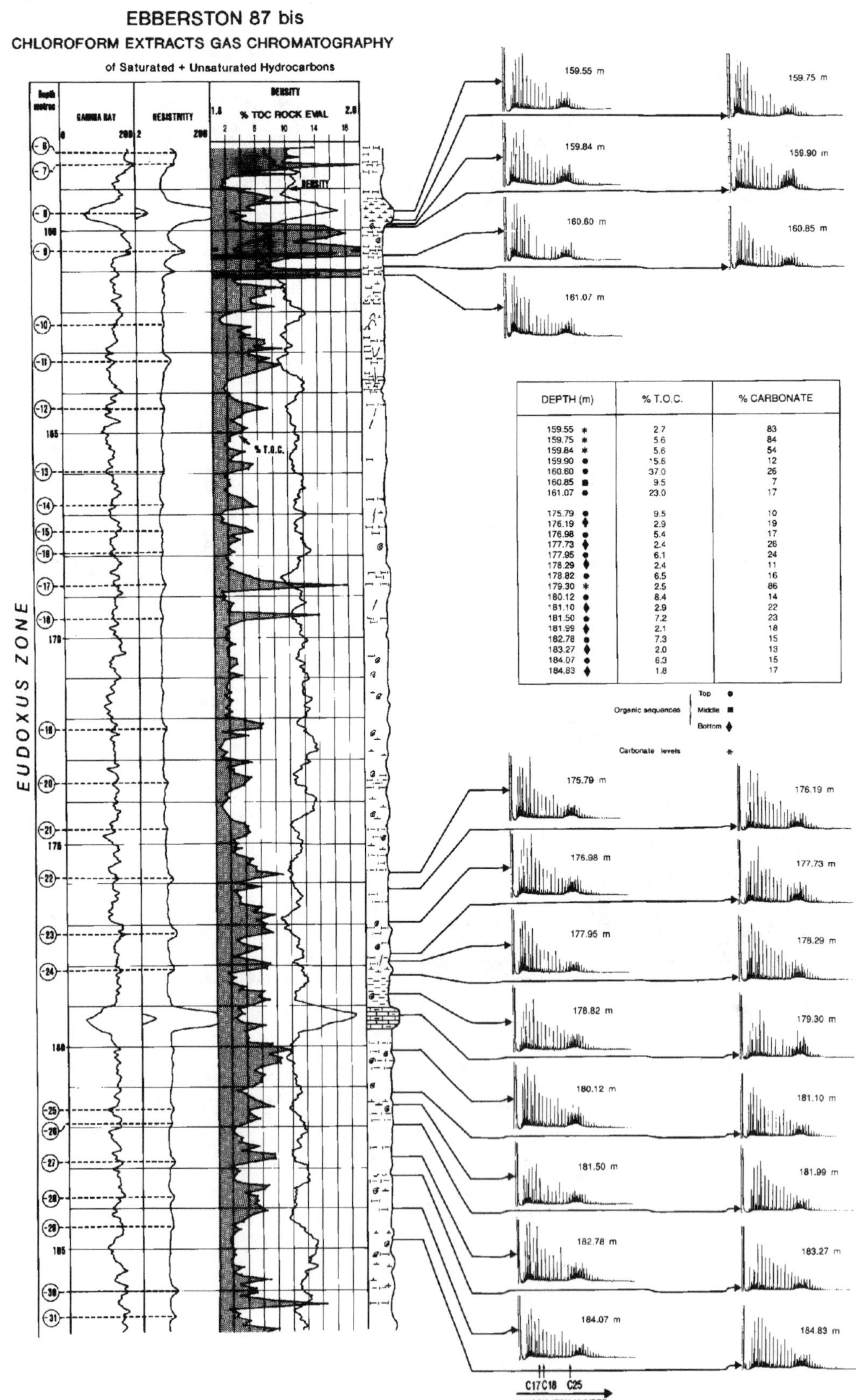

DEPTH (m)	% T.O.C.	% CARBONATE
159.55 *	2.7	83
159.75 *	5.6	84
159.84 *	5.6	54
159.90 ●	15.6	12
160.60 ●	37.0	26
160.85 ■	9.5	7
161.07 ●	23.0	17
175.79 ●	9.5	10
176.19 ♦	2.9	19
176.98 ●	5.4	17
177.73 ♦	2.4	26
177.95 ●	6.1	24
178.29 ♦	2.4	11
178.82 ●	6.5	16
179.30 *	2.5	86
180.12 ●	8.4	14
181.10 ♦	2.9	22
181.50 ●	7.2	23
181.99 ♦	2.1	18
182.78 ●	7.3	15
183.27 ♦	2.0	13
184.07 ●	6.3	15
184.83 ♦	1.8	17

Figure 14 (continued). B—Samples from the top of the Eudoxus Zone from 159.55 to 184.83 m in the Ebberston 87 bis borehole (corresponding to Figure 12B).

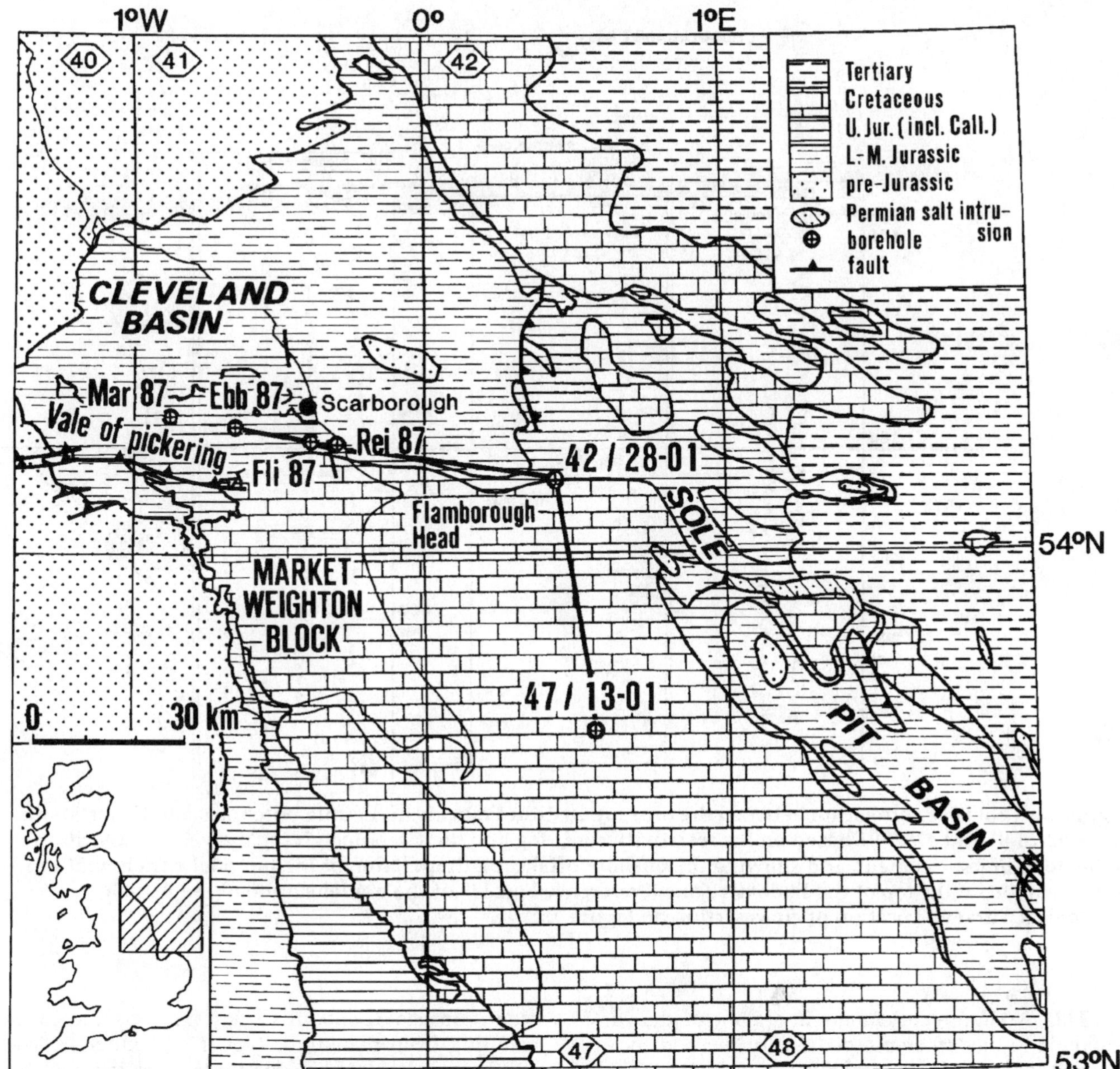

Figure 15. Geological sketch map showing the position of the IFP boreholes and the two offshore wells, 42/28-01 and 47/13-01, allowing extrapolation from the Cleveland basin to the Sole Pit Basin in the North Sea (from B. M. Cox et al., 1987). Location of cross section corresponds to Figure 16.

In these wells, the Kimmeridge Clay log signature appears similar to that of the onshore sequence with peaks related to cementstones and organic-rich levels (Figure 16). The correlation between the two offshore oil wells is rather good (marker beds 1 to 13). These boreholes also show the transition between the lower Kimmeridgian and the Callovian-Oxfordian which was not proved in the Vale of Pickering boreholes, while the upper part of the Kimmeridge Clay is partially eroded beneath lower Cretaceous strata in 47-13/1 (as it is in Reighton 87).

The combined use of the geophysical logs (sonic and resistivity) and the total organic carbon distribution resulting from the Carbolog suggests a possible correlation between the onshore and the offshore boreholes, which are separated by a distance of more than 100 km. The distribution of the organic matter includes fluctuations of rich and poor horizons which allows the identification of the large scale cycles. The bimodal enrichment which characterizes the top of the Eudoxus Zone (Figure 7) can be extended for example, to the offshore area (C, D, E peaks on Figure

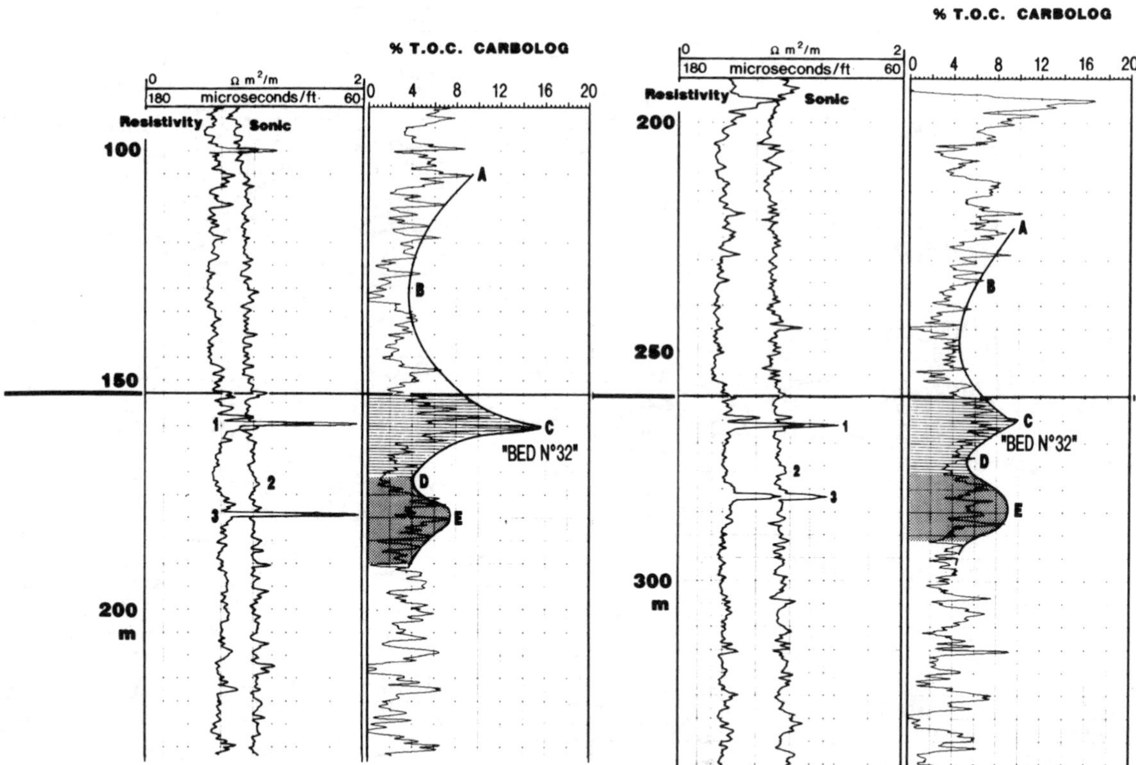

Figure 16. Tentative correlation between the onshore Institut Français du Pétrole boreholes and the offshore wells 42/28-01 and 47/13-01 based on the resistivity/sonic log and the Carbolog giving the vertical distribution of the total organic content, and showing the onshore/offshore extrapolation of the bimodal enrichment in organic matter at the top of the Eudoxus Zone ("reference bed 32" of the nomenclature of the British Geological Survey). Location of cross section on Figure 15.

16). This maximum enrichment in organic matter corresponds to the "reference bed 32" of the nomenclature of the British Geological Survey, which, as stated previously, can be identified from Yorkshire to Dorset (Gallois, 1988).

In the Norwegian–Danish Basin, Rawson and Riley (1982) mention also the existence of the "hot shales" facies at the base of the "Tau Formation," approximately the equivalent of the Kimmeridgian clays of Yorkshire. These shales correspond precisely to the "oil shales" facies of the Eudoxus Zone. According to Doré et al. (1985), this event results from a broad process of transgression and gives rise to lithostratigraphic sequence boundaries that may extend to regions other than the Norwegian–Danish Basin.

CONCLUSIONS

Quantitative mineralogy shows that, on the whole, the sediments of the area examined are distinguished by horizontal and vertical homogeneity indicating the permanence of the conditions of deposition. The sedimentary material, deposited in a low-energy environment, consists essentially of clay minerals and calcite resulting from planktonic biogenic production (coccoliths). Intercalated in the mass of claystones, are carbonate beds, about 1 m thick, which can be correlated between the boreholes. They consist chiefly of dolomite resulting from the diagenetic transformation (more complete in the west than in the east) of layers formed nearly exclusively of coccoliths. This dolomitization appears to have been controlled by the degree of abundance of the organic matter.

The study of the variation of the distribution of the organic matter in the Kimmeridgian sediments of Yorkshire reveals the orderly (nonrandom) character of the TOC distribution in space and in time. The correlation of the organic sequences in space displays an enrichment of up to 50% from east to west. The accumulations that appear synchronously in the "basin" area and in the "platform" area are richer and more prominent (TOC contents and thicknesses) in deeper paleobathymetric zones which show higher sedimentary accumulation rates (Marton 87). Because the synchroneity of the deposits is established between the two areas lying only 35 km apart, and because the

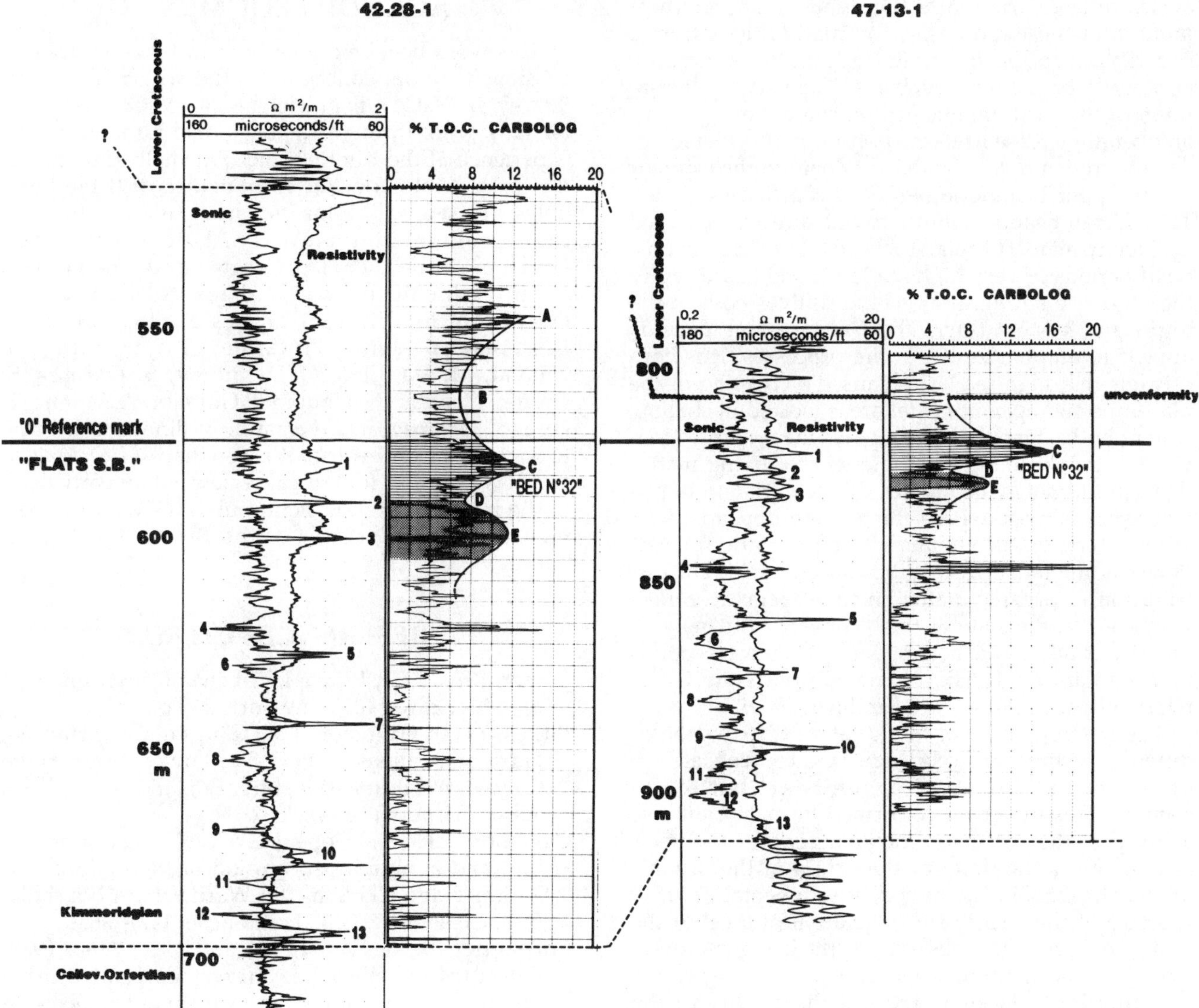

Figure 16 (continued).

enrichments are systematically greater in the "basin," variation in primary planktonic productivity alone over the entire period investigated (6.5 m.y.) cannot justify the existence of this constant gradient between the two areas during so long a period of time. The enrichment in the "basin" area as compared with the "platform" area (corresponding to the sedimentary high of Market Weighton) can be explained partly by an environment that was always relatively more depleted in dissolved oxygen (more favorable to the accumulation of organic matter) in the deeper bathymetric areas (at the same time a calmer, lower-energy environment), and also by better preservation associated with rapid burial there. Such a model previously established by Müller and Suess (1979) has been described for the Kimmeridge Clay Formation of the Central Graben of the North Sea (Demaison et al., 1983). Thus, this model of the distribution of the organic matter must be valid over a large area.

Further, the widespread geographical extension of both organic and calcareous sedimentological events (over the 400 km from Yorkshire to Dorset) confirms that factors controlling sedimentary accumulation must be geodynamically driven. Their origin cannot be sought in the presence of local factors, but must result from global changes, as for example in eustatic sea level fluctuations, and related to the paleogeography of source rocks.

Variation in time is characterized by a cyclicity of alternating organic-poor and organic-rich levels. The shortest scale of cycle observed appears to have a periodicity of about 25,000 years. In fact many deposits, both recent and ancient, display such regular alternations. During Kimmeridgian time the process appears to be well developed on the Eastern England Platform with a baseline at 2 wt.% TOC punctuated by pulses of organic matter deposition leading to some 40 wt.% TOC. In these deposits the cycles reflect variation of the depositional environment, and the hypothesis of climatic fluctuation, a repetitive process of broad geographic extension, appears to be the most probable. The vertical distribution of the total organic carbon content also reveals an overall increase from the bottom part of the lower Kimmeridgian (*s.a.*) (Cymodoce Zone) to the top part of the upper Kimmeridgian (*s.a.*) (Pectinatus Zone). This overall change (similar to one already observed in Cenomanian/Turonian rocks) could be characteristic of periods of very high sea level, and could typify the sequence stratigraphy of the source rocks in the highstand system tract. The transgressive periods appear to emphasize the cyclic processes. It is conceivable that high sea levels caused a covering of the platforms and epicontinental areas located in shallow water depths. Since the ocean plays the role of a "converter" (transforming solar energy into living matter that can ultimately be fossilized), the variation in primary productivity within this system depends, *inter alia*, on temperature (climatic) and eustatic fluctuations leading to increase in the collector area. The distribution of organic matter in time seems to reflect variations in the state of this "converter." It may be that this stratigraphic heterogeneity even corresponds to a regulation mechanism that supports the GAIA hypothesis (Lovelock, 1989; Kirchner, 1989).

The heterogeneity of quantitative distribution in different parts of the cycle (high TOC contents at "top of sequence," low TOC contents at "bottom of sequence") is moreover confirmed by the qualitative analysis of the kerogens. The analysis of the extracts also suggests the unity of the origin of the organic matter throughout the period of sedimentation, irrespective of the terms of the hierarchical level of the cycle. The differences observed in the kerogens consequently reflect a difference in the state of preservation affecting both the quantity and the quality of the same organic matter (zoo- and phytoplankton). The vertical distribution of the type II organic matter in the cycles can be consequently interpreted in terms of a fluctuation of the oxygen deficiency, whose variations (dysaerobic, anaerobic or anoxic) are recorded in time and in space. The study of the Kimmeridgian rocks of the Dorset area (Huc et al., 1991) reveals fluctuation in primary productivity in layers very rich in organic carbon, with the "organic productivity" going hand-in-hand with a "mineral productivity." In the Yorkshire area such examples exist at the boundary of the Hudlestoni and Pectinatus zones, with organic matter contents up to 40 wt.% and located above and below the "White Stone Band," which itself contains over 70 wt.% coccoliths (Figure 6A). Furthermore, this enrichment corresponds to one of the pulses that can be correlated with a period of maximum transgression. Two factors therefore appear to coexist: climatic—regulating the sedimentary cycles, and eustatic—producing an environment with a tendency toward oxygen deficiency in areas of shallow submergence. These combine to produce alternating levels, richer or poorer in organic matter, on the platforms.

ACKNOWLEDGMENTS

This work benefited greatly from very helpful discussions with our colleagues of the British Geological Survey: B. M. Cox, B. N. Fletcher, R. W. Gallois, and J. Hallam, who, through their regional familiarity and experience of the Kimmeridge Clay, helped us to put this project into concrete form and to drill the boreholes. We thank ELF for its participation in funding the borehole campaign (Ford de Soutien aux Hydrocarbunes, FSH program) and the Yorkim Group, which contributed at Institut Français du Pétrole to the different phases of the project: G. Bessereau, B. Buton, G. Caillet, B. Carpentier, B. Cheval, M. Da Silva, C. Delaunay, B. Doligez, S. Drouet, J. Dufau, N. Goubin, E Guimiot, A. Y. Huc, T. Lesage, F. Papagni, G. Pichaud, B. Pinoteau, J. Priol, L. Sage, and C. Schwartz. We also thank G. Demaison and R. W. Gallois for critical reviews of an early draft of the manuscript. I. E. Penn publishes with the permission of the Director of the British Geological Survey.

REFERENCES CITED

Berner, R.A., 1971, Principles of chemical sedimentology: McGraw–Hill, New York, 240 p.

Birkelund, T., Callomon, J.H., Clausen, C.K., Hansen, H.N., and Salinas, I., 1983, The Lower Kimmeridge Clay at Westbury, Wiltshire, England, Proc. Geol. Assoc., London, 94 (4), 289–309.

Callomon, J. H., and Cope, J.C.W., 1971, The stratigraphy and ammonite succession of the Oxford and Kimmeridge Clays in the Warlingham borehole: Bull. Geol. Surv. G.B., London, 36, 147–176.

Carpentier, B., Bessereau, G., and Huc, A.Y., 1989, Diagraphies et roches mères. Estimation des teneurs en carbone organique par la méthode CARBOLOG. Revue IFP, vol. 44, n. 6, 699–720.

Claret, J., Jardiné, S., and Robert, P., 1981, La diversité des roches mères pétrolières: aspects géologiques et implications économiques à partir de quatre exemples. Bull. Rech. SNEA(P). Vol 5, 383–417.

Cope, J.C.W., 1967, The palaeontology and stratigraphy of the lower part of the Upper Kimmeridge Clay of Dorset. Bull. Br. Mu. Nat. Hist. (Geol.), London, 15 (1), 1–79.

Cope, J.C.W., 1974a, New information of the Kimmeridge Clay of Yorkshire. Proc. Geol. Assoc., London, 85 (2), 211–221.

Cope, J.C.W., 1974b, Upper Kimmeridgian ammonite faunas of the Wash area and a subzonal scheme for the lower part of the Upper Kimmeridgian. Bull. Geol. Surv. G.B., London, 47, 29–37.

Cope, J.C.W., 1978, The ammonite faunas and stratigraphy of the upper part of the Upper Kimmeridge Clay of Dorset. Palaeontology, London, 21 (3), 469–533.

Cope, J.C.W., 1980, Kimmeridgian correlation chart, *in* Cope, J.C.W., ed., A correlation of Jurassic rocks in the British Isles. Part 2: Middle and Upper

Jurassic. Geol. Soc. Lond., Spec. Rep., London, 15, 76–85.

Cornford, C., 1984, Source Rocks and Hydrocarbon of the North Sea, *in* Glennie, K. W., ed., Introduction to Petroleum of the North Sea. Blackwell Sci. Pub., Oxford, U.K., 171–204.

Cox, B.M., and Gallois, R.W., 1981, The stratigraphy of the Kimmeridge Clay of the Dorset type area and its correlation with some other Kimmeridgian sequences. Rep. Inst. Geol. Sci., London, 80/4, 1–44.

Cox, B.M., Lott, G.K., Thomas, J.E., and Wilkinson, I.P., 1987, Upper Jurassic stratigraphy of four shallow cored boreholes in the U. K. sector of the Southern North Sea. Proceedings of the Yorkshire Geological Society, Vol. 46, Part 2, 97–109.

Dean, W.E., and Gardner, I.V., 1982, Origin and geochemistry of redox cycles of Jurassic to Eocene age, Cape Verde Basin (DSDP Site 367), continental margin of North-West Africa, *in* Nature and Origin of Cretaceous Carbon Rich Facies, edited by S.O. Schlanger and M.B. Cita, 55–78.

Demaison, G., Holck, A.J.J., Jones, R.W., and Moore, G. T., 1983, Predictive source bed stratigraphy; a guide to regional petroleum occrrence. North Sea Basin and Eastern North American Continental Margin. Proceedings 11th World Petroleum Congress, London, 2, 17-29.

Deroo, G., Herbin, J.P., and Huc, A. Y., 1984, Organic geochemistry of Cretaceous Black Shales from Deep Sea Drilling Project Site 530, Leg 75, Eastern South Atlantic—Initial Reports of the DSDP, Vol. 75, 983– 999.

Doré, A.G., Vollset, J., and Hamar, G.P., 1985, Correlation of the offshore sequences referred to the Kimmeridge Clay formation. Relevance to the Norwegian sector, *in* Petroleum geochemistry in exploration of the Norwegian shelf, B.M. Thomas et al., eds., Norwegian Petroleum Society, 27–37.

Espitalié, J., Deroo, G., and Marquis, F., 1985/1986, La pyrolyse Rock-Eval et ses applications. Revue de l'Institut Français du Pétrole Vol 40 n. 5, 563-580, Vol. 40 n. 6, 755-784, Vol. 41 no. 1, 73–90.

Fritz, P., and Smith, D.G.W., 1970, The isotope composition of secondary dolomites. Geochimica and Cosmochimica Acta, 34, 1161–1173.

Gallois, R.W., and Cox, B.M., 1974, Stratigraphy of the Upper Kimmeridge Clay of the Wash Area. Bull. Geol. Surv. G.B., London, 47, 1–28.

Gallois, R.W., and Cox, B.M., 1976, The Stratigraphy of the Lower Kimmeridge Clay of Eastern England. Proc. Yorkshire Geol. Soc., Leeds, 41 (1), 13–26.

Gallois, R.W., 1976, Coccolith blooms in the Kimmeridge Clay and the origin of North Sea Oil. Nature 259, 473–475.

Gallois, R.W., 1979, Oil Shale Resources in Great Britain. Internal report of the Institute of Geological Sciences—2 volumes, 300 p.

Gallois, R. W., 1988, Geology of the country around Ely. Memoir for 1:50000 geological sheet 173. British Geological Survey. 116 p.

Grace, J. D., and Hart, G. F., 1986, Giant gas fields of Northern West Siberia: AAPG Bulletin Vol. 70, No. 7, 830–852.

Grant, A. C., Jansa, L. F., McAlpine, K.D., and Edwards, A., 1988, Mesozoic–Cenozoic Geology of the Eastern Margin of the Grand Banks and its relation to Galicia Bank, Proc. ODP, Vol. 103, 787–807.

Haq, B.U., Hardenbol, J., and Vail, P.R., 1987, Chronology of fluctuating sea levels since the Triassic. Science Vol. 235, 1156–1167.

Herbin, J.P., Deroo, G., and Roucaché, J., 1984, Organic geochemistry of lower Cretaceous sediments from Site 535, Leg 77, Florida straits—Initial reports of the deep sea drilling project. Vol. 77, 459–475.

Herbin, J.P., Montadert, L., Muller, C., Gomez, R., Thurow, J., and Wiedmann, J., 1986, Organic-rich sedimentation at the Cenomanian–Turonian boundary in oceanic and coastal basins in the North Atlantic and Tethys. North Atlantic Palaeoceanography, Summerhayes, C.P., Shackleton, N.J., eds. Geological Society Special Publication, N 21, 389–422.

Herbin, J.P., Masure, E., and Roucaché, J., 1987, Cretaceous formations from the lower continental rise off Cape Hatteras: organic geochemistry, dinoflagellate cysts, and the Cenomanian/Turonian Boundary Event at Sites 603 (Leg 93) and 105 (Leg 11). Initial Reports of DSDP, volume 93, 1139–1162.

Herbin, J.P., Geyssant, J.R., Mélières, F., Müller, C., Penn, I.E. et le groupe Yorkim, 1991, Hétérogénéité quantitative et qualitative de la matière organique dans les argiles du Kimméridgien du Val de Pickering (Yorkshire, U.K.), cadre sédimentologique et stratigraphique. Revue de l'Institut Français du Pétrole, Vol. 46, No. 6, 1-39.

House, M.R., 1987, Are Jurassic sedimentary microrhythms due to orbital forcing? Proc. Usher Soc. 6, 299–311.

Huc, A.Y., Lallier, Verges E., Bertrand, P., Carpentier, B., and Hollander, D.G., 1991, Organic matter response to change of depositional environment in Kimmeridgian shales, Dorset, UK, in Organic matter: Productivity, accumulation and preservation in recent sediments. J. Whelan, J. Farrington, eds., Columbia University Press.

Irwin, H., and Curtis, C., 1977, Isotopic evidence for source of diagenetic carbonates formed during burial of organic-rich seidments. Nature, 269, 209–213.

Irwin, H., 1980, Early diagenetic carbonate precipitation and pore fluid migration in the Kimmeridge Clay of Dorset, England. Sedimentology, 27, 577–591.

Kirby, G.A., and Swallow, P.W., 1987, Tectonism and sedimentation in the Flamborough Head region of north-east England. Proc. Yorks. Geol. Soc. 46, 301–309.

Kirchner, J.W., 1989, The GAIA hypothesis: can it be tested? Reviews of Geophysics, 27, 2, 223–235.

Lovelock, J.E., 1989, Geophysiology, the science of GAIA. Reviews of Geophysics, 27, 2, 215–222.

Mann, A.L., and Myers, K.J., 1989, The effect of climate on the geochemistry of the Kimmeridge Clay Formation, *in* Biomarkers in Petroleum, Memorial Symposium for W. Seifert, Dallas Meeting—Division of Petroleum Chemistry—American Chemical Society, 139–142.

Müller, P.J., and Suess, E., 1979, Productivity, sedimentation rate and sedimentary organic matter in the oceans. Organic carbon preservation. Deep Sea Research, Vol. 26 A, 12A, 1347–1362.

Oschmann, W., 1988a, Upper Kimmeridgian and Portlandian Marine Macrobenthic Associations from Southern England and Northern France. Facies, 18, 49–84.

Oschmann, W., 1988b, Kimmeridge Clay sedimentation—a new cyclic model. Palaeogeogr., Palaeoclimatol., Palaeoecol., 65, 217–251.

Oschmann, W., 1990, Environmental cycles in the late Jurassic northwest European epeiric basin: interaction with atmospheric and hydrospheric circulations. Sedimentary Geology, 69, 313–332.

Pelet, R., 1984, A model for the biological degradation of recent sedimentary organic matter. Organic Geochemistry. V. 6, 167–180.

Penn, I.E., and Abbott, M.A.W., 1989, Geological context of the Kimmeridge Clay of Eastern England and the adjacent Southern North Sea. Internal report BGS no. DO/89/1, 25 p.

Rawson, P.F., and Riley, L.A., 1982, Latest Jurassic–Early Cretaceous events and the "Late Cimmerian unconformity" in the Northern Sea area: AAPG Bulletin, 66 (12), 2628–2648.

Schlanger, S.O., and Jenkyns, H. C., 1976, Cretaceous oceanic anoxic events, causes and consequences. Geologie en Mijnbouw 55, (3-4), 179–184.

Tissot, B., 1979, Effects on prolific petroleum source rocks and major coal deposits caused by sea-level changes. Nature, 277, No. 5696, 463–465.

Tyson, R.V., Wilson, R.C., and Downie, C., 1979, A stratified water column environmental model for the type Kimmeridge Clay. Nature, 277, No. 5695, 377–380.

Tyson, R.V., 1987, The genesis and palynofacies characteristics of marine petroleum source rocks. Marine Petroleum Source Rocks. Geological Society Special Publication, 26, 47–67.

Waterhouse, H.K., 1990, Quantitative palynofacies analysis of Milankovitch cyclicity. International Symposium to celebrate 25 years of Palynology in the North Sea Basin. Nottingham April 1990 (Summary).

Whittaker A., (ed.), 1985, Atlas of Onshore Sedimentary Basins in England and Wales: Post Carboniferous Tectonics and Stratigraphy. Blackie, Glasgow, 68 pp.

Wignall, P.B., and Ruffell, A.H., 1990, The influence of a sudden climatic change on marine deposition in the Kimmeridgian of northwest Europe. Journ. of the Geolog. Soc. London, Vol. 147, 365–371.

Wignall, P.B., 1990, Benthic palaeocology of the late Jurassic Kimmeridge Clay of England. Spec. Papers in Palaeontology, No. 43, 75 p.

Williams, P.F.V., 1986, Petroleum geochemistry of the Kimmeridge Clay of onshore southern and eastern England. Marine Petroleum Geology, 3, 258–281.

Chapter 7

Characterization of the Source Horizons Within the Late Cretaceous Transgressive Sequence of Egypt

Vaughn D. Robison
Texaco Inc.
Exploration and Production Technology Department
Houston, Texas, USA

Michael H. Engel
University of Oklahoma
School of Geology and Geophysics
Norman, Oklahoma, USA

ABSTRACT

Source rocks were deposited in northeastern Africa during a major Late Cretaceous transgression. The stratigraphic section begins with a series of phosphorites and oil shales and progresses up into an organically lean chalk. These deposits represent increasing water depths and depositional isolation from the continental landmass. Within the source sequences of the Duwi and Dakhla formations in the eastern desert of Egypt, variations in the organic matter are recorded. Understanding the systematics of this variability and predicting the geochemical attributes of the source system away from control can be important to resource assessment. Even though a general increase in both organic richness and liquid hydrocarbon generation potential is observed from the base to the top of the source section in Egypt, no clear, predictable trends are evident. The variability observed in source quality is best explained in a sequence stratigraphic framework. In Egypt, three third order eustatic cycles can be identified. The best oil-prone source rocks occur within the condensed sections at the top of the transgressive systems tracts of the three sequences. Little to no oil-prone source potential can be identified in the highstand systems tracts, where the bulk of the organic matter exhibits gas-prone attributes.

INTRODUCTION

A major Late Cretaceous and early Tertiary transgression flooded a broad marine shelf across northeast Africa and led to the deposition of numerous organic-rich, potential source rocks. These units extend today across broad regions of Egypt, Israel, Jordan, Saudi Arabia, and Iraq, and are characterized by phosphorite and "oil shale" horizons (Bein and Amit, 1982; Kempler and Zimmerle, 1983; Troger,

Table 1. Organic carbon and pyrolysis data.

Sample ID[1]	Stratigraphic Position[2]	TOC (%)	S_1 (mg/g)	S_2 (mg/g)	S_3 (mg/g)	HI	OI	T_{max}	System Tract[3]
GD-01	17.6 m above A	8.5	1.8	41.9	6.8	492	80	418	T
GD-02	14.6 m above A	2.7	0.6	14.1	2.2	520	83	419	T
GD-03	13.1 m above A	7.8	1.6	41.6	5.8	532	74	414	T
GD-04	9.1 m above A	7.8	1.4	37.2	6.3	476	81	419	T
GD-05	5.8 m above A	1.4	0.5	8.5	7.4	606	77	418	T
GD-06	3.4 m above A	1.8	0.5	7.1	2.3	396	128	417	T
GD-07	1.7 m above A	1.1	0.2	1.2	1.0	111	91	420	T
GD-08	0.6 m above A	0.9	0.1	0.8	0.8	84	88	412	T
YN-01	20.1 m above A	4.3	0.2	16.0	1.9	371	45	412	T
YN-02	18.2 m above A	5.7	0.3	19.8	1.4	347	25	411	T
YN-03	16.5 m above A	5.2	0.7	27.3	1.5	524	29	412	T
YN-04	15.6 m above A	4.3	0.6	20.8	1.5	484	34	415	T
YN-05	8.2 m above A	16.2	4.0	68.9	5.8	425	36	412	T
YN-06	6.4 m above A	20.8	4.1	78.6	6.0	377	29	412	T
YN-07	1.3 m above A	1.6	0.6	8.0	0.7	498	41	415	T
YN-08	0.3 m above A	4.2	1.0	19.5	1.5	465	35	413	T
YN-09	2.3 m above A	2.1	0.4	9.2	0.6	439	28	414	T
YN-10	40.6 m above A	7.6	1.4	48.6	5.0	639	66	417	T
YN-11	37.5 m above A	5.0	0.9	30.3	4.1	605	82	418	T
YN-12	34.6 m above A	5.6	1.2	35.6	4.3	635	77	416	T
YN-13	27.6 m above A	6.6	1.2	35.2	1.5	533	23	419	T
YN-14	21.8 m above A	5.5	0.8	29.3	2.3	532	42	422	T
YN-15	20.4 m above A	5.6	1.1	32.5	3.9	580	69	416	T

Six samples collected from the Oyster Limestone in the Abu Shigaila and Mohammad mines were organically lean, containing <0.5% TOC. Pyrolysis data was not generated on these samples.

[1]GD—Gebel Duwi mine, YN—Younis North mine, AS—Abu Shigaila mine, H—Hamrawein mine, M—Mohammad mine, W—Wasif mine.
[2]Meters above or below base of phosphorite unit. A—A group phosphorites, B—B group phosphorites, C—C group phosphorites.
[3]Interpreted position of sample in a sequence stratigraphic framework. T—Transgressive systems tract, H—Highstand systems tract.

1984; Germann et al., 1985; Mikbel and Abed, 1985; Notholt, 1985; Abed and Al-Agha, 1989). The association of phosphorites, carbonates, cherts, and organic-rich shales and marls indicates that a zone of high primary productivity occurred well inboard on the shallow shelves and that conditions favorable for organic matter preservation varied through time.

Deposition of these organic-rich, Late Cretaceous units was controlled by a major second order (~5–50-m.y. cycles) transgressive–regressive cycle (Lewy, 1990). Within this second order cycle, several sequences can be delineated which appear to correspond to the third order eustatic cycles (~0.5–5-m.y. cycles) defined by Haq et al. (1987). These sequences and their geochemical characteristics are described in this chapter for an organic-rich, Late Cretaceous section from the eastern desert of Egypt. Here, the organic-rich portion of the Late Cretaceous transgression is represented by the Duwi and Dakhla formations (Kerdany and Cherif, 1990). Both formations locally contain good to excellent hydrocarbon source facies as manifested by the organic enrichment and generation potential (Table 1; Figure 1) and the elevated hydrogen index values for many of the samples (Figure 2). The Duwi and Dakhla formations were examined to determine the relationships among organic enrichment and organic matter type to interpretations of the sequence stratigraphic framework of

Table 1 (continued).

Sample ID[1]	Stratigraphic Position[2]	TOC (%)	S_1 (mg/g)	S_2 (mg/g)	S_3 (mg/g)	HI	OI	T_{max}	System Tract[3]
YN-16	16.8 m above A	5.3	1.0	31.8	2.9	600	55	419	T
AS-01	10.2 m above A	3.2	0.3	12.4	3.3	387	102	428	T
AS-02	7.8 m above A	2.4	0.6	9.4	1.9	391	80	423	T
AS-03	1.8 m above B	2.4	0.3	9.9	1.8	412	74	416	T
AS-04	1.4 m above B	2.6	0.5	10.5	2.2	404	83	419	T
AS-05	13.8 m above C	3.3	0.2	4.1	3.2	123	97	423	H
AS-06	13.6 m above C	2.1	0.3	2.5	2.0	119	93	426	H
AS-07	13.1 m above C	3.8	0.2	3.3	2.8	86	74	421	H
AS-08	12.2 m above C	4.4	0.6	4.3	3.3	97	75	422	H
H-01	6.7 m above A	7.6	0.8	24.7	4.3	325	56	417	T
H-02	6.5 m above A	3.8	0.2	13.9	1.4	365	36	419	T
H-03	6.2 m above A	4.1	0.3	14.0	2.0	341	48	424	T
H-05	5.4 m above A	6.2	0.8	31.1	1.7	501	28	423	T
H-06	5.4 m above A	8.1	1.6	43.5	3.0	536	37	426	T
H-07	5.1 m above A	7.0	1.1	32.8	2.5	468	35	422	T
H-08	4.6 m above A	8.4	1.2	35.9	3.4	427	41	424	T
H-09	3.2 m above B	1.2	0.2	1.3	0.5	111	38	422	T
M-01	6.4 m above B	3.7	0.4	12.6	1.5	341	41	423	T
M-03	5.3 m above B	5.6	0.7	18.0	2.4	321	43	421	T
M-04	4.2 m above B	1.1	0.1	4.1	0.3	370	29	425	T
M-05	1.2 m above B	1.9	0.2	6.3	0.9	331	45	423	T
M-06	13.3 m above C	3.8	0.3	15.7	2.6	413	68	426	H
M-07	12.1 m above C	2.3	0.4	10.6	1.4	460	62	423	H
M-08	8.8 m above C	1.6	0.1	7.2	1.2	448	76	421	H
W-01	8.3 m above C	7.8	0.9	9.4	4.9	121	63	417	H
W-02	6.6 m above C	7.0	0.9	7.1	4.6	102	66	419	H
W-03	6.1 m above C	3.5	0.4	2.9	2.4	84	69	424	H
W-04	4.6 m above C	5.8	0.3	5.7	4.2	98	73	419	T
W-05	4.5 m above C	3.4	0.5	6.1	1.5	179	45	423	T
W-06	2.5 m above C	5.1	0.5	6.3	4.5	124	88	416	T

these potential source rocks. The geochemical attributes of these units are interpreted in light of these stratigraphic relationships. Samples from the Duwi and Dakhla formations were collected throughout Egypt but exposures in the eastern desert of Egypt are emphasized in this work (Figure 3). Data reported in this work were collected from the Gebel Duwi, Younis North, Abu Shigaila, Hamrawein, Mohammad, and Wasif phosphate mines in the eastern desert of Egypt. Equivalent strata in the Gulf of Suez, where mature, may be a source of that region's produced oil (Rohrback, 1983). The Duwi Formation is also one of the largest phosphate deposits in the world (Notholt, 1985). Interpretation of the sequence stratigraphic framework is based on previous interpretations (Ganz et al., 1990; Glenn, 1990; Glenn and Arthur, 1990).

EXPERIMENTAL METHODS

Pre-washed samples were ground to pass 100 mesh prior to analysis. Organic carbon content was determined using a Leco CS-244 carbon analyzer after the samples were treated with HCl to remove carbonate minerals. Rock-Eval analysis, as described by Espitalié et al. (1977), was used to further characterize the organic matter. Approximately 50 grams of each sample were refluxed for 24 hours with methylene

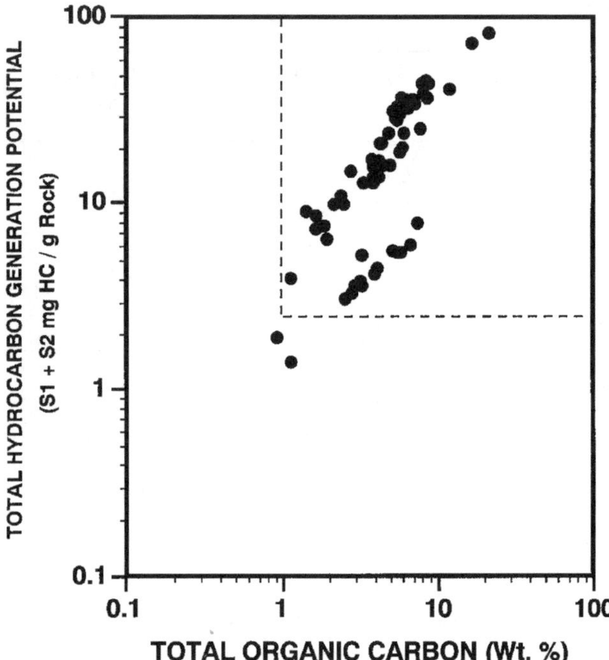

Figure 1. Organic carbon contents and generation potential for Duwi and Dakhla formations samples from the eastern desert of Egypt. Based on visual kerogen data (Table 2), samples on the upper trend contain primarily oil-prone organic matter while samples from the lower trend contain abundant terrestrial organic matter. Samples from the lower trend are primarily from the lower Duwi Formation, equivalent to the C group phosphorites in the Abu Shigaila and Mohammad mines (Figure 3). Dashed lines represent minimum organic carbon content and generation potential necessary to be considered a potential hydrocarbon source rock (Bissada, 1982).

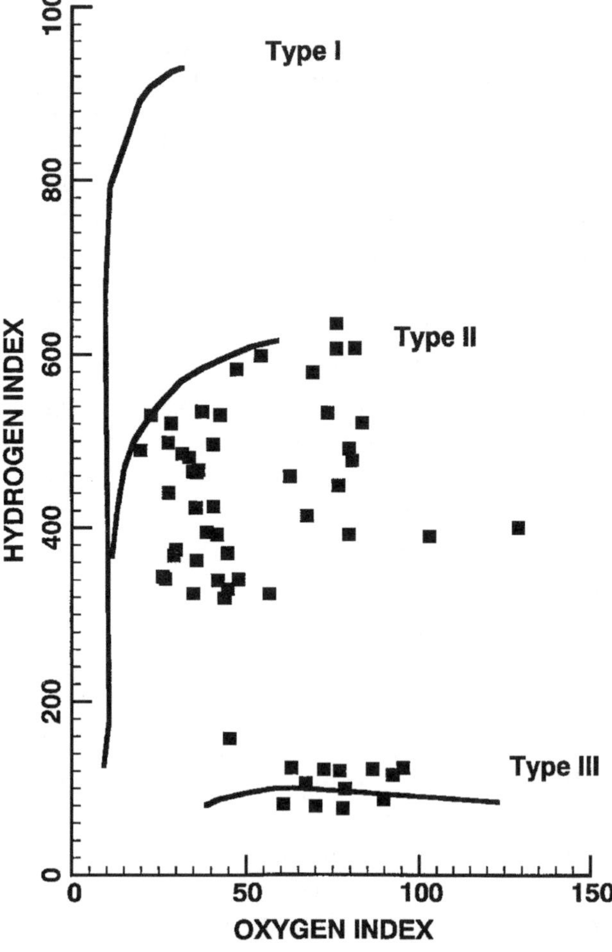

Figure 2. Modified van Krevelen diagram showing the moderate to excellent hydrogen enrichment and oil-prone nature of samples from the upper portion of the Duwi and the Dakhla Formations (Type II). Samples from the base of the Duwi Formation (Type III) commonly contain abundant terrestrial organic matter, primarily vitrinite, resulting in low hydrogen indices. Type I, II, and III kerogens after Espitalié et al. (1977).

chloride in a soxhlet extraction apparatus to recover the bitumen. The bitumens were separated into saturated hydrocarbons, aromatic hydrocarbons, resins, and asphaltenes (Nolte and Colling, 1989) and the saturated hydrocarbons were further separated into normal paraffins, isoprenoids, and naphthenes (Nolte, 1991). The saturated hydrocarbon fractions were analyzed using a Hewlett-Packard 5840 gas chromatograph equipped with a 30-m capillary column coated with a DB-1 stationary phase equivalent to SE-30. The isoprenoid and naphthene fractions were recombined and analyzed using a Hewlett-Packard 5980 gas chromatograph with a 5970B mass selective detector at 70 eV electron impact ionization. The column was a 50-m capillary coated with DB-5 stationary phase equivalent to SE-54. Additional details of analytical conditions can be found in Robison (1986).

Separate aliquots of samples were treated with HCl and HF to isolate the kerogens for elemental and/or visual analysis. Elemental analysis was performed on a Carlo Erba 1106 elemental analyzer. Visual kerogen analysis was performed in transmitted and fluorescent light using standard petrographic techniques.

REGIONAL GEOLOGY

During the Late Cretaceous the African continent was rotating counterclockwise and moving toward Asia (Condie, 1975; Le Pichon et al., 1976), resulting in a gradual closing of the Tethys Sea. Also during this time a broad downwarping of the northeast margin of Africa occurred, causing a major transgression that covered portions of northeast Africa (Figure 4) during which a series of phosphorites and organic-rich source rocks were deposited.

An upwelling zone has been postulated to have existed along the northeast African shelf during

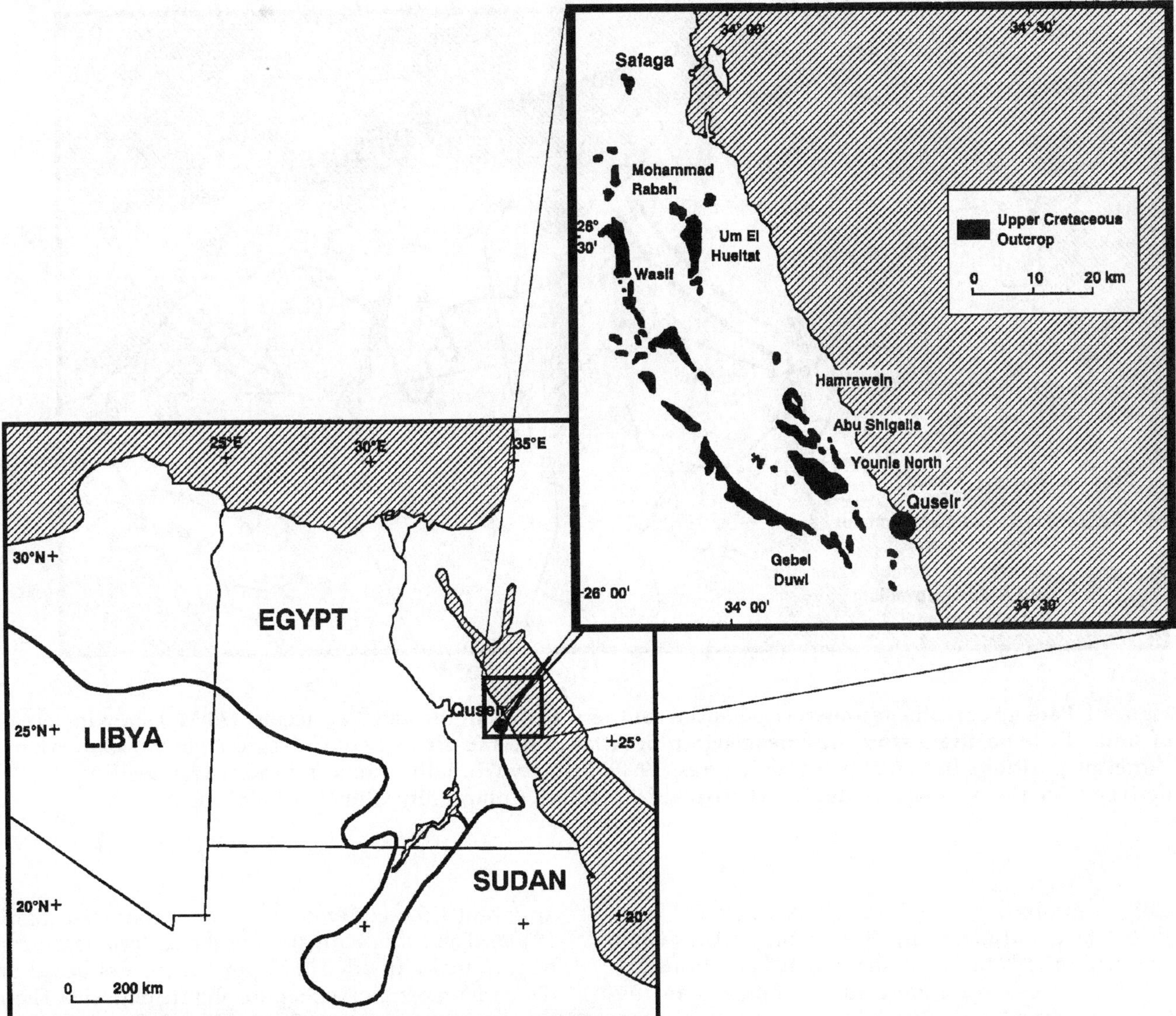

Figure 3. Locality map for the eastern desert of Egypt showing location of phosphate mines where sampling occurred. Maximum extent of Maastrichtian transgression (thick line, lower box) from Klitzsch and Squyres (1990).

deposition of these units (Ganz, 1984; Germann et al., 1985). Upwelling, which is often suggested as a primary mechanism in supplying nutrients to the shallow photic zone, would have supplied an area with high levels of primary productivity along the shelf (Sheldon, 1980). Glenn and Arthur (1990), however, argue that upwelling is not necessary to provide the nutrients to the African shelf. Krajewski (1989) also suggests that Jurassic phosphorite and black shales of the Spitsbergen shelf were not deposited beneath an upwelling zone. Glenn and Arthur (1990) suggest that fluvial input from a deeply weathered continental land mass could have supplied much of the needed phosphorus to enhance surface primary productivity on the northeastern African shelf.

Danian shorelines of the southern Tethys Sea are believed to have been lined by mangrove swamps (Gregor and Hahn, 1982). Pollen data obtained on formations in the eastern desert of Egypt indicate that these mangrove swamps may have existed since the Late Cretaceous (Schrank, 1984a). Spore and pollen data also suggest that a region of heavy rainfall existed along the Cretaceous shoreline (Schrank, 1984b). The drainage system for this area of heavy rainfall could have supplied the nutrients to drive primary productivity as Glenn and Arthur (1990) suggest.

The association of organic-rich and phosphatic horizons indirectly suggests that high levels of surface organic productivity were coupled with benthic low oxygen levels (Schlanger and Jenkyns, 1976; Waples, 1982). Several authors have suggested that phosphate formation takes place through inorganic precipitation of apatite within pore waters of anoxic sediments near the upper or lower boundary of an

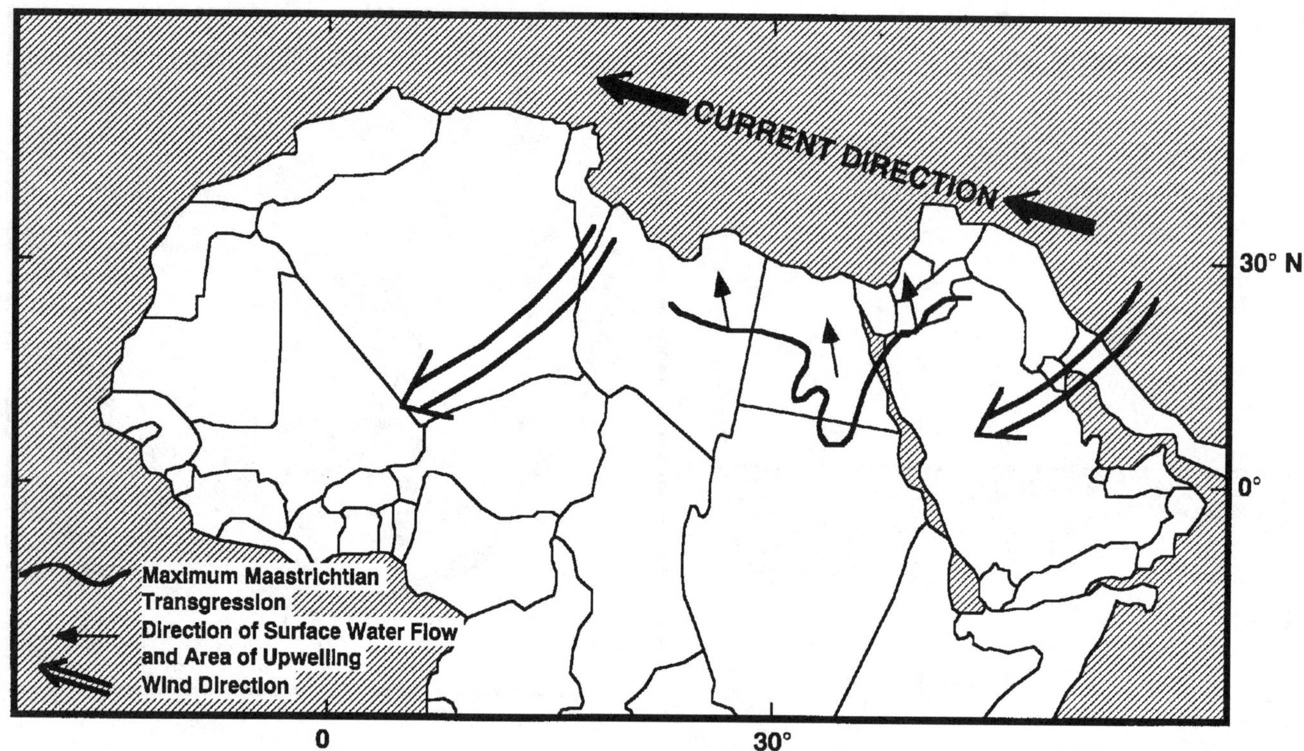

Figure 4. Paleogeographic reconstruction of the southern Tethys in the Late Cretaceous (70 Ma) showing area of study. Plate positions from the Paleogeographic Atlas Project (University of Chicago, 1990). Maastrichtian shoreline positions from Klitzsch and Squyres (1990). Primary wind directions and inferred upwelling derived from the National Center for Atmospheric Research Community Climate Model CCM0.

oxygen minimum zone (Burnett 1974, 1977; Cook, 1976; Glenn and Mansour, 1979; Baturin, 1982; Glenn and Arthur, 1990). Prevailing anoxic conditions are also favorable sites for preservation of organic matter (Demaison and Moore, 1980).

The favorable conditions that existed for source rock development in Egypt during the Late Cretaceous do not represent an isolated event on the southern Tethys margin. Similar, age-equivalent depositional belts of phosphorite accumulations and shales with elevated organic carbon content are described across this region of the Tethys (Kolodny, 1980; Bein and Amit, 1982; Kemper and Zimmerle, 1983; Mikbel and Abed, 1985; Abed and Al-Agha, 1989). An upwelling system has also been invoked to explain the equivalent of these deposits in northeast Jordan, northwest Iraq, and Syria (Kemper and Zimmerle, 1983).

STRATIGRAPHIC SECTION AND SOURCE ROCK POTENTIAL

Initial deposits associated with the transgression are characterized by phosphorites, bioclastic carbonates, and organic-rich shales and marls of the Duwi Formation (Figure 5). The Duwi Formation represents the first fully marine conditions in Egypt, and has no apparent time-equivalent section landward (Said, 1990a). The Duwi Formation in the eastern desert can be informally divided into three members based on the cyclic repetition of its phosphorite deposits. These are referred to here as the A, B, and C group phosphorites (Soliman and Amer, 1972).

A late Campanian to early Maastrichtian age of the main phosphorites of Egypt has been established in the Wadi Qena by Hendriks and Luger (1987) and Luger and Groschke (1989). Ammonite assemblages from the lower C group and middle B group phosphorites are of early late and late late Campanian age, respectively (Luger and Groschke, 1989). Glenn (1980) indicated that in Gebel Duwi (Figure 3) the Campanian–Maastrichtian stage boundary occurs in the uppermost portion of the Duwi Formation. Ganz et al. (1990) suggest an early Maastrichtian age for the upper phosphorite horizon.

The Duwi Formation exhibits considerable variability in its organic enrichment (Table 1; Figure 6). Organic carbon contents range from <0.5 to >7 wt.%. The lowest values occur in the Oyster Limestone samples from the Abu Shigaila mine where the organic carbon content is <0.5 wt.%. The highest organic carbon contents measured in the Duwi Formation occur in samples from the Wasif mine. The Duwi Formation has an average organic carbon content of 3 wt.% for all samples.

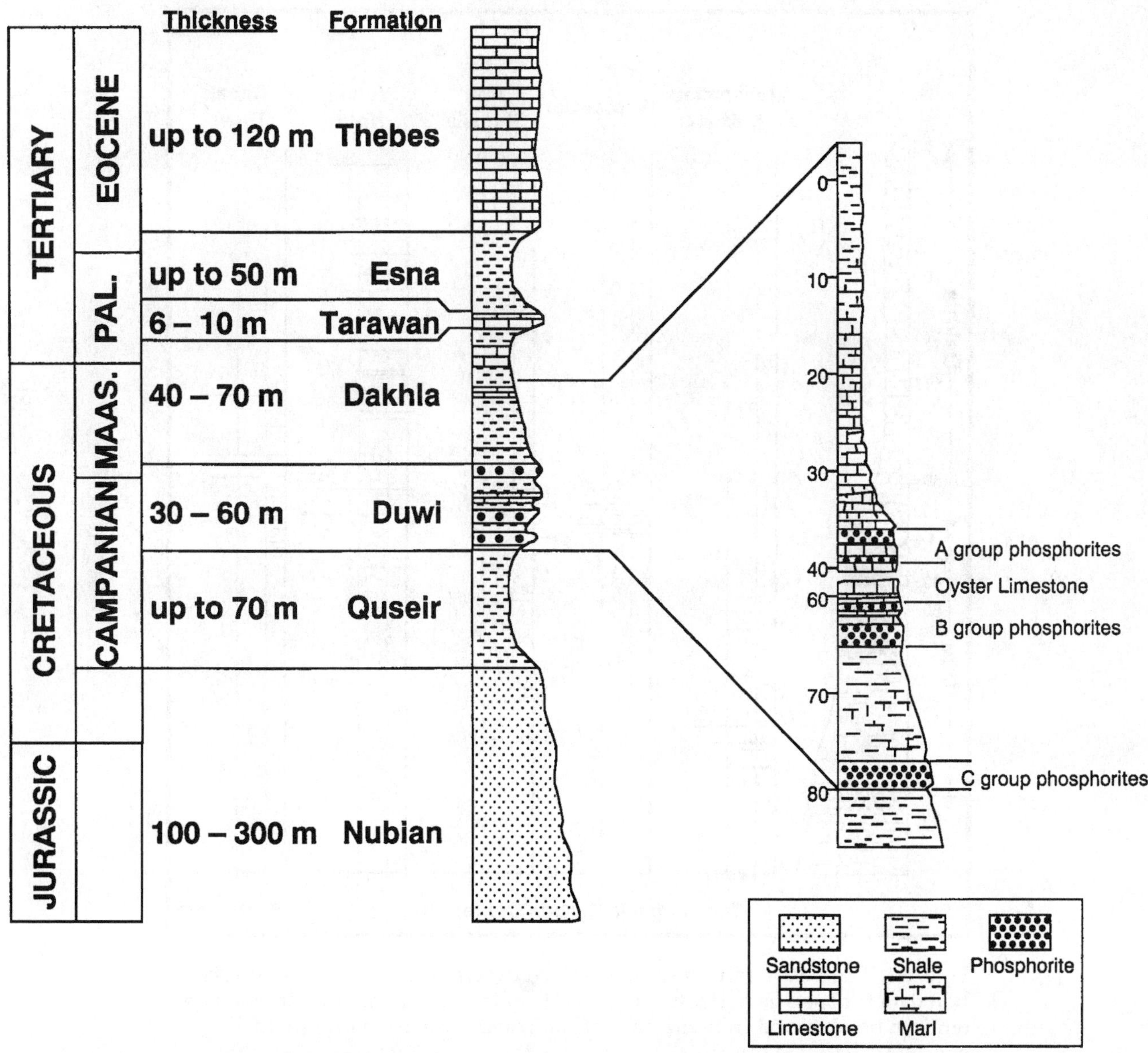

Figure 5. Idealized stratigraphic section for the eastern desert of Egypt. A, B, and C group phosphorites are informal designations used by the local phosphate companies (Soliman and Amer, 1972). Note that for simplification ~20 m of bioclastic limestone is not shown in the Duwi Formation.

Kerogen type is not directly related to organic enrichment in the Duwi Formation. For example, six samples from the Wasif mine have an average organic carbon content of 5.4% with an average hydrogen index from Rock-Eval pyrolysis of 118 (Table 1). In contrast, eight samples from the Mohammad mine have a lower average organic carbon content, 3.1%, but are more hydrogen-enriched with an average hydrogen index of 376. In general, the lower part of the Duwi Formation has the highest organic carbon content (Figure 6) while the organic matter in the upper part of the Duwi Formation is more hydrogen-enriched (Figure 7). Vitrinite reflectance measurements (Table 2) indicate that the section is immature with respect to hydrocarbon generation and expulsion and that the measured organic carbon content and hydrogen indices have not been affected by heating. T_{max} values from pyrolysis (Table 1) are consistent with the vitrinite reflectance measurements.

Visual kerogen analyses (Table 2) indicate that vitrinite is present only in samples from the Duwi Formation. Also, the only significant content of nonfluorescent organic matter is found in the Duwi Formation. Nonfluorescent organic matter as used here includes degraded terrestrial (humic) material and any unidentifiable organic material. Both visual kerogen and Rock-Eval analysis suggest that input of terrestrial organic matter was greatest during deposition of the lower portion of the Duwi Formation. A decrease over time in input of higher plant material

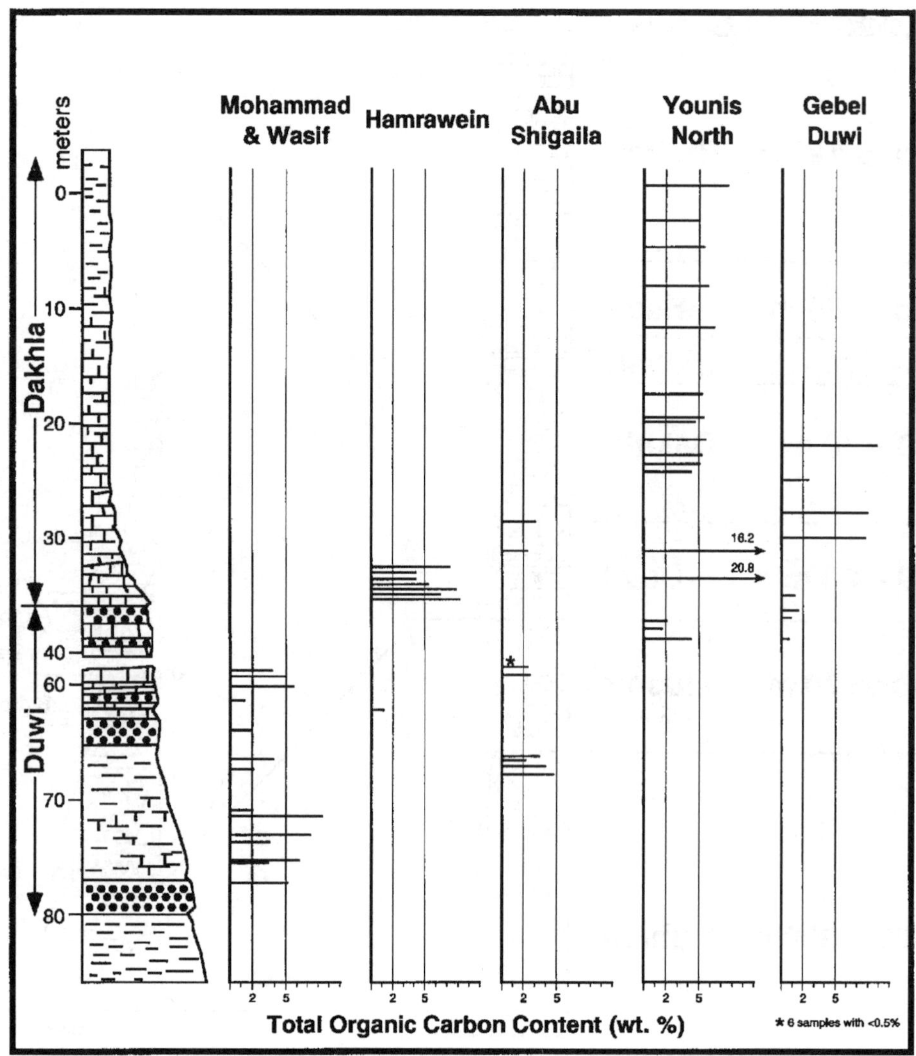

Figure 6. Composite section for the eastern desert of Egypt with organic carbon data for specific horizons. In general, an increase in organic carbon content can be observed from the base of the Duwi Formation up into the overlying Dakhla Formation. When placed into a sequence stratigraphic framework, the transgressive systems tracts are organically enriched relative to the highstand systems tracts (see Figures 10, 11). Patterns from Figure 5.

relative to marine phytoplankton is evident from lower Duwi Formation time on up into deposition of the Dakhla Formation.

Conformably overlying the Duwi Formation is the Dakhla Formation, thought to represent deposition in slightly deeper water (Said, 1990a). The lower Dakhla Formation in most areas exhibits significant organic enrichment and is often classified as an oil shale (Troger, 1984). Depositional conditions were probably very similar between the Duwi and Dakhla formations, with the primary difference being water depth. The base of the Dakhla Formation is assigned to the early Maastrichtian (Ganz et al., 1990).

In the eastern desert, total organic carbon content in the Dakhla Formation ranges from less than 1 to greater than 20 wt.% (Table 1; Figure 6). Organic carbon contents > 15 wt.% have also been reported for the Late Cretaceous in the eastern desert of Egypt by Ganz et al. (1990). In this study, samples from the Dakhla Formation have an average organic carbon content > 5.5 wt.% and an average hydrogen index of 459, indicating a significant potential for liquid hydrocarbon generation with sufficient thermal maturity. In the Gulf of Suez, equivalent horizons to the Dakhla and upper Duwi formations may have sourced much of the region's reservoired oil (Rohrback, 1983).

Samples from the Late Cretaceous Ghareb Formation in Israel also exhibit significant organic enrichment, with organic carbon contents over 20%

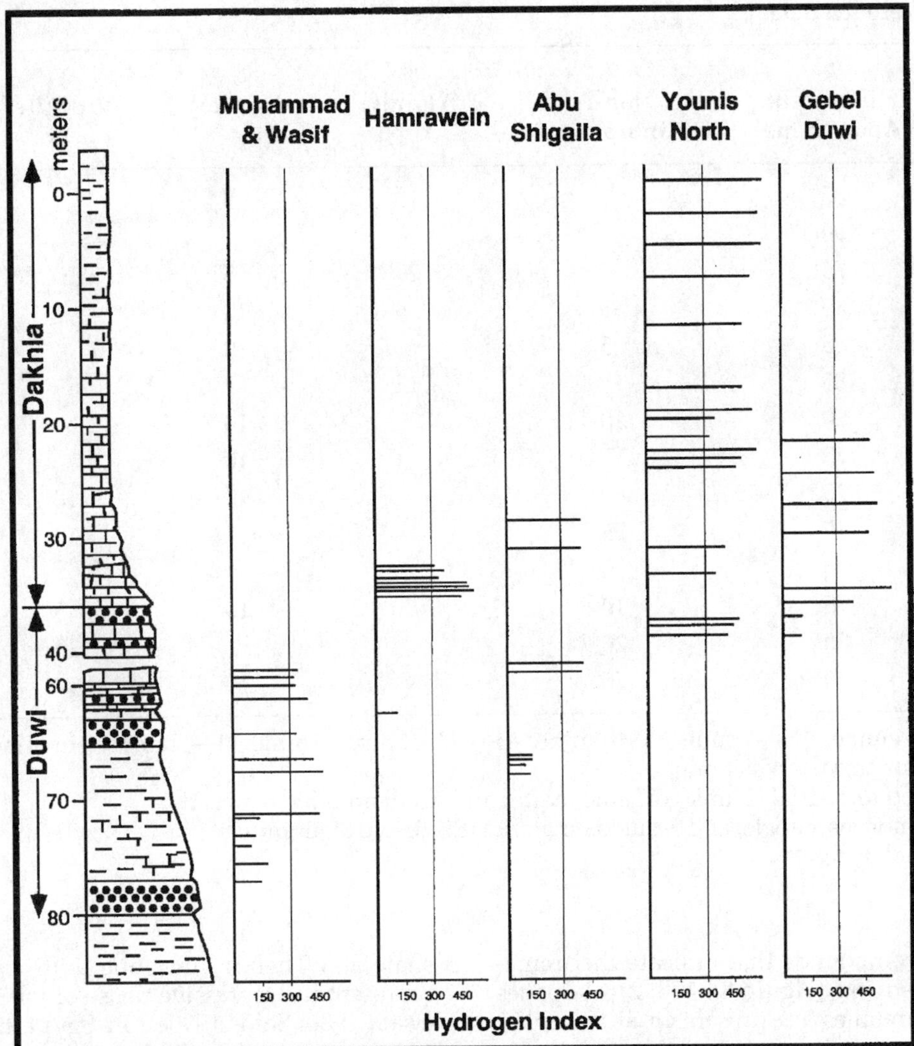

Figure 7. Composite section for the eastern desert of Egypt with hydrogen index values for specific horizons. The oil-prone nature of the upper Duwi and Dakhla formations is evident by the elevated hydrogen index values. The increased organic carbon content observed in the transgressive systems tracts relative to the highstand systems tracts occurs in conjunction with an increase in the oil proneness of the section (see also Figures 10, 12). Patterns from Figure 5.

being reported (Shirav and Ginzburg, 1983). The organic matter consists primarily of amorphous algal debris that exhibits significant generation potential based on elemental analysis of the kerogen. In most areas, these source rocks are immature to moderately mature due to insufficient burial (Tannenbaum and Aizenshtat, 1984). However, several authors (Spiro et al., 1983; Tannenbaum and Aizenshtat, 1985; Rullkötter et al., 1985) have suggested that many of the biodegraded tars in the area can be correlated to the Ghareb Formation.

Little evidence for input of terrestrial organic matter from the nearby shoreline is evident in the Dakhla Formation in the eastern desert. Visual kerogen data (Table 2) indicate the Dakhla Formation contains primarily fluorescent, amorphous organic matter with minor amounts of alginite and other exinites. Vitrinite is sparse through most of the section, suggesting little detrital input from the nearby shoreline.

Typical saturated hydrocarbon chromatograms from oil-prone facies of the Duwi and Dakhla formations are dominated by isoprenoids, steranes, and terpanes (Figure 8) consistent with the immature nature of the section. Pristane/phytane ratios are consistently less than 1, suggesting reducing conditions at the time of deposition (Didyk et al., 1978).

The m/z 217 chromatograms exhibit simple patterns that are dominated by the regular 5α,14α,17α (20R) sterane configuration (Figure 8), indicating the section is immature. The m/z 191 chromatogram also

Table 2. Visual kerogen data.

Sample ID[1]	% Fluorescent Amorphous	% Non-Fl. Amorphous[2]	% Alginite	% Exinite[3]	% Vitrinite	Vitrinite Reflectance % in oil
GD-02	85		10	5		
GD-06	90		10			
YN-03	75		20	5		
YN-05	80	5	5	10		
YN-09	85	5		10		
YN-10	85		15			
YN-15	80	10		10		
YN-16	75	15		10		
AS-05	20	60		10	10	.33
AS-08	15	75			10	.39
H-04	85	10	5			
M-02	75	10		10	5	.34
M-08	70	20			10	.29
W-03	10	70		10	10	.44

[1]GD—Gebel Duwi mine, YN—Younis North mine, AS—Abu Shigaila mine, H—Hamrawein mine, M—Mohammad mine, W—Wasif mine.
[2]Includes degraded terrestrial (humic) organic matter and unidentifiable material.
[3]Includes palynomorphs, cuticle, and resinous material. Exclusive of alginite.

exhibits several compounds that indicate the immature nature of the rocks (Figure 8); 17β, 21β hopanes and 17β, 21α moretanes are present in all samples. Tricyclic terpanes are in low concentrations relative to pentacyclic triterpanes.

No significant differences are apparent between the type of organic matter contained in the Dakhla Formation compared with the upper part of the Duwi Formation. The primary difference is both the higher organic carbon content and relatively greater hydrogen enrichment in the Dakhla Formation (Figures 6, 7). Since organic matter input does not appear to change significantly between the upper Duwi and lower Dakhla formations, the primary difference between the two formations is the lack of phosphorite hardpans within the Dakhla Formation, indicating deposition in slightly deeper water, probably below storm wave base (Said, 1990a). The lower Dakhla Formation is primarily an organic-rich marl in the eastern desert. In contrast, upper portions of the Dakhla Formation are typically not organic-rich and are more of a true hemipelagic shale with little carbonate content.

Overlying the Dakhla Formation is the Paleocene Tarawan chalk. To the north, in the Cretaceous unstable shelf areas in Egypt, the Dakhla grades laterally into chalk, which likely represents the maximum extent of the transgression during the marine phase (Said, 1990b). Above the Tarawan chalks are the Esna shale and Thebes limestone units (Figure 5), which represent the regressive phase of the section (Soliman et al., 1986; Said, 1990c). In Egypt, the stratigraphic sequence that spans the Duwi through Thebes formations represents a several-million-year transgressive–regressive cycle.

DEPOSITIONAL ENVIRONMENT

Potential source rocks were deposited on a shallow shelf that developed across the northeast African margin and the Arabian shield as the result of the Late Cretaceous transgression. Deposition of the source units was likely controlled by local tectonics. Minor sags that developed along the shelf resulted in the development of restricted and commonly anoxic conditions during the initial stages of the transgression (Bartov and Steinitz, 1977; Kolodny, 1980; Schroter, 1986; Bock, 1987; Abed and Al-Agha, 1989). These troughs were the site of deposition of significant quantities of oil-prone organic matter. During Duwi and Dakhla Formation times, these troughs occurred in restricted coastal environments, e.g., lagoonal to shallow shelf settings (Ganz et al, 1990).

During the early stages of Late Cretaceous deposition, water depths were still shallow enough to allow periodic reworking of the sediments due to wave action and/or periodic storms (Glenn and Arthur, 1990). This reworking led to the development of

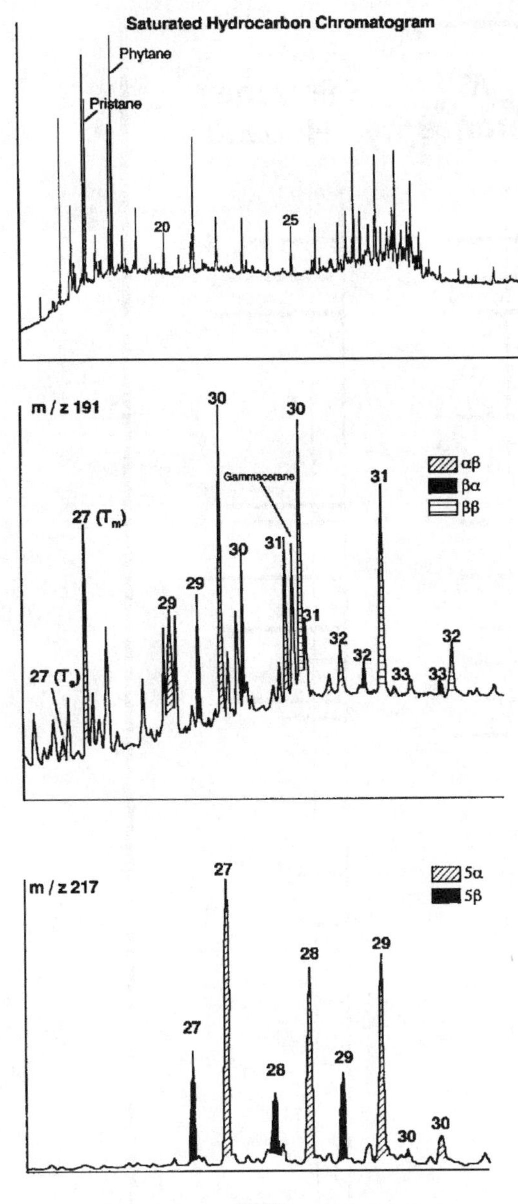

Figure 8. Typical saturated hydrocarbon, m/z 191 and m/z 217 chromatograms from the lower portion of the Dakhla Formation (sample GD-03, Table 1). The immature nature of the samples is exhibited in these chromatograms. On the m/z 191 fragmentogram, αβ = 17α,21β; βα = 17β,21α; and ββ = 17β,21β hopane configuration. On the m/z 217 fragmentogram, 5α = 5α,14α,17α (20R) and 5β = 5β,14α,17α (20R) sterane configuration.

phosphorite lag deposits typical of the Duwi Formation. As the transgression continued and water depth increased, input of both organic and inorganic fractions did not change significantly from the lower sections (Robison, 1986). However, water depths had increased to the point where lag deposits are not recorded.

Several geochemical parameters exhibit significant changes from the base to the top of the sampled stratigraphic interval, representing the change from a mixed marine–terrestrial organic input to a primarily marine organic assemblage. These include the kerogen maceral distribution and the hydrogen indices from Rock-Eval type pyrolysis (Figure 7). In addition, both the C_{27}/C_{29} 5α,14α,17α (20R) sterane ratio and the sterane/hopane ratio increase with increasing marine algal/bacterial influence (Figure 9). A similar sequence of phosphorite to black shale with terrestrial organic matter to shale and marlstone with phytoplankton-derived organic matter has been described from the Jurassic of the Spitsbergen shelf (Krajewski, 1989).

SOURCE VARIABILITY IN THE EASTERN DESERT

Even though general trends in the source quality can be observed through the section from the base of the Duwi Formation through the Dakhla Formation, no clear predictable pattern in source rock quality emerges from a traditional approach to stratigraphic interpretation. Prediction of the variability in source quality is important as it can impact resource assessment, source to oil correlations, and the prediction of the relative abundance of oil and gas (Katz et al., 1993). When the geochemistry is placed into the context of sequence stratigraphy, predictable associations of organic matter type and source rock quality are observed (Pasley et al., 1991; Bohacs, 1993). Relying on the sequence stratigraphic framework developed by Glenn (1990) and Glenn and Arthur (1990), the geochemical attributes of the Duwi and Dakhla formations are examined below.

The stratigraphy of the Duwi and Dakhla formations indicates that depositional conditions were not stagnant in the troughs along the shallow shelf. The deposition of organic-rich marls and the formation of phosphorite lag deposits suggests alternating anoxic and oxic conditions. Significant variability can be observed in the organic matter analyzed in the Duwi and Dakhla formations from the eastern desert of Egypt. In general, total organic carbon content and the hydrocarbon generation potential as determined by Rock-Eval pyrolysis tend to increase up section (Figures 6, 7).

Samples analyzed from the upper Duwi and lower Dakhla formations contain greater quantities of organic matter that is hydrogen-enriched compared to the lower Duwi Formation. In the upper Duwi and Dakhla formations, the quantity and quality of the organic matter may be primarily related to preservation, since both visual kerogen and chemical analyses suggest that organic input did not vary significantly. Throughout this portion of the section, the biomarker compounds present and their distribution remain relatively constant. Some variability is observed that likely reflects minor differences in reworking and degradation of the organic matter. If the type of

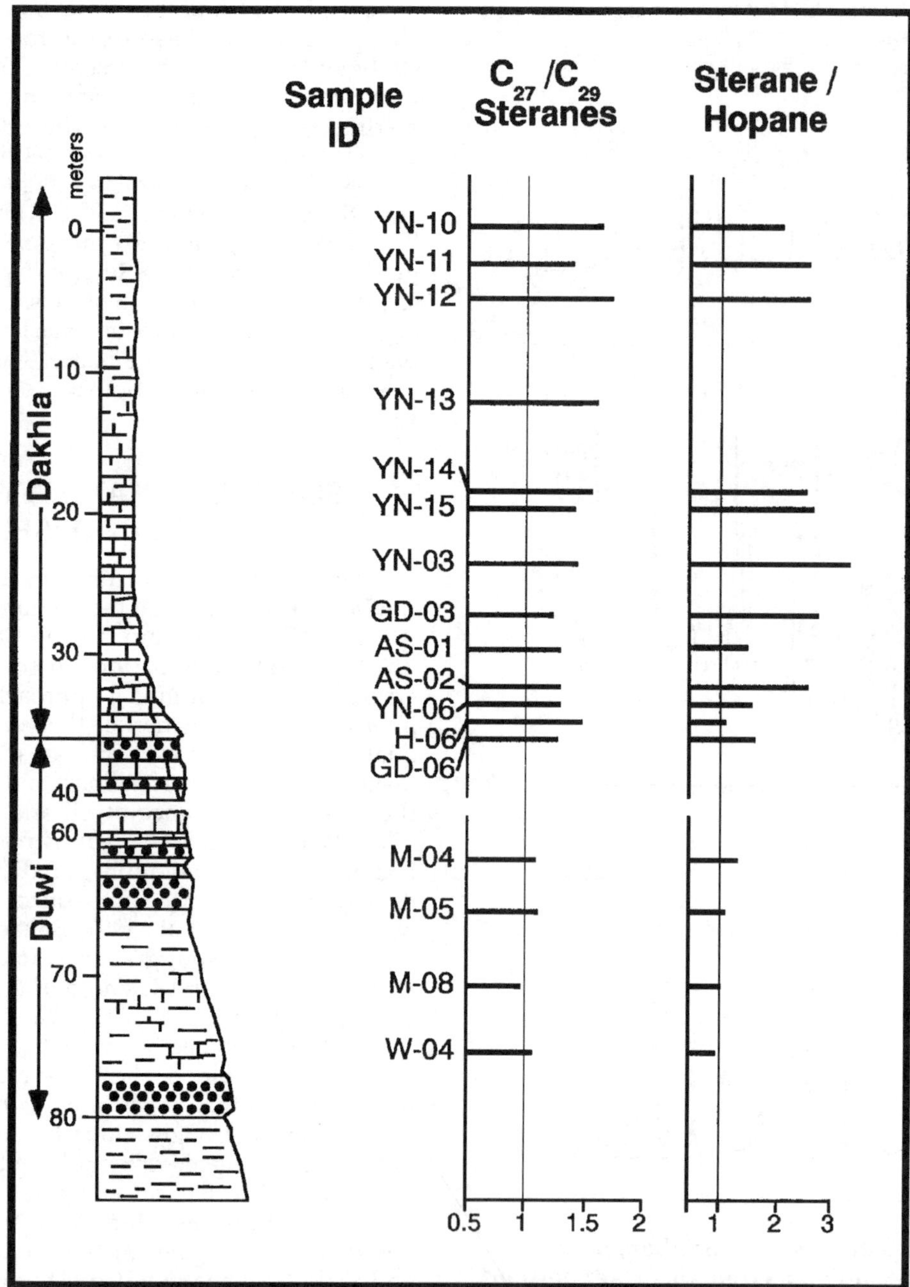

Figure 9. Sterane/Hopane and C_{27}/C_{29} sterane ratios for selected samples. Both ratios tend to increase with increasing hydrogen index values (Table 1) and increasing content of fluorescent amorphous organic matter (Table 2). Peak areas from the m/z 217 and m/z 191 fragmentograms were used for calculations. Patterns from Figure 5.

C_{27}/C_{29} steranes = C_{27}/C_{29} 5α,14α,17α (20R) sterane ratio

$$\text{Sterane/Hopane} = \frac{C_{27} + C_{28} + C_{29}\ \ 5\beta + 5\alpha,14\alpha,17\alpha\ \text{(20R) steranes}}{C_{30}\ \ 17\alpha,21\beta + C_{30}\ \ 17\beta,21\alpha + C_{30}\ \ 17\beta,21\beta\ \ \text{hopanes}}$$

organic matter being input to the system and primary productivity were relatively constant, the primary factor controlling the amount and quality of the kerogen observed in the rock record would be the preservation potential of the section. However, samples from the base of the section are leaner in organic matter and hydrogen-depleted relative to samples from higher in the section, due to significant quantities of terrestrial organic matter found there. This suggests that the low organic carbon content and low hydro-

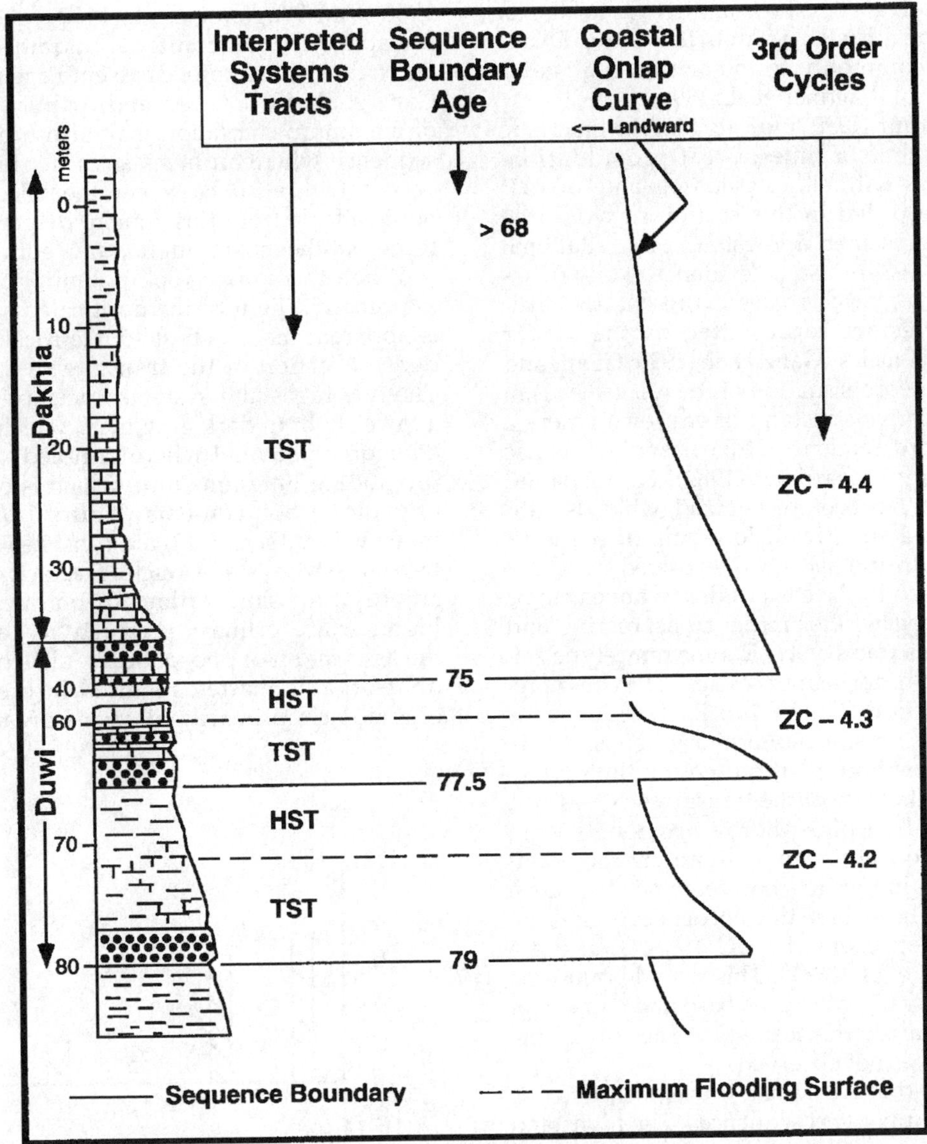

Figure 10. Composite section for the eastern desert of Egypt with interpretative correlation to the sequences of Haq et al. (1987). Three sequences can be identified in the organic-rich portions of the Upper Cretaceous of the eastern desert of Egypt based on the distribution of systems tracts. Due to the shallow shelf depositional environment (Glenn and Arthur, 1990) only transgressive and highstand systems tracts are recognized. Patterns from Figure 5.

gen enrichment may be due to a combination of factors, including the type of organic matter input to the environment, as well as the preservation potential of the system.

The variability observed in organic richness and generation potential can be explained in a sequence stratigraphic framework. Glenn and Arthur (1990) first attempted to explain the distribution of stratal patterns observed in the Upper Cretaceous of Egypt using sequence stratigraphic concepts. Adapting their convention yields the interpretation presented in Figure 10.

A significant deviation from their designations is that their lower unit in the Duwi Formation is interpreted to represent a highstand systems tract rather than a lowstand progradational wedge. Although the two may display similar geometry, the sequence under investigation here is interpreted to be inboard of the shelf-slope break, such that the lowstand systems tract is not developed (Van Wagoner et al., 1990). The Late Cretaceous, north Africa Tethyan margin of the eastern desert, where the Duwi and Dakhla formations were deposited, has been interpreted as a broad, stable, relatively shallow, epicontinental basin (Said, 1990a). Low-angle clinoforms, as observed in the bioclastic debris of the Duwi Formation, are commonly found within highstand systems tracts (Van Wagoner et al., 1990).

Glenn and Arthur (1990) indicate that the bioclastic limestone represents a time of reef progradation. Erosional surfaces with either paleosols and/or root zones are not recorded in this section as would be expected if it represented a lowstand progradational wedge (Weimer, 1988). Also no evidence of any fresh-water deposits on marine shales is present; the shallow marine facies are represented by the oyster limestone "reefal" facies (Ganz et al., 1990; Glenn and Arthur, 1990). The progradation of the reef facies is in response to a relative stillstand of sea level where a coarsening upward sequence of bioclastic debris has been described by Richardson (1982). No lowstand surface of erosion has been recognized, which is critical to document a significant lowering of sea level (Weimer, 1988). In the setting interpreted here, the sequences recorded in the eastern desert appear to be represented solely by alternating transgressive and highstand systems tracts as typical for ramp-type settings. Three third order sequences appear to be represented in the section (Figure 10). Each sequence is marked at its base by a phosphorite lag deposit (A, B, and C group phosphorites) representing the ravinement surfaces at the base of the transgressive systems tract. The individual phosphorite horizons within each group of phosphorites may represent the flooding surface of an individual parasequence or fourth order sequence. These three third order sequences are interpreted to represent the ZC 4.2, 4.3, and 4.4 sequences of Haq et al. (1987). This would bound the age of the section as 79 Ma at the base and somewhat older than 68 Ma for the top, since the top of the upper sequence was not observed.

As defined in this work, the lowest sequence is thinner than the upper two sequences. The highstand tract in the lowest sequence appears to be represented by a fine-grained, light gray to brown shale with interbedded thin sand lenses. The transgressive systems tract in the middle sequence is overlain by a thick bioclastic limestone, which is named locally the Oyster limestone. This bioclastic limestone represents the progradational highstand systems tract of the middle sequence. The highstand systems tract in the upper sequence cannot clearly be defined. The A group phosphorites grade upward into the dark, marly, organic-rich Dakhla Formation. It is not clear from the available data where the boundary for the sequence tract would be placed. It appears that all of this upper portion of the measured section is part of the transgressive systems tract.

Glenn and Arthur (1990) estimated that the rates of sediment accumulation were approximately 20 m/m.y. for the entire sequence. The transgressive surface is marked in each cycle by ravinement surfaces at the base of the transgressive systems tracts. The condensed sections, which occur in the upper part of the transgressive systems tracts (Loutit et al., 1988; Van Wagoner et al., 1990), exhibit associated phosphorite, glauconite, abundant organic matter, and a diverse microfossil assemblage (Schrank, 1984b; Ganz et al., 1990; Glenn and Arthur, 1990). A surface of maximum starvation may also occur in the upper sequence where an increase in detrital shale content occurs (Figure 10). However, the paleontologic data is not available from this zone to determine if this does represent the maximum flooding surface.

When the source rock potential of the section is examined in light of the designated systems tracts, it is apparent that the best source rocks are in the condensed section of the transgressive systems tracts. The two highstand systems tracts are represented by a lower light to dark gray marl, which has only limited source potential where sampled, and a bioclastic prograding limestone unit, which is organically lean. Organic carbon contents (Figure 11) and hydrogen index values (Figure 12) are consistently higher in the transgressive systems tracts. The lack of detrital input and resulting slow sedimentation rates, together with high surface primary productivity, as indicated by the association of phosphorites, glauconite, and abundant organic matter, resulted in the deposition of organic-rich, primarily oil-prone organic matter. The

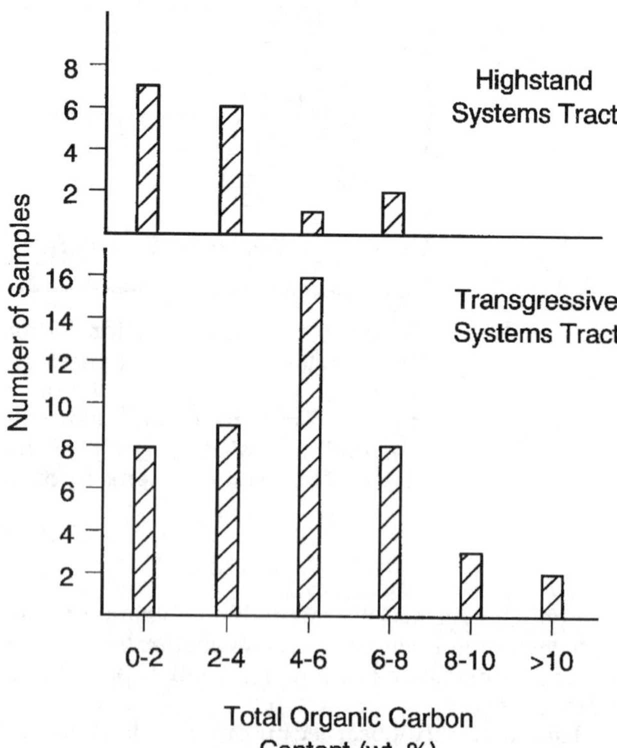

Figure 11. Distribution of organic carbon content relative to the interpreted systems tracts. Organic carbon content in the transgressive systems tracts is significantly higher than in the highstand systems tracts.

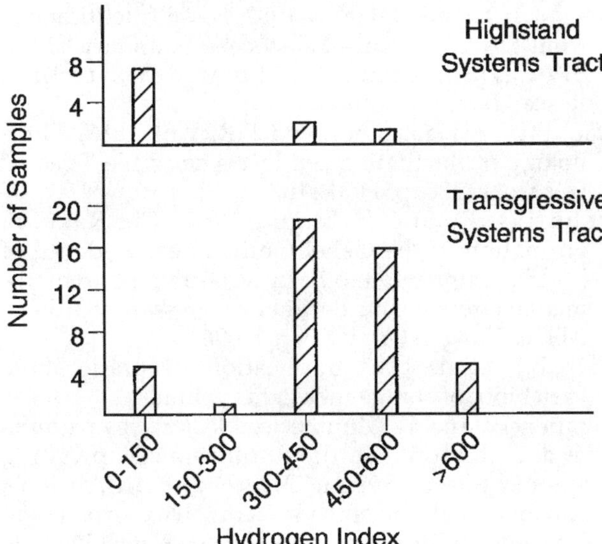

Figure 12. The type of organic matter, as indicated by the hydrogen index values, is significantly more oil-prone in the transgressive systems tracts relative to the highstand systems tracts. Organic carbon contents are also considerably higher in sections with elevated hydrogen index values (see also Figures 6, 7, 11).

association of abundant oil-prone organic matter with the transgressive systems tracts appears to be consistent. During the transgressions, the amount of detrital material supplied to the deep troughs would be greatly reduced, allowing a clastic-starved, low-energy, environment to persist resulting in the formation of a condensed section. Progradation of the shoreline during the highstand supplied coarser clastics to the trough in the form of thin sands in the lower sequence and coarse bioclastic debris in the middle sequence. This higher energy regime would not favor the deposition and preservation of fine-grained, organic-rich sediments.

SUMMARY

A Late Cretaceous transgression that flooded a broad marine shelf in Northeast Africa deposited organic-rich, potential source rocks of the Duwi and Dakhla formations. These formations exhibit considerable variability in their source quality. The lower Duwi Formation is relatively organic-rich and hydrogen-lean. The upper Duwi section is not as organically enriched but exhibits significant hydrogen enrichment, while the overlying Dakhla Formation exhibits the best oil-prone, hydrocarbon source rock potential.

Even though trends in hydrocarbon source quality are observed, no obvious pattern is apparent. Projection of the variability in source quality away from points of control can be important for volumetric calculations, oil to source correlation considerations, and the prediction of the type of hydrocarbon product to be encountered in the reservoir (Katz et al., 1993).

The variability observed in organic richness, generation potential, and organic matter type in the Duwi and Dakhla formations can be explained in a sequence stratigraphic framework. The best oil-prone, hydrocarbon source rocks occur within the condensed sections of the transgressive systems tracts. In general, the highstand systems tracts are organically lean and/or contain abundant terrestrial, primarily gas-prone organic matter.

The association of abundant oil-prone organic matter with the transgressive systems tracts is in response to reduction in supply of detrital material, allowing a clastic-starved, low-energy, condensed section to persist. Progradation of the shoreline during the highstand systems tracts supplied detrital material to these depositional sites and the resulting higher energy regime was not as favorable for the deposition and preservation of fine-grained, organic-rich sediments.

ACKNOWLEDGMENTS

Vaughn D. Robison thanks Texaco Inc. for permission to publish. Also special thanks to Joe Curiale, Lou Liro, and an anonymous reviewer for their suggestions for improving an earlier version of this manuscript. Analytical support was provided by the geochemical laboratories at the Texaco Exploration and Technology Department.

REFERENCES CITED

Abed, A.M., and M.R. Al-Agha, 1989, Petrography, geochemistry and origin of NW Jordan phosphorites: Journal of the Geological Society, London, v. 146, p. 499-506.

Bartov, Y., and G. Steinitz, 1977, The Judea and Mount Scopus Groups in the Negev and Sinai with trend surface analysis of the thickness data: Israel Journal of Earth Sciences, v. 26, p. 119-148.

Baturin, G.N., 1982, Phosphorites on the sea floor; origin, composition and distribution: Developments in Sedimentology 33, Amsterdam, Elsevier, 345 p.

Bein, A., and O. Amit, 1982, Depositional environments of the Senonian chert, phosphorite and oil shale sequence in Israel as deduced from their organic matter composition: Sedimentology, v. 29, p. 81-90.

Bissada, K.K., 1982, Geochemical constraints on petroleum generation and migration—a review: Proceedings Association of South East Asian Nations Council on Petroleum '81, p. 69-87.

Bock, W.D., 1987, Geochemie und Genese der oberkretazischen Phosphorite Agyptens. Berliner geowissenschaftliche Abhandlungen (A), v. 82, 138 p.

Bohacs, K.M., 1993, Source quality variations tied to sequence development in the Monterey and associated formations, southwestern California; this volume.

Burnett, W.C., 1974, Phosphorite deposits from the sea floor off Peru and Chile; radiochemical and geochemical investigations concerning their origin: Hawaii Institute of Geophysics Report, HIG-74-3, 164 p.

Burnett, W.C., 1977, Geochemistry and origin of phosphorite deposits off Peru and Chile: Bulletin of the Geological Society of America, v. 88, no. 6, p. 813-823.

Condie, K.G., 1975, Plate tectonics and crustal evolution: New York, Pergamon, 346 p.

Cook, P.Y., 1976, Sedimentary phosphate deposits, *in* K.H. Wolf, ed., Handbook of stratabound and strataform ore deposits, v. 7, p. 505-535.

Demaison, G.J., and G.T. Moore, 1980, Anoxic environments and oil source bed genesis: Organic geochemistry v. 2, p. 9-31.

Didyk, B.M., B.R.T. Simoneit, S.C. Brassell, and G. Eglinton, 1978, Organic geochemical indicators of paleoenvironmental conditions of sedimentation: Nature, v. 272, p. 216-222.

Espitalié, J., J.L. Laporte, M. Madec, F. Marquis, P. Leplat, J. Poulet, and A. Boutefeu, 1977, Methode rapide de caracterisation des roches meres de leur potentiel petrolier et de leur degre d'evolution: Revue de l'Institut Francais du Petrole, v. 32, p. 23-42.

Ganz, H., 1984, Organic-geochemical and palynological studies of a Dakhla shale profile in southeast Egypt: Berliner geowissenschaftliche Abhandlungen (A), v. 50, p. 363-374.

Ganz, H.H., P. Luger, E. Schrank, P.W. Brooks, and M.G. Fowler, 1990, Facies evolution of Late Cretaceous black shales from southeast Egypt, *in* A.Y. Huc, ed., Deposition of Organic Facies: AAPG Studies in Geology 30, p. 217-229.

Germann, K., W.D. Bock, and T. Schroter, 1985, Properties and origin of Upper Cretaceous Campanian phosphorites in Egypt: Sciences Geologiques–Memoires, v. 77, p. 23-33.

Glenn, C.R., 1980, Stratigraphy, petrology and sedimentology of the Duwi formation (Late Cretaceous) eastern Egypt: Unpublished MS thesis, University of California, Santa Cruz, 269 p.

Glenn, C.R., 1990, Depositional sequences of the Duwi, Sibaiya and Phosphate Formations, Egypt: phosphogenesis and glauconitization in a Late Cretaceous epeiric sea, *in* A.J.G. Notholt and I. Jarvis, eds., Phosphorite Research and Development: Geology Society Special Publication No. 52, p. 205-222.

Glenn, C.R., and M.A. Arthur, 1990, Anatomy and origin of a Cretaceous phosphorite-greensand giant, Egypt: Sedimentology, v. 37, p. 123-154.

Glenn, C.R., and S.E.A. Mansour, 1979, Reconstruction of the depositional and diagenetic history of phosphorites and associated rock of the Duwi Formation (Late Cretaceous) eastern desert, Egypt: Annals of the Geological Survey of Egypt, v. 9, p. 388-407.

Gregor, H.J., and H. Hahn, 1982, Fossil fructifications from the Cretaceous–Palaeocene boundary of the SW-Egypt (Danian, Bir Abu Mangar): Tertiary Research, v. 4, p. 121-147.

Haq, B.U., J. Hardenbol, and P.R. Vail, 1987, Chronology of fluctuating sea levels since the Triassic: Science, v. 235, p. 1156-1167.

Hendriks, F., and P. Luger, 1987, The Rakhiya Formation of the Gebel Qreiya area: evidence of Middle Campanian to Early Maastrichtian synsedimentary tectonism: Berliner geowissenschaftliche Abhandlungen (A), v. 75, p. 83-96.

Katz, B.J., et al., 1993, Implications of stratigraphic variability of source rocks; this volume.

Kemper, E., and W. Zimmerle, 1983, Facies patterns of a Cretaceous/Tertiary subtropical upwelling system (Great Syrian desert) and an Aptian/Albian boreal upwelling system (NW Germany), *in* J. Thiede and E. Suess, eds., Coastal upwelling—Its sediment record, Part B. Sedimentary records of ancient coastal upwelling, New York, Plenum Press, p. 501-533.

Kerdany, M.T., and O.H. Cherif, 1990, Part 4: Discussion; Mesozoic, *in* R. Said, ed., The Geology of Egypt, Rotterdam, Balkema, 734 p.

Klitzsch, E.H., and C.H. Squyres, 1990, Paleozoic and Mesozoic geological history of northeastern Africa based upon new interpretation of Nubian strata: AAPG Bulletin, v. 74, p. 1203-1211.

Kolodny, Y., 1980, Carbon isotopes and depositional environment of a high productivity sedimentary sequence—The case of the Mishash–Ghareb Formations, Israel: Israel Journal of Earth Sciences, v. 29, p. 147-156.

Krajewski, K.P., 1989, Organic geochemistry of a phosphorite to black shale transgressive succession: Wilhelmoya and Janusfjellet Formations (Rhaetian-Jurassic) in Central Spitsbergen, Arctic Ocean: Chemical Geology, v. 74, p. 249-263.

Le Pichon, X., J. Francheteau, and J. Bonnin, 1976, Plate tectonics: New York, Elsevier, 241 p.

Lewy, Z., 1990, Transgressions, regressions and relative sea level changes on the Cretaceous shelf of Israel and adjacent countries, A critical evaluation of Cretaceous global sea level correlations: Paleoceanography, v. 5, no. 4, p. 619-637.

Loutit, T.S., J. Hardenbol, P.R. Vail, and G.R. Baum, 1988, Condensed sections: The key to age determinations and correlation of continental margin sequences, *in* C.K. Wilgus, B.S. Hastings, C.G.St.C. Kendall, H.W. Posamentier, C.A. Ross, and J.C. Van Wagoner, eds., Sea-level changes: An integrated approach: Society of Economic Paleontologists and Mineralogists Special Publication Number 42, p. 183-225.

Luger, P., and M. Groschke, 1989, Late Cretaceous ammonites from the Wadi Qena area in the Egyptian eastern desert: Paleontology, v. 32, p. 355-407.

Mikbel, S., and A.M. Abed, 1985. Discovery of large phosphate deposits in NW Jordan: Dirasat, v. 12, p. 125-136.

Nolte, D.G., 1991, Separation of a mixture of normal paraffins branched chain paraffins and cyclic paraffins. United States Patent 4,982,052.

Nolte, D.G., and E.L. Colling, Jr., 1989, Separation of oil into fractions of asphaltenes, resins, aromatics, and saturated hydrocarbons. United States Patent 4,865,741.

Notholt, A.J.G., 1985, Phosphorite resources in the Mediterranean (Tethyan) phosphogenic province: A progress report: Sciences Geologiques–Memoires, v. 77, p. 9-21.

Pasley, M.A., W.A. Gregory, and G.F. Hart, 1991, Organic matter variations in transgressive and regressive shales: Organic geochemistry, v. 17, p. 483-509.

Richardson, M., 1982, A depositional model for the Cretaceous Duwi (phosphate) Formation, south of Quseir, Red Sea coast, Egypt: Unpublished MS thesis, University of South Carolina, 395 p.

Robison, V.D., 1986, Organic geochemical characterization of the Late Cretaceous–Early Tertiary transgressive sequence found in the Duwi and Dakhla Formations, Egypt: Unpublished Ph.D. thesis, University of Oklahoma, 176 p.

Rohrback, B.G., 1983, Crude oil geochemistry of the Gulf of Suez, *in* M. Bjoroy, ed., Advances in Organic Geochemistry, 1981, Wiley, v. p. 39-48.

Rullkötter, J., B. Spiro, and A. Nissenbaum, 1985, Biological marker characteristics of oils and asphalts from carbonate source rocks in a rapidly subsiding graben, Dead Sea, Israel: Geochemica et Cosmochimica Acta, v. 49, p. 1357-1370.

Said, R., 1990a, Part 4: Discussion; Cretaceous paleogeographic maps, *in* R. Said, ed., The Geology of Egypt, Rotterdam, Balkema, 734 p.

Said, R., 1990b, Part 3: Geology of selected areas; Red Sea coastal plain, *in* R. Said, ed., The Geology of Egypt: Rotterdam, Balkema, 734 p.

Said, R., 1990c, Part 4: Discussion; Cenozoic, *in* R. Said, ed., The Geology of Egypt: Rotterdam, Balkema, 734 p.

Schlanger, S.O., and H.C. Jenkyns, 1976, Cretaceous oceanic anoxic events—causes and consequences: Geologica Mijnbouw, v. 55, p. 179-184.

Schrank, E., 1984a, Organic-walled microfossils and sedimentary facies in the Abu Tartur phosphates (Late Cretaceous, Egypt): Berliner geowissenschaftliche Abhandlungen (A), v. 50, p. 177-187.

Schrank, E., 1984b, Organic-geochemical and palynological studies of a Dakhla Shale profile (Late Cretaceous) in southeast Egypt: Part A: Succession of microfloras and depositional environment: Berliner geowissenschaftliche Abhandlungen (A), v. 50, p. 189-207.

Schroter, T., 1986, Die lithofazille Entwicklung der oberkretazischen Phosphatgesteine Agyptens—ein Beitrag zur Genese der Tethys–Phosphorite der Ostsahara: Berliner geowissenschaftliche Abhandlungen (A), v. 67, 105 p.

Sheldon, R.P., 1980, Episodicity of phosphate deposition and deep ocean circulation, A hypothesis, *in* Y.K. Bentor, ed., Marine Phosphorites: Special publication of the Society of Economic Paleontologists and Mineralogists, v. 29, 239-247.

Shirav, M., and D. Ginzburg, 1983, Geochemistry of Israeli oil shales—A review: Symposium on geochemistry and chemistry of oil shale.

Soliman, S.M., and K.M. Amer, 1972, Petrology of the phosphorite deposits, Qusseir area, Egypt: Arabian Mineralogy and Petrology Association Transactions, v. 27, no. 1, p. 17-48.

Soliman, M.A., M.E. Habib, and E.A. Ahmed, 1986, Sedimentologic and tectonic evolution of the Upper Cretaceous–Lower Tertiary succession at Wadi Qena, Egypt: Sedimentary Geology, v. 46, p. 111-133.

Spiro, B., D.H. Welte, J. Rullkötter, and R.G. Schaefer, 1983, Asphalts, oils, and bituminous rocks from the Dead Sea area—A geochemical correlation study: AAPG Bulletin, v. 67, no. 7, p. 1163-1175.

Tannenbaum, E., and Z. Aizenshtat, 1984, Formation of immature asphalt from organic-rich carbonate rock—II. Correlations of maturation indicators: Organic geochemistry, v. 6, p. 503-511.

Tannenbaum, E., and Z. Aizenshtat, 1985, Formation of immature asphalt from organic-rich rocks—I. Geochemical correlation: Organic geochemistry, v. 8, no. 2, p. 181-192.

Troger, U., 1984, The oil shale potential of Egypt: Berliner geowissenschaftliche Abhandlungen (A), v. 50, p. 375-380.

Van Wagoner, J.C., R.M. Mitchum, K.M. Campion, and V.D. Rahmanian, 1990, Siliciclastic sequence stratigraphy in well logs, cores, and outcrops: AAPG Methods in Exploration 7, 55 p.

Waples, D.W., 1982, Phosphate-rich sedimentary rocks: Significance for organic facies and petroleum exploration: Journal of Geochemical Exploration, v. 16, p. 135-160.

Weimer, R.J., 1988, Record of relative sea-level changes, Cretaceous of Western Interior, USA, *in* C.K. Wilgus, B.S. Hastings, C.G.St.C. Kendall, H.W. Posamentier, C.A. Ross, and J.C. Van Wagoner, eds., Sea-level changes: An integrated approach: Society of Economic Paleontologists and Mineralogists Special Publication Number 42, p. 285-288.

University of Chicago, 1990, Paleogeographic atlas project, A.M. Ziegler.

Chapter 8

Types and Thermal Maturity of Organic Matter Accumulated During Early Cretaceous Subsidence of the Exmouth Plateau, Northwest Australian Margin

Philip A. Meyers
Department of Geological Sciences
The University of Michigan
Ann Arbor, Michigan, USA

Lloyd R. Snowdon
Institute of Sedimentary and Petroleum Geology
Sedimentary and Marine Geoscience Branch
Geological Survey of Canada
Calgary, Alberta, Canada

ABSTRACT

Coring done on the Exmouth Plateau of the northwest Australian margin during Ocean Drilling Program Leg 122 encountered a sequence of organic-carbon-rich Lower Cretaceous siltstones and shales. The types of organic matter contained in these stratigraphic units have been assessed by Rock-Eval pyrolysis, extractable saturated hydrocarbon and biomarker analyses, pyrolysis GC/MS, and carbon isotope analysis. Rock-Eval, pyrolysis GC/MS, and carbon isotope results indicate predominance of continental plant organic matter throughout the sequence. Distributions of the saturated hydrocarbons and the biomarkers, however, reveal a progressively decreasing dominance of land-plant contribution in younger strata. Comparison of stratigraphic sequences, sediment provenances, and organic matter types shows that these undersea strata are thermally immature equivalents of the onshore Neocomian Barrow Formation and Aptian Muderong Shale and represent the distal portions of a subsiding wedge of fluviodeltaic sediments which accumulated on this passive margin during the Early Cretaceous.

INTRODUCTION

The stratigraphic sequences of the Exmouth Plateau, on the passive northwestern Australian continental margin, were investigated during Ocean Drilling Program (ODP) Legs 122 and 123 in 1988. The plateau is underlain by rifted and deeply subsided continental crust which is isolated from the Australian shelf by the Kangaroo Syncline (Exon et al., 1982; von Rad et al., 1989; Lorenzo et al., 1991).

Prior to rifting in the Jurassic, a series of shallow-water lagoonal and reefal sedimentary sequences were deposited (von Rad et al., 1990). This was followed by deeper-water clastic sedimentary sequences, totaling as much as 10 km in thickness, which accumulated as part of the fluviodeltaic Barrow Formation during subsidence. Strata from both the shallow-water and deeper environments typically contain 1 to 2 wt.% organic carbon, mostly derived from land plants, although occasional coaly deposits have as much as 40 wt.% organic carbon (Barber, 1982; Campbell et al., 1984; Cook et al., 1985; Snowdon and Meyers, 1992). Accumulation of terrigenous sediments became limited after tectonic isolation of the plateau from Australia in the Early Cretaceous. The predominantly biogenous hemipelagic sediments that have deposited since then contain very little organic matter (Snowdon and Meyers, 1992).

The deeper sections of the Barrow Formation have generated oil and gas and are the sources of commercial production in northwestern Australia. Noncommercial gas shows have been found by exploration drilling on the Exmouth Plateau (cf. Barber, 1982). Commercial quantities of hydrocarbons exist not far from the drill sites of Legs 122 and 123, both on the continental shelf and on Barrow Island (cf. Campbell et al., 1984).

ODP Sites 762 and 763 (Figure 1) were selected to investigate the postrift stratigraphic sequence which was deposited on the Exmouth Plateau as this passive margin subsided. Drilling at both sites was limited to the uppermost portions of the Barrow Delta Formation to avoid penetrating the possible petroleum reservoir caprocks indicated from earlier drilling at the Vinck and Eendracht exploration wells (Figure 1). The stratigraphic sequences at the two sites differ in thickness of the various strata, each recording different intervals better than or not as well as the other. Combining the two sections provides a composite Early Cretaceous history of terrigenous sedimentation on this fluviodeltaic feature. Continued subsidence during this time progressively changed the depositional environment to a more marine, deep-water, sediment-starved setting (Exon et al., 1982; Haq et al., 1990). As a contribution to the investigation of this stratigraphic sequence, the sources and thermal maturities of organic matter were characterized in the organic-carbon-rich Lower Cretaceous strata.

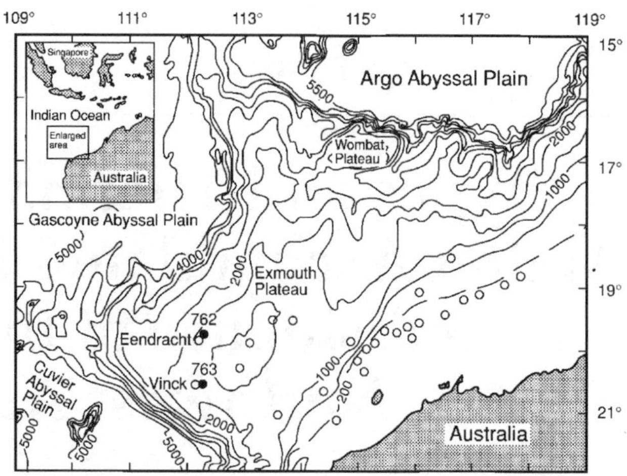

Figure 1. Locations of Sites 762 and 763 on the Exmouth Plateau, northwest Australian margin. Positions of exploratory wells are given as open circles; well logs from Vinck and Eendracht helped plan ODP drilling. Water depths are in meters.

SAMPLING AND ANALYSIS

Samples

Samples were routinely selected for determination of total organic carbon (TOC) and for Rock-Eval analysis as drilling of Sites 762 and 763 proceeded. A total of 131 samples was collected and analyzed over the 940 m of the sedimentary sequence drilled at Site 762; it was 268 samples over the 1064 m drilled at Site 763. Sample selection was based partly on subbottom depth and partly on stratigraphic boundaries visible in the recovered cores (Haq et al., 1990).

Five samples of Lower Cretaceous sediment remaining after pore waters had been squeezed were selected for more detailed study of organic matter. These samples were chosen because of their stratigraphic significance; relatively large amounts of sediment could be obtained for organic analyses, and additional information about the inorganic compositions of the sediments would ultimately become available from post-cruise studies. The five samples were from the stratigraphic equivalents of the onshore Aptian Muderong Formation (one sample), the Valanginian Barrow Group B (three samples), and the Berriasian Barrow Group C (one sample).

Analysis

Total organic carbon (TOC) concentrations were determined onboard the *JOIDES Resolution* as part of the Rock-Eval analysis of samples using a Girdel Rock-Eval II instrument equipped with a TOC module. Samples were pyrolyzed by heating at a rate of 25°C/min between 300°C to 600°C, yielding the amount of volatile hydrocarbons (S_1), the amount of thermogenic hydrocarbons (S_2), and the amount of CO_2 released during pyrolysis to 390°C (S_3). The TOC module combusts the residue of the temperature-programmed sample in air at 600°C and algorithmically determines the carbon content of this oxidation peak as well as the carbon in the pyrolysis peaks to give the TOC. These values provide the information necessary to calculate the hydrogen index (HI = 100 × S_2/TOC, or mg hydrocarbons/g organic carbon) and the oxygen index (OI = 100 × S_3/TOC, or mg CO_2/g organic carbon). The temperature of maximum hydrocarbon release during pyrolysis (T_{max}, °C) is also obtained and can be interpreted as a measure of organic matter thermal maturity (cf. Espitalié et al., 1977).

Carbon and oxygen isotope measurements were done in the Stable Isotope Laboratory at The University of Michigan. For carbonate analyses, powdered samples were roasted in vacuo at 200°C for 2 hr to remove volatile organic matter prior to treatment with anhydrous phosphoric acid to convert the carbonate minerals to CO_2. This treatment was performed in an automated carbonate extraction system (Kiel Device) coupled to the Finnigan MAT Model 251 mass spectrometer which measured the $^{18}O/^{16}O$ and $^{13}C/^{12}C$ ratios of the evolved CO_2. Organic carbon $^{13}C/^{12}C$ ratios of samples which had been treated with 3N HCl to destroy carbonate minerals were measured after reacting the samples with CuO_2 in evacuated and sealed quartz tubes for 3 hr at 800°C. The CO_2 produced by oxidation of the organic matter was analyzed with a Finnigan Delta S mass spectrometer. National Bureau of Standards carbon isotope standards were routinely used to calibrate the instruments. Results are reported relative to the Peedee belemnite standard.

Extractions of the five squeezed samples were done onboard the *JOIDES Resolution*. The freeze-dried samples were extracted with chloroform/methanol (80/20) by using ultrasound for 15 min. Solvents were evaporated from the extracted material, and the dried residue was redissolved in hexane for transfer to a silica gel chromatography column. The saturated hydrocarbon fraction was eluted from the column in hexane and analyzed later onshore by capillary gas chromatography.

Gas chromatographic analysis of the extractable hydrocarbons was done at the Geological Survey of Canada in Calgary using a Varian 3700 FID gas chromatograph fitted with a splitter injection port and a 30m DB-1 fused silica column. Combined gas chromatography-mass spectrometry (GC-MS) was done using a Kratos MS-80 mass spectrometer interfaced with a Carlo Erba gas chromatograph and a DS-90 data system. Column type and conditions were the same as for the gas chromatograph, and the GC-MS system was operated in both full scan and specific ion modes (m/z 191 and m/z 217).

Extractable biomarkers typically constitute a minor, albeit significant, fraction of the total organic matter. For this reason, it was decided to examine the geochemical composition of bulk organic matter in samples after the biomarker contents had been extracted and identified.

A pyrolysis-gas chromatography-mass spectrometry procedure was used to identify molecular fragments in the pyrolysate of the non-extractable kerogen matrix. Four of the extracted pore-water squeeze-cake samples were examined by the Pyran pyroanalytical system developed by Ruska Laboratories in Houston (e.g., Imbus et al., 1988). In this procedure, a thermal chromatograph-mass spectrometer pyrolyzes organic matter over a carefully controlled temperature range. The molecular fragments released are isolated and analyzed. Pyrolysis is achieved by heating from 330°C to 600°C at 30°C/min, and the quantity of the total pyrolysates produced over this temperature range is measured and cryogenically trapped. The total yield is then separated into molecular fragments by temperature-programmed capillary gas chromatography-mass spectrometry. The capillary column employed in this study was a 30m by 0.25 mm DB-5 fused silica column. It was operated over the temperature range of –35°C to 315°C at a heating rate of 5°C/min.

RESULTS AND DISCUSSION

The stratigraphic sequence encountered by drilling at both Site 762 and Site 763 consisted of an upper section composed of Quaternary, Tertiary, and Upper Cretaceous pelagic oozes and chalks, an early Aptian dark-colored hemipelagic calcareous claystone lithologically equivalent to the onshore Muderong Shale, and a series of Neocomian black-colored distal deltaic siltstones lithologically equivalent to the onshore Barrow Formation (Haq et al., 1990). This sequence represents stages in the progressive subsidence of the Exmouth Plateau. Organic carbon content is generally inversely related to the carbonate carbon content in these strata (Rullkötter et al., 1992; Snowdon and Meyers, 1992). The pelagic carbonate oozes and chalks generally contain less than 0.1 wt.% TOC, whereas the claystone and siltstone facies usually contain more than 0.5 wt.% TOC (Figure 2). The silt-

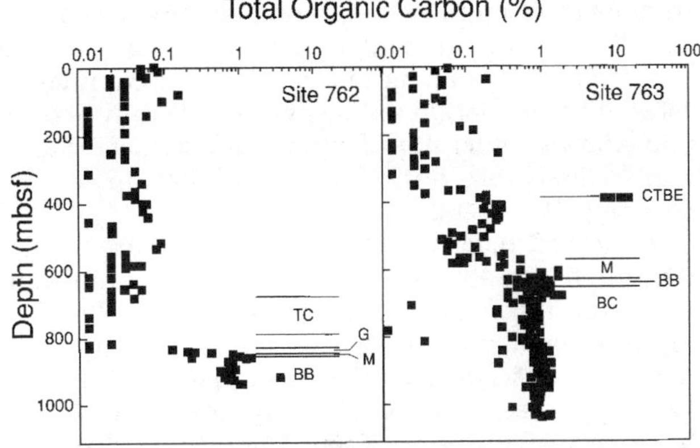

Figure 2. Total organic carbon vs. depth (in meters below seafloor, mbsf) for Sites 762 and 763 on the Exmouth Plateau. Rock units equivalent to strata identified on northeastern Australia are indicated: Toolonga Calcilutite (TC); Gearle Siltstone (G); Muderong Formation (M); Barrow "B" Formation (BB); Barrow "C" Formation (BC). An unnamed unit exists between 780 and 815 mbsf in the Site 762 sedimentary column. The Cenomanian–Turonian Boundary Event (CTBE) appears with its characteristic TOC enhancement at 380 mbsf in the Site 763 column.

stones equivalent to the Barrow Formation typically contain about 1 wt.% TOC (Figure 2), with a few samples having up to 1.5 wt.% at Site 762 and up to 2 wt.% TOC at Site 763.

Two thin layers (4 cm and 12 cm) of organic-carbon-rich black claystones were encountered at the Cenomanian–Turonian boundary at Site 763. Total organic carbon values as high as 26 wt.% in the thinner of these layers and 9 wt.% in the thicker one have been reported by Rullkötter et al., 1992. Rock-Eval pyrolysis results indicate that these layers contain Type II, marine organic matter which has been partially oxidized (Rullkötter et al., 1992; Snowdon and Meyers, 1992). A vitrinite reflectance of 0.2 wt.% indicates that the organic matter in the layer containing 9 wt.% organic carbon is thermally immature (Rullkötter et al., 1992). These two layers deviate from the overall concentration trends of increasing $CaCO_3$ and decreasing TOC as the Exmouth Plateau subsided through Cretaceous and Tertiary time and as water paleodepths increased. Cenomanian–Turonian boundary sediments were also recovered at Site 762; this boundary section, however, is carbonate-rich and contains ~0.01 wt.% TOC (Haq et al., 1990).

Organic carbon concentrations vary within the Neocomian siliciclastic fluviodeltaic unit corresponding to Barrow Group B in the upper part of the Barrow Formation (Figure 2). The down-core decrease and then increase in the TOC content may reflect variable oxidation of organic matter in these sections at Sites 762 and 763. Increased levels of oxidative degradation could be related to a coarser grain size and concomitant increase in the circulation of oxygen-bearing water through the sediment. Alternatively, the section having lower TOC concentrations might represent a period of diminished deltaic progradation and consequent diminished accumulation of organic matter. Higher accumulation rates of ocean margin sediments are generally accompanied by a higher flux of organic debris. Clastic dilution is therefore not a likely explanation for the observed TOC trend.

Rock-Eval data from Sites 762 and 763 are similar (Figure 3), yet a significant difference exists in samples from the two locations. Rock-Eval results from samples containing less than 0.25 wt.% TOC are omitted from these presentations to avoid possible artifacts reported in rocks which are low in TOC and high in $CaCO_3$ (e.g., Katz, 1983; Snowdon and Meyers, 1992). Samples from Site 762 with 0.5 wt.% TOC or more show a constant HI of between 50 to 100, typical of Type III, higher land plant organic matter. Samples from Site 763, however, show a small but consistent increase in HI with increasing TOC content, and TOC concentrations are somewhat higher in sediments from this site. The difference between the two sites, which had different sedimentation rates in the Neocomian, is consistent with differential preservation being an important control of the TOC concentration: where sedimentation rates are higher, degradation is less and preservation of hydrogen-rich

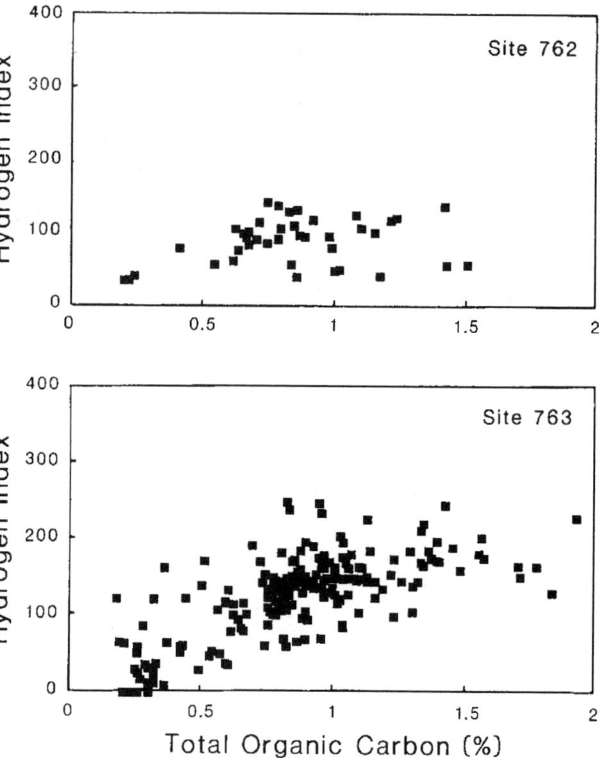

Figure 3. Rock-Eval hydrogen index vs. total organic carbon for Sites 762 and 763 on Exmouth Plateau. Organic matter is predominantly Type III, continentally derived material.

lipid material is improved. The essentially constant and low HI values of the samples from the stratigraphic unit at Site 762 equivalent to the onshore Barrow Formation probably resulted from the extensive degradation of organic matter which would accompany a low sedimentation rate. Lengthened exposure time within the bioturbated sediment–water interface allows preservation of only the most refractory portions of the deposited organic debris (cf. Emerson and Hedges, 1988).

The level of thermal maturity indicated by the Rock-Eval T_{max} parameter for the deltaic Barrow Formation siltstones is immature to marginally mature with respect to oil generation. T_{max} values for Site 762 are about 410°C to 425°C (equivalent to a vitrinite reflectance of 0.4 to 0.5% R_o), with no discernible trend over the relatively short depth interval from the Barrow Formation at this location (Haq et al., 1990). In contrast, T_{max} values increase at Site 763 over a 400-m interval from about 422°C to about 430°C (equivalent to about 0.45% to 0.60% R_o). This trend may represent a true increase in thermal maturation with depth, or alternatively it may reflect some change in the nature of the sedimentary organic matter. Thermal maturation modeling done at the Geological Survey of Canada using the Institut Français du Pétrole software package MATOIL indicates that the heat flow at this site would need to

Table 1. Descriptions and bulk organic matter characteristics of rock samples from selected Lower Cretaceous sequences in ODP holes 762C, 763B, and 763C, Exmouth Plateau.

Sample and Description	Depth (mbsf)	$CaCO_3$ (%)	TOC (wt.%)	$\delta^{13}C$ (‰) Inorg.	$\delta^{13}C$ (‰) Organic	Rock-Eval* HI	Rock-Eval* OI	Rock-Eval* T_{max}
122-763B-45X-3, 140–150 cm Aptian black silty claystone (Muderong Formation equivalent)	602.9	12.7	1.70	–0.9	–28.4	167	51	431ºC
122-763C-6R-1, 140–150 cm Valanginian dark gray silty claystone (Barrow Group B equivalent)	662.0	12.4	0.89	–8.0	–28.3	144	46	424ºC
122-762C-82X-3, 140–150 cm Valanginian dark gray silty claystone (Barrow Group B equivalent)	857.9	11.7	1.14	–9.4	–26.7	137	35	426ºC
122-762C-89X-4, 140–150 cm Valanginian dark gray silty claystone (Barrow Group B equivalent)	916.5	14.2	0.77** (0.15)	–19.1	–27.5	96** (19)	149** (53)	424ºC** (7)
122-763C-41R-4, 140–150 cm Berriasian dark gray clayey siltstone (Barrow Group C equivalent)	985.5	3.3** (0.9)	0.98* (0.09)	–0.4	n.d.	178** (22)	110** (23)	433ºC** (2)

*HI = hydrogen index (mg HC/g TOC); OI = oxygen index (mg CO_2/g TOC)
**not determined on this sample, mean value (s.d.) of samples within 10 m is substituted.
n.d. = not determined

have exceeded 120 mW/m^2 since the Neocomian in order to attain a vitrinite reflectance in excess of 0.5% R_o. Physical properties measurements (Haq et al., 1990) indicate a contemporary heat flow of about 75 mW/m^2 for the uppermost 150 m of Site 763, which is far too low to have achieved the thermal maturity indicated by the Rock-Eval T_{max} data. Instead of in situ thermal maturation, the increasing T_{max} values might record the presence of an increasingly larger contribution of thermally mature, recycled detrital organic matter in the progressively deeper deltaic siltstones. A larger proportion of land-derived detrital organic matter is consistent with sedimentological evidence that the older siltstones were deposited in shallower waters closer to the ancient shoreline (Haq et al., 1990) as the delta prograded and this passive margin subsided. If the terrigenous organic matter were derived from erosion of ancient sedimentary rocks, some fraction of thermally mature material could have been delivered to the Neocomian strata as they deposited.

Rock-Eval and $\delta^{13}C$ data in Table 1 indicate that most of the organic matter contained in these Lower Cretaceous clastic strata originated from continental vegetation. Rock-Eval oxygen indices are low, but most importantly the hydrogen indices are also low, generally ca. 150 mg hydrocarbons/g TOC. The organic carbon stable isotope ratios are uniformly typical of those of modern C3 land plants, ca. –27‰. Inorganic carbon isotope values, in contrast, have a surprisingly broad range, from essentially 0‰ to –19‰. The light values are unexpected, inasmuch as most carbonates have ratios close to modern marine carbonate values, about 0 to +2‰. The isotopically light carbonate values may result from oxidation of organic carbon and incorporation of the CO_2 so produced into carbonates. This type of organic matter recycling is quite possible on the Exmouth Plateau, where large volumes of thermogenic methane appear to have migrated through the Cretaceous sediments from deeper Jurassic sources (Snowdon and Meyers, 1992). A parallel example of methane oxidation and precipitation of isotopically light carbonate has also been postulated in Miocene hemipelagic sediments at ODP Sites 767 and 768 in the Celebes-Sulu Sea (von Breymann and Berner, 1990).

Chromatograms of the saturated hydrocarbon contents extracted from samples of the Lower Cretaceous strata show that long-chain C_{25}, C_{27}, and C_{29} n-alkanes diagnostic of the epicuticular waxes of higher land plants (e.g., Rieley et al., 1991) are dominant. Examples from Berriasian, Valanginian, and Aptian samples are shown in Figure 4 and are similar to chromatograms presented in Rullkötter et al. (1992). This dominance is particularly evident in the four samples from Exmouth Plateau sequences equivalent to parts of the Neocomian Barrow Formation in northwestern Australia (Table 2). In Sample 122-763B-

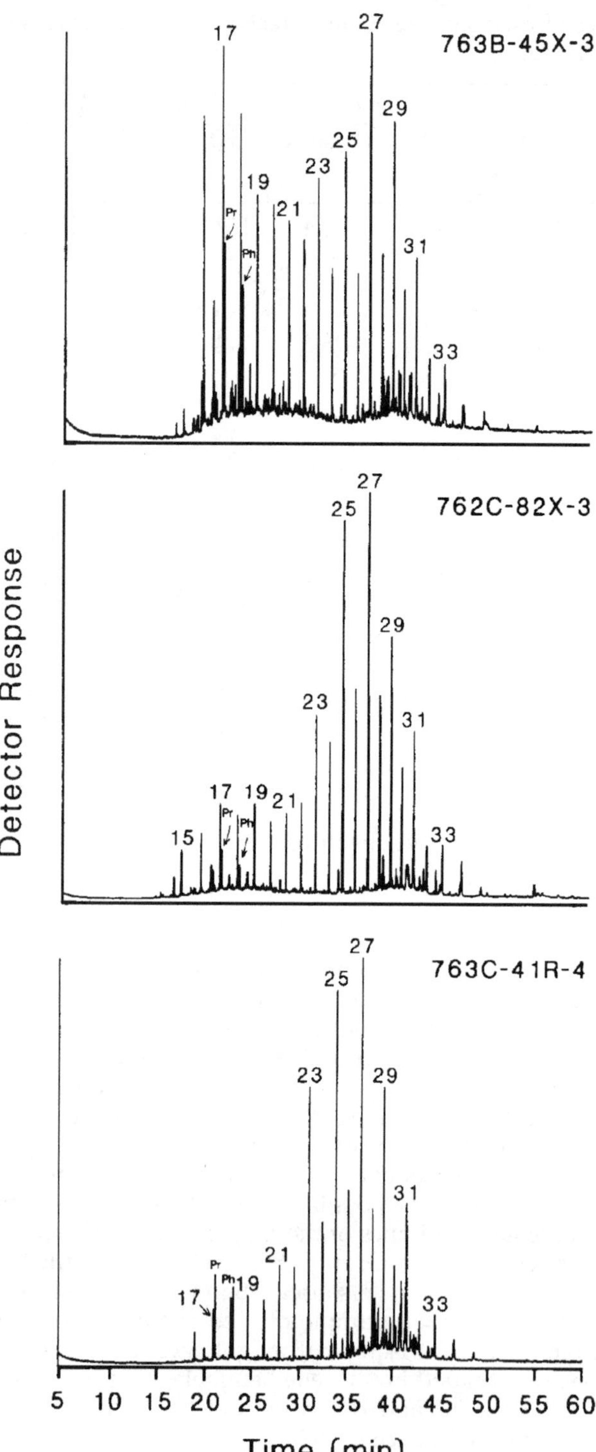

Figure 4. Chromatograms of extractable hydrocarbon fractions from representative Lower Cretaceous samples from Sites 762 and 763. Major n-alkanes are identified by their carbon numbers; the isoprenoid hydrocarbons pristane and phytane are identified by Pr and Ph, respectively. Long-chain n-alkanes representative of land plant contributions are dominant except in the chromatogram of Sample 122-763B-45X-3 from the offshore equivalent of the Aptian Muderong Formation on Australia.

45X-3, 140–150 cm, the contributions of marine lower molecular weight n-alkanes maximizing at n-C_{17} and terrigenous n-alkane biomarkers are equivalent. This Aptian claystone corresponds to the Muderong Formation onshore and was evidently deposited under deeper water conditions than the older siltstones which contain a larger proportion of continental lipid matter.

The n-alkane odd/even ratios calculated from n-C_{16} to n-C_{33} are high relative to oil source rocks (Table 2), although substantially lower than in land plant waxes (e.g., Rieley et al., 1991). The odd/even ratios agree with the low thermal maturity of bulk organic matter indicated by Rock-Eval T_{max} values, which range from 425°C to 435°C (Haq et al., 1990). The ratios also indicate changes in delivery of organic matter to this sequence of sediments. The values for the Neocomian samples are notably higher than for the Aptian sample, averaging 1.92 as opposed to 1.32. This difference cannot be caused by greater thermal maturity of the Aptian sediments than the older and more deeply buried Neocomian strata. Inspection of the chromatograms (Figure 4) shows that it is the land-derived long-chain n-alkanes that have a strong odd/even predominance. The shorter-chain marine hydrocarbons comprising about half of the distribution in the Aptian claystone lack a strong odd/even character; these may include bacterial components, which typically have low odd/even signatures, as well as algal hydrocarbons. The lower proportions of these lower molecular weight hydrocarbons in the Neocomian siltstones show that a change in the type of organic matter is responsible for the difference in odd/even n-alkane ratios.

The pristane/phytane ratios are >1 in all the samples, a common observation in clastic sedimentary rocks. This value also suggests that methanogenic bacteria have not been active in these sedimentary settings, inasmuch as such bacteria typically synthesize phytane precursor compounds rather than pristane precursors (Risatti et al., 1984). Rullkötter et al. (1992) note that bacterial hopanoid biomarkers are at very low concentrations in extracts of Site 763 sediments, which is consistent with little bacterial activity. Diminished microbial activity is typical of settings in which the organic matter is dominated by detrital land-derived material, rather than more easily metabolized algal material. The two isoprenoid compounds make smaller contributions to the total hydrocarbons than the C_{17} and C_{18} n-alkanes in all of the samples except Sample 122-763C-41R-4, 140–150 cm, equivalent to the Berriasian Group C siltstone of the Barrow Formation. This difference further suggests that marine contributions of extractable hydrocarbons are smaller in this stratigraphic unit than in the others.

Selected ratios and distributions of sterane and pentacyclic triterpane compounds are presented in Table 3. The general appearances of both the m/z 191 and m/z 217 total ion chromatograms (Figure 5), specifically the high abundance of $\beta\beta$ pentacyclic terpanes and the virtual absence of $\alpha\beta\beta$ steranes, reflect

Table 2. Descriptions and extractable hydrocarbon characteristics of rock samples from selected Lower Cretaceous sequences in ODP holes 762C, 763B, and 763C, Exmouth Plateau. Long/short ratio is the sum of the percent contributions of n-C_{17} + n-C_{19} + n-C_{21} divided by the sum of the contributions of n-C_{27} + n-C_{29} + n-C_{31}. Odd/even n-alkanes calculated over the range C_{17} to C_{30}. Pr = Pristane; Ph = phytane.

Sample and Description	Long/short	Odd/even	Pr/Ph	Pr/n-C_{17}	Ph/n-C_{18}
122-763B-45X-3, 140–150 cm Aptian black silty claystone (Muderong Formation equivalent)	1.07	1.32	1.43	0.48	0.36
122-763C-6R-1, 140–150 cm Valanginian dark gray silty claystone (Barrow Group B equivalent)	5.45	1.92	1.14	0.76	0.60
122-762C-82X-3, 140–150 cm Valanginian dark gray silty claystone (Barrow Group B equivalent)	3.17	1.63	1.44	0.76	0.36
122-762C-89X-4, 140–150 cm Valanginian dark gray silty claystone (Barrow Group B equivalent)	4.54	2.05	1.71	0.66	0.37
122-763C-41R-4, 140–150 cm Berriasian dark gray clayey siltstone (Barrow Group C equivalent)	3.77	2.08	1.21	1.65	1.17

Table 3. Sterane and triterpane compositions of extracts from Early Cretaceous rocks, ODP sites 762 and 763, Exmouth Plateau, expressed as percentages of αααR steranes and of αβ triterpanes.

Sample	Age	Sterane Percentage				Triterpane Percentage			Hopane Ratios	
		C_{27}	C_{28}	C_{29}	C_{30}	C_{29}	C_{30}	C_{31}	30ββ/αβ	31ββ/αβ
122-763B-45X-3, 140–150 cm (Muderong Formation equivalent)	Aptian	34	34	28	5	30	32	38	0.46	0.34
122-763C-6R-1, 140–150 cm (Barrow Group B equivalent)	Valanginian	36	12	44	7	32	34	34	0.72	0.75
122-762C-82X-3, 140–150 cm (Barrow Group B equivalent)	Valanginian	21	13	63	4	32	30	38	0.98	0.93
122-762C-89X-4, 140–150 cm (Barrow Group B equivalent)	Valanginian	28	11	56	5	21	40	39	1.11	1.31
122-763C-41R-4, 140–150 cm (Barrow Group C equivalent)	Berriasian	26	18	41	14	26	44	29	1.33	1.13

the low level of thermal maturity of these strata. The classic biomarker thermal ααα sterane ratios may not be particularly useful at this level of maturity because either the peaks are so small or the specific compounds of interest have not yet been generated. Similarly, 18α(H)-22,29,30-trisnorneohopane is either absent or present in very low concentrations in these extracts. This nonbiological hopanoid compound is formed during thermal maturation of organic matter; its absence indicates immaturity relative to onset of petroleum generation (e.g., Seifert and Moldowan, 1986).

A progressive increase in the ββ/αβ hopane ratio (Table 3) and the odd/even ratio (Table 2) as the age of the sediment layers increases is opposite to the trends expected if these parameters were being controlled by thermal maturity. ββ-hopanes are highly sensitive to thermal maturation and are generally

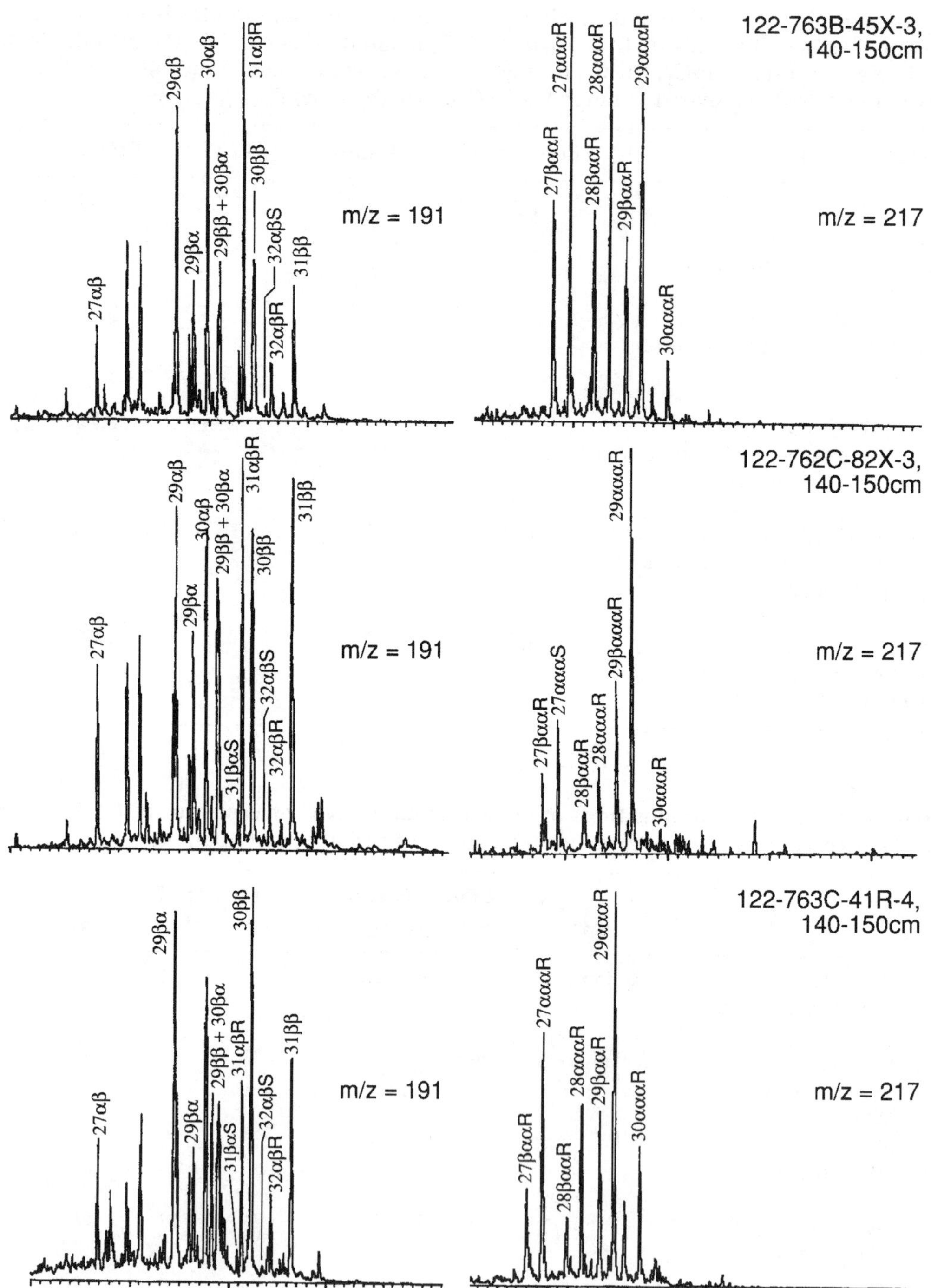

Figure 5. Total ion chromatograms of triterpanes (m/z 191) and steranes (m/z 217) extracted from representative Lower Cretaceous samples from Sites 762 and 763. Progressively larger contributions of C_{27} marine steranes appear as strata ages decrease from Berriasian to Valanginian to Aptian (upwards).

converted to the αβ form in the earliest stages of oil generation (e.g., Seifert and Moldowan, 1986). The existence of the anomalous trends of increasing values of the ββ/αβ and odd/even ratios with depth indicates progressively larger contributions of thermally immature, continental organic matter in the deeper strata of this distal deltaic sequence. The small increase observed in Rock-Eval T_{max} (Haq et al., 1990) cannot be related to in situ thermal maturation effects, nor can it be caused by the presence of a larger proportion of recycled, eroded organic matter. It must be related instead to the larger contribution of continental organic matter in the deeper sections of this deltaic sedimentary sequence.

The relative abundance of C_{27} relative to C_{29} steranes (m/z 217, Figure 5) confirms the change in organic matter sources. A dominance of land-derived organic matter exists in the extracts from the Neocomian strata, whereas subequal contributions of marine and continental components are found in the Aptian facies. This pattern of source difference contained in the sterane compositions agrees with that of the extractable n-alkanes (Figure 4, Table 2).

Total yields from temperature-programmed pyrolysis of the organic matter contents of samples of the Aptian and Valanginian strata are shown in Figure 6. The organic matter in these Early Cretaceous rocks pyrolyzes at a fairly narrow and high temperature range (ca. 450°C). This pattern presumably results from condensation of low-molecular-weight organic matter into high-molecular-weight kerogen in the ancient sediments. Some lower molecular weight fragments appear in the pyrogram of the Aptian sample at a pyrolysis time of 15 min, corresponding to a temperature of 300°C. As noted from the extractable biomarker distributions shown in Figure 4 and summarized in Table 2, this sample differs from the Neocomian samples in containing a larger proportion of marine components. Differences evidently exist in bulk kerogen as well, although these differences are not evident from $\delta^{13}C$ values (Table 1). The small fraction of marine organic matter may be more susceptible to pyrolysis than the predominant continental organic matter.

The results of capillary gas chromatography-mass spectrometry of the pyrolysates show that a large proportion of cyclic molecules are produced from the kerogen of these Lower Cretaceous strata. Continental plant matter, rich in cellulose and lignin, would yield such fragments on pyrolysis. These data are consistent with the biomarker, Rock-Eval, and isotope data in indicating that most of the organic matter in the strata of this paleodelta is derived from land plants.

Comparisons of two types of cyclic molecular fragments were done to investigate possible differences in the kerogen of the different stratigraphic units. One comparison was of aromatic hydrocarbons to their organosulfur analogs, and the other was of polyaromatic hydrocarbons of the same molecular weights but different structures. The eight molecular fragments selected from the Pyran data are illustrated in Figure 7, and the ratios derived from the percent contribution each molecule made to the total ion current are given in Table 4.

Several changes with stratigraphic depth appear in the ratios of the cyclic hydrocarbons (Table 4). The ratios of benzene/thiophene and naphthalene/benzothiophene increase as sediment depth increases. These changes could possibly result from diagenetic changes in kerogen compositions with increasing time or thermal maturation, but aromatic hydrocarbons and heterocompounds are generally resistant to alterations at low temperatures and all measures indicate that thermal maturity is low in these strata. The

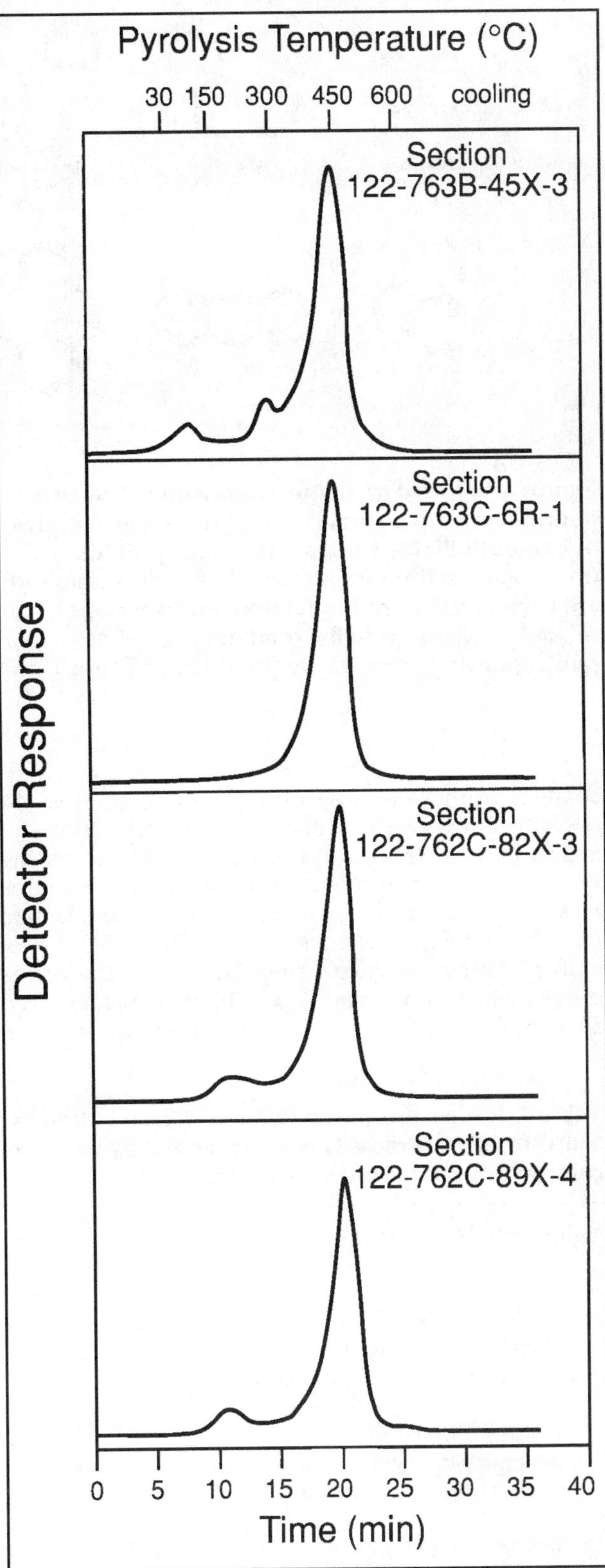

Figure 6. Temperature-programmed Pyran pyrolysis of organic matter from Lower Cretaceous samples from Sites 762 and 763. Samples were heated from 30°C to 600°C at 30°C/min. Volatiles were measured with a flame ionization detector; the detector response is shown as heating progressed.

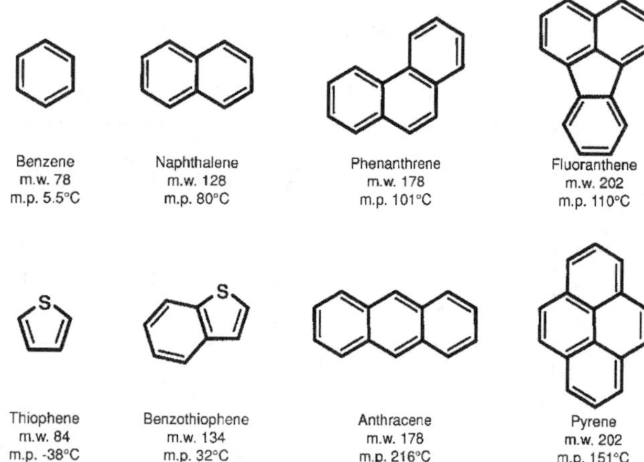

Figure 7. Selected aromatic and organosulfur compounds generated from Pyran pyrolysis of samples of Exmouth Plateau strata. Molecular weights (m.w.) and melting points (m.p.) of each compound are shown. Ratios of the relative contributions of related compounds to the total ion current of pyrolysate mass spectra are provided in Table 4.

changes in these two pyrolysate ratios may instead result from progressively smaller proportions of organosulfur compounds in older strata. As the Barrow delta prograded, older sections may have been deposited in shallower settings with brackish to freshwater conditions, thus limiting the availability of sulfur. Younger sections were laid down under the deeper water, more marine conditions which evolved as the Exmouth Plateau continued to subside. Sinninghe-Damste et al. (1989a,b) have found that incorporation of sulfur into organic matter occurs during the early stages of diagenesis and that the salinity of the waters of the depositional environment is important in affecting the amounts of organosulfur compounds that will form. The progressive subsidence of this ocean margin sequence thus may have controlled the sulfur content of its organic matter.

The second change is general increase in the ratios of phenanthrene/anthracene and fluoranthene/pyrene with greater age. These ratios represent the compound with lower melting point over the analog with a higher melting point. The generally higher values of these two ratios in older strata from the Exmouth Plateau may suggest that compounds like phenanthrene and fluoranthene are more sensitive to temperature–time factors than compounds like anthracene and pyrene. More likely, the greater proportions of recycled continental organic matter in the deeper parts of the sedimentary sequence control these ratios.

SUMMARY AND CONCLUSIONS

Our organic geochemical data consist of a large number of Rock-Eval and organic carbon analyses and a small number of isotopic and extractable hydrocarbon and biomarker analyses. Characterization of lithologic sequences with a limited number of samples can be unreliable. We nonetheless observe consistent patterns in organic carbon concentrations and in Rock-Eval results that give us confidence in our interpretations of the limited amount of extract data. Furthermore, samples from within a given lithologic unit yield similar hydrocarbon and biomarker distributions which are consistent with petrographic

Table 4. Ratios of selected Pyran pyrolysis products from Exmouth Plateau samples. Pyrolysis was done from 330°C to 600°C. Products were separated by gas chromatography from –35°C to 315°C on a 30m DB-5 capillary column prior to mass spectrometry.

Sample and Description	Depth (mbsf)	Benzene / Thiophene	Naphthalene / Benzothiophene	Phenanthrene / Anthracene	Fluoranthene / Pyrene
122-763B-45X-3, 140–150 cm Aptian black silty claystone (Muderong Formation equivalent)	602.9	1.86	1.67	2.00	0.89
122-763C-6R-1, 140–150 cm Valanginian dark gray silty claystone (Barrow Group B equivalent)	662.0	2.68	2.86	1.96	1.00
122-762C-82X-3, 140–150 cm Valanginian dark gray silty claystone (Barrow Group B equivalent)	857.9	2.81	2.90	3.71	1.53
122-762C-89X-4, 140–150 cm Valanginian dark gray silty claystone (Barrow Group B equivalent)	926.5	6.07	8.79	4.33	1.44

maceral analyses of Rullkötter et al. (1992), further bolstering our conclusions as to organic matter sources and maturity in this stratigraphic sequence.

The stratigraphic sequence at ODP Sites 762 and 763 on the Exmouth Plateau represents progressive stages in the termination of deltaic sedimentation during the Early Cretaceous as this passive margin continued to subside and eventually became sediment-starved. Strata encountered in Sites 762 and 763 are equivalent to units present in the onshore Barrow basin of northwestern Australia. Organic matter in Lower Cretaceous units is dominated by type III, continental organic matter. The concentration of sedimentary organic carbon decreased as the depositional environment subsided and the contribution of continental organic matter diminished. Marine productivity evidently provided little organic matter to sediments on this margin. The Aptian claystone unit lithologically equivalent to the Muderong Formation nonetheless contains material somewhat richer in marine organic matter than is found in the Neocomian siltstones, which are equivalent to the Barrow Formation. The consequent hydrogen enrichment may confer limited potential to generate and to expel liquid as well as gaseous hydrocarbons in portions of this unit. This potential is constrained by the fact that organic matter in the Exmouth Plateau Lower Cretaceous strata has never been deeply buried and is therefore immature to marginally mature with respect to peak oil generation.

ACKNOWLEDGMENTS

We are grateful for the experience (which we actually enjoyed!) of being at sea for nine weeks during Ocean Drilling Program Leg 122. Comments from J. Palacas and an anonymous reviewer helped improve this contribution. We thank Sneh Achal, Marg Northcott, and Ron Fanjoy of the geochemistry research laboratory of the Geological Survey of Canada, Calgary, for their analytical assistance, and Dale Austin of the University of Michigan Department of Geological Sciences for his graphic services. Part of this research was supported by a Joint Oceanographic Institutes grant of National Science Foundation funds to Philip A. Meyers.

REFERENCES CITED

Barber, P.M., 1982, Palaeotectonic evolution and hydrocarbon genesis of the central Exmouth Plateau: Australian Petroleum Exploration Association Journal: v. 22, p. 131-144.

Campbell, I.R., A.M. Tait, and R.F. Reiser, 1984, Barrow Island oilfield, revisited: Australian Petroleum Exploration Association Journal, v. 24, p. 289-298.

Cook, A.C., M. Smyth, and R.G. Vos, 1985, Source potential of Upper Triassic fluvio-deltaic systems of the Exmouth Plateau: Australian Petroleum Exploration Association Journal, v. 25, p. 204-215.

Emerson, S., and J.I. Hedges, 1988, Processes controlling the organic carbon content of open ocean sediments: Paleoceanography, v. 3, p. 621-634.

Espitalié, J., J.L. Laporte, M. Madec, F. Marquis, P. Leplat, J. Paulet, and A. Boutefeu, 1977, Méthode rapide de caractérisation des roches mères, de leur potential pétrolier et de leur degré d'évolution: Revue de l'Institut Français du Pétrole, v. 32, p. 23-42.

Exon, N.F., U. von Rad, and U. von Stackelberg, 1982, The geological development of the passive margins of the Exmouth Plateau off northwest Australia: Marine Geology, v. 47, p. 131-152.

Haq, B.U., U. von Rad, S. O'Connell, and Shipboard Scientific Party, 1990, Proceedings of the Ocean Drilling Program, Initial Reports, v. 122: College Station, Ocean Drilling Program.

Imbus, S.W., M.H. Engel, R.D. Elmore, and J.E. Zumberge, 1988, The origin, distribution and hydrocarbon generation potential of organic-rich facies in the Nonesuch Formation, Central North American Rift System: A regional study: Organic Geochemistry, v. 13, p. 207-219.

Katz, B.J., 1983, Limitations of "Rock-Eval" pyrolysis for typing organic matter: Organic Geochemistry, v. 4, p. 195-199.

Lorenzo, J.M., J.C Mutter, R.L. Larson, and Northwest Australia Study Group, 1991, Development of the continent–ocean transform boundary of the southern Exmouth Plateau: Geology, v. 19, p. 843-846.

Rieley, G., R.J. Collier, D.M. Jones, and G. Eglinton, 1991, The biogeochemistry of Ellesmere Lake, U.K.—I: source correlation of leaf wax inputs to the sedimentary record: Organic Geochemistry, v. 17, p. 901-912.

Risatti, J.B., S.J. Rowland, D.A. Yon, and J.R. Maxwell, 1984, Stereochemical studies of acyclic isoprenoids—XII. Lipids of methanogenic bacteria and possible contributions to sediments: Organic Geochemistry, v. 6, p. 93-104.

Rullkötter, J., R. Littke, U. Disko, B. Horsfield, and J. Thurow, 1992, Petrography and geochemistry of organic matter in Triassic and Cretaceous deep-sea sediments from the Wombat and Exmouth Plateaus and nearby abyssal plains off northwest Australia: *in* U. von Rad, B.U. Haq, S. O'Connell, and Shipboard Scientific Party, eds., Proceedings of the Ocean Drilling Program, Scientific Results: College Station, Ocean Drilling Program, v. 122, p. 317-333.

Seifert, W.K., and J.M. Moldowan, 1986. Use of biological markers in petroleum exploration: *in* R.B. Johns, ed., Biological Markers in the Sedimentary Record: Amsterdam, Elsevier, p. 261-290.

Sinninghe-Damste, J.S., W.I.C. Rijpstra, A.C. Kock-van Dalen, J.W. de Leeuw, and P.A. Schenck, 1989a, Quenching of labile functionalized lipids by inorganic sulphur species: Evidence for the formation of sedimentary organic sulphur compounds at the early stages of diagenesis: Geochimica et Cosmochimica Acta, v. 53, p. 1343-1355.

Sinninghe-Damste, J.S., W.I.C. Rijpstra, J.W. de Leeuw, and P.A. Schenck, 1989b, The occurrence and identification of series of organic sulphur compounds in oils and sediment extracts: II. Their presence in samples from hypersaline and non-hypersaline palaeoenvironments and possible application as source, palaeoenvironmental and maturity parameters: Geochimica et Cosmochimica Acta, v. 53, p. 1323-1341.

Snowdon, L.R., and P.A. Meyers, 1992, Source and maturity of organic matter in sediments and rocks from Sites 759, 760, 761, and 764 (Wombat Plateau) and Sites 762 and 763 (Exmouth Plateau): *in* U. von Rad, B.U. Haq, S. O'Connell, and Shipboard Scientific Party, eds., Proceedings of the Ocean Drilling Program, Scientific Results: College Station, Ocean Drilling Program, v. 122, p. 309-315.

von Breymann, M., and U. Berner, 1990, Diagenetic dolomite formation in the Sulu and Celebes seas from continentally derived turbidite deposits rich in organic matter: EOS, Transactions American Geophysical Union, v. 71, p. 1392 (abstract).

von Rad, U., J. Thurow, B.U. Haq, F. Gradstein, and J. Ludden, 1989, Triassic to Cenozoic evolution of the NW Australian continental margin and the birth of the Indian Ocean (preliminary results of ODP Legs 122 and 123): Geologische Rundshau, v. 78, p. 1189-1210.

von Rad, U., M. Schott, N.F. Exon, J. Mutterlose, P.G. Quilty, and J.W. Thurow, 1990, Mesozoic sedimentary and volcanic rocks dredged from the northern Exmouth Plateau: petrography and microfacies: Journal of Australian Geology and Geophysics, v. 11, p. 449-472.

Chapter 9

Sea Level Changes, Anoxic Conditions, Organic Matter Enrichment, and Petroleum Source Rock Potential of the Cretaceous Sequences of the Cauvery Basin, India

Kuldeep Chandra
Institute of Management Development
Oil and Natural Gas Commission
Dehra Dun, India

D. S. N. Raju and P. K. Mishra
Paleontology Laboratory
Oil and Natural Gas Commission
Dehra Dun, India

ABSTRACT

The Cauvery basin is a pericratonic rift basin along the east coast of India. A thick succession of sedimentary rocks of Early Cretaceous to Holocene age are known in the basin. The Cretaceous sedimentary sequences are the dominant habitat for petroleum in the basin.

The relationships among sea level changes, anoxic conditions, organic matter richness, and petroleum source rock potential have been investigated, especially for the Cretaceous sediments of the basin. Foraminiferal data have been used to evaluate sea level changes. Anoxic conditions have been assessed based on (1) localized occurrence of agglutinated foraminifera, (2) presence of pyritized tests, and (3) absence of benthonic foraminifera and presence of planktonic foraminifera or very high planktonic–benthonic ratios. Organic contents (TOC), hydrogen index (HI), and oxygen index (OI) are utilized to evaluate source potential. The results of thermal maturation studies based on T_{max} and vitrinite reflectance (R_o) are included to qualify "immature" (T_{max} less than 430°C and R_o less than 0.5%) and "mature" (T_{max} of 430–465°C and R_o of 0.5 to 1.3%) source rock sequences.

Five major cycles of sea level changes are described for the Cauvery basin. In each cycle usually two or more microcycles of sea level changes are observed. Anoxic conditions are found to be a characteristic usually of sediments deposited in a paleobathymetric regime of >80 m. The observed anoxic conditions do not coincide necessarily with the Global Anoxic Events of Jenkyns (1980) at any of the studied locales in the basin.

Organic matter enrichment and petroleum source rock development are observed in each transgressive event. The sediments associated with basal transgressions, having no distinct foraminiferal evidence of anoxic conditions, show relatively better organic matter enrichment compared with the post-basal transgressive phases having foraminiferal evidence of anoxic conditions. Peaking of organic matter enrichment often is observed to be coincident with the peaking of the transgression. At any given locale in the basin, the hydrogen richness of organic matter, as revealed by hydrogen indices, is generally higher (HI=1.5 to 2 times higher) for the sediments of the very first transgressive event from pre-Albian to Santonian ages.

Relatively better organic matter richness coupled with hydrogen enrichment in the early transgressive events is due to higher organic matter flux in the paleobathymetric regimes of 50 to 80 m, as well as better preservation of organic matter. The decrease in hydrogen richness in terms of hydrogen indices in most of the younger transgressive events associated with anoxic conditions at each studied locale is most likely due to lesser organic matter flux in the paleobathymetric regime of >80 m, preburial alteration, and early diagenetic transformations.

The anoxic conditions and organic matter enrichment in any single transgressive event within any major chronostratigraphic sequence are not observed to be synchronous in different structural locales within sub-depressions. Variations in paleobathymetric regimes are considered the likely factors to have controlled the level of organic enrichment. The distribution and volume of shallow marine transgressive pre-Albian–Santonian sediments may be of use in prioritizing generative depocenters in the basin.

INTRODUCTION

The Cauvery basin is situated on the eastern flank of peninsular India, in the state of Tamil Nadu, along the Palk Strait and Coromandal Coast (Figure 1a). The basin came into existence in Late Jurassic time during the breakup of eastern Gondwana (Rangaraju et al., 1991). It is a pericratonic rift basin having horsts and grabens aligned northeast–southwest (Figure 1b) in consonance with Archean basement morphology (Prabhaker and Zutshi, 1993).

The sedimentary fill in the subsurface consists dominantly of clastics with a succession of thick limestones at various locales both in the Cretaceous and the Tertiary (Sastri et al., 1973; Kumar, 1983). In the deeper parts of the basin, maximum thickness of the sedimentary column is inferred to be more than 6000 m (Rangaraju et al., 1991). The biostratigraphy of the basin has been described by a number of workers in recent times (Banarji and Radha, 1970; Raju, 1970). The synrift Lower Cretaceous sediments are nonmarine. The oldest marine sediments known in the Cauvery basin are of Barremian–Aptian age. The post-rift Cretaceous sediments are marine sequences deposited under varying paleoclimatological conditions across various paleolatitudes during the northward journey of India. The Cretaceous sediments form the dominant habitat of petroleum.

Govindan (1982, 1991), following Jenkyns (1980) and Schlanger and Jenkyns (1976), came to the conclusion that there are two to three distinct stratigraphic levels during which oxygen-depleted water should have transgressed onto the shelf and coincided with the Global Oceanic Anoxic Events (OAE 1, 2, 3). Such a transgression should result in preservation of organic matter and deposition of good regional source rocks. Over the years a large geochemical data base on organic matter content, type, and matu-

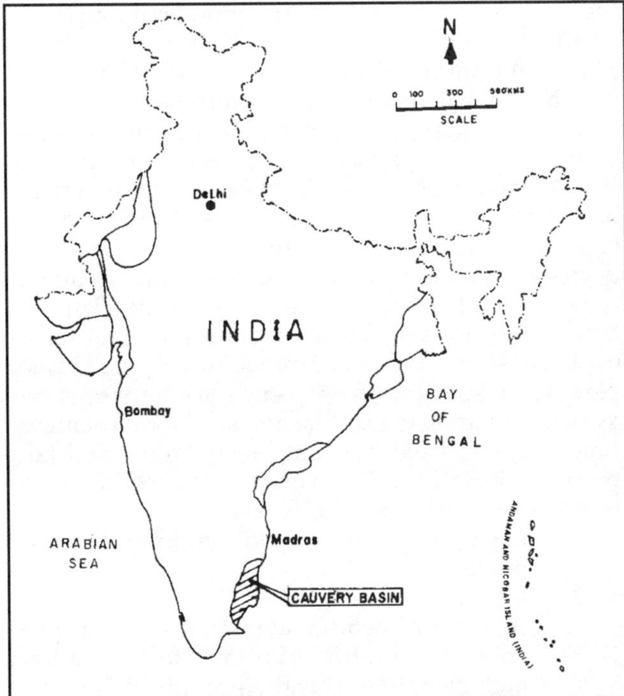

Figure 1. Maps of Cauvery basin. A—Location map.

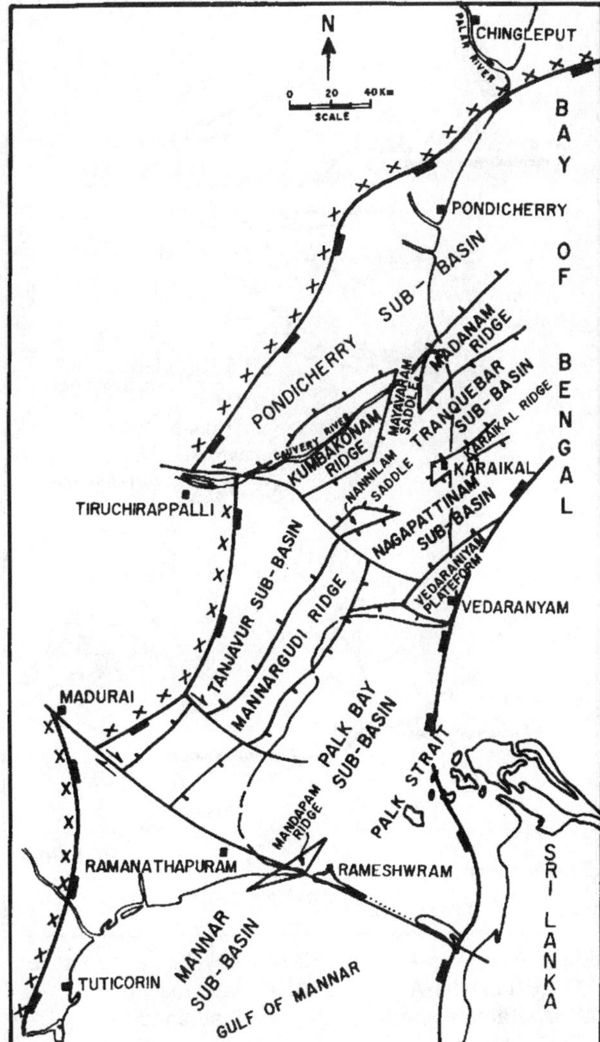

Figure 1 (continued). B—Tectonic map (after Prabhakar and Zutshi, in press).

rity has been built up for the basin. Some of the major results have been published by Chandra et al. (1991), Philip and Neeraja (1991), and Thomas et al. (1991). Thomas et al. (1991) concluded that in the Cauvery basin, the major source rocks are in the Cretaceous sequences. The oil and gas in the Cauvery basin belong to different genetic species within one single depression as well as in different subbasins, implying multiplicity of source rocks. Philip and Neeraja (1991), in their evaluation of source rock potential of a part of the Cauvery basin, came to the conclusion that the main hydrocarbon-generating centers are located around Tirukadiyur in the Tranquebar depression, Orathanadu in the Tanjore depression, and Thevur in the Nagapattinam depression. Chandra et al. (1991) concluded that the development of very good source rocks with type III kerogens was favored during the global anoxic events associated with the pre-Albian–Cenomanian major marine transgression with good source rocks developing during Coniacian transgressive sedimentation.

Parallel to the geochemical studies, micropaleontological data were generated both on the outcrops and subsurface strata. Knowledge on the distribution of ancient foraminifera has paved the way for the zonation, correlation, and calibration of the numerical geochronometric scale for reconstruction of relative sea level changes and the periods of anoxic conditions in the Cauvery basin. Raju and Ravindran (1990) recognized five major cycles of sea level changes from the exposed sections of the Ariyalur-Pondicherry depression. Since then much additional data have been compiled and a revised sea level curve has been prepared by Raju et al. (1992).

The present work aims to (1) recognize sea level changes in a few key exploratory wells in the Cauvery basin, (2) identify anoxic conditions from foraminiferal criteria, (3) relate organic matter enrichment and source potential with transgressive events and anoxic conditions, and (4) identify generalized trends for development of good source rocks in the stratigraphic framework of the basin.

FORAMINIFERAL AND GEOCHEMICAL STUDIES

The key exploratory structural locales included in the present study are (1) Nannilam, (2) Adiyakamangalam, (3) Tirukallar, (4) Pondicherry offshore PY-3, (5) Palk Bay PH-11, (6) Pappanaccheri, (7) Kamalapuram, (8) Orathanadu, and (9) Bhuvanagiri. Their locations are shown in Figure 2.

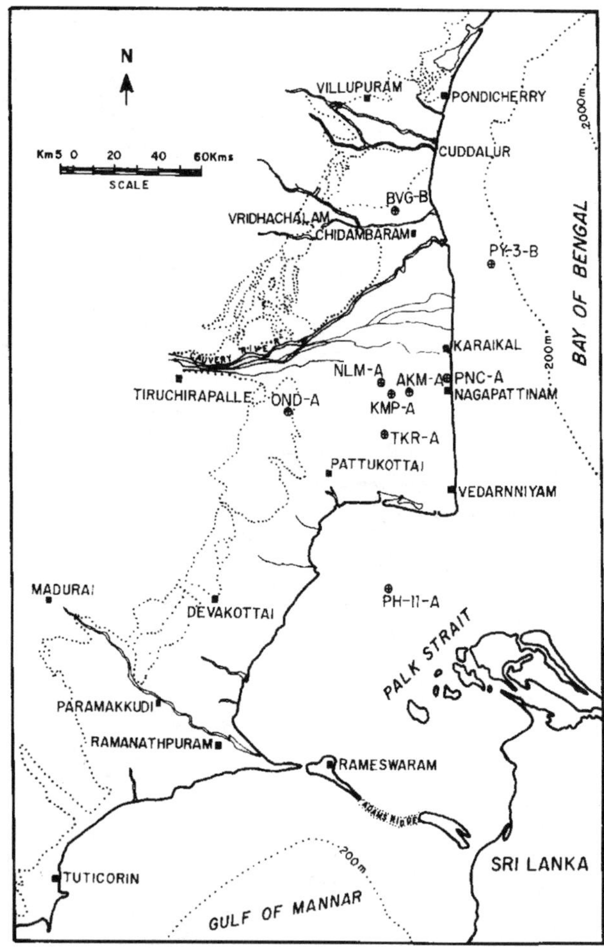

NLM-A=Nannilam-A, AKM-A=Adiyakamangalam-A, TKR-A=Tirukallar-A, PNC-A=Pappanaccheri-A, KMP-A=Kamalapuram-A, OND-A=Orathanadu-A, BVG-B=Bhuvanagiri-B

Figure 2. Location map of the studied wells in the Cauvery basin.

Standard paleontological methods were used to process samples and to sort and identify foraminifera. The sample control and results for the selected key wells from the different exploratory structural locales are shown in Figures 3 to 11.

The paleobathymetries inferred from the foraminiferal assemblages have been quantified and their range is shown against the sample studied. The paleobathymetric curve is the result of joining the midpoints of the individual paleobathymetric lines.

Identification of sea level changes is based on work of Raju et al. (1992). Table 1 summarizes the cycles of sea level changes in the Cauvery basin.

The foraminiferal criteria employed to infer anoxic conditions are based on an earlier approach of paleontologists of Robertson Research (1987). This approach is qualitative and cannot independently distinguish suboxic conditions having dissolved oxygen concentration in an aquatic regime >0.5‰ to 8‰, as per Demaison et al. (1983), and anoxic conditions having dissolved oxygen concentration in an aquatic regime of <0.1‰ to nil. The foraminiferal criteria are (i) absence of benthonic foraminifera but presence of planktonic foraminifera, or very high planktonic/benthonic ratios, (ii) localized occurrence of *Chilostomella* spp. or *Globobulimina* spp., (iii) localized occurrence of "thick uvigerines," (iv) localized occurrence of bolivines, (v) localized occurrence of agglutinated forms, (vi) pyritized tests, (vii) tendency toward dwarfed specimens, (viii) tendency to decrease ornament and to increase size and frequency of pores, (ix) localized increase of elongate/flattened, tapered flaring and flattened planispiral test forms, (x) local decrease in occurrence of angular, keeled planktonic species, (xi) presence of pteropods, (xii) associated laminated sediments, i.e., lack of bioturbation, (xiii) localized gypsum precipitation, and (xiv) presence of otoliths. The main criteria used in the present work, in order of priority, are:

1. Localized occurrence of agglutinated foraminifera.
2. Pyritized tests.
3. Absence of benthonic foraminifera but presence of planktonic foraminifera, or a very high planktonic/benthonic ratio.

Among the 14 criteria mentioned above, criteria (ii), (iii), (iv), (xi), and (xiv) are inapplicable due to absence or very rare presence of characteristic fauna. The criteria (vii), (viii), (ix), and (xiii) could not be utilized due to inadequacy of the desired level of observational data. The criteria (x) and (xii) were considered whenever possible.

The geochemical data analyzed in this work were taken from unpublished petroleum source rock evaluation reports of the Oil and Natural Gas Commission (ONGC) and incorporated in the earlier work of Philip and Neeraja (1991), Thomas et al. (1991), and Chandra et al. (1991). The geochemical data are also depicted in Figures 3 to 11. The qualitative results on organic matter maturation included in Figures 3 to 11 are also based on the recent study of Balan et al. (1991). Sedimentary sequences having Rock-Eval T_{max} less than 430°C are qualified as "immature." Sediments having vitrinite reflectance (R_o) ranging from 0.5 to 1.3% and Rock-Eval T_{max} ranging from 430 to 465°C are qualified as "mature" and within the oil window.

Structure

Nannilam Structure

The Nannilam structure is situated on the "saddle" at the juncture of the Nagapattinam depression and the Tanjore–Tranquebar depression and between the Karaikal and the Manargudi ridges (Figure 1b). The oldest sediments in NLM-A well (Figures 2, 3) are of Albian age. The Cretaceous sediments also include younger sediments of Cenomanian to Santonian and Campanian to Maastrichtian ages. The sediments are exclusively clastic, comprising sand and shale alternations in the Albian to the middle part of the Santonian, and silt and sand up to the Maastrichtian.

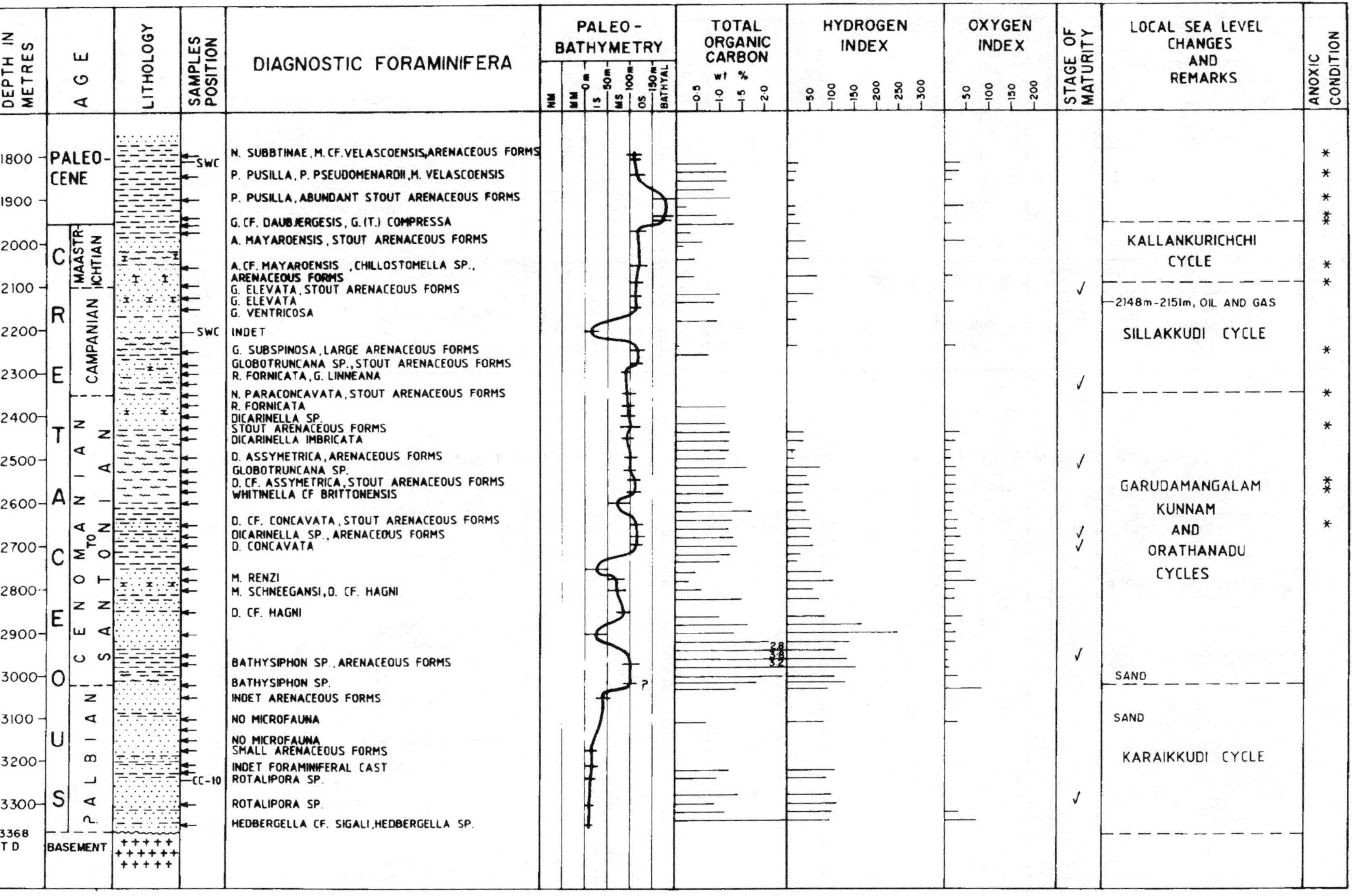

Figure 3. Stratigraphic succession, depositional environments, and geochemical logs in the Cretaceous–Paleocene of well NLM-A.

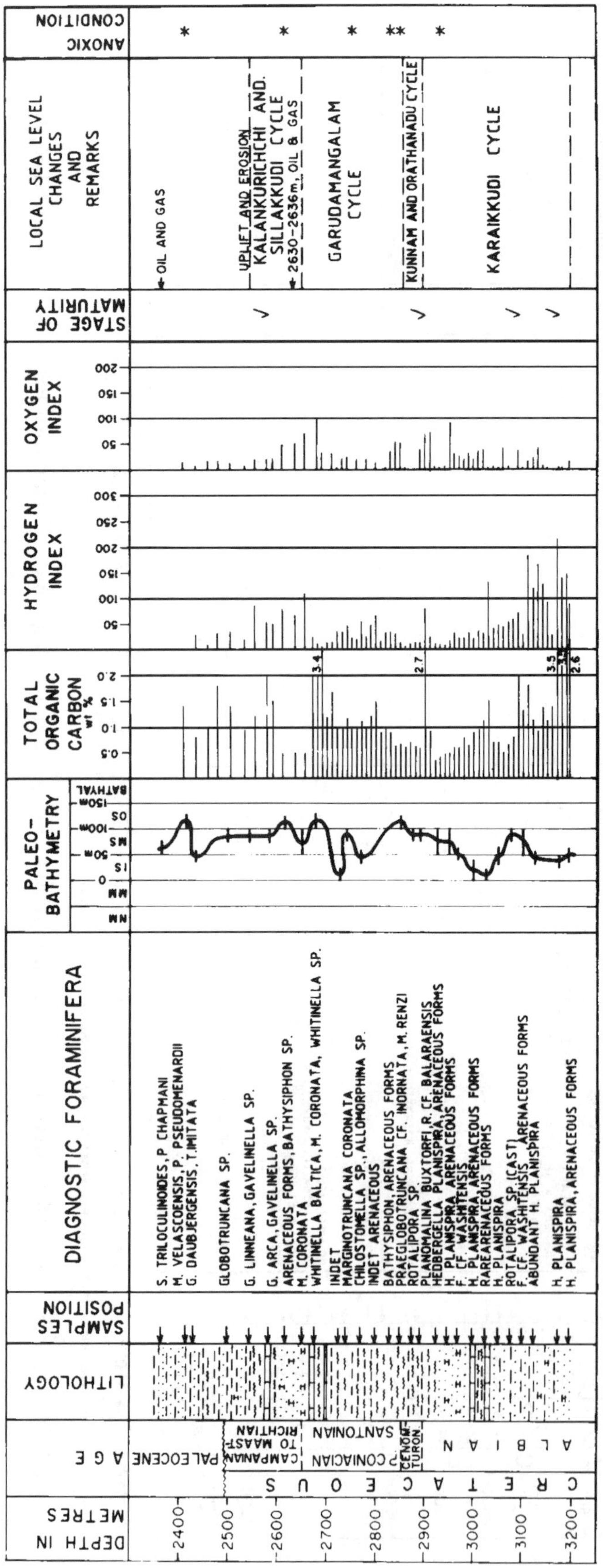

Figure 4. Stratigraphic succession, depositional environments, and geochemical logs in the Cretaceous–Paleocene of well AKM-A.

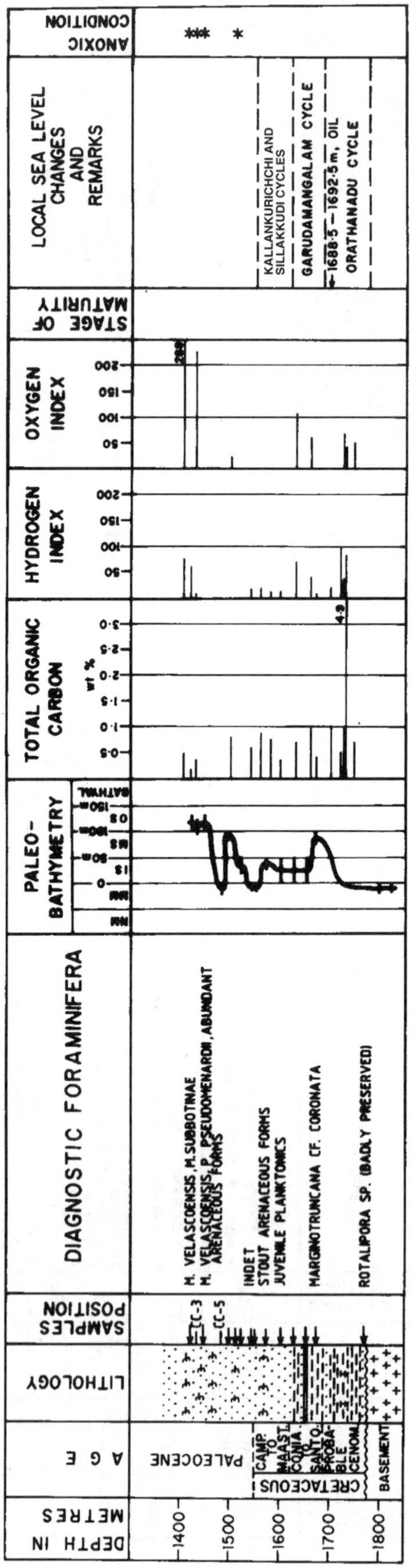

Figure 5. Stratigraphic succession, depositional environments, and geochemical logs in the Cretaceous–Paleocene of well TKR-A.

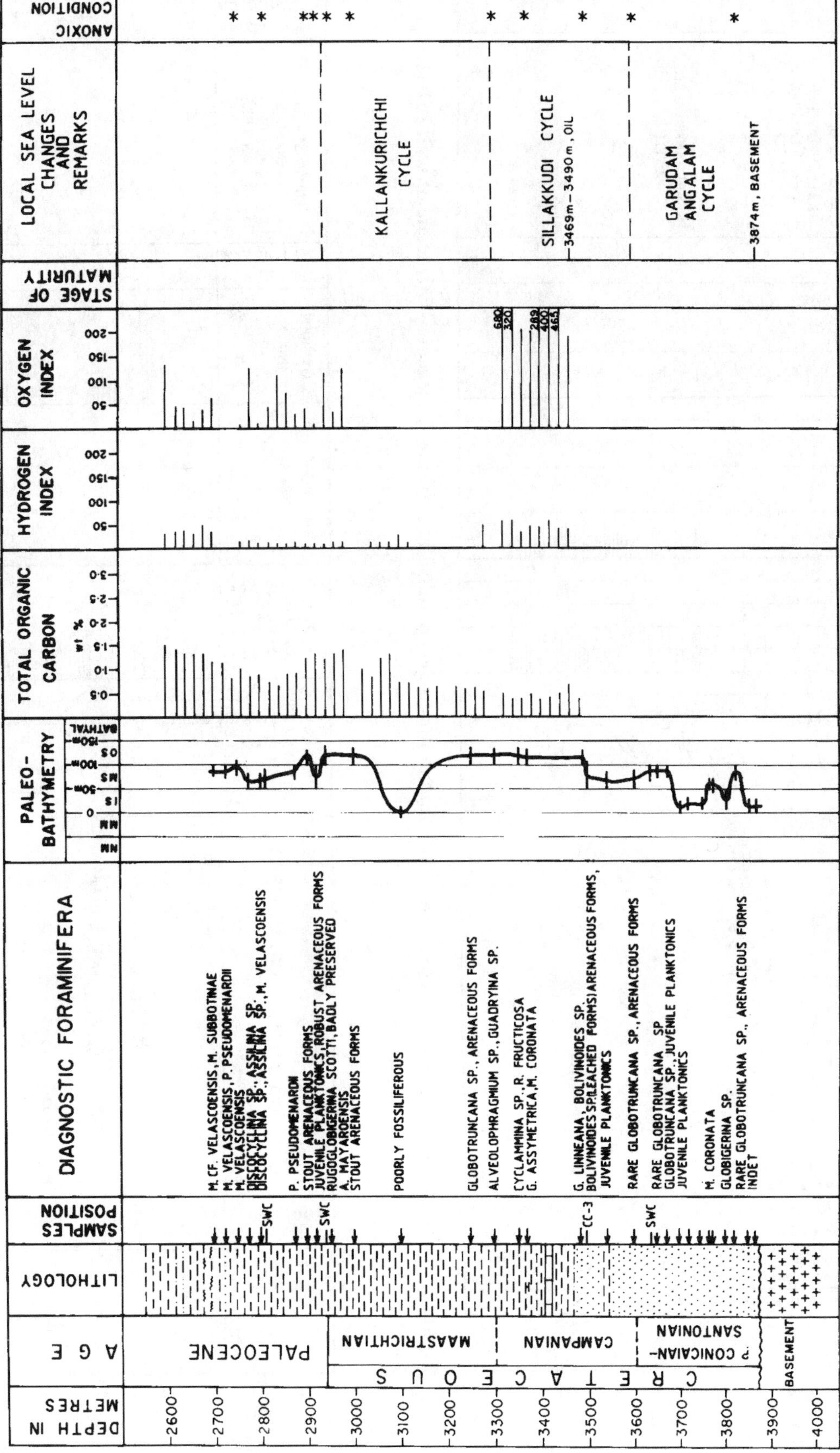

Figure 6. Stratigraphic succession, depositional environments, and geochemical logs in the Cretaceous–Paleocene of well PY-3B.

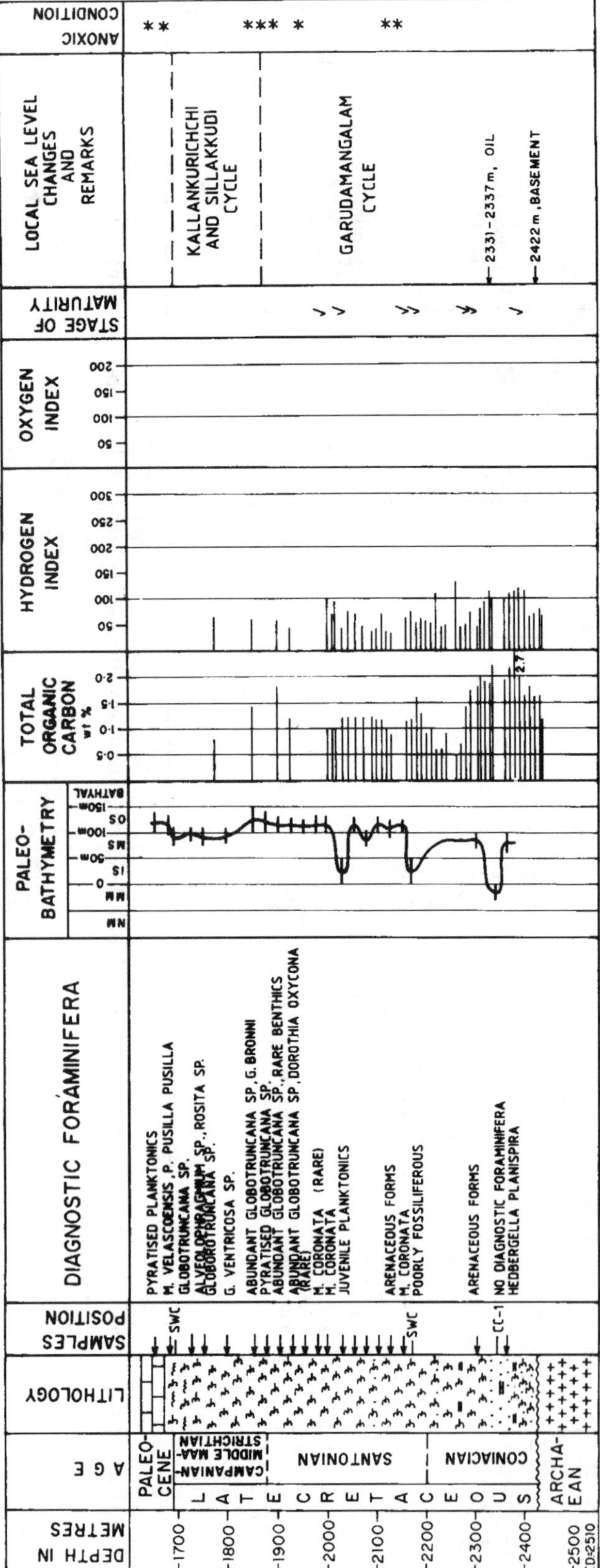

Figure 7. Stratigraphic succession, depositional environments, and geochemical logs in the Cretaceous–Paleocene of well PH-11A.

Figure 8. Stratigraphic succession, depositional environments, and geochemical logs in the Cretaceous–Paleocene of well PNC-A.

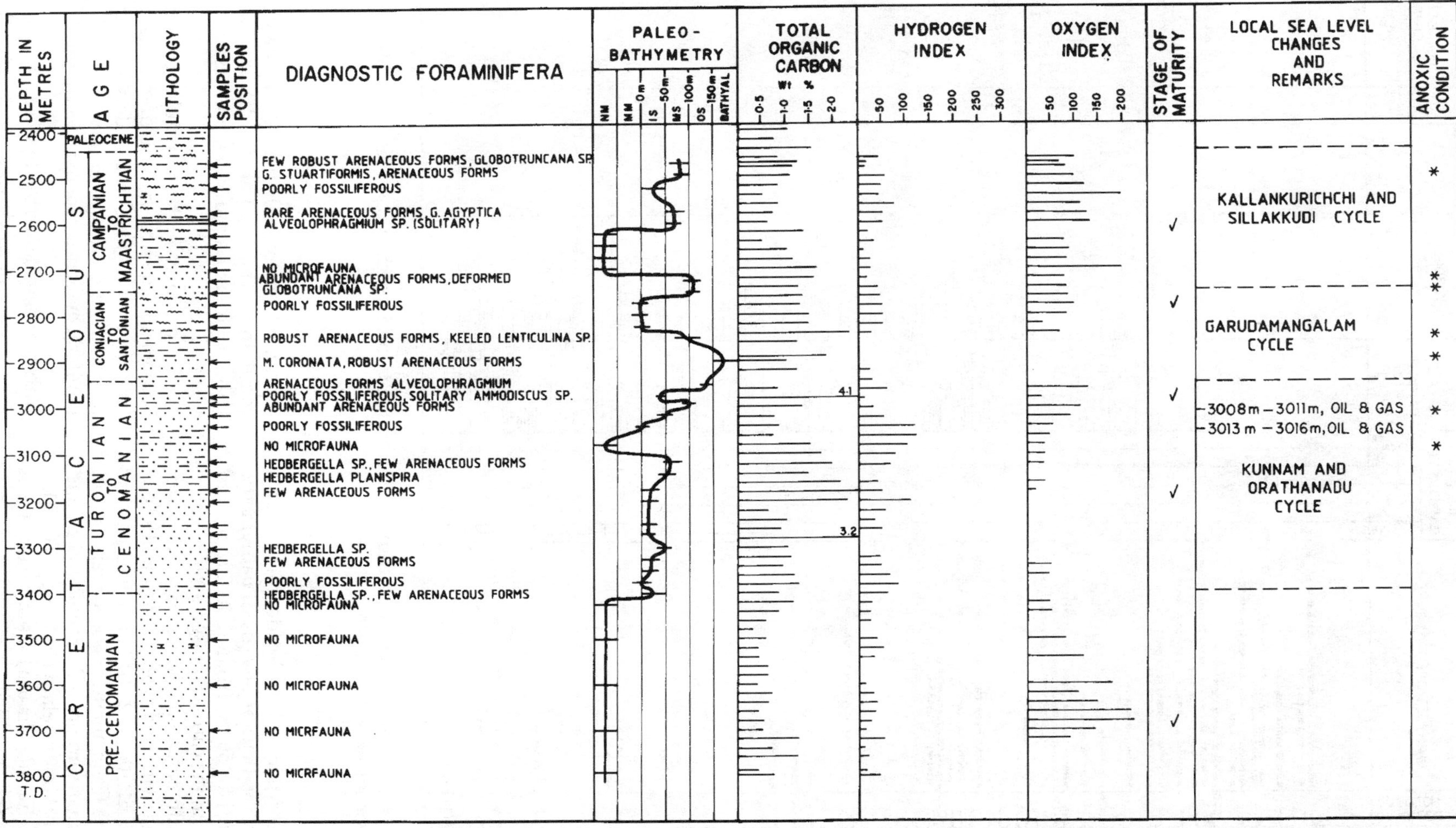

Figure 9. Stratigraphic succession, depositional environments, and geochemical logs in the Cretaceous–Paleocene of well KMP-A.

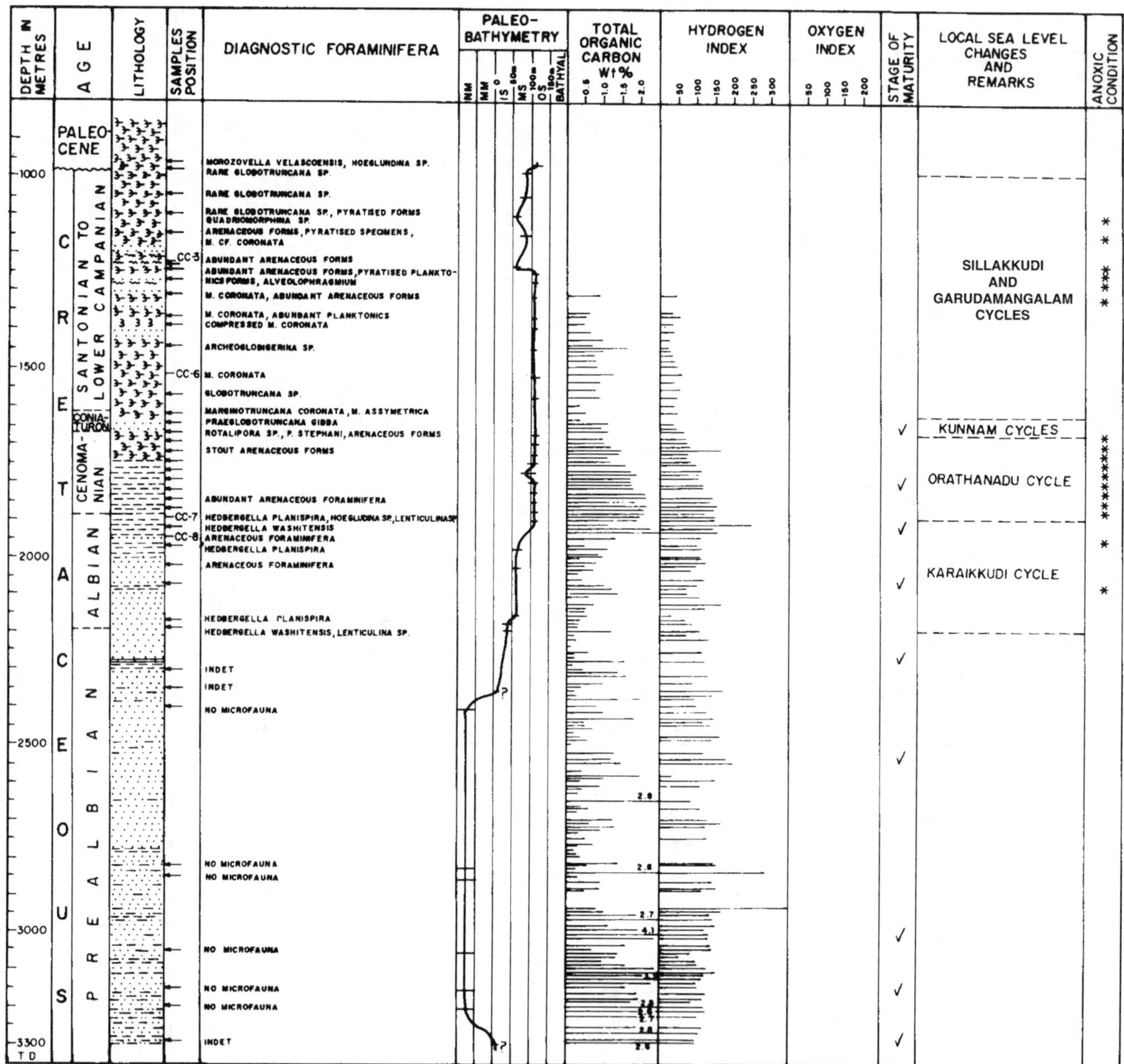

Figure 10. Stratigraphic succession, depositional environments, and geochemical logs in the Cretaceous–Paleocene of well OND-A.

One of the observations that is made from the data shown in Figure 3 is that for any given bathymetric regime, i.e., less than 50 m, from 50 to 100 m, and greater than 100 m, the earliest transgressive event in the Cenomanian to Santonian shows organic matter enrichment along with hydrogen enrichment (TOC 1.0 to 3.8%, and HI 100 to 250). Major bathymetric fluctuations in the Campanian–Maastrichtian are not seen. The bathymetry remained around 100 m and TOC content and HI are relatively lower (TOC 1% maximum; HI 50 maximum) compared with Cenomanian–Santonian sediments.

Anoxic conditions prevailed in the younger sequences where the bathymetry is greater or around 100 m, but the organic matter and hydrogen enrichment show that anoxic conditions have not contributed to any significant preservation of organic matter.

The organic matter enrichment in Albian to Santonian sequences may be due to the deposition of land-derived organic matter deposited in these shallow marine environments with no obvious foraminiferal evidence of anoxic conditions.

Adiyakamangalam Structure

The Adiyakamangalam structure is located in the northern part of the Nagapattinam depression. The AKM-A well (Figures 2, 4) penetrated sediments

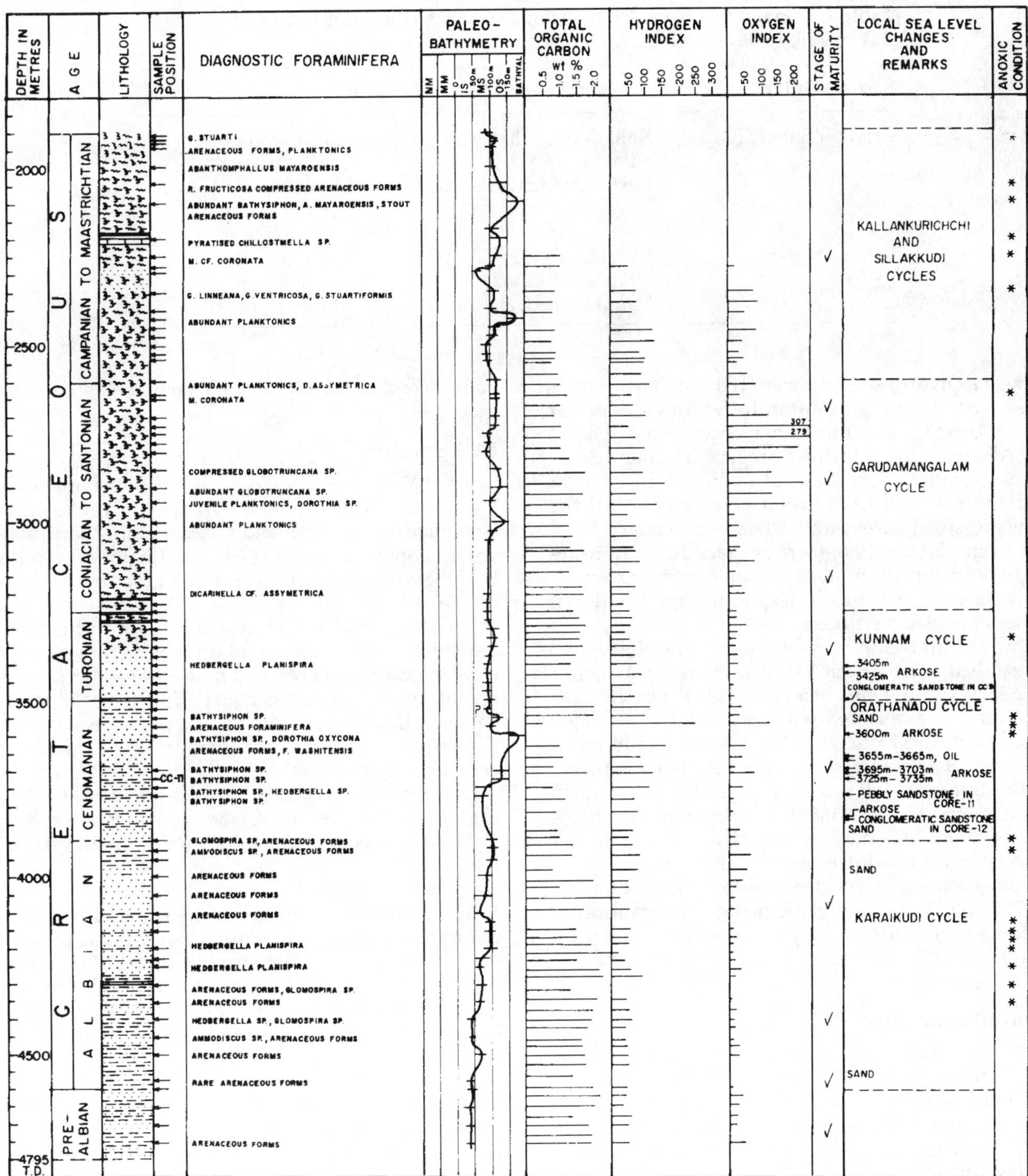

Figure 11. Stratigraphic succession, depositional environments, and geochemical logs in the Cretaceous–Paleocene of well BVG-B.

including a fairly thick Albian, a condensed Cenomanian–Turonian, and moderately thick Santonian to lower Maastrichtian section. The lithology consists of shale and sand from upper Albian to Santonian, and sand, silt, and calcareous layers in the Campanian to Maastrichtian.

Fluctuations in the paleobathymetry from 20 to 120 m are observed in the Albian to Santonian sections and a relatively constant bathymetry (80 m) is seen in the Campanian to Maastrichtian sequences.

In the bathymetric regime around 50 m in the Albian, organic matter enrichment (TOC 2.6 to 3.5%),

Table 1. Cycle of Sea Level Changes in Cauvery basin (after Raju and Ravindran, 1990, and Raju et al., 1992).

Second Order Cycle	Third Order Cycle	European Stages
Ariyalur Super Cycle	Kallankurichchi Cycle Sillakkudi Cycle Garudamangalam Cycle	Maastrichtian Campanian Santonian Coniacian
Uttatur Super Cycle	Kunnam Cycle Orathanadu Cycle Karaikkudi Cycle	Turonian Cenomanian Albian

as well as hydrogen enrichment (HI 210 maximum) is observed. There is no foraminiferal evidence for anoxic conditions in this bathymetric regime. Also in the Albian, in the bathymetric regime of around 80 m, organic matter (TOC 2%) and hydrogen enrichment (HI 180 maximum) are relatively poor. With the transgressive phase in the Albian (bathymetry varying from <50 to +100), there is practically no meaningful enrichment in organic matter or improvement in hydrogen enrichment despite foraminiferal evidence of anoxic conditions.

In the Coniacian to Santonian in the regressive phase (bathymetry from 100 to 50 m), there is again a tendency of slight enrichment in organic matter (TOC around 1%) associated with very minor hydrogen enrichment (HI 25 to 50), though some foraminiferal evidence of anoxic conditions does exist in this phase. In the transgressive phase in the earlier Campanian (bathymetry >100 m), there is enrichment of organic matter (TOC 3.4% maximum) which is, however, not associated with hydrogen enrichment (HI less than 25) despite foraminiferal evidence of anoxic conditions.

The conditions of organic matter preservation in sediments of transgressive phases progressively deteriorate along the stratigraphic column from Albian to early Campanian.

Tirukallar Structure

The Tirukallar structure is located in the Nagapattinam depression. The entire Cretaceous succession is greatly condensed (225 m) in the TKR-A well (Figures 2, 5). Within this thickness Campanian to Maastrichtian sequences are observed. The lithology consists of sand and shale alternations with occasional calcareous layers.

From the initial transgression (bathymetry 80 m maximum) in the probable Cenomanian, the shale layers contain high TOC (4.9% maximum) content and low HI values (100 maximum), with no clear-cut foraminiferal evidence of an anoxic event. The sand development in the regressive cycle of Campanian–Maastrichtian contains poor organic matter and shows low hydrogen index and higher oxygen index.

In the example of this well it is also seen that the bathymetric regime of <100 m, with no evidence of anoxic conditions, shows some source rock development.

PY-3 Structure

The PY-3 structure is located in the offshore extension of the Tranquebar depression. Coniacian to Santonian, Campanian and Maastrichtian successions are encountered in the PY-3B well (Figures 3, 6). The lithology of Coniacian to lower Campanian is predominantly sandstone.

The transgressive zone from the middle Campanian to early Maastrichtian, characterized by a bathymetric regime of >100 m, is poorly enriched in organic matter (TOC less than 0.75%) and also poorly enriched in hydrogen (HI around 50). However, this transgressive phase shows foraminiferal evidence of anoxic conditions.

The regressive phase in the Maastrichtian displays slightly better levels of organic enrichment (TOC 0.75 to 1.5%) and is extremely depleted in hydrogen (HI less than 10). The regressive phase is not associated with anoxic conditions.

The transgressive phase in the late Maastrichtian (bathymetry >100 m) exhibits fairly good organic matter enrichment (TOC maximum of 1.5%) but is poorly enriched in hydrogen (HI less than 10). However, this late Maastrichtian transgressive phase is associated with foraminiferal evidence of anoxic conditions.

In this well the bathymetric regime of >100 m does not form good source rocks. The earliest transgressive phase of the Campanian to middle Maastrichtian, however, may be seen as containing one of the best developed source rocks in the stratigraphic column of PY-3B (Figure 6).

PH-11 Structure

The PH-11 structure is situated in the northern part of Palk Bay. A moderately thick succession of Coniacian, Santonian, Campanian, and middle Maastrichtian sequences are present in the PH-11A well (Figures 2, 7). The lithology consists of claystone and sand.

The earliest Coniacian, basal transgression (bathymetry around 100 m) displays elevated levels of organic matter enrichment (TOC maximum 2.7%)

and low levels of hydrogen enrichment (HI 50–110) with, however, no foraminiferal evidence of anoxic conditions. The transgression from middle Coniacian to early Santonian (bathymetry 50–80 m) is associated with fluctuating organic matter concentration (TOC 0.5 to 2.0%) and fluctuating hydrogen enrichment (HI 50–120). However, in this zone the sediments deposited close to the 50 m bathymetry do show relatively higher organic matter enrichment (TOC 1.2 to 1.6%) with little improvement in the hydrogen enrichment (HI close to 50). Such sediments are not associated with anoxic events in the late Coniacian.

The significant transgressive (bathymetry 120 m) zone from early to middle Santonian is practically indistinguishable from the preceding transgressive sequence and also shows organic matter enrichment (TOC 1%) with no conspicuous enrichment in hydrogen richness (HI 25–50). However, there is some evidence of anoxic conditions in the early part of the transgressive zone.

Yet another regressive zone in the Santonian (bathymetry 25–50 m) is indistinguishable from the preceding transgressive zone. The sediments in other transgressive zones from middle Santonian to Maastrichtian (bathymetry 100–120 m) show moderate levels of organic matter enrichment (TOC maximum 1.7%) and have rather poor hydrogen enrichment (HI 25–50). This transgressive zone is also associated with foraminiferal evidence of anoxic conditions.

In this example, only the first basal transgression in the Coniacian is associated with development of the best source rock, and all other, younger transgressions are associated with low levels of organic matter and hydrogen richness despite being associated with foraminiferal evidence of anoxic conditions.

The deeper bathymetric regimes (>80 m around) have been associated with comparatively poor organic matter preservation and poor hydrogen enrichment.

Pappanaccheri Structure

The Pappanaccheri structure is located in the northernmost part of the Nagapattinam depression. Only 115 m of condensed section of Campanian to middle Maastrichtian, besides Cenozoic, are drilled in the PNC-A well (Figures 2, 8). The Archean basement is not penetrated. The lithology consists of shale, sand, and minor limestone.

Very limited fractions of Cretaceous sediments from Campanian to Maastrichtian have been studied. The transgressive sequences of Maastrichtian having bathymetry >100 m have moderate organic matter enrichment (TOC 0.75–1.0%) and very poor hydrogen enrichment (HI close to 5). However, this transgressive zone is associated with foraminiferal evidence of anoxic conditions.

Kamalapuram Structure

The Kamalapuram structure is located in the western flank of the Karaikal ridge. The KMP-A well (Figures 2, 9) penetrated into the pre-Cenomanian sequence, which is devoid of microfauna. The interval 3425–3800 m did not yield any microfauna to date the section precisely.

The first major transgressive zone of interest commences in the Turonian to Cenomanian, with a maximum bathymetry of 50 m. This transgressive zone is rich in organic matter (TOC maximum 3.4%); however, the hydrogen enrichment is poor (HI 25–100). This zone also is not associated with foraminiferal evidence of anoxic conditions. The intervening regressive phase is practically indistinguishable from the preceding transgressive phase.

The succeeding transgressive phase in the Turonian to Cenomanian, with bathymetry ranging up to 120 m, has moderate organic matter enrichment (TOC 1.0 to 1.2%) and poor to fair hydrogen indices (HI around 100). This zone is associated with anoxic conditions. The succeeding Coniacian to Santonian transgressive sequence is also moderately rich in organic matter (TOC 1.0 to 2.5%), but is extremely poor in hydrogen enrichment (HI 0–50) and is associated with foraminiferal evidence of anoxic conditions.

Yet another interesting transgressive sequence (bathymetry up to 120 m) in the Campanian to Maastrichtian is moderately rich in organic matter (TOC 1.0 to 1.5%) but has poor hydrogen enrichment (HI around 50) and is again associated with foraminiferal evidence of anoxic conditions. A still younger transgressive zone in the upper Campanian to Maastrichtian sequence is practically indistinguishable from the preceding transgressive zone and is associated with anoxic conditions.

In this example, the first important transgression during the Cenomanian to Turonian which displays little evidence of anoxic conditions seems to have the best source rock potential. The younger transgressive zones with bathymetry of more than 100 m are poorer both in organic matter richness and hydrogen richness, despite their evidence of anoxic conditions.

Orathanadu Structure

The Orathanadu structure is located in the Tanjavur subbasin. A 3200-m-thick Cretaceous succession of pre-Albian, Albian, Cenomanian, condensed Coniacian to Turonian, and Santonian to lower Campanian sediments is present in the OND-A well (Figures 2, 10).

Lithology is predominantly sand with minor shale in the pre-Albian, and Albian, shale and claystone in the Cenomanian, a thin layer of sand in Turonian–Coniacian, and claystone with minor sand in the middle Santonian to lower Campanian.

The pre-Albian is dominantly nonfossiliferous and has good source rock development in the lowermost section. The major transgressive zone (bathymetry >100 m) of interest with fine-grained shales begins in the upper Albian to upper Campanian. The earliest transgressive zone with fine clastics is in the upper Albian to Coniacian, and this zone is rich in organic matter (TOC 1 to 2%), moderately rich in hydrogen

(HI between 70 and 250), and associated with foraminiferal evidence of anoxic conditions. The younger transgressive sequences of Coniacian to lower Campanian are moderately rich in organic material (TOC 1.2 to 2.5%) that is poor in hydrogen enrichment (HI 50 to less than 50) and there is no evidence of anoxic conditions. Geochemical data are inadequate in the lower Campanian sequence in the bathymetry regime of 50 to 100 m, while there is evidence of anoxic conditions.

Also in this example the sediments in the earliest transgression of interest are abundantly rich in organic matter, as in all other cases, and the organic matter is also rich in hydrogen.

Bhuvanagiri Structure

The Bhuvanagiri structure is located in the Ariyalur Pondicherry depression. A very thick (2900 m) and nearly continuous succession of marine sediments from pre-Albian to Maastrichtian is present in the BVG-B well (Figures 2, 11). The lithology consists predominantly of sandstone/minor shale alternations in the interval of pre-Albian, Albian, and Cenomanian. In the lower to middle part of the Turonian, sandstone occurs with claystone and thin limestone beds occasionally in the upper part of Turonian. In the Coniacian–Santonian and lower part of Campanian, claystone is the predominant lithology. Some sandstones occur in the upper part of the Campanian claystone and siltstone alternations occur in the Maastrichtian. Conglomerates are also noted in a few intervals in the Cenomanian and lower Turonian.

The Cretaceous sediments of BVG-B are deposited in a transgressive regime with a number of fluctuations between 40–100 m and occasionally between 100–150 m. The first transgressive sequence of the pre-Albian to Albian sediments (bathymetric regime 40–100 m) is rich in organic matter (TOC 0.75 to 2.0%). The sediments are poor in hydrogen (HI 35–50) and associated with foraminiferal evidence of anoxic events.

The Cenomanian to Turonian sediments in the bathymetric regime of 50–100 m are poorer in organic matter (TOC 0.5 to 1.5%) and slightly better in hydrogen richness (HI 50–200) and are associated with anoxic conditions.

The Turonian transgressive sequence (bathymetry 100 m) is rich in organic matter (TOC around 1.5%), relatively poor in hydrogen enrichment (HI 50 or less than 50), and associated with anoxic conditions.

The Coniacian to Santonian transgressive sequence (bathymetry 50 to more than 100 m) is equally rich in organic matter and is indistinguishable from the Turonian sequences. However, it is not associated with evidence of anoxic conditions.

The Campanian to Maastrichtian sequences (bathymetry 50–100 m) are moderately rich in organic matter (TOC 1.0 to 1.5%), relatively better in hydrogen enrichment (HI 50–100), and are occasionally associated with anoxic conditions.

In the example of the BVG-B well, the fine-grained transgressive phase sediments in the pre-Albian to Albian having a bathymetric regime of 40–100 m are the best source rocks. The apparently poor hydrogen enrichment, as evidenced from the hydrogen indices in the pre-Albian and Albian, is due to the maturation associated with a high degree of subsidence for the sequence (more than 4000 m).

DISCUSSION

The results have demonstrated quite amply that:

1. There have been wide fluctuations in sea level in the Cretaceous of the Cauvery basin.
2. The sea level changes within one major time stratigraphic unit are spatially time-transgressive.
3. Anoxic conditions are observed in each time stratigraphic sequence. There seems to be some relationship between anoxic conditions and bathymetry. The sediments deposited in water depths of 50 m have sporadic foraminiferal evidence of anoxic conditions, and sediments deposited between 80 m and more than 100 m are generally associated with anoxic conditions. Anoxic conditions are more than abundantly prevalent in the Santonian and younger sequences. There is no basin-wide association of anoxic conditions with the three Oceanic Anoxic Events (OAE's of Jenkyns, 1980), the first being from the upper Barremian to the Albian, the second one close to the Cenomanian–Turonian boundary, and the third one covering the Coniacian to the Santonian.
4. In all the key wells studied, stratigraphically, the first major transgressive zones associated with fine-grained sediments have the highest organic matter enrichment and good source rock potentials. Generally the sediments in first major transgression have paleobathymetric regimes of 40–100 m.

The present study in the Cauvery basin covers the post-rift Cretaceous sedimentary sequences. In this phase of basin evolution the sedimentation and subsidence history was associated with stresses related to the movement of the Indian Plate from lower latitudes toward higher latitudes (Rangaraju et al., 1991). These stresses resulted in varying degrees of differentiation of depositional floor at different times in the basin's evolutionary history. Additionally, the basin floor differentiation was compounded by the regional movements along the Archean basement lineaments (Prabhakar and Zutshi, 1993) apart from the nuances of movement along the basement fault system in the various grabens in the basin. The changing differentiation in depositional floor entailed varying bathymetric regimes laterally, even within one major time stratigraphic unit, and also caused the time-transgressive nature of peak transgressive events in different parts of the basin.

The anoxic conditions have been observed to correlate with the bathymetric regimes and thus are not laterally extensive even within a single time stratigraphic unit within the Cretaceous.

Sediments in the deeper (>80 m) bathymetric regimes with anoxic conditions are found to be poorer in organic matter content (maximum around 2 times) compared with the sediments in the shallower bathymetric regimes with or without anoxic conditions. This depletion in organic matter concentration in the sediments of deeper bathymetric regimes seems to be related to preburial oxidation. The observation of a concomitant decrease in hydrogen richness of the organic matter in these deeper bathymetric regime sediments to some extent supports this preburial oxidation theory.

Very often, the generation of anoxic conditions occurs due to early diagenetic transformation of organic matter involving sulfate-reducing bacteria and formation of hydrogen sulfide and pyrite. In these early diagenetic transformations, a fraction of sedimentary organic matter is consumed in various microbiological and chemical processes. The relatively lower organic matter enrichment as well as hydrogen enrichment in the sediments of deeper bathymetric regimes having anoxic conditions may be ascribed to these early diagenetic transformations (Fisher and Hudson, 1987). Additionally, the lower amounts of organic matter in the sediments of deeper bathymetric regimes may be related to decreased availability of land-derived organic matter due to the greater distance from the shoreline.

In contrast, organic matter richness and relatively better hydrogen enrichment of organic matter in the sediments of the shallower bathymetric regimes (40 to 80 m), even in the apparent absence of anoxic conditions, is related to abundant influx of terrestrial organic matter, less preburial oxidation, and limited loss of organic matter in early diagenetic transformations. Simultaneous occurrence of organic matter richness and hydrogen richness in organic matter and anoxic conditions are observed in the sediments of the OND-A well, and such exceptionally favorable situations for good source rock development seem to have involved an optimum rate of sedimentation that facilitated better preservation of organic matter (Pedersen and Calvert, 1990, 1991).

CONCLUSIONS

1. Foraminiferal characteristics indicate that the anoxic conditions prevailed in the Cauvery basin throughout the Cretaceous at one place or another but were not limited to the three Oceanic Anoxic Events that have been discussed so much in the last decade. Anoxic conditions are abundantly evident in the Cretaceous sediments deposited in the paleobathymetric regime with water depths >80 m, and are rarely evident in water depths as shallow as 50 m.
2. Limited lateral continuity of anoxic conditions is due to varying intensities of transgression related to depositional floor differentiation in response to the tectonic evolutionary history of the basin.
3. The sediments deposited in the basal transgressive phase or during the peaks of transgression contain abundant organic matter in the Cauvery basin. The hydrogen richness of the organic matter is, however, favored in sediments deposited in paleobathymetric regimes of usually <100 m. Thus the Albian to Santonian sediments are the preferred chronostratigraphic units for the development of good quality source rocks.
4. In most cases, the poor organic matter enrichment in the anoxic Santonian–Maastrichtian sequences may be due to probable preburial alteration (oxidation), decreased contribution of terrestrial organic matter in a deeper bathymetric regime (>100 m), and loss of organic matter in early diagenetic transformations.

ACKNOWLEDGMENTS

The authors wish to thank Sh. L.L. Bhandri M(E&D), Sh. S.K. Manglik, M(P), and Dr. S.K. Biswas, director, KDMIPE, for encouragement and permission to publish this chapter, and Dr. Jagdish Pandey and Dr. K.S. Soodan for providing the facilities. We also thank Sh. P. Kumar, chief geologist, for critical review of the manuscript and Sh. Achal Singh and Sh. Sadhu Ram for drafting the figures and preparing the plates. The authors also thank Mrs. N.J. Thomas, Sh. V.N. Sharma, Dr. Prabhu, Sh. M.N. Pande, and Sh. P. Sridharan of the Geochemistry Division for their cooperation in the collection of data.

REFERENCES CITED

Balan, K.C., S.M. Bandopadhyay, N.J. Thomas, K.B. Shilpkar, M.V.N. Chari, L.N. Pati, A. Neog, and R.P. Sharma, 1991, Quantitative genetic modeling of Cauvery basin: Dehra Dun, unpublished ONGC (KDMIPE) report.

Banarji, R.K., and Mohan Radha, 1970, Foraminiferal biostratigraphy of Meso–Cenozoic sequences of the Cauvery basin, South India: Journal, Geological Society of India, v. 11, n. 4, p. 348-357.

Chandra, K., P.C. Phillip, P. Sridharan, V.S. Chopra, B. Rao, and P.K. Saha, 1991, Petroleum source rock potential of the Cretaceous transgressive regressive sedimentary sequences of the Cauvery basin: Journal of South East Asian Earth Sciences, v. 1-4, p. 367-371.

Demaison, G.J., A.J.J. Holk, R.W. Jones, and G.T. Moore, 1983, Predictive source bed stratigraphy; a guide to regional petroleum occurrence, *in* London, 11th World Petroleum Congress panel discussion, PDI paper 2.

Fisher, I. St. John, and J.D. Hudson, 1987, Pyrite formation in Jurassic shales of contrasting biofacies, *in* J. Brooks and J.D. Hudson, eds., Marine Petroleum Source Rocks: Geological Society, Special Publication 26, p. 69-78.

Govindan, A., 1982, Imprint of global Cretaceous anoxic events in east coast basins of India, and their implication: Dehra Dun, Bulletin, Oil and Natural Gas Commission, v. 19, n. 2.

Govindan, A., 1991, Cretaceous anoxic events, sea level changes and microfauna in Cauvery basin, India: Dehra Dun, abstract, Second Seminar, Petroliferous Basins of India, p. 14.

Jenkyns, H.C., 1980, Cretaceous anoxic events from continents to oceans: Journal of the Geological Society of London, v. 137, p. 171-188.

Kumar, S.P., 1983, Geology and hydrocarbon prospects of Krishna–Godavari and Cauvery basins: Petroleum Asia Journal, Dehra Dun, a Himachal Times group publication, v. 6, p. 57-65.

Pedersen, T.F., and S.E. Calvert, 1990, Anoxia vs. productivity; what controls the formation of organic-carbon-rich sediments and sedimentary rocks?: AAPG Bulletin, v. 74, n. 4, p. 454-466.

Pedersen, T.F., and S.E. Calvert, 1991, Anoxia vs. productivity; what controls the formation of organic-carbon-rich sediments and sedimentary rocks? Reply 1: AAPG Bulletin, v. 75, n. 3, p. 500-501.

Philip, P.C., and P. Neeraja, 1991, Source rock development in Tanjore and Nagapattinam subbasins: J. Pandey and V. Banerjee, eds., Dehra Dun, Proceedings Conference on Integrated Exploration Research, Achievement, and Perspectives, KDM Institute of Petroleum Exploration, p. 449-454.

Prabhakar, K.N., and P.L. Zutshi, 1993, Evolution of southern part of Indian east coast basins: Bangalore, Journal of the Geological Society of India, v. 41, n. 3, p. 215-230.

Raju, D.S.N., 1970, Zonal distribution for selected foraminifera in the Cretaceous and Cenozoic sediments of Cauvery basin, and some problems of Indian biostratigraphy classification: Chandigarh, Punjab University, Publication CAS in Geology, n. 8, p. 85-110.

Raju, D.S.N., and C.N. Ravindran, 1990, Cretaceous sea level changes and transgressive/regressive phase in India; a review: A. Sahni and A. Jolly, eds., Chandigarh, Contribution from the seminar cum workshop, IGCP 216, p. 38-46.

Raju, D.S.N, C.N. Ravindran, and R. Kalyanasundar, 1992, Cretaceous cycle of sea level changes in Cauvery basin, India, a first revision manuscript.

Rangaraju, M.K., A. Agarwal, and K.N. Prabhakar, 1991, Techno-stratigraphy, structural styles, evolutionary model, and hydrocarbon habitat, Cauvery and Palar basins: Dehra Dun, Abstracts, Second Seminar, Petroliferous Basins of India, p. 3.

Robertson Research, 1987, Krishna–Godavari Basin stratigraphy, petroleum geochemistry, and petroleum geology: Dehra Dun, ONGC unpublished report.

Sastri, V.V., R.N. Sinha, G. Singh, and K.V.S. Murti, 1973, Stratigraphy and tectonics of sedimentary basins on east coast of peninsular India: AAPG Bulletin, v. 57, p. 655-678.

Schlanger, S.O., and H.C. Jenkyns, 1976, Cretaceous oceanic anoxic events, causes and consequences: Geologie en Mijnbouw, v. 55 (3-4), p. 179-184.

Thomas, N.J., AK. Pandey, and A.K. Samant, 1991, Geochemistry in petroleum exploration in Krishna Godavari and Cauvery basins: J. Pandey and V. Banerjie, eds., Dehra Dun, Proceedings Conference on Integrated Exploration Research, Achievements, and Perspectives, KDM Institute of Petroleum Exploration, p. 483-494.

Chapter 10

Petroleum Source Rock Potential of Mesozoic Condensed Section Deposits of Southwest Alabama

Ernest A. Mancini
University of Alabama and Geological Survey of Alabama
Tuscaloosa, Alabama, USA

Berry H. Tew
Robert M. Mink
Geological Survey of Alabama
Tuscaloosa, Alabama, USA

ABSTRACT

Condensed section deposits in carbonates and siliciclastics are generally fine-grained rocks that commonly contain relatively high concentrations of organic matter; therefore, these rocks may have the potential to be petroleum source rocks if buried under conditions favorable for thermogenic hydrocarbon generation. Of the condensed section deposits in the Mesozoic strata of southwest Alabama, only the Upper Jurassic Smackover Formation carbonate mudstones from the condensed section of the lower Zuni A Gulf Coast-4.1 depositional cycle have sufficient organic carbon and were subjected to burial and thermal conditions in which this potential has been realized. These condensed section and transgressive carbonate mudstones contain total organic carbon contents of algal and amorphous kerogen of as much as 2.19% and exhibit thermal alteration indices of 2^- to 3^+. The laminated carbonate mudstones of the Smackover Formation have apparently served as the hydrocarbon source for the majority of Mesozoic reservoirs throughout southwest Alabama. The Upper Cretaceous Tuscaloosa Group marine claystones from the condensed section of the upper Zuni A Gulf Coast-2.5 depositional cycle are rich (total organic carbon values of as much as 2.91%) in herbaceous and amorphous organic matter but have not been subjected to burial and thermal conditions favorable for thermogenic hydrocarbon generation. These claystones exhibit thermal alteration indices of 1^+ to 2^-. The Jurassic Norphlet shales of the condensed section of the lower Zuni A Gulf Coast-3.1 depositional cycle are low in total organic carbon content (0.1%). These rocks have experienced burial and thermal conditions favorable for thermogenic hydrocarbon generation, but depositional conditions have limited their potential as source rocks because of the paucity of organic carbon preserved in these deposits. No well-developed condensed sections are recognized in the Upper Jurassic Haynesville lower Zuni A Gulf Coast-4.2 depositional cycle or the Upper Cretaceous Tuscaloosa upper Zuni A Gulf Coast-2.3 and upper Zuni A Gulf Coast-2.4 depositional cycles. Although condensed sections within depositional sequences, in general, should have the highest source rock potential, specific environmental, preservational and/or burial and thermal history conditions within a particular basin dictate whether or not this potential is realized. This relationship is shown by the condensed sections of the Mesozoic depositional sequences in southwest Alabama. Therefore, petroleum geologists can use sequence stratigraphy as a tool to help identify stratigraphic intervals that might have potential to contain hydrocarbon source rocks; however, only through geochemical analyses can the actual source rock potential be determined.

INTRODUCTION

Condensed section deposits in carbonates and siliciclastics are generally fine-grained rocks that commonly contain relatively high concentrations of organic matter (Loutit et al., 1988; Sarg, 1988); therefore, these strata have the potential to be petroleum source rocks if buried under conditions favorable for thermogenic hydrocarbon generation. In addition to the richness (total organic carbon content) of source rocks, the type of kerogen and thermogenic conditions (time and temperature effects) determine the petroleum generating capacity of source rocks. Generally, siliciclastics having a minimum of 0.5% total organic carbon and carbonate rocks containing as much as 0.3% total organic carbon are considered to be potential source rocks (Tissot and Welte, 1984). In the Mesozoic deposits of southwest Alabama, the condensed section deposits which have the potential to be petroleum source rocks are Jurassic shales associated with the Norphlet Formation, Jurassic carbonate mudstones of the Smackover Formation, and Upper Cretaceous marine claystones of the Tuscaloosa Group. The Smackover carbonate mudstones and the Tuscaloosa marine claystones were deposited during widespread transgressive events that resulted in sea level and depositional conditions conducive to the accumulation and preservation of significant amounts of organic matter in these strata. The magnitude of these sea level events also is recognized in the global coastal onlap charts and eustatic sea level curves of Haq et al. (1988). The purpose of this chapter is to evaluate the petroleum source rock potential of the Mesozoic condensed section deposits in southwest Alabama and to present general conclusions about the source rock potential of these deposits in this area.

GEOLOGIC SETTING

The Mesozoic strata of southwest Alabama were deposited as part of a seaward-dipping and -thickening wedge of sediment that accumulated in the Gulf of Mexico Basin, a differentially subsiding depositional basin on the passive margin of the North American continent. Major structural elements that have altered the general orientation of these strata include basement features associated with continental collision and suturing or with continental rifting, features formed due to halokinesis of Jurassic salt, and intrusions resulting from igneous processes.

The major positive basement features that influenced the distribution and nature of Mesozoic deposits are the Wiggins arch complex, which includes the Wiggins arch and the Baldwin high, the Choctaw Ridge complex, and the Conecuh Ridge complex (Figure 1). These structural elements are, in part, associated with the Appalachian fold and thrust structural trend that was formed in the late Paleozoic by tectonic events resulting from the convergence of the North American and Afro–South American continental plates. Paleotopography had a significant impact on the distribution of sediments, and positive areas within basins and along basin margins provided major sources for Mesozoic terrigenous siliciclastic sediments (Mancini et al., 1985).

The Mississippi interior salt basin is a major negative structural feature in southwest Alabama (Figure 1). This salt basin was an actively subsiding depocenter throughout the Mesozoic and into the Cenozoic. Based on gravity data, Wilson (1975) interpreted the salt basin to be an area of attenuated granitic continental crust; crustal thinning resulted from tectonic extension of the lithosphere during the rifting of the Gulf in the Triassic and Jurassic periods. This attenua-

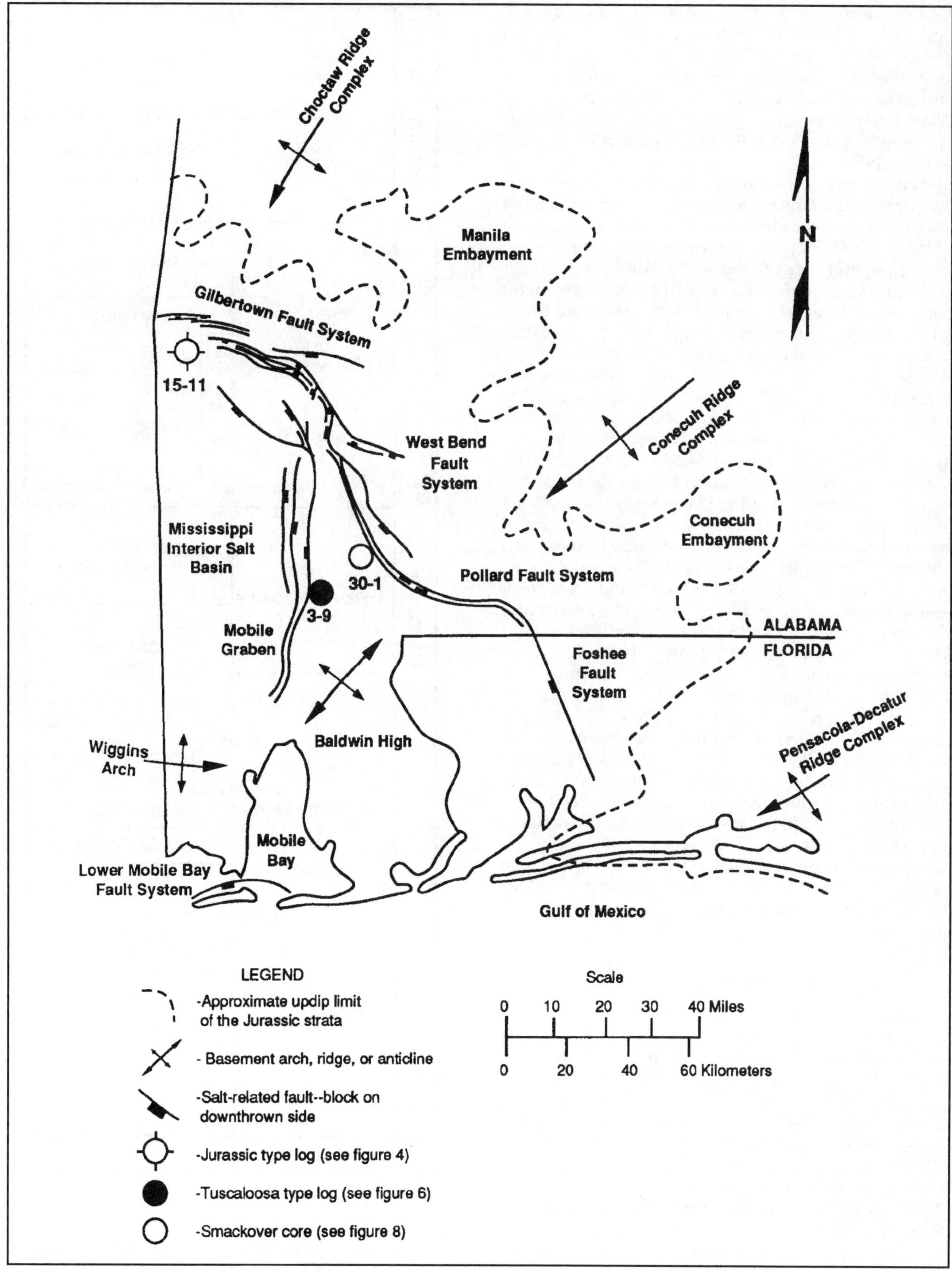

Figure 1. Map of southwest Alabama, showing the major structural features and the approximate location of the type-log wells.

tion of the crust established a subsiding basin cratonward of the rifted and elevated continental margin (Wood and Walper, 1974). The Conecuh and Manila embayments were also significant Mesozoic depocenters (Mancini and Benson, 1980). These embayments have been interpreted to have originated as rift grabens associated with the breakup of Pangea (Miller, 1982).

Halokinesis of the Jurassic Louann Salt (Figure 2) has produced an intricate complex of salt-movement related structural elements in southwest Alabama (Martin, 1978). Salt-related structures include diapirs, anticlines, and extensional fault and graben systems. Structural elements resulting from salt movement include the regional peripheral fault trend, the lower Mobile Bay fault system, the Mobile graben, and numerous salt domes and anticlines (Figure 1).

The regional peripheral fault trend is comprised of a group of genetically related, generally en echelon extensional faults which are associated with salt movement. In the area of study, this trend is composed of the Gilbertown, West Bend, Pollard, and Foshee fault systems (Figure 1). The faults of the regional peripheral fault trend are generally parallel or subparallel to regional strike, and the trend approximates the updip limit of thick Jurassic salt (Martin, 1978). Most of the faults in the trend are normal, down-to-the-basin or antithetic faults that form grabens that are generally 5–8 mi (8–13 km) across (Murray, 1961). The faults are listric and the dips of the fault surfaces range from 35° to 70°. Displacements on major faults in the trend range from 200 ft (61 m) to more than 2000 ft (610 m) in the Jurassic section (Mancini et al., 1985). The Mobile graben, which is considered to be the eastern limit of the Mississippi interior salt basin, represents a more mature stage of halokinesis as evidenced by an association with diapiric features. The lower Mobile Bay fault system has characteristics similar to the regional peripheral fault trend; however, most of these faults terminate upward into or just above the Haynesville Formation, with the exception of the main down-to-the-basin fault in the system, which extends upward into Cretaceous strata.

STRATIGRAPHY

Sedimentation in southwest Alabama was associated with rifted continental margin tectonics resulting from the breakup of Pangea and the opening of the Gulf of Mexico Basin. Triassic graben-fill red beds of the Eagle Mills Formation (Figure 2) were deposited locally as the oldest Mesozoic strata above Paleozoic basement during the early stages of extension and rifting (Tolson et al., 1983). The Jurassic Werner Formation and Louann Salt are evaporite deposits that formed during the initial transgression of marine water into the Gulf of Mexico Basin (Salvador, 1987). The upper part of the Louann Salt is the Pine Hill Anhydrite Member. The Upper Jurassic Norphlet Formation, which overlies the Louann Salt, includes basal intertidal shale, alluvial fan conglomeratic sandstone, alluvial, fluvial, and wadi red beds, and upper eolian and marine shoreface quartzose sandstone (Mancini et al., 1985). The Upper Jurassic Smackover Formation overlies the Norphlet Formation and con-

Erathem	System/ Series		Rock Unit
Cenozoic	Tertiary		undifferentiated Tertiary strata
Mesozoic	Upper Cretaceous		Selma Group
			Eutaw Formation
		Tuscaloosa Group	"upper" Tuscaloosa
			"marine" Tuscaloosa
			"lower" Tuscaloosa
	Lower Cretaceous		Lower Cretaceous undifferentiated
			Cotton Valley Group
	Upper Jurassic		Haynesville Formation
			Buckner Anhydrite Member
			Smackover Formation
			Norphlet Formation
			Pine Hill Anhydrite Member
	Middle Jurassic		Louann Salt
			Werner Formation
	Lower Jurassic/ Upper Triassic		Eagle Mills Formation
Paleozoic			undifferentiated Paleozoic "basement" rocks

Figure 2. Mesozoic stratigraphy for southwest Alabama.

sists of a lower unit of intertidal to subtidal, laminated carbonate mudstone and intraclastic, peloidal and oncoidal wackestone and packstone, a middle unit of subtidal, laminated carbonate mudstone interbedded with peloidal and skeletal wackestone and packstone, and an upper unit of peritidal, oolitic, oncoidal, and peloidal grainstone and packstone interbedded with laminated carbonate mudstone (Mancini and Benson, 1980; Moore, 1984; Benson, 1988). The Upper Jurassic Haynesville Formation includes the subaqueous to subaerial, massive anhydrites of its lower member, the Buckner Anhydrite Member, and restricted shelf to littoral anhydritic shale and sandstone in the upper part (Tolson et al., 1983; Mann, 1988). The Upper Jurassic to Lower Cretaceous Cotton Valley Group overlies the Haynesville Formation and includes fluvial-deltaic and delta destructive rocks, including barrier bar and strandplain sandstones and conglomerates (Moore, 1983; Tolson et al., 1983).

Lower Cretaceous rock units are not differentiated in southwest Alabama, but generally consist of fine- to medium-grained fluvial to deltaic sandstones with minor nodular limestones and thin anhydrite beds (Eaves, 1976). The Upper Cretaceous Series is comprised of the Tuscaloosa Group, the Eutaw Formation, and the Selma Group. In the subsurface of the study area, the Tuscaloosa Group is composed of the informal "lower Tuscaloosa," which consists of wave-dominated, high destructive deltaic and barrier–shoal sandstones, the "marine Tuscaloosa," a unit of deeper water marine shales, and the "upper Tuscaloosa," comprised of regressive marine shelf to marginal marine sands (Mancini et al., 1987). The "lower Tuscaloosa" can be further subdivided into a lower "Massive sandstone interval" and an upper "Pilot sandstone interval." Three units compose the "Massive sandstone interval." These are, in ascending order, the "basal sandstone unit," the "interbedded sandstone and claystone unit," and the "Massive sandstone unit." The "Pilot sandstone interval" includes a lower "claystone unit," overlain by the "Pilot sandstone interval." The foregoing subsurface stratigraphic terminology for the Tuscaloosa Group is from Mancini et al. (1987), who adapted and modified the usage of Winter (1954).

The Eutaw Formation consists of glauconitic, fine-grained marine-shelf sands, and the Selma Group includes massive marine-shelf chalks interbedded with marls and calcareous clays (Mancini and Mink, 1985). Tertiary units in the study area represent alternating marine-shelf marls and limestones and marginal marine to fluvial-deltaic sands, carbonaceous clays, and lignites (Mancini and Tew, 1991).

SEQUENCE STRATIGRAPHY

Middle and Upper Jurassic

Three unconformity-bounded depositional sequences that are based on regional stratigraphic and sedimentologic data and which resulted from relative changes in sea level and coastal onlap during the Middle and Upper Jurassic (Callovian, Oxfordian and Kimmeridgian) have been recognized for southwest Alabama (Mancini et al., 1990; Tew et al., 1991). These sequences are designated the LZAGC-3.1 (lower Zuni A Gulf Coast), the LZAGC-4.1, and the LZAGC-4.2 (Figure 3). The Gulf Coastal Plain sequences of Mancini et al. (1990) are correlated to the LZA-3.2 sequence (155.5 to 150.5 Ma), a composite of the LZA-4.1 through LZA-4.4 sequences (150.5 to 144 Ma), and the LZA-4.5 sequence (144 to 142 Ma) in the global system of Haq et al. (1988), respectively.

The LZAGC-3.1 depositional sequence (Figure 4) consists of a basal unconformity, transgressive deposits (Werner and Louann evaporites), condensed section deposits (Pine Hill anhydrites and Norphlet shales), and progradational, highstand regressive deposits (Norphlet continental deposits). The basal unconformity is a composite surface that is underlain by Paleozoic igneous, metamorphic, or sedimentary rocks, or Late Triassic–Early Jurassic rift-graben fill continental sedimentary rocks and intrusive igneous rocks, depending upon location. The Werner and Louann evaporite deposits represent the initial (and probably sporadic) influx of marine water into the basin during rifting of the Gulf of Mexico basin. The subaqueous anhydrites of the Pine Hill Anhydrite Member of the Louann Salt and basal intertidal shales of the Norphlet Formation accumulated during maximum transgression within the cycle and, thus, in part, represent the condensed section deposits of the sequence. Relative sea level fall after accumulation of these deposits is indicated by the progradation of Norphlet continental sediments southward from the eroding Appalachian highlands bordering the developing Gulf of Mexico Basin; this highland front served as a source area for these terrigenous sediments.

The LZAGC-4.1 depositional sequence (Figure 4) includes a basal type 2 sequence boundary, shallow-water, shelf margin deposits (Norphlet marine sandstone), transgressive deposits (lower Smackover intertidal to subtidal carbonate mudstone, wackestone, and packstone), condensed section deposits (middle Smackover subtidal carbonate mudstone), and progradational, regressive highstand deposits (upper Smackover subtidal to supratidal, shoaling-upward carbonate mudstone to grainstone cycles, Buckner peritidal shoaling-upward evaporite cycles, and Haynesville peritidal carbonate, sandstone, shale, and evaporite). Eolian sandstone of the upper part of the Norphlet Formation was reworked and deposited in a marine shoreface environment with the onset of sea level rise. The initiation of Smackover carbonate deposition occurred during a major Jurassic transgressive event in the Gulf of Mexico Basin as indicated by the widespread distribution of this marine unit throughout the region. The lower Smackover, which includes a sequence of intraclastic, peloidal and oncoidal wackestone and packstone and laminated carbonate mudstone, represents the lower part of this transgressive phase. These lithologies are overlain gradationally by a thick sequence of laminated car-

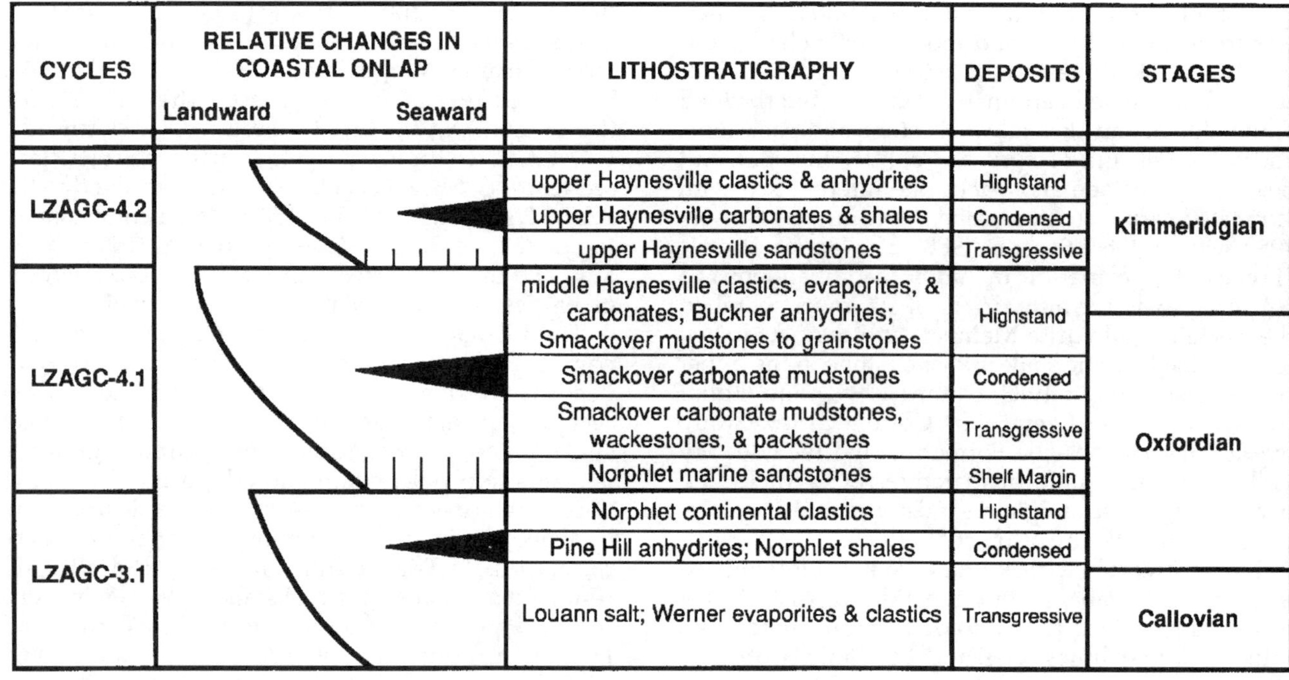

Figure 3. Sequence stratigraphy of Jurassic strata in southwest Alabama. LZAGC = lower Zuni A Gulf Coast.

bonate mudstone which is included in the middle Smackover. This stratigraphic succession indicates that, initially, Smackover sedimentation, in part, occurred in an intertidal to subtidal, moderately high-energy environment, followed by rapid relative sea level rise and concomitant increase in sediment accommodation, which produced the thick, upward-deepening, carbonate mud-dominated subtidal interval of the middle Smackover. Smackover rocks that accumulated during maximum transgression of the LZAGC-4.1 cycle are the organic-rich, laminated carbonate mudstone and interbedded peloidal and skeletal wackestone and packstone of the upper portion of the middle Smackover. The middle Smackover is, in part, a condensed section within the depositional sequence. Condensed sections in carbonate sequences generally consist of mudstone with thin interbeds of wackestone (Sarg, 1988) and condensed intervals often exhibit concentrations of organic matter (Loutit et al., 1988). Aggradational to progradational, shallowing-upward parasequences of peritidal, oolitic, oncoidal, peloidal packstone and grainstone compose the lower part of the progradational, highstand regressive deposits of the LZAGC-4.1 depositional sequence (Mancini et al., 1990). These rocks are interbedded with intertidal to supratidal, laminated, carbonate mudstone that commonly exhibits mud cracks, fenestral fabrics, dissolution fabrics, and exposure surfaces in association with anhydrite.

The Buckner Anhydrite Member and middle part of the Haynesville Formation constitute the uppermost highstand deposits of the LZAGC-4.1 sequence. The Buckner is a subaqueous to subaerial evaporite sequence that was deposited in a series of upward-shallowing cycles (Mann, 1988). The interbedded carbonate, sandstone, shale, and anhydrite of the middle part of the Haynesville accumulated in subtidal to supratidal environments. Lagoonal/sabkha evaporite lithofacies commonly occur in the progradational, highstand regressive deposits of carbonate depositional systems (Sarg, 1988).

The LZAGC-4.2 depositional sequence (Figure 4) includes a basal type 2 sequence boundary, transgressive deposits (intertidal to subtidal sandstone of the upper Haynesville), condensed section deposits (subtidal shale and limestone of the upper Haynesville), and progradational, regressive highstand deposits (subtidal to continental interbedded sandstone, shale, and anhydrite of the upper Haynesville). The transgressive upper Haynesville sandstone was deposited in environments ranging from tidal flats to offshore marine bars. Marine shale and limestone of the upper Haynesville accumulated during maximum transgression within the relative sea level cycle. These condensed section deposits as recognized from geophysical well logs do not constitute a significant stratigraphic thickness. Sea level fall is indicated by interbedded sandstone, shale, and anhydrite in the upper part of the upper Haynesville; these rocks were deposited in subtidal to continental environments. Marginal marine to continental deposits of the Cotton Valley Group disconformably overlie the upper Haynesville. The sequence stratigraphy of the Cotton

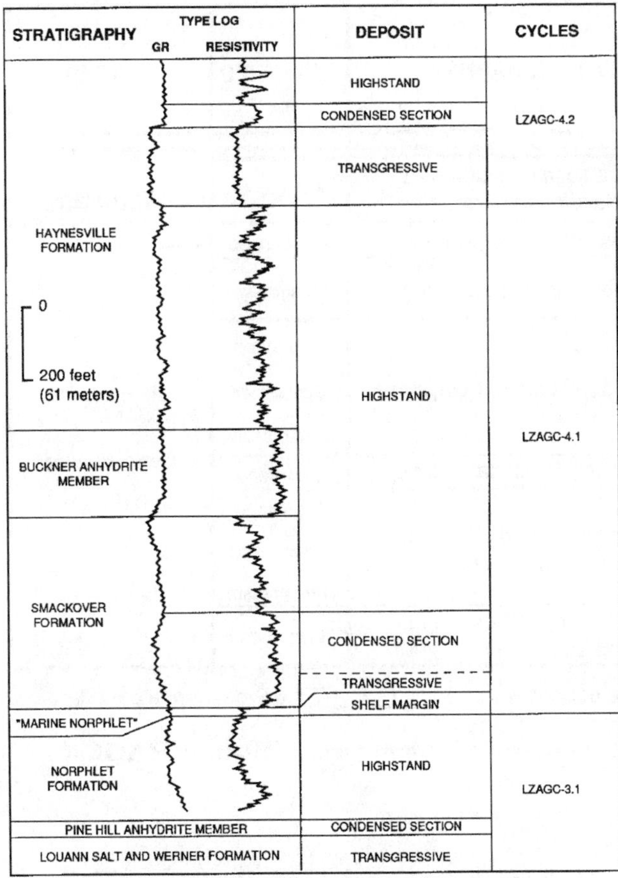

Figure 4. Type log of Jurassic strata in southwest Alabama (Exxon, USA, # 1 H. W. Smith Lumber Company 15-11). See Figure 1 for location of type-log wells. GR = gamma ray.

Valley Group and the Lower Cretaceous undifferentiated in southwest Alabama has not been studied.

Upper Cretaceous

Based on regional stratigraphic, sedimentologic, petrophysical, and paleontologic data, three unconformity-bounded depositional sequences reflecting relative changes in sea level and coastal onlap are recognized for the Upper Cretaceous (Cenomanian and Turonian) Tuscaloosa Group in the subsurface of southwest Alabama. The depositional sequences probably correspond to the UZA-2.3 (95.5 to 94 Ma), UZA-2.4 (94 to 93 Ma), and UZA-2.5 (93 to 91 Ma) global sequences of Haq et al. (1988) and are here designated the UZAGC-2.3 (upper Zuni A Gulf Coast), UZAGC-2.4, and UZAGC-2.5 depositional sequences (Figure 5).

The UZAGC-2.3 depositional sequence (Figure 6) is underlain by undifferentiated Lower Cretaceous terrigenous siliciclastic deposits and consists of strata of the lower part of the "lower Tuscaloosa." The sequence includes a basal type 2 sequence boundary, shelf margin deposits ("basal sandstone unit," marginal marine sandstones), transgressive and condensed section deposits ("interbedded sandstone and claystone unit" marine shelf sandstones and claystones) and progradational, regressive highstand deposits ("interbedded sandstone and clay unit," marine shelf to strandplain sandstones and claystones). The shelf margin and transgressive deposits represent progressive relative sea level rise and transgression during the early Cenomanian following a long period of Early Cretaceous regression. No discrete, well-developed condensed section is recognized from geophysical well logs in this sequence. The regressive highstand deposits of the sequence record a relative sea level fall in the middle Cenomanian, culminating in a type 1 sequence boundary.

The UZAGC-2.4 depositional sequence (Figure 6) consists of strata of the upper part of the "lower Tuscaloosa" (exclusive of the "Pilot sandstone unit"). The sequence includes a basal type 1 unconformity, lowstand deposits ("Massive sandstone unit," aggradational coastal barrier sandstones), transgressive and condensed section deposits ("claystone unit," marine shelf sandstones and claystones) and progradational, regressive highstand deposits ("claystone unit," strandplain to lagoonal sandstones and claystones). The initiation of relative sea level rise in the Cenomanian is indicated by a sequence of aggradational lowstand deposits that overlie the regressive, highstand deposits of the underlying sequence. These sandstones are overlain by a regressive sequence of marine-shelf strata. As with the underlying UZAGC-2.3 depositional sequence, no well-developed condensed section is recognized from geophysical well logs in the UZAGC-2.4 sequence. Relative sea level fall within the cycle is represented by progradational nearshore to marginal marine strata. The sequence is capped by a type 2 sequence boundary.

The UZAGC-2.5 depositional sequence (Figure 6) consists of the "Pilot sandstone unit" of the "lower Tuscaloosa," the "marine Tuscaloosa," and the "upper Tuscaloosa." The sequence includes a basal Type 2 unconformity, transgressive deposits ("Pilot sandstone unit," marine shelf sandstones), condensed section deposits ("marine Tuscaloosa," marine shelf claystones), and progradational, regressive highstand deposits ("upper Tuscaloosa," marine shelf to marginal marine sandstones and claystones). This sequence reflects a major sea level rise beginning in the late Cenomanian and continuing into the Turonian. Maximum transgression within the sequence is represented by the open-marine-shelf clays of the "marine Tuscaloosa." This unit, which is rich in organic matter, contains a diverse macroinvertebrate faunal assemblage, including ammonites, gastropods, inoceramids, and other bivalves, as well as a rich microfossil assemblage of planktonic foraminifera and calcareous nannofossils (Mancini et al., 1987). Relative sea level fall in the Turonian is reflected in the deposition of the regressive strata of the "upper Tuscaloosa."

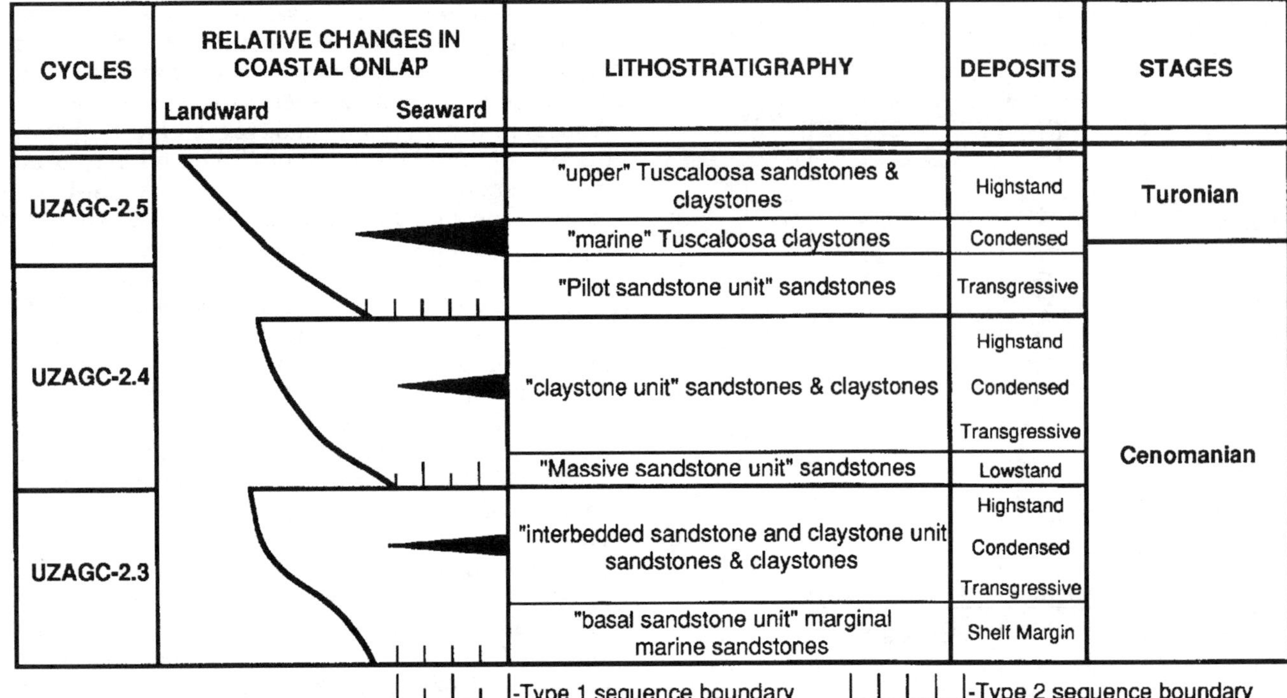

CYCLES
RELATIVE CHANGES IN COASTAL ONLAP
Landward
Seaward
LITHOSTRATIGRAPHY
DEPOSITS
STAGES
UZAGC-2.5
UZAGC-2.4
UZAGC-2.3
"upper" Tuscaloosa sandstones & claystones
"marine" Tuscaloosa claystones
"Pilot sandstone unit" sandstones
"claystone unit" sandstones & claystones
"Massive sandstone unit" sandstones
"interbedded sandstone and claystone unit sandstones & claystones
"basal sandstone unit" marginal marine sandstones
Highstand
Condensed
Transgressive
Lowstand
Shelf Margin
Turonian
Cenomanian
-Type 1 sequence boundary
-Type 2 sequence boundary

Figure 5. Sequence stratigraphy of Upper Cretaceous Tuscaloosa strata in southwest Alabama. UZAGC = upper Zuni A Gulf Coast.

SOURCE ROCK ANALYSIS

Organic geochemical analyses of 77 core samples representing Mesozoic transgressive, condensed section, and highstand deposits (Figure 7) are summarized by rock unit, geologic age, sequence component, location, and burial depth in Table 1. The Tuscaloosa samples (13), which are from transgressive, condensed section, and highstand deposits, have the highest total organic carbon content as a group (as much as 2.91% and an average of 1.53%). The eight Norphlet samples (from condensed section deposits) and 17 Haynesville samples (from highstand deposits) analyzed are low in total organic carbon, having an average of 0.10% and 0.12%, respectively (Table 1). These total organic carbon percentages are consistent with those reported by Sassen et al. (1987) for five Norphlet and 12 Haynesville samples from the Mississippi interior salt basin, which had an average of 0.09% and 0.26% total organic carbon, respectively. The Smackover samples (39), representing transgressive, condensed section, and highstand deposits, cover a wide range of total organic carbon content (<0.1% to 2.19%). Smackover samples from transgressive and condensed section deposits average 0.81% total organic carbon and those from highstand deposits average 0.29%. Sassen et al. (1987) reported ranges of 0.23% to 1.56% total organic carbon for lower and middle Smackover rocks. For 149 samples analyzed for the lower and middle Smackover (transgressive and condensed section deposits) in Mississippi, Alabama, and Florida, Sassen et al. (1987) recorded an average of 0.48% total organic carbon, and for the 193 upper Smackover samples (highstand deposits) from this area, they reported an average of 0.25% total organic carbon. Oehler (1984) indicated that lower and middle Smackover rocks (transgressive and condensed section deposits) of the Mississippi, Alabama, and Florida area have total organic carbon contents of 0.05% to 2.52% and average about 0.48%.

The relationship of total organic carbon content with the depositional sequence component for the Smackover as determined from core samples and geophysical log characteristics is shown in Figure 8. In this particular well (Alabama Oil and Gas Board Permit Number 3648, Clarke County, Alabama), the entire Smackover section was cored. In general, higher total organic carbon percentages occur in the condensed section deposits (2.01% to 2.19%) and lower percentages are present in the highstand component of the section (0.20% to 0.64%). This observation is consistent with the total organic carbon analysis data reported by Sassen et al. (1987) for the Smackover Formation.

The Jurassic samples analyzed show moderate to advanced degrees of conversion of organic matter to hydrocarbons and samples of more deeply buried rocks show advanced degrees of hydrocarbon destruction and loss (Figure 9). The dominant kerogen types in the Smackover are algal and amorphous. Algal and amorphous kerogen, enriched in hydrogen,

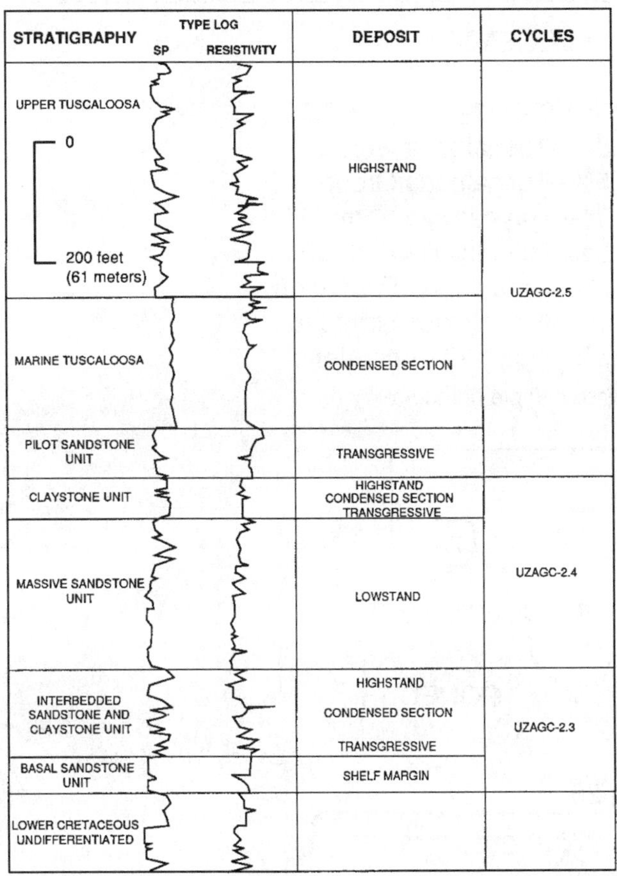

STRATIGRAPHY
TYPE LOG
SP
RESISTIVITY
DEPOSIT
CYCLES
UPPER TUSCALOOSA
0
200 feet
(61 meters)
HIGHSTAND
UZAGC-2.5
MARINE TUSCALOOSA
CONDENSED SECTION
PILOT SANDSTONE UNIT
TRANSGRESSIVE
CLAYSTONE UNIT
HIGHSTAND
CONDENSED SECTION
TRANSGRESSIVE
MASSIVE SANDSTONE UNIT
LOWSTAND
UZAGC-2.4
INTERBEDDED SANDSTONE AND CLAYSTONE UNIT
HIGHSTAND
CONDENSED SECTION
TRANSGRESSIVE
UZAGC-2.3
BASAL SANDSTONE UNIT
SHELF MARGIN
LOWER CRETACEOUS UNDIFFERENTIATED

Figure 6. Type log of Upper Cretaceous Tuscaloosa strata in southwest Alabama (Belden and Blake Corp., Wall et al. Unit # 3-9). See Figure 1 for location of type-log wells. SP = spontaneous potential.

are more oil prone than herbaceous kerogen, which is deficient in hydrogen (Tissot and Welte, 1984). Oehler (1984) and Sassen et al. (1987) also reported that algal-derived amorphous organic matter is the dominant kerogen type in the Smackover. These samples exhibit thermal alteration indices of 2^- to 3^+ (Table 1). Oehler (1984) recorded thermal alteration indices of 3^- to $>3^+$ for Smackover samples from this area, while Sassen et al. (1987) reported indices of 2^- to 4 for the Smackover of Mississippi, Alabama, and Florida. The generation of crude oil from potential source rocks in the Gulf of Mexico area is believed by Nunn and Sassen (1986) to have been initiated at a level of thermal maturity equivalent to a mean vitrinite reflectance of 0.55% (equivalent thermal alteration index of 2^- to 2) and concluded at a level of thermal maturity equivalent to a mean vitrinite reflectance of about 1.5% (equivalent to thermal alteration index of 3^- to 3^+).

The Tuscaloosa rocks show low degrees of conversion to hydrocarbons. The thermal alteration indices of these rocks are 1^+ to 2^-. The dominant kerogen types are herbaceous and amorphous.

DISCUSSION

Sarg (1988) and Loutit et al. (1988) have demonstrated that condensed section deposits in carbonates and siliciclastics are generally fine-grained rocks that often contain relatively high concentrations of organic matter. In carbonate sequences, transgressive deposits associated with a catch-up system (Sarg, 1988) also consist predominantly of fine-grained carbonate rocks containing significant quantities of organic matter. A catch-up system is characterized by a relatively slow rate of sediment accumulation (Sarg, 1988). In addition, transgressive deposits associated with siliciclastics commonly consist of relatively fine-grained sediment containing high concentrations of organic matter (Pasley, 1991).

In the Mesozoic stratigraphic succession of southwest Alabama, the shales of the Norphlet, the stromatolitic carbonate mudstones of the Smackover, and the claystones of the Tuscaloosa accumulated, in part, as condensed sections associated with maximum transgression within genetically-related, unconformity-bounded depositional sequences. Condensed sections are deposited during relative sea level maxima which result in active sedimentation moving landward and in relatively low sedimentation or sediment starvation in more seaward positions on the shelf. Because of relatively low sedimentation rates during deposition, the potential exists for condensed section strata to contain high concentrations of organic matter due to the lack of sediment dilution of these materials which would result from more rapid sediment accumulation; this organic matter can be preserved if oceanographic conditions of the overlying water column are favorable (Loutit et al., 1988). Thus, these deposits have the potential to be hydrocarbon source rocks if subsequent burial and thermal conditions are favorable for thermogenic hydrocarbon generation (Claypool and Mancini, 1989). In addition, all but the predominantly grain-rich rocks in the lowermost phase of the transgressive systems tract of the LZAGC-4.1 depositional cycle were deposited in a catch-up carbonate regime. Consequently, rocks above the basal grain-rich interval are predominantly fine-grained and contain a significant amount of organic carbon (generally more than 0.3%); thus, these rocks are potential hydrocarbon source rocks.

The results of the source rock analyses performed as part of this study, in combination with those of Oehler (1984) and Sassen et al. (1987), indicate that the petroleum source rock potential of the lower and middle Smackover carbonate mudstones of the LZAGC-4.1 depositional cycle has been optimized by the combination of favorable conditions of deposition, preservation, and subsequent burial and thermal histories (Figure 9). These rocks contain ample amounts of algal and algal-derived amorphous kerogens which have been subjected to favorable burial and thermal history conditions, including adequate time-temperature effects, conducive to thermogenic petroleum generation. The Norphlet intertidal shales of the LZAGC-3.1 cycle are low in total organic car-

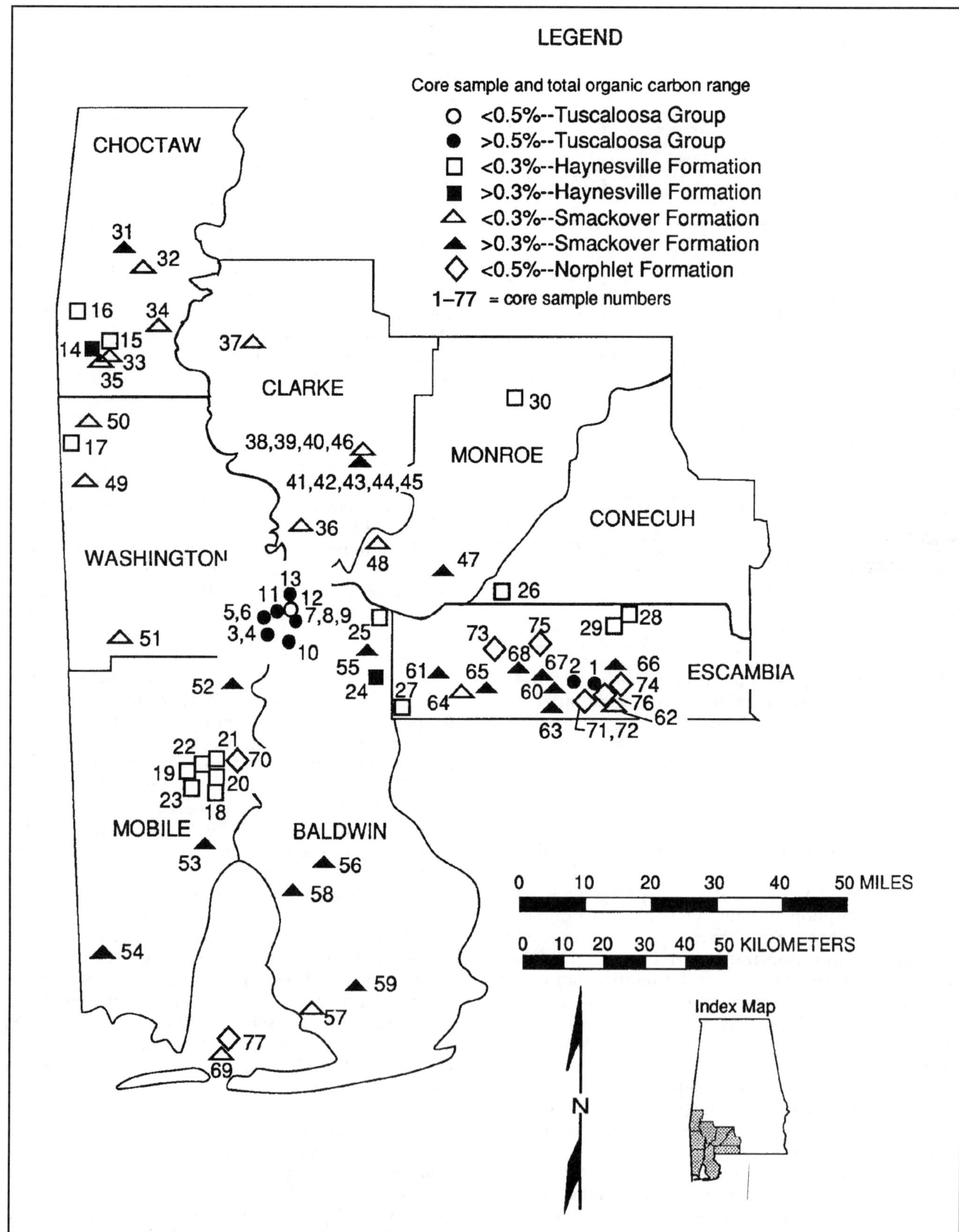

Figure 7. Location of core samples analyzed in this study. Numbers refer to core sample numbers (Table 1). Siliciclastics having a minimum of 0.5% total organic carbon and carbonate rocks containing as much as 0.3% total organic carbon are considered potential hydrocarbon source rocks.

bon (Table 1). These Norphlet rocks have experienced burial and thermal conditions favorable for thermogenic hydrocarbon generation, but depositional conditions resulting in low rates of accumulation and/or preservation of organic carbon have limited their potential as source rocks (Table 1). It should be noted that some workers (Marzano et al., 1988) have speculated that offshore facies of the Norphlet may have served as a hydrocarbon source for natural gas being produced from that unit in the offshore area. The Tuscaloosa marine-shelf claystones of the UZAGC-2.5 cycle are rich in herbaceous and herbaceous- and algal-derived amorphous kerogen (Table 1), but have not been subjected to burial and thermal conditions favorable for thermogenic petroleum generation.

Based on the results of this study, the organic-rich, laminated, lower to middle Smackover carbonate mudstones of the LZAGC-4.1 depositional sequence are the principal source rocks for the hydrocarbon resources of southwest Alabama. Carbonate mudstones also occur in the highstand systems tract of the sequence, but these strata are lower in total organic carbon, containing about three times less total organic carbon than the laminated carbonate mudstones of the transgressive systems tract and associated condensed section.

The differences in Smackover organic facies are attributed to different conditions of deposition of the various component systems tracts within the depositional sequence. The lower and middle Smackover lithofacies represent low-energy, shallow-water, hypersaline, anoxic intertidal to subtidal depositional conditions that occurred during progressive marine transgression. Lithofacies relationships indicate that transgression outpaced carbonate productivity. This led to the establishment of a catch-up carbonate depositional system which persisted through the deposition of the lower and middle parts of the Smackover (transgressive system tract and associated condensed section); these strata contain the organic-rich, stromatolitic carbonate mudstones. Mud-rich parasequences, such as those observed in the lower and middle Smackover, are typical of catch-up carbonate systems (Sarg, 1988). Catch-up systems are characterized by relatively slow rates of sediment accumulation which may result in anaerobic and hypersaline environmental conditions (Sarg, 1988). The low-energy Smackover depositional conditions favored the growth of algal mats and the hypersaline and oxygen-poor conditions precluded the establishment of large populations of grazing herbivores in this environment. These factors enhanced the preservation of the organic material.

In the Manila and Conecuh embayment areas (Figure 1), the carbonate mudstones of the Smackover catch-up system contain a significant percentage of kerogen derived from terrestrial sources (Claypool and Mancini, 1989; Sassen, 1989). These carbonate mudstones accumulated in embayments adjacent to the Appalachian structural trend to the north and east and therefore experienced greater input of clay minerals and terrigenous organic matter than the Mississippi interior salt basin, where hypersaline, anoxic conditions dominated during the Smackover marine transgression. These differences in depositional conditions and resulting organic facies have produced variations in the molecular and isotopic compositions (Figure 10) of the Mesozoic oils from southwest Alabama (Claypool and Mancini, 1989).

The upper Smackover lithofacies represent higher energy, normal, open-marine intertidal to subtidal, upward-shoaling mudstone to grainstone parasequence sets that accumulated as part of a progradational, keep-up carbonate system. The carbonate mudstones associated with this lithofacies are highly bioturbated unlike the majority of carbonate mudstones in the lower and middle Smackover (Benson, 1988; Sassen and Moore, 1988). Grain-rich, mud-poor parasequences are typical of keep-up carbonate systems (Sarg, 1988). Such a system is characterized by a relatively rapid rate of sediment accumulation which results in shallow, normal open-marine conditions (Sarg, 1988). In these higher-energy, normal marine environments, organic matter, particularly algal mats, is not as abundant. High energy impedes the growth of these mats and normal marine conditions support populations of algal-ingesting herbivores. The high degree of bioturbation observed in these mudstones records the destructive ability of these invertebrates.

As evidenced by the large petroleum resources of southwest Alabama (more than 147.2 million barrels (MMB) of oil, 162 MMB of condensate, and 1.42 trillion ft^3 (40 billion m^3) of natural gas produced from Jurassic-sourced reservoirs as of May, 1991), the hydrocarbon generating capacity of the lower and middle Smackover laminated carbonate mudstones was substantial. The source rock capability of these rocks was enhanced by a number of factors. First, pressure solution acted to concentrate algal organic matter along stylolitic surfaces (Sassen et al., 1987). Second, the Smackover was subjected to elevated temperatures during the early stages of its burial history due to initial rapid sediment accumulation and to tectonic processes associated with early rifting of the Gulf of Mexico basin, and these elevated temperatures resulted in the generation and migration of liquid hydrocarbons from the Smackover source rocks during the Cretaceous (Nunn and Sassen, 1986). Further, diagenetic processes such as dissolution and fracturing in association with salt movement provided enhanced migration pathways, and impermeable evaporites overlying Smackover reservoirs acted as effective roof rocks (Sassen, 1989).

Koons et al. (1974), in their study of the composition of oils from the Tuscaloosa reservoirs in Mississippi and Alabama, concluded that two different oil families were represented. One family was an indigenous lower Tuscaloosa oil largely confined to a region in Mississippi where the Tuscaloosa was buried to depths of 10,000 feet or greater. The other oil family they recognized, which included oils from fields in southwest Alabama, was thought to have

Table 1. Kerogen analyses for core samples, southwest Alabama.

Core sample number[1]	Well permit number	County/ area[2]	Rock unit[3]	Depth (m)	Total organic carbon %	Kerogen type[4]	TAI 1–5 scale	Sequence component[5]
1	355	Esc	Tus	1772.1	1.18	Am	2^-	CS
2	427	Esc	Tus	1853.2	2.63	Am	2^-	CS
3	199	Cla	Tus	1592.0	0.76	Herb	1^+	CS
4	199	Cla	Tus	1607.8	2.90	Am	1^+	CS
5	243	Cla	Tus	1602.0	1.79	Am	1^+	CS
6	243	Cla	Tus	1645.9	1.10	Woody	1^+	HS
7	245	Cla	Tus	1600.2	0.88	Herb	1^+	CS
8	245	Cla	Tus	1604.8	2.91	Am	1^+	CS
9	245	Cla	Tus	1652.0	0.54	Woody	1^+	T
10	257	Cla	Tus	1661.2	0.92	Herb	2^-	T
11	269	Cla	Tus	1612.4	1.14	Herb	2^-	CS
12	269	Cla	Tus	1649.0	0.44	Herb	2^-	T
13	2182	Cla	Tus	1606.6	2.75	Am	2^-	CS
14	1582	Cho	Hay	4035.2	0.32	—	—	HS
15	1659	Cho	Hay	4156.9	0.11	—	—	HS
16	2054	Cho	Hay	3732.9	0.09	—	—	HS
17	1910	Was	Hay	4943.9	0.14	—	—	HS
18	2073	Mob	Hay	5561.1	0.07	—	—	HS
19	2098	Mob	Hay	5668.7	0.13	—	—	HS
20	2101	Mob	Hay	5601.3	0.05	—	—	HS
21	2207	Mob	Hay	5564.7	0.23	—	—	HS
22	2250	Mob	Hay	5611.1	0.12	—	—	HS
23	2383	Mob	Hay	5641.8	0.07	—	—	HS
24	1584	Bal	Hay	4944.8	0.35	—	—	HS
25	3299	Bal	Hay	4572.6	0.05	—	—	HS
26	2212	Con	Hay	4264.5	0.07	—	—	HS
27	1672	Esc	Hay	4960.3	0.06	—	—	HS
28	4925	Esc	Hay	4094.1	0.12	—	—	HS
29	8790	Esc	Hay	4414.1	0.07	—	—	HS
30	1352	Mon	Hay	2810.6	0.04	—	—	HS
31	1330	Cho	Smk	3210.2	0.30	Am(Al)	2^-	HS
32	1412	Cho	Smk	3233.9	0.07	—	—	HS
33	1659	Cho	Smk	4183.1	0.09	—	—	HS
34	1875	Cho	Smk	3732.3	0.24	Am(Al)	2^-	HS
35	2205	Cho	Smk	4160.8	0.09	—	—	HS

[1]See Figure 7 for location of core samples.
[2]Bal—Baldwin, Cho—Choctaw, Cla—Clarke, Esc—Escambia, Mob—Mobile, Mon—Monroe, Was—Washington, M Bay—Mobile Bay.
[3]Tus—Tuscaloosa, Hay—Haynesville, Smk—Smackover, Nor—Norphlet.
[4]Am—amorphous, Al—algal, Herb—herbaceous.
[5]HS—highstand, CS—condensed section, T—transgressive.

been derived from a deeper source, such as Upper Jurassic Smackover rocks, and to have migrated into the Tuscaloosa reservoirs along faults. Claypool and Mancini (1989) demonstrated not only the similarities in molecular and isotopic characteristics of Jurassic and Cretaceous oils in southwest Alabama, but also the similarities in composition of probable Upper Jurassic Smackover source rocks and Jurassic and Cretaceous oils (Figure 10). Based on these similarities, Claypool and Mancini (1989) concurred with other workers, including Oehler (1984), Sassen et al. (1987), Sofer (1988), and Sassen (1989), that the Smackover was the main source of Mesozoic oils in southwest Alabama. These Mesozoic oils are enriched in aromatic hydrocarbons, have pristane/phytane ratios of less than one, are enriched in sulfur, and have saturated and aromatic hydrocarbon fractions with carbon isotope differences of almost zero. These

Table 1. Continued

Core sample number[1]	Well permit number	County/ area[2]	Rock unit[3]	Depth (m)	Total organic carbon %	Kerogen type[4]	TAI 1–5 scale	Sequence component[5]
36	735	Cla	Smk	3400.0	0.29	Am(Al)	2^-	HS
37	1438	Cla	Smk	3346.7	0.11	Am(Al)	2^-	T
38	3648	Cla	Smk	4014.5	0.20	—	—	HS
39	3648	Cla	Smk	4020.0	0.28	—	—	HS
40	3648	Cla	Smk	4034.3	0.28	—	—	HS
41	3648	Cla	Smk	4048.0	0.64	—	—	HS
42	3648	Cla	Smk	4064.2	2.01	—	—	CS
43	3648	Cla	Smk	4102.0	2.19	—	—	CS
44	3648	Cla	Smk	4106.0	0.36	—	—	T
45	3648	Cla	Smk	4124.2	1.10	—	—	T
46	3648	Cla	Smk	4131.3	0.25	—	—	T
47	1592	Mon	Smk	4341.9	0.54	Am	2^+	HS
48	4673	Mon	Smk	4448.9	0.05	—	—	HS
49	1596	Was	Smk	4927.7	0.19	Am	3	HS
50	2478	Was	Smk	4919.2	0.17	—	—	CS
51	2645	Was	Smk	5900.6	0.25	Am	3	CS
52	2126	Mob	Smk	5275.5	0.81	Am(Al)	2^+	CS
53	2178	Mob	Smk	5580.6	0.89	Am	3^-	HS
54	3735	Mob	Smk	5856.7	1.74	—	—	CS
55	1584	Bal	Smk	4945.4	0.42	Am	2^-	HS
56	2075	Bal	Smk	5588.5	0.49	Am	3^-	CS
57	2587	Bal	Smk	6053.3	0.20	Am(Al)	3^+	CS
58	2621	Bal	Smk	5629.7	1.17	Am	3	T/CS
59	2915	Bal	Smk	5915.9	0.88	Am	3^+	T/CS
60	1460	Esc	Smk	4664.7	0.33	—	—	T
61	1674	Esc	Smk	4877.7	0.32	Am	2^+	T
62	1766	Esc	Smk	4671.4	0.26	Am(Al)	2^+	CS
63	1770	Esc	Smk	4766.2	0.99	Am	2^+	CS
64	1837	Esc	Smk	4760.7	0.17	—	—	HS
65	1895	Esc	Smk	4758.2	0.91	Am(Al)	2^+	CS
66	2041	Esc	Smk	4493.4	1.35	Am	2	CS
67	3402	Esc	Smk	4728.7	1.05	—	—	T/CS
68	3900	Esc	Smk	4663.7	0.91	Am	2^+	T/CS
69	3632	M Bay	Smk	6003.0	0.25	—	—	HS
70	1978	Mob	Nor	5615.0	0.09	—	—	CS
71	2991	Esc	Nor	4720.4	0.07	—	—	CS
72	2991	Esc	Nor	4723.2	0.17	—	—	CS
73	4183	Esc	Nor	4695.7	0.08	—	—	CS
74	4395	Esc	Nor	4545.8	0.07	—	—	CS
75	4543	Esc	Nor	4590.0	0.12	—	—	CS
76	4693	Esc	Nor	4536.0	0.07	—	—	CS
77	3632	M Bay	Nor	6238.0	0.11	—	—	CS

characteristics indicate that these oils have been derived from algal kerogen associated with marine carbonates that accumulated in a reducing environment (Claypool and Mancini, 1989). This observation is consistent with the findings of this study. Of the Mesozoic condensed section deposits studied, only the Smackover carbonate mudstones of the condensed section deposits of the LZAGC-4.1 depositional sequence have had their petroleum source rock potential realized.

SUMMARY

Although condensed sections within depositional sequences, in general, should have the highest source rock potential, specific environmental, preservational and/or burial and thermal history conditions within a particular basin will dictate whether or not this potential is realized. This relationship is shown by the condensed sections of the Mesozoic depositional sequences in southwest Alabama. Smackover con-

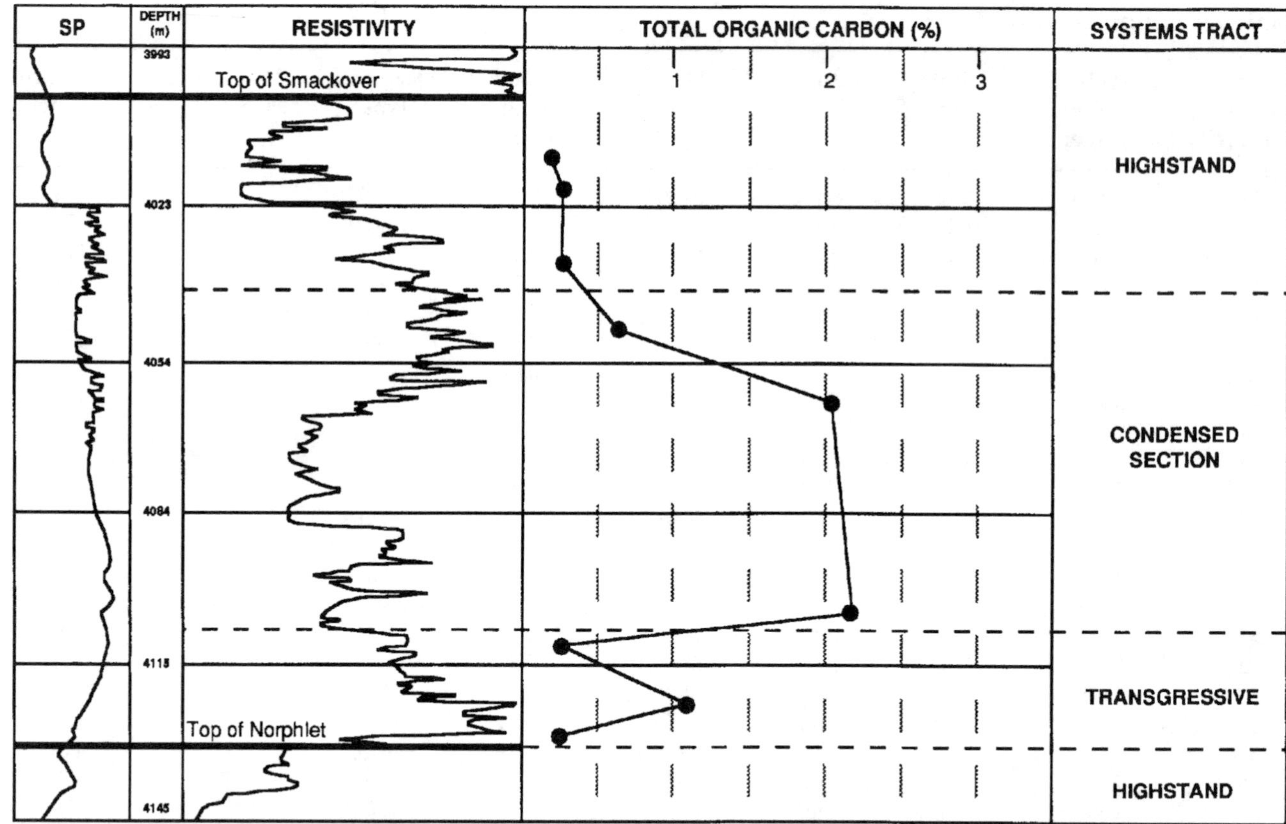

Figure 8. Spontaneous potential (SP)–resistivity log of Shell Oil Co., #1 Neal et al. Unit 30-1 (Alabama Oil and Gas Board permit number 3648) well showing relationship of total organic carbon content to Smackover depositional sequence components. See Figure 1 for location of well.

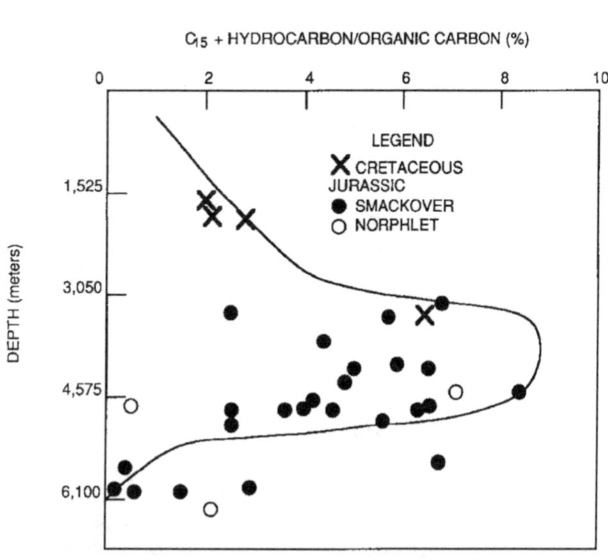

Figure 9. Extractable (C_{15+}) hydrocarbon/organic carbon ratio of core samples for southwest Alabama plotted against burial history (modified from Claypool and Mancini, 1989).

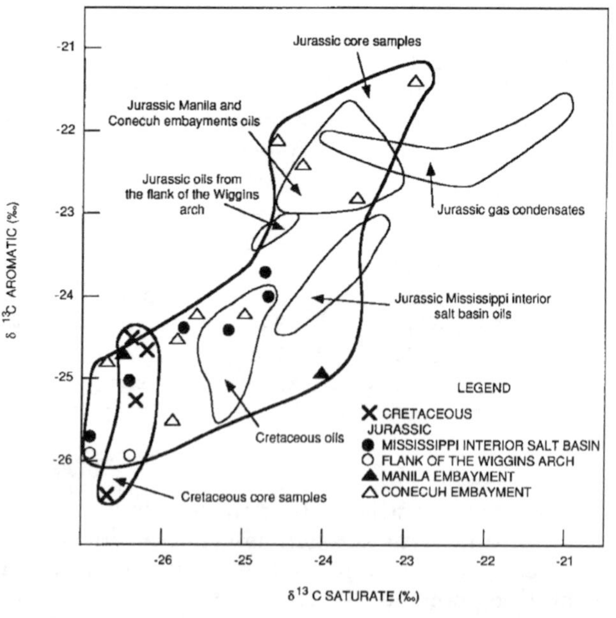

Figure 10. Carbon isotope ratios of C_{15+} saturated and aromatic hydrocarbon fractions of rock extracts from Cretaceous and Jurassic core samples from southwest Alabama (modified from Claypool and Mancini, 1989).

densed section and transgressive laminated carbonate mudstones deposited in a catch-up carbonate system served as the petroleum source for the majority of Mesozoic reservoirs in southwest Alabama. Petroleum geologists can use sequence stratigraphy as a tool to help identify stratigraphic intervals that might have potential to contain hydrocarbon source rocks; however, only through geochemical analyses can the actual source rock potential be determined.

ACKNOWLEDGMENTS

The authors would like to thank Drs. Wayne M. Ahr and Dudley D. Rice for their constructive reviews of this manuscript. Their thoughtful comments greatly improved the final product.

REFERENCES CITED

Benson, D.J., 1988, Depositional history of the Smackover Formation in southwest Alabama: Gulf Coast Association of Geological Societies Transactions, v. 38, p. 197-205.

Claypool, G.E., and E.A. Mancini, 1989, Geochemical relationships of petroleum in Mesozoic reservoirs to carbonate source rocks of Jurassic Smackover Formation, southwestern Alabama: American Association of Petroleum Geologists Bulletin, v. 73, p. 904-924.

Eaves, E., 1976, Citronelle oil field, Mobile County, Alabama: American Association of Petroleum Geologists Memoir 24, p. 259-275.

Haq, B.U., J. Hardenbol, and P.R. Vail, 1988, Mesozoic and Cenozoic chronostratigraphy and cycles of sea-level change, *in* C.K. Wilgus, B.S. Hastings, C.G.St.C. Kendall, H.W. Posamentier, C.A. Ross, and J.C. Van Wagoner, eds., Sea-level changes: an integrated approach: Society of Economic Paleontologists and Mineralogists Special Publication 42, p. 71-108.

Koons, C.B., J.G. Bond, and F.L. Peirce, 1974, Effects of depositional environment and postdepositional history on chemical composition of lower Tuscaloosa oils: American Association of Petroleum Geologists Bulletin, v. 58, p. 1272-1280.

Loutit, T.S., J. Hardenbol, P.R. Vail, and G.R. Baum, 1988, Condensed sections: The key to age determination and correlation of continental margin sequences, *in* C.K. Wilgus, B.S. Hastings, C.G.St.C. Kendall, H.W. Posamentier, C.A. Ross, and J.C. Van Wagoner, eds., Sea-level changes: an integrated approach: Society of Economic Paleontologists and Mineralogists Special Publication 42, p. 184-213.

Mancini, E.A., and D.J. Benson, 1980, Regional stratigraphy of Upper Jurassic Smackover carbonates of southwest Alabama: Gulf Coast Association of Geological Societies Transactions, v. 30, p. 151-165.

Mancini, E.A., and R.M. Mink, 1985, Petroleum production and hydrocarbon potential of Alabama's coastal plain and territorial waters, *in* B.F. Perkins and G.B. Martin, eds., Habitat of oil and gas in the Gulf Coast: Proceedings of the Fourth Annual Research Conference, Gulf Coast Section—Society of Economic Paleontologists and Mineralogists Foundation, p. 43-60.

Mancini, E.A., and B.H. Tew, 1991, Relationships of Paleogene stage and planktonic foraminiferal zone boundaries to lithostratigraphic and allostratigraphic contacts in the eastern Gulf Coastal plain: Journal of Foraminiferal Research, v. 21, p. 48-66.

Mancini, E.A., B.H. Tew, and R.M. Mink, 1990, Jurassic sequence stratigraphy in the Mississippi interior salt basin of Alabama: Gulf Coast Association of Geological Societies Transactions, v. 40, p. 521-529.

Mancini, E.A., R.M. Mink, B.L. Bearden, and R.P. Wilkerson, 1985, Norphlet Formation (Upper Jurassic) of southwestern and offshore Alabama: Environments of deposition and petroleum geology: American Association of Petroleum Geologists Bulletin, v. 69, p. 881-898.

Mancini, E.A., R.M. Mink, J.W. Payton, and B.L. Bearden, 1987, Environments of deposition and petroleum geology of Tuscaloosa Group (Upper Cretaceous), South Carlton and Pollard fields, southwestern Alabama: American Association of Petroleum Geologists Bulletin, v. 71, p. 1128-1142.

Mann, S.D., 1988, Subaqueous evaporites of the Buckner Member, Haynesville Formation, northeastern Mobile County, Alabama: Gulf Coast Association of Geological Societies Transactions, v. 38, p. 187-196.

Martin, R.G., 1978, Northern and eastern Gulf of Mexico continental margin: stratigraphic and structural framework, *in* A.H. Bouma, G.T. Moore, and J.M. Coleman, eds., Framework, facies, and oil-trapping characteristics of the upper continental margin: American Association of Petroleum Geologists Studies in Geology 7, p. 21-42.

Marzano, M.S., G.M. Pense, and P. Andronaco, 1988, A comparison of the Jurassic Norphlet Formation in Mary Ann field, Mobile Bay, Alabama to onshore regional Norphlet trends: Gulf Coast Association of Geological Societies Transactions, v. 38, p. 85-100.

Miller, J.A., 1982, Structural control of Jurassic sedimentation in Alabama and Florida: American Association of Petroleum Geologists Bulletin, v. 66, p. 1289-1301.

Moore, C.H., 1984, The upper Smackover of the Gulf rim: depositional systems, diagenesis, porosity evolution and hydrocarbon production, *in* W.P.S. Ventress, D.G. Bebout, B.F. Perkins, and C.H. Moore, eds., The Jurassic of the Gulf rim: Proceedings of the Third Annual Research Conference, Gulf Coast Section, Society of Economic Paleontologists and Mineralogists, p. 283-307.

Moore, T., 1983, Cotton Valley depositional systems of Mississippi: Gulf Coast Association of Geological Societies Transactions, v. 33, p. 163-167.

Murray, G.E., 1961, Geology of the Atlantic and Gulf

coastal province of North America: New York, Harper and Brothers, 692 p.

Nunn, J.A., and R. Sassen, 1986, The framework of hydrocarbon generation and migration, Gulf of Mexico continental slope: Gulf Coast Association of Geological Societies Transactions, v. 36, p. 257-262.

Oehler, J.H., 1984, Carbonate source rocks in the Jurassic Smackover trend of Mississippi, Alabama, and Florida, *in* J.G. Palacas, ed., Petroleum geochemistry and source rock potential of carbonate rocks: American Association of Petroleum Geologists Studies in Geology 18, p. 63-69.

Pasley, M.A., 1991, Organic matter variations in a depositional sequence: Implications for use of source rock data in sequence stratigraphy (abs.): American Association of Petroleum Geologists Bulletin, v. 75, p. 650.

Salvador, A., 1987, Late Triassic-Jurassic paleogeography and origin of Gulf of Mexico basin: American Association of Petroleum Geologists Bulletin, v. 71, p. 419-451.

Sarg, J.F., 1988, Carbonate sequence stratigraphy, *in* C.K. Wilgus, B.S. Hastings, C.G.St.C. Kendall, H.W. Posamentier, C.A. Ross, and J.C. Van Wagoner, eds., Sea-level changes: an integrated approach: Society of Economic Paleontologists and Mineralogists Special Publication 42, p. 155-181.

Sassen, R., 1989, Migration of crude oil from the Smackover source rock to Jurassic and Cretaceous reservoirs of the northern Gulf rim: Organic Geochemistry, v. 14, p. 51-60.

Sassen R., and C.H. Moore, 1988, Framework of hydrocarbon generation and destruction in eastern Smackover trend: American Association of Petroleum Geologists Bulletin, v. 72, p. 649-663.

Sassen, R., C.H. Moore, and F.C. Meendsen, 1987, Distribution of hydrocarbon source potential in the Jurassic Smackover Formation: Organic Geochemistry, v. 11, p. 379-383.

Sofer, Z., 1988, Stable carbon isotope compositions of crude oils: application to source depositional environments and petroleum alteration: American Association of Petroleum Geologists Bulletin, v. 68, p. 31-49.

Tew, B.H., E.A. Mancini, and R.M. Mink, 1991, Jurassic sequence stratigraphy of the eastern Gulf Coastal Plain (abs.): Applications to hydrocarbon exploration: American Association of Petroleum Geologists Bulletin, v. 75, p. 680.

Tissot, B.P., and D.H. Welte, 1984, Petroleum formation and occurrence (2nd ed.): New York, Springer–Verlag, 699 p.

Tolson, J.S., C.W. Copeland, and B.L. Bearden, 1983, Stratigraphic profiles of Jurassic strata in the western part of the Alabama Coastal Plain: Alabama Geological Survey Bulletin 122, 425 p.

Wilson, G.V., 1975, Early differential subsidence and configuration of the northern Gulf Coast basin in southwest Alabama and northwest Florida: Gulf Coast Association of Geological Societies Transactions, v. 25, p. 196-206.

Winter, C.V., Jr., 1954, Pollard field, Escambia County, Alabama: Gulf Coast Association of Geological Societies Transactions, v. 4, p. 121-142.

Wood, M.L., and J.L. Walper, 1974, The evolution of the interior Mesozoic basin and the Gulf of Mexico: Gulf Coast Association of Geological Societies Transactions, v. 24, p. 31-41.

Chapter 11

Internal Stratigraphy and Organic Facies of the Devonian–Mississippian Chattanooga (Woodford) Shale in Oklahoma and Kansas

Michael W. Lambert
Kansas Geological Survey
Lawrence, Kansas, USA[1]

ABSTRACT

The Devonian–Mississippian Chattanooga (Woodford) Shale and equivalent formations are widely distributed in North America. Deposited under generally anoxic conditions during the Kaskaskian marine transgression, they are known to be important petroleum source rocks in many intracratonic basins of the Midcontinent.

The formation can be divided into several members, including the basal Misener Sandstone, and informally designated lower, middle, and upper shale members. The middle shale member is the most radioactive part of the formation, although the radioactivity of all three members decreases to the north. Isopach maps of the shale members in northwestern Oklahoma and Kansas show that the middle shale member has the greatest areal extent and thickness, indicating that it was deposited when the Kaskaskia I transgression was at its maximum. Total organic carbon content of the three shale members decreases to the north, but the middle shale member is the most organic-rich part of the formation and has the greatest component of hydrogen-rich type I and II organic matter, making it the part of the Chattanooga (Woodford) Shale likely to be the best petroleum source rock where the formation is thermally mature.

The shale members of the Chattanooga (Woodford) Shale represent a third-order depositional sequence that is bounded below and above by unconformities. The distribution and organic geochemistry of the shale members suggest that the lower, middle, and upper shale members are the transgressive and early and late highstand systems tracts of the sequence.

INTRODUCTION

Organic-rich shales were deposited over a large part of what is now North America during the Late Devonian. Shale of this age in Nebraska, Kansas, and eastern Oklahoma is called the Chattanooga Shale (Carlson, 1963; Goebel, 1968; Amsden, 1980). However, in the Permian basin of western Texas and southeastern New Mexico, and in the Anadarko basin of western Oklahoma, correlative beds are called the Woodford Shale (Ellison, 1950; Amsden, 1975). Other equivalent formations include the Bakken Formation in the Williston basin of North and South Dakota,

[1]*Current address: Naval Research Laboratory, Stennis Space Center, Mississippi, USA*

Montana, Saskatchewan, and Manitoba (Meissner, 1978), and shales within the New Albany Shale Group in the Illinois basin of Illinois, Indiana, and western Kentucky (Conant and Swanson, 1961).

The Chattanooga Shale has also been called the "Kinderhook Shale" (Ver Wiebe, 1946), a usage that is still common among drillers in Kansas, but this name was based on the erroneous assumption that the shale is entirely Mississippian in age, and its use should be discontinued (Lee, 1956; Lambert, in press). In this chapter, the formation is called the Chattanooga Shale when the discussion is limited to Kansas, and the Woodford Shale when the discussion is limited to western Oklahoma. When the formation in the Midcontinent as a whole is discussed, it is called the Chattanooga (Woodford) Shale. Comer (1991) preferred the name Woodford Formation because in western Texas and southeastern New Mexico both shale and siltstone are major lithologies, but the formation is predominately shale in Oklahoma and Kansas.

Amsden and Klapper (1972) used conodont biostratigraphy to date the basal Misener Sandstone of the Woodford Shale in north-central Oklahoma as Middle to Late Devonian (Givetian to early Famennian) in age. Hass and Huddle (1965) and Over and Barrick (1990), also using conodont biostratigraphy, determined the age of the upper Woodford Shale in south-central Oklahoma to be mostly Late Devonian (Famennian), although the uppermost 2 ft (0.6 m) is Early Mississippian (Kinderhookian).

It has been estimated that 8% of the world's original petroleum reserves were generated from the Chattanooga (Woodford) Shale and equivalent formations (Fritz, 1991). It is believed to be a major petroleum source rock in the Permian basin (Wright, 1963), and geochemical comparison of sedimentary organic matter (kerogen) with produced oil indicates that it is also an important source rock in the Anadarko, Williston, and Illinois basins (Fertl and Chilingarian, 1990; Williams, 1974; Dow, 1974; Bethke et al., 1991). In this chapter, it will be shown that shale members within the Chattanooga (Woodford) Shale of the Midcontinent differ in the amount and kind of kerogen that they contain, suggesting the presence of organic facies. This in turn controls the effectiveness of the formation as a petroleum source rock and has sequence stratigraphic implications.

DEPOSITIONAL SETTING

North America in the Late Devonian was located in the tropics (Woodrow et al., 1973), possibly in low southerly latitudes (Heckel and Witzke, 1979; Witzke and Heckel, 1988; Streel et al., 1990). Structural features that were present in the Midcontinent during deposition of the Chattanooga (Woodford) Shale are illustrated in Figure 1. Topographically positive areas included the ancestral central Kansas uplift of northwestern Kansas (a southern extension of the Transcontinental arch) and the Chautauqua arch of southeastern Kansas (a part of the Ozark dome), features that were connected by the central Kansas arch (Merriam, 1963). Negative areas that were present included the Iowa basin of northeastern Kansas and the Oklahoma basin of southwestern Kansas and western Oklahoma (Johnson, 1989). The depocenter of the Oklahoma basin was the southern Oklahoma aulacogen, the axis of which was in the position of the modern Anadarko basin.

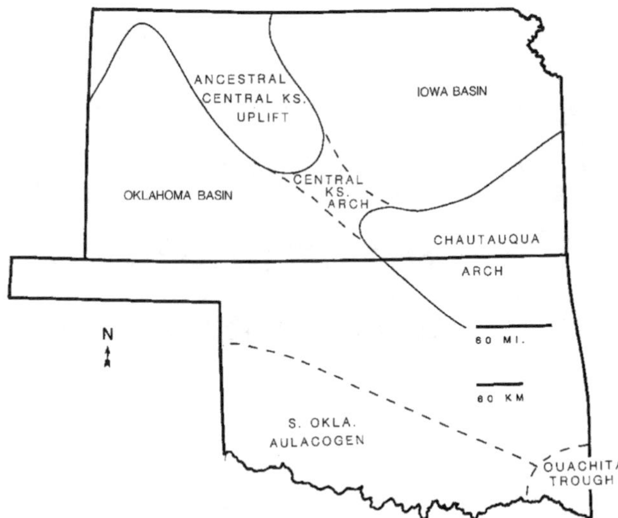

Figure 1. Map showing Late Devonian structures in Kansas and Oklahoma.

The Chattanooga (Woodford) Shale and its equivalents form the base of the Kaskaskia sequence in the Midcontinent (Sloss, 1963). A period of regional exposure and erosion prior to the deposition of these rocks resulted in the incision of major stream channels in pre-Kaskaskian strata, and was followed by a marine transgression that advanced onto the craton from the south (Bunker et al., 1988). A minor regression at the end of Chattanooga (Woodford) Shale deposition marks the upper boundary of the Kaskaskia I subsequence of Sloss (1988) in the study area.

Many of the Devonian shales of North America have a high total organic carbon (TOC) content, and Twenhofel (1939) believed that deposition in anoxic water would preserve the organic matter in such rocks from oxidation or ingestion by burrowing organisms. However, Tourtelot (1979) suggested high rates of biological productivity during deposition as an alternative explanation for the origin of organic-rich shales. Demaison and Moore (1980) argued for anoxia as the cause, stating that in modern oxic environments the bulk sedimentation rate increases with organic content of the sediment, up to the point where clastic dilution actually decreases TOC. Only in anoxic settings, where organic matter would be preserved in any case, does high biological productivity in the oxygenated surface layer of the ocean create organic-rich sediment.

Ettensohn (1985) believed that restricted connections with the open ocean would have promoted density and thermal stratification of the Devonian epicontinental sea in North America, preventing mixing with oxygenated surface waters and causing oxygen-depletion in the lower depths. Many researchers (Byers, 1977; Cluff, 1980; Ettensohn and Elam, 1985) have used the modern Black Sea as a model for anoxic Paleozoic epicontinental seas. Byers defined three general zones of oxygenation in the Black Sea, with well-oxygenated water above a depth of 164 ft (50 m), dysoxic water between depths of 164 to 492 ft (50 to 150 m), and anoxic water beneath 492 ft (150 m). Because the Black Sea is the largest modern anoxic sea, Byers cautioned that this depth zonation may represent maximum values for the three zones. An organic-rich shale deposited in the deep, anoxic bottom waters of an epicontinental sea should have a much more widespread distribution than one deposited in a shallower anoxic basin, such as an anoxic lagoon (Cluff, 1981; McGhee and Bayer, 1985). The Chattanooga (Woodford) Shale and its equivalents are distributed over much of North America, implying a deep-water origin.

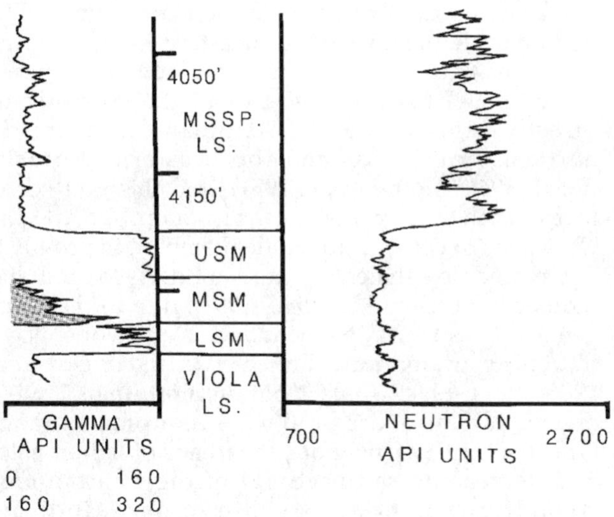

Figure 2. Geophysical log for the Phoenix #1 Orme well, located in Kingman County, south-central Kansas (4-28S-6W). Depth below surface (feet) is shown in central column. All three shale members of the Chattanooga (Woodford) Shale are present. LSM = lower shale member, MSM = middle shale member, USM = upper shale member.

INTERNAL STRATIGRAPHY OF THE CHATTANOOGA (WOODFORD) SHALE

Misener Sandstone Member

Throughout the Midcontinent, a sandstone is sometimes found at the base of the Chattanooga (Woodford) Shale. It is called the Sylamore Sandstone in eastern Oklahoma and northwestern Arkansas (Amsden, 1980; Manger, 1985), but in Kansas and western Oklahoma it is called the Misener Sandstone (Moore et al., 1951; Amsden, 1975). The Misener has an erratic geographic distribution, and where present it is often only 3.3 ft (1 m) or less in thickness (Lee, 1956; Hilpman, 1967; Amsden and Klapper, 1972; Newell, 1989). It apparently developed where Ordovician-age Simpson Group sandstones subcropped beneath the pre-Kaskaskian erosional surface, becoming a source of sand-sized clastic detritus in the early stages of the Kaskaskian transgression. As previously stated, the Misener Sandstone of north-central Oklahoma has been dated as Givetian to early Famennian (Amsden and Klapper, 1972). However, the Misener in Kansas may be younger than this because of the discontinuous nature of the member and the fact that the Kaskaskian transgression advanced from south to north across the craton.

Lower, Middle, and Upper Shale Members

Ellison (1950) and Comer (1991) used geophysical logs to distinguish three shale units within the Woodford Shale of the Permian basin. Hester et al. (1988) found the same three units in the Woodford of northwestern Oklahoma and informally named them the lower, middle, and upper shale members, nomenclature adopted by Lambert (1992) for the Chattanooga Shale of Kansas. Figure 2 is a geophysical log for the Phoenix #1 Orme well in Kingman County, south-central Kansas, showing the highly radioactive middle shale member and less radioactive lower and upper shale members. All three are present in Kansas, although the lower shale member is restricted to the south-central part of the state, and the upper shale member is not present in extreme northern Kansas.

The radioactivity of each shale member decreases to the north. Gamma-ray log response for the lower shale member locally is more than 300 API units in Oklahoma, decreasing to 136 in south-central Kansas. The response for the middle shale member can be more than 320 API units in Oklahoma, decreasing to 240 to 320 in southern Kansas, and 110 to 120 in northern Kansas. Similarly, the radioactivity of the upper shale member is as much as 213 API units in Oklahoma, decreasing to 140 to 150 in southern Kansas, and to 80 to 90 in northern Kansas.

Lee (1956) mentioned a progressive color change in subsurface samples of the Chattanooga Shale of Kansas, from black in the southern part of the state to lighter colors in the north. Examination of cores and well-cutting descriptions confirms that the color change does occur, but it is not the same for each shale member (Lambert, 1992). All three shale members are commonly black shales in Oklahoma, but in south-central Kansas the middle shale member is black, while the upper shale member is gray. The middle shale member in northeastern Kansas is gray to green, with a few darker gray beds. The name "Boice Shale" has been given to a light-colored shale directly beneath the Mississippian carbonates in northeastern Kansas and southeastern Nebraska (Reed, 1946; Moore et al., 1951), and this unit correlates with the middle shale member of the Chattanooga (Woodford) Shale (Lambert, in press).

Organic-rich shales typically are dark-colored and radioactive (Schmoker, 1980, 1981), due to the kerogen acting as both a pigment and as a site of concentration for uranyl ions (Potter et al., 1980; Doveton, 1986). Such black shales contain more than 1 wt.% organic carbon, and 2 to 10 wt.% is a common range (Tourtelot, 1979). Therefore, the trends of lighter color and decreasing radioactivity of the Chattanooga (Woodford) Shale from south to north indicate decreasing organic content to the north. In addition, values that Hester et al. (1989, 1990) calculated from density log curves for wells in northwestern Oklahoma show that the middle shale member has a higher TOC than the other shale members (5.5 wt.% vs. 3.2 and 2.7 for the lower and upper shale members, respectively), demonstrating both vertical and areal zonation of organic content.

Byers (1977), Cluff (1980), and Ettensohn and Elam (1985) reasoned that Devonian black shales were deposited in relatively deep water (deeper than 492 ft or 150 m) by analogy with the modern Black Sea. By reference to the same Black Sea model, the fact that the shale members in Oklahoma are all black shales implies that deep water existed there throughout Chattanooga (Woodford) Shale deposition. Deep water can also be inferred to have been present in south-central Kansas during deposition of the black shale of the middle shale member, but water depths must have been relatively shallow by the time the gray shale of the upper shale member was deposited. Progressively shallower water to the north (the direction of the Transcontinental arch) during middle shale member deposition is suggested by the transition from black shale in south-central Kansas to gray and green shale in northeastern Kansas.

DISTRIBUTION OF CHATTANOOGA (WOODFORD) SHALE MEMBERS

Lower Shale Member

Figure 3 is an isopach map of the lower shale member of the Chattanooga (Woodford) Shale in

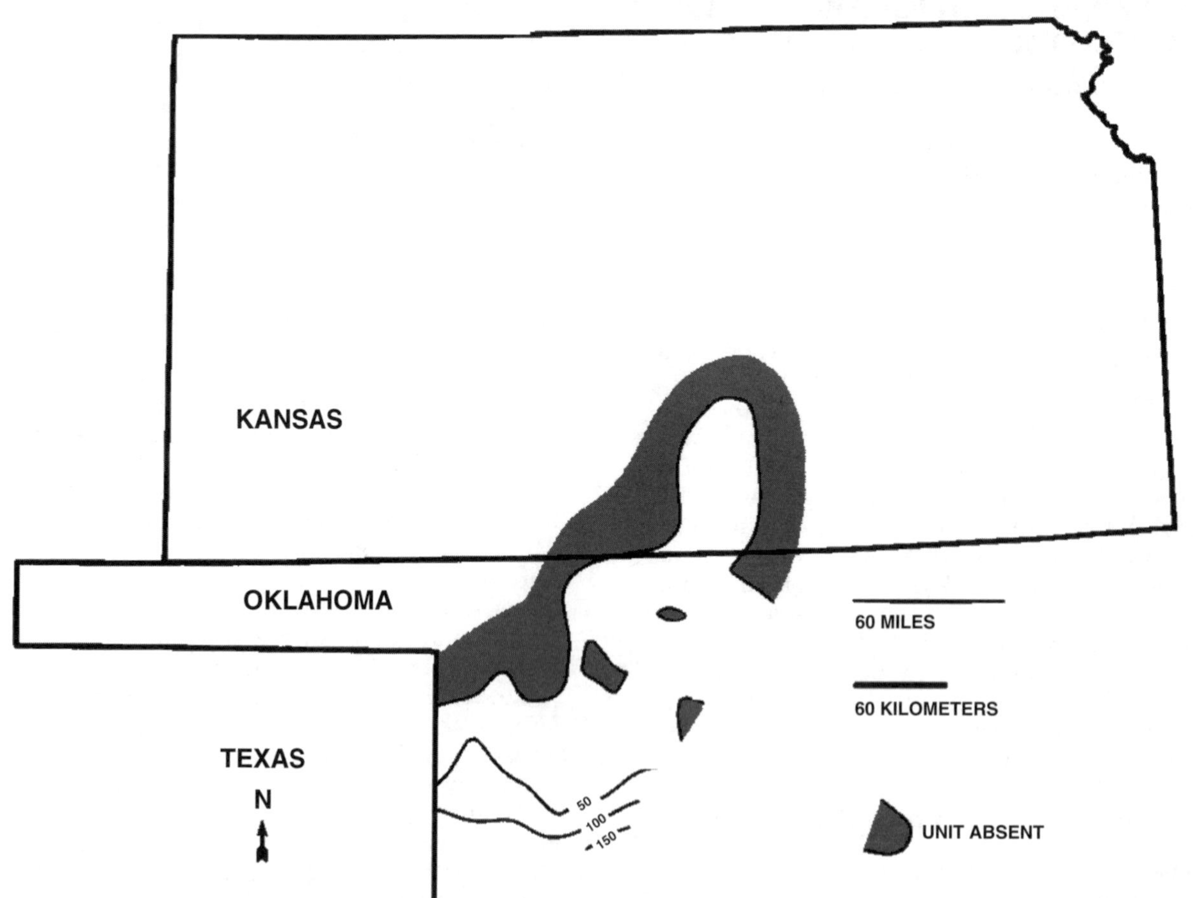

Figure 3. Isopach map of the lower shale member of the Chattanooga (Woodford) Shale. Contour interval is 50 ft (15.2 m). Figure after Lambert (this paper) and Hester et al. (1990).

Kansas and northwestern Oklahoma. The isopach maps included in this chapter incorporate mapping by Hester et al. (1990) for Oklahoma and by the present author for Kansas. The zero edge of the Chattanooga Shale in Kansas is from the sub-Chattanooga subcrop map of Newell (1987) for northern Kansas, and from a Kansas Geological Survey open-file report by Wilson and Berendsen (1988) for southeastern Kansas.

The least areally widespread of the shale members, the lower shale member is present only in the southern part of the mapped area. The Kaskaskia I transgression had apparently not yet reached its greatest northward extent on the craton by the time this shale member was deposited. The greatest thickness of the lower shale member (about 150 ft or 45.7 m) is found in the southern Oklahoma aulacogen. However, it is commonly less than 50 ft (15.2 m) thick over most of the study area. A transmitted light photomicrograph of the lower shale member is shown in Figure 4. Quartz silt (light) is relatively abundant in the clay matrix, and glauconite silt (gray) is also present.

Middle Shale Member

The middle shale member of the Chattanooga (Woodford) Shale has the greatest areal extent of any shale member (Figure 5), and is found throughout northwestern Oklahoma and eastern Kansas. Its widespread distribution indicates that the Kaskaskia I transgression was at its maximum when this member was deposited. The thickest part of the Chattanooga (Woodford) Shale, the middle shale member is about 200 ft (60.9 m) thick in the central Iowa basin of northeastern Kansas and in the McPherson Valley area of central Kansas. The McPherson Valley was first described by Lee (1940, 1956), who believed it to be an east–west stream channel incised onto the pre-Chattanooga (Woodford) surface along the southwestern margin of the Iowa basin. In contrast, the middle shale member in southeastern Kansas is thin (less than 50 ft or 15.2 m thick) due to the presence of the Chautauqua arch during deposition. Post-depositional erosion in central and northeastern Kansas has removed the middle shale member over the modern central Kansas uplift and the Nemaha anticline (Lee, 1956).

Thin limestone and dolomite beds (usually less than 3.3 ft or 1 m thick) are found within the upper part of the middle shale member on geophysical logs and in well-cutting descriptions of wells in northeastern Kansas and southeastern Nebraska (Lambert, in press). Assuming that these carbonate beds were deposited in the photic zone, this suggests that the relatively shallow depositional environment in the north may have become even shallower late in the deposition of the middle shale member. An additional indication of this is the fact that hematite oolites are present near the top of the middle shale member in northeastern Kansas.

Figure 6 is a transmitted light photomicrograph of the middle shale member. Pyrite grains appear in the lower left, with bedding-parallel wisps of organic matter contorted around them. Fecal pellets that have been flattened by compaction appear in the upper right. Only a few grains of quartz silt are present in this sample from south-central Oklahoma, but dolomite silt becomes common in the middle shale member in northeastern Kansas.

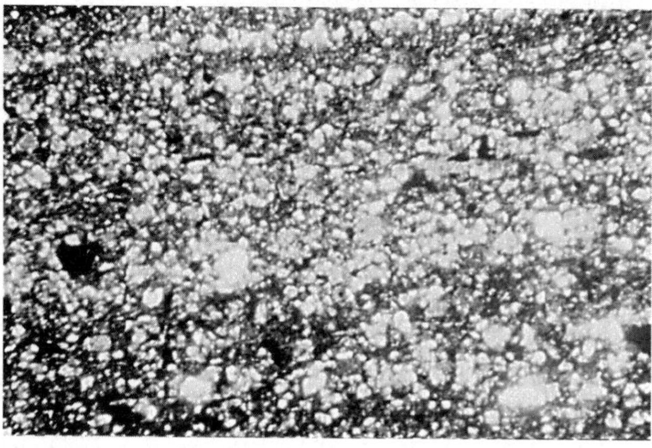

Figure 4. Transmitted light photomicrograph of lower shale member sample from the Gulf #1 Dyer well in McClain County, south-central Oklahoma (20-6N-3W). Scale bar = 0.5 mm.

Upper Shale Member

The upper shale member isopach (Figure 7) shows that it is not as areally widespread as the middle shale member, indicating that the Kaskaskia I transgression was past its maximum extent by the time the upper shale member was deposited. It is thickest (about 150 ft or 45.7 m) in the McPherson Valley area of central Kansas. Dolomite silt is abundant in the upper shale member (Figure 8).

Geophysical logs of wells in the McPherson Valley area reveal that a lenticular limestone bed as much as 40 ft (12.2 m) thick is present near the base of the upper shale member, and may be the same bed that Lee (1956) described for this location and called the Chattanooga limestone lentil (Lambert, in press). Its presence could indicate deposition within the photic zone early in upper shale member deposition. Limestone and dolomite beds less than 3.3 ft (1 m) thick are found in the upper part of the upper shale member on geophysical logs and in well-cutting descriptions for wells in northeastern Kansas, south of similar beds in the middle shale member, and may indicate a shallowing of water depth in the northern part of the study area late in the deposition of the upper shale member.

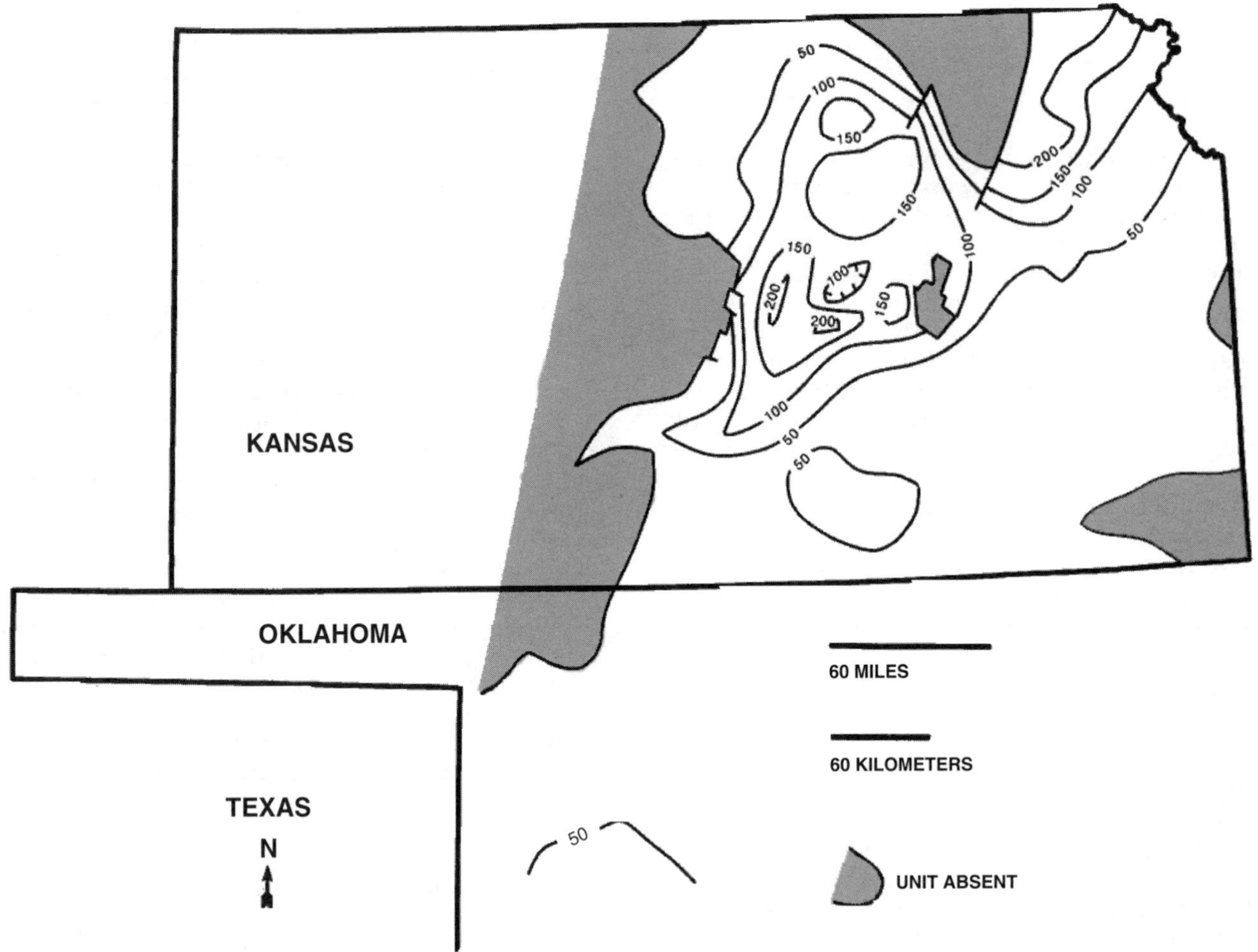

Figure 5. Isopach map of the middle shale member of the Chattanooga (Woodford) Shale. Contour interval is 50 ft (15.2 m). Figure after Lambert (this paper) and Hester et al. (1990).

Figures 3, 5, and 7 show that the depocenter shifted during deposition of the Chattanooga (Woodford) Shale. The lower shale member depocenter was in the southern Oklahoma aulacogen. The middle shale member had two depocenters, both located in Kansas. The upper shale member had only one depocenter, in the McPherson Valley area of central Kansas.

Well-cutting descriptions for wells along the periphery of the Chattanooga (Woodford) Shale in Kansas mention that red and purple shale is found at the top of the formation, beneath the Mississippian carbonates. This is suggestive of exposure and reworking, and may mark the unconformity separating the Kaskaskia I and II subsequences.

ORGANIC GEOCHEMISTRY OF THE CHATTANOOGA (WOODFORD) SHALE MEMBERS

Total Organic Carbon

Table 1 lists the location and depth of 21 Chattanooga (Woodford) Shale core samples whose kerogen content was analyzed for this study. In Oklahoma these include one upper shale member sample, six middle shale member samples, and two lower shale member samples. Three upper shale member and eight middle shale member samples are from Kansas; one middle shale member sample is from southeastern Nebraska, several miles north of the Kansas state line. TOC content, hydrogen and oxygen indices (HI and OI, respectively), kerogen type, and T_{max} values for the samples are presented in Table 2. All of the geochemical analyses referenced in this chapter are courtesy of Atlantic Richfield's Plano Research Center.

The average TOC for the lower, middle, and upper shale member core samples from Oklahoma, Kansas, and southeastern Nebraska are reported in Table 2, and confirm the finding of Hester et al. (1989, 1990) that the middle shale member is always the most organic-rich part of the formation. In addition, the data show that average TOC values for the middle and upper shale members are lower in Kansas and southeastern Nebraska than in Oklahoma, which corresponds with the northward color change and decrease in radioactivity.

Shales with a TOC of more than 0.5 wt.% are considered sufficiently organic-rich to be possible petro-

Figure 6. Transmitted light photomicrograph of middle shale member sample from the Shell #1 Farr well in Greenwood County, southeastern Kansas (7-20S-12E). Crossed nicols. Scale bar = 0.5 mm.

leum source rocks (Tissot and Welte, 1984). By this criterion, the Chattanooga (Woodford) Shale could be considered a petroleum source rock over most of the study area. However, vitrinite reflectance data for the Woodford Shale in western Oklahoma indicate that north of the Anadarko basin, the formation is not thermally mature enough to have generated oil or gas (Lambert, 1982; Cardott and Lambert, 1985). Somewhat more encouraging are the T_{max} values reported in Table 2 for the Chattanooga Shale in Kansas and southeastern Nebraska, which range from 424°C to 445°C (and average 439°C), placing the Chattanooga Shale at the onset of oil and gas generation according to Waples (1985).

Kerogen Type

The hydrogen- and oxygen-richness of kerogen in Chattanooga (Woodford) Shale core samples from Oklahoma have been plotted on the modified van Krevelen diagram in Figure 9. Two lower shale member samples contain kerogen with relatively little hydrogen, plotting as type IV. This is believed to be reworked or highly oxidized organic matter, and has essentially no capacity for generating hydrocarbons (Waples, 1985). The six middle shale member samples contain type I and type II kerogens, and the one upper shale member sample contains type II kerogen. Type I and type II kerogens are of probable marine origin, and were protected from oxidation by deposition in anoxic water. Upon thermal maturation they will generate mostly oil.

Figure 10 is a modified van Krevelen diagram showing the hydrogen- and oxygen-richness of kerogen in Chattanooga (Woodford) Shale core samples from Kansas. The hydrogen- and oxygen-richness of kerogen in the one core sample from southeastern Nebraska has also been plotted on this diagram. The nine middle shale member samples (eight from Kansas and one from southeastern Nebraska) contain kerogen with compositions that range from type II to type III. Terrestrial land plants are believed to be the source for type III kerogen, which generates mostly gas upon thermal maturation. Kerogen analyses for three samples of the upper shale member in Kansas are also plotted, and they range from type III to type IV in composition.

As hydrocarbons are generated, sedimentary organic matter loses hydrogen and the composition of the residual kerogen changes in such a way that its position on a modified van Krevelen diagram will follow a maturation pathway, down and to the left along the curve representing its kerogen type (Tissot and Welte, 1984). However, the hydrogen indices for the Chattanooga (Woodford) Shale samples listed in Table 2 do not consistently decrease with increasing T_{max}. This suggests that the hydrogen indices, and thus the position of kerogen samples plotted in Figures 9 and 10, represent original compositional differences of the kerogen rather than differences in thermal maturity.

The lower shale member contains oxidized kerogen and therefore may not be a petroleum-source rock. The upper shale member contains terrestrial kerogen that will be likely to generate gas upon thermal maturation. The part of the Chattanooga (Woodford) Shale that has the greatest potential for generating oil is the middle shale member, and this is especially true in Oklahoma, where the formation is also thermally mature enough to generate hydrocarbons. In Kansas and southeastern Nebraska, the middle shale member contains more terrestrial kerogen than it does in Oklahoma, perhaps due to the proximity of the emergent Transcontinental arch during deposition. The fact that the middle shale member is the part of the Chattanooga (Woodford) Shale with the greatest marine influence in its organic composition is another indication that it was deposited when the Kaskaskia I transgression was at its maximum.

Organic Facies

An organic facies has been defined as a mappable subdivision of a stratigraphic unit, distinguished by

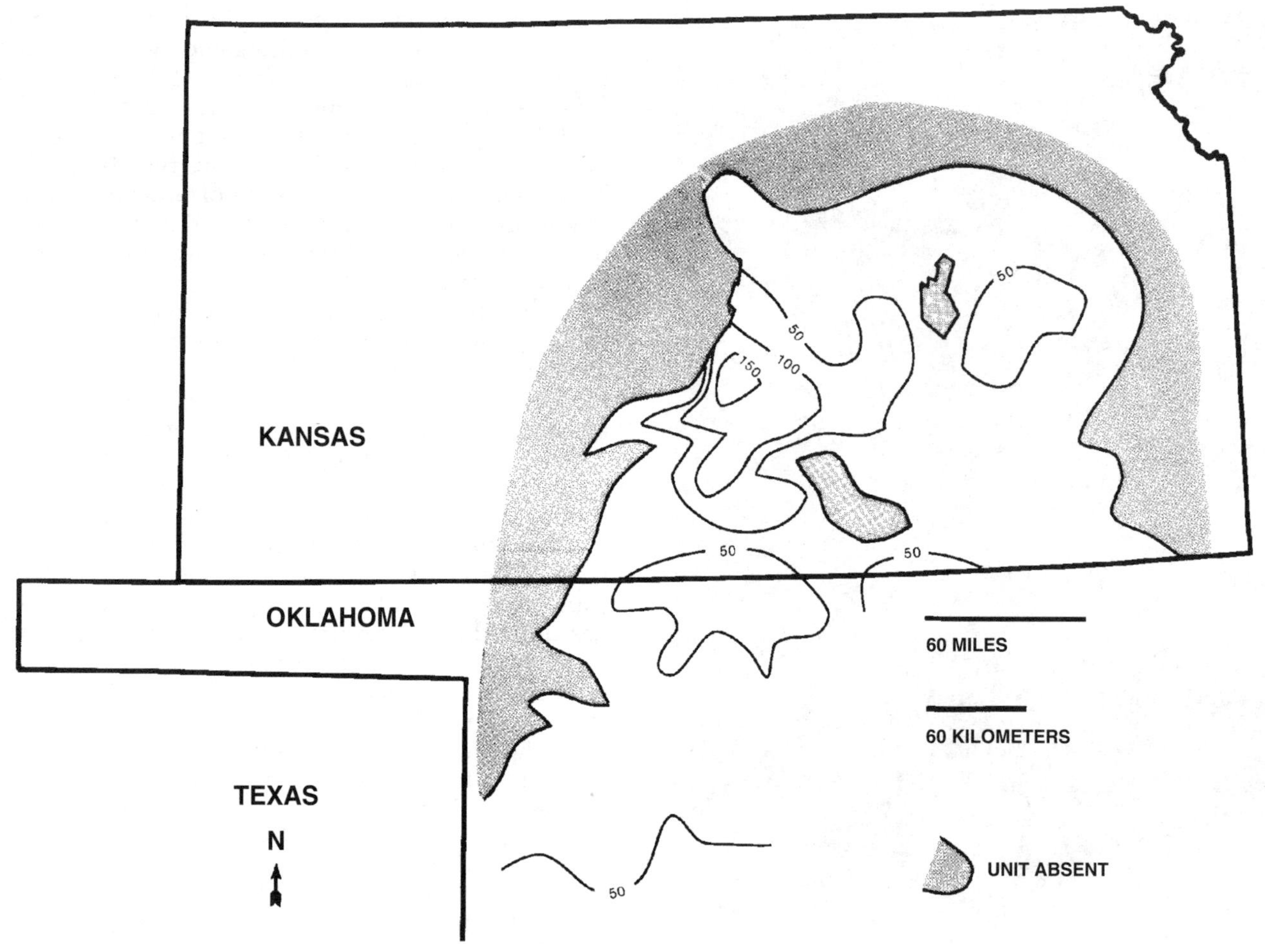

Figure 7. Isopach map of the upper shale member of the Chattanooga (Woodford) Shale. Contour interval is 50 ft (15.2 m). Figure after Lambert (this paper) and Hester et al. (1990).

the nature of its organic constituents (Jones and Demaison, 1980). Several workers have used the organic content of the Chattanooga (Woodford) Shale to recognize subdivisions of the formation.

Urban (1960) examined an outcrop of the Woodford Shale at a single locality in south-central Oklahoma, and used paleoecological inferences drawn from palynology to distinguish three zones within the formation. Urban found both a lower and an upper zone that appeared to have been deposited close to shore, and a middle zone that was deposited far from shore.

Sullivan (1985) used the n-alkane distribution of the extractable organic matter in Woodford Shale cores and well-cuttings from western Oklahoma to distinguish two groups of wells. One group had organic matter with marine precursors in an upper zone and terrestrial precursors in a lower zone. The other group had an upper zone with marine and terrestrial precursors, and a lower zone with mainly marine precursors.

The results presented in this chapter are comparable with those of Urban and Sullivan. The lower and upper shale members have relatively low TOC content, and their kerogen is less hydrogen-rich and more terrestrial in origin than that of the middle shale member. Both the middle and upper shale members have greater amounts of terrestrial kerogen in Kansas and southeastern Nebraska than they do in Oklahoma, indicating that organic facies are present within the shale members. Urban apparently examined a section that included all three shale members of the Chattanooga (Woodford) Shale. Sullivan, in contrast, may have examined sections that contained only the middle and lower shale members or the upper and middle shale members.

SEQUENCE STRATIGRAPHY OF THE CHATTANOOGA (WOODFORD) SHALE

Cunningham and Newell (1991) examined closely spaced Misener Sandstone cores from the Zenith oil field in Stafford county, south-central Kansas. They divided the Chattanooga Shale in that area into two depositional sequences. The oldest, which they desig-

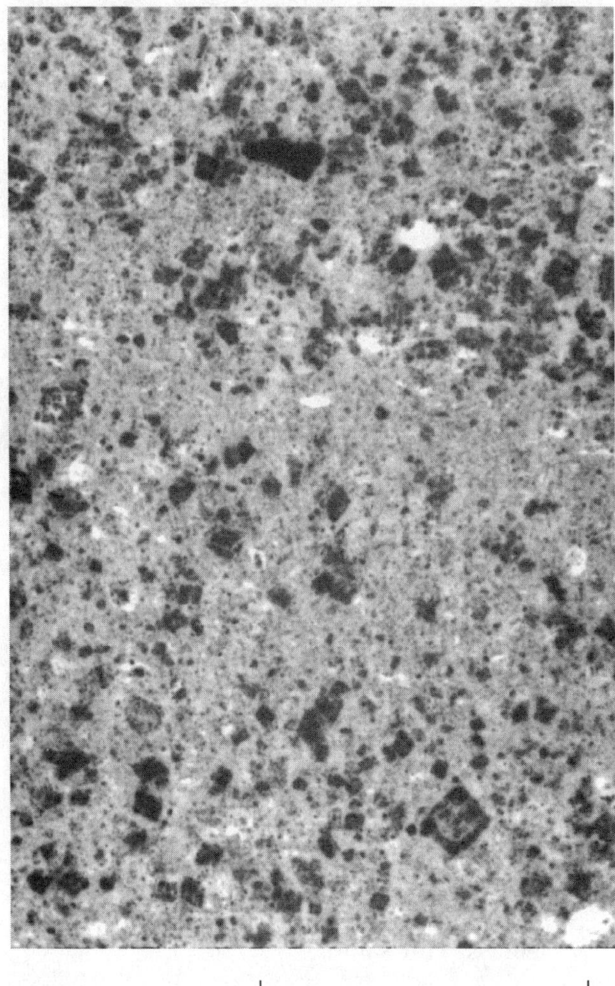

0.5 mm

Figure 8. Transmitted light photomicrograph of upper shale member sample from the Western #1A Brown well in Stafford County, south-central Kansas (30-24S-14W). Scale bar = 0.5 mm.

nated DS1, includes the lower Misener sandstone submember and the overlying Misener limestone submember. The lower Misener sandstone submember is the transgressive systems tract of DS1, and the Misener limestone submember is the highstand systems tract. DS2, separated from DS1 by an unconformity, comprises the upper Misener sandstone submember and overlying mudrocks of the formation. Cunningham and Newell were uncertain if the upper Misener sandstone submember is the lowstand or transgressive systems tract of DS2.

As previously stated, the Misener Sandstone Member has an erratic distribution in the Midcontinent. The lower, middle, and upper shale members constitute depositional sequence DS2 over most of the study area. It is proposed that DS1 be called the Misener sequence, and that DS2 be called the Chattanooga (Woodford) sequence.

The Chattanooga (Woodford) sequence is bounded at its base and top by unconformities, and was deposited during a cycle of sea level rise and fall that occurred over a period of several million years, making it a third-order depositional sequence in the terminology of Mitchum and Van Wagoner (1991). Glacial eustasy and tectonism have both been proposed as the causal mechanism for sea level change during the formation of this type of sequence (Cloetingh, 1988; Vail et al., 1977).

Berry and Wilde (1978) cited glacial eustasy as the cause of the sea level rise that took place during the Late Devonian and at other times of black shale deposition during the Paleozoic, and Veevers and Powell (1987) have correlated Late Devonian transgressive–regressive sequences in North America with glacial events in Gondwana. Evidence of cyclic sedimentation with Milankovich periodicities was found within the Woodford Shale of south-central Oklahoma by Roberts (1988), but the alternating black shale and chert beds that he described are only centimeters thick, and may represent more rapid changes in sea level than those discussed here.

Bond and Kominz (1991) and Kominz and Bond (1991) have suggested that the sea level rise that took place between Middle Devonian and earliest Mississippian time had a tectonic cause. They stated that mantle downwelling beneath North America during the initial assembly of Pangea may have caused the craton to subside, reactivating interior basins and raising relative sea level.

According to Creaney et al. (1991), vertical sections through marine mudrocks that are single sequences have the highest TOC values in the center of the section, a point which probably represents the time of maximum flooding. An idealized prograding cycle would have a lowstand systems tract that contains type III to IV kerogen, a transgressive systems tract with the highest TOC content and type II kerogen, and a highstand systems tract with type III to IV kerogen. Bohacs and Isaksen (1991) used organic geochemistry to define depositional environments in sequences within Triassic mudrocks of the Barents Sea. They found evidence of oxic environments in lowstand systems tracts, and dysoxic to anoxic environments in transgressive and early highstand systems tracts.

The lower shale member of the Chattanooga (Woodford) Shale has the most restricted areal distribution of any shale member. It also contains type IV kerogen, which would indicate deposition in oxygenated (and therefore relatively shallow) water during the initial phase of the Kaskaskia I transgression. However, type IV kerogen may be reworked from older sediment, and the presence of glauconite in the lower shale member suggests an anoxic depositional setting. Therefore, the lower shale member is considered to be the transgressive systems tract of the Chattanooga (Woodford) sequence.

Table 1. Chattanooga (Woodford) Shale core samples analyzed for this study.

Member*	Sample Number**	Location	Depth ft	(m)	Well Name
USM	OC1	21-27N-15W	6189.5	(1887)	Calvert #2 Bloyd
	OC2	11-06S-06E	4030.5	(1228)	Texas #1 Gipson
	OC3	16-13N-06W	8619.5	(2627)	Universal #2-16 Dannehl
MSM	OC4	21-15N-01W	5627.3	(1715)	Gomoco #1-26 Gaffe
	OC5	11-21N-14W	8507.5	(2593)	Tenneco #1-11 Edwards
	OC6	21-26N-11W	6084.5	(1855)	Huber #1 Cherokee Methodist
	OC7	21-27N-15W	6199.5	(1890)	Calvert #2 Bloyd
LSM	OC8	01-14N-16W	14267.5	(4349)	GHK #1 Hoffman
	OC9	06-19N-24W	14332.5	(4368)	Lone Starr #1 Hannan
	KC1	14-26S-09W	4118	(1255)	Shell #1 Mauck
USM	KC2	05-22S-03W	3476.6	(1060)	Shell #1 Duerkson
	KC3	33-19S-08W	3362.6	(1025)	Sutton #3 Bell
	KC4	27-32S-14W	4961	(1512)	Olson-Skelly #1 Elsea
	KC5	33-31S-01W	4096	(1249)	Terra Resources #1 Peasel
	KC6	25-30S-01E	3612.2	(1101)	APC-DEI #1 Goyer
	KC7	34-25S-04W	4096	(1249)	Mid-Cont. #1 N. Hilger Unit
MSM	KC8	15-23S-12E	2170	(661)	ERDA #1 Bock
	KC9	07-20S-12E	2427	(740)	Shell #1 Farr
	KC10	07-17S-03W	3353.9	(1022)	Dalmac #1 Allen
	KC11	33-13S-10E	2978.5	(908)	Carter #2A Davis
	NC1	20-01N-16E	2196	(669)	Harper #1 Sibberson

*USM = upper shale member, MSM = middle shale member, LSM = lower shale member.
**OC = Oklahoma core sample, KC = Kansas core sample, NC = Nebraska core sample.

The middle shale member, apparently deposited when the Kaskaskia I transgression was at its maximum extent, has the widest areal distribution of the shale members and is also the thickest. It is the part of the formation with the highest TOC, although the middle shale member in Kansas and southeastern Nebraska is less organic-rich than it is in Oklahoma (see Table 2). During this time, relatively deep, anoxic water was present in Oklahoma, where kerogen in the middle shale member is of predominately marine origin (types I and II). Shallower, more oxygenated water was present closer to the Transcontinental arch in Kansas and southeastern Nebraska, where organic matter in the middle shale member has a greater terrestrial component (kerogen types I, II, and III). This shift in organic facies may indicate the presence of parasequences within the middle shale member, which forms the early highstand systems tract of the Chattanooga (Woodford) sequence.

The upper shale member is the late highstand systems tract of the Chattanooga (Woodford) sequence. It has a somewhat more restricted areal distribution than the middle shale member, and appears to have been deposited when sea level had begun to fall following the Kaskaskia I transgression. Kerogen in the upper shale member in Oklahoma is marine (type II), while kerogen in the upper shale member in Kansas is more terrestrial and oxidized (types III and IV), indicating the same trend to more oxidizing conditions in the northern part of the study area that was noted for the middle shale member. However, the change from marine to more terrestrial organic facies in the upper shale member occurs farther to the south than does the corresponding change in the middle shale member.

CONCLUSIONS

The Chattanooga (Woodford) Shale is an organic-rich shale that is an important source of hydrocarbons in many of the intracratonic basins of the Mid-continent. In addition to the basal Misener Sandstone, informally named lower, middle, and upper shale members were deposited during the Kaskaskia I transgression. The distribution and organic geochemistry of these shale members suggest that the middle shale member was deposited during maximum transgression. The middle shale member has the highest TOC and contains hydrogen-rich organic matter of marine origin, making it the part of the formation that

Table 2. Geochemical data for Chattanooga (Woodford) Shale samples.

Sample Number	TOC (wt.%)	HI	OI	Kerogen Type	T_{max} (°C)
OC1	1.7	366	23	II	448
OC2	1.2	514	38	II	442
OC3	5.3	352	9	I–II	446
OC4	7.0	551	8	I	445
OC5	6.8	347	5	I–II	447
OC6	4.4	397	8	I–II	445
OC7	5.2	443	9	I	446
	$\bar{x}$ = 5.0				
OC8	4.1	20	1	IV	486
OC9	0.4	10	13	IV	363
	$\bar{x}$ = 2.3				
KC1	1.8	285	24	III	440
KC2	0.2	91	43	IV	429
KC3	0.3	117	38	III–IV	424
	$\bar{x}$ = 0.8				
KC4	2.8	261	21	III	441
KC5	2.3	336	8	II	443
KC6	2.2	384	7	II	440
KC7	2.0	420	13	I–II	445
KC8	4.3	373	17	I–II	441
KC9	4.4	394	13	I–II	442
KC10	0.5	167	20	III	439
KC11	2.2	242	25	III	441
NC1	2.8	494	22	II	438
	$\bar{x}$ = 2.6				

is most likely to generate oil where it is thermally mature. Bounded on the bottom and top by unconformities, the three shale members constitute a third-order depositional sequence. The lower, middle, and upper shale members are the transgressive and early and late highstand (respectively) depositional systems tracts of this sequence.

ACKNOWLEDGMENTS

This chapter represents a part of my dissertation research at the University of Kansas, and I wish to thank Sigma Xi, the Scientific Research Society, for providing me with a Grant-in-Aid of Research during my studies. I also wish to thank the state geological surveys of Oklahoma, Kansas, and Nebraska for providing access to cores of the Chattanooga (Woodford) Shale, and Joseph Senftle of Atlantic Richfield's Plano Research Center for geochemical analyses of those cores. Finally, John Comer and James Schmoker were very helpful in their reviews of an early version of this chapter.

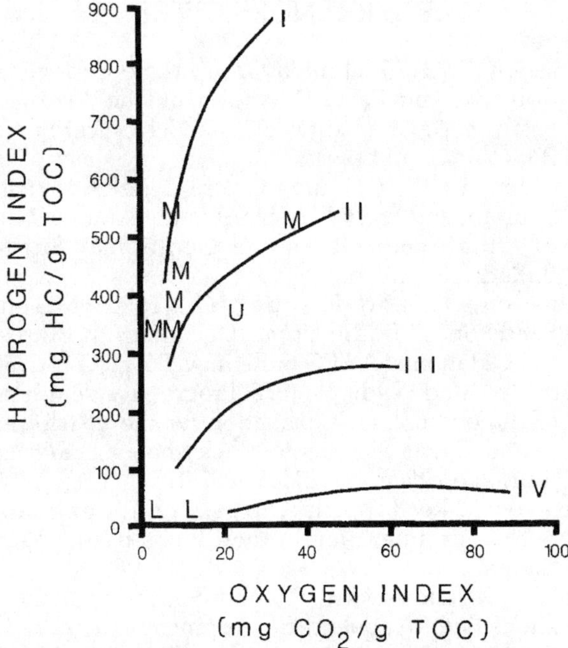

Figure 9. Modified van Krevelen diagram for core samples of the Chattanooga (Woodford) Shale in Oklahoma. Compositional curves for type I, II, III, and IV kerogen are shown. L = lower shale member, M = middle shale member, and U = upper shale member. Hydrogen and oxygen indices for these samples are listed in Table 2.

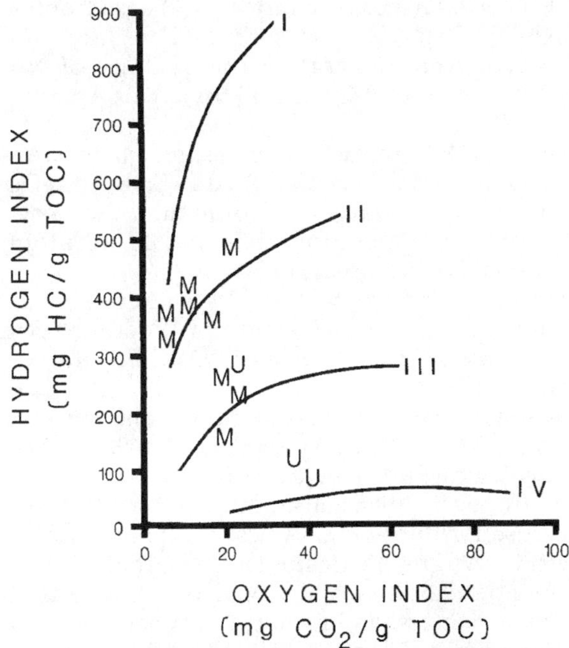

Figure 10. Modified van Krevelen diagram for core samples of the Chattanooga (Woodford) Shale in Kansas and southeastern Nebraska. Compositional curves for type I, II, III, and IV kerogen are shown. M = middle shale member, and U = upper shale member. Hydrogen and oxygen indices for these samples are listed in Table 2.

REFERENCES CITED

Amsden, T., 1975, Hunton Group (Late Ordovician, Silurian, and Early Devonian) in the Anadarko basin of Oklahoma: Oklahoma Geological Survey Bulletin 121, 214 p.

Amsden, T., 1980, Hunton Group (Late Ordovician, Silurian, and Early Devonian) in the Arkoma basin of Oklahoma: Oklahoma Geological Survey Bulletin 129, 136 p.

Amsden, T., and Klapper, G., 1972, Misener Sandstone (Middle-Upper Devonian), north-central Oklahoma: AAPG Bulletin, v. 56, p. 2323-2334.

Berry, W., and Wilde, P., 1978, Progressive ventilation of the oceans—an explanation for the distribution of the Lower Paleozoic black shales: American Journal of Science, v. 278, p. 257-275.

Bethke, C., Reed, J., and Oltz, D., 1991, Long-range petroleum migration in the Illinois basin: AAPG Bulletin, v. 75, p. 925-945.

Bohacs, K., and Isaksen, G., 1991, Source quality variations tied to sequence development: integration of physical and chemical aspects, Lower to Middle Triassic, western Barents Sea (abstract): AAPG Bulletin, v. 75, p. 544.

Bond, G., and Kominz, M., 1991, Disentangling middle Paleozoic sea level and tectonic events in cratonic margins and cratonic basins of North America: Journal of Geophysical Research, v. 96, no. B-4, p. 6619-6639.

Bunker, B., Witzke, B., Watney, W., and Ludvigson, G., 1988, Phanerozoic history of the midcontinent, United States, *in* L. Sloss, ed., Sedimentary cover—North America craton: the geology of North America, v. D-2: Geological Society of America, p. 243-260.

Byers, C., 1977, Biofacies patterns in euxinic basins: a general model, *in* H. Cook and P. Enos, eds., Deep-water carbonate environments, Society of Economic Paleontologists and Mineralogists Special Publication 25, p. 5-17.

Cardott, B., and Lambert, M., 1985, Thermal maturation by vitrinite reflectance of Woodford Shale, Anadarko basin, Oklahoma: AAPG Bulletin, v. 69, p. 1982-1998.

Carlson, M., 1963, Lithostratigraphy and correlation of the Mississippian System in Nebraska: Nebraska Geological Survey Bulletin 21, 46 p.

Cloetingh, S., 1988, Intraplate stresses: a tectonic cause for third-order cycles in apparent sea level, *in* C. Wilgus, B. Hastings, C. Kendall, H. Posamentier, C. Ross, and J. Van Wagoner, eds., Sea level changes: an integrated approach: Society of Economic Paleontologists and Mineralogists Special Publication 42, p. 19-29.

Cluff, R., 1980, Paleoenvironment of the New Albany Shale Group (Devonian–Mississippian) of Illinois: Journal of Sedimentary Petrology, v. 50, p. 767-780.

Cluff, R., 1981, Mudrock fabrics and their significance—reply: Journal of Sedimentary Petrology, v. 51, p. 1027-1029.

Comer, J., 1991, Stratigraphic analysis of the Upper Devonian Woodford Formation, Permian basin, west Texas and southeastern New Mexico: Texas Bureau of Economic Geology, Report of Investigation 201, 63 p.

Conant, L., and Swanson, V., 1961, Chattanooga Shale and related rocks of central Tennessee and nearby areas: United States Geological Survey Professional Paper 357, 91 p.

Creaney, S., Passey, Q., and Allan, J., 1991, Use of well logs and core data to assess the sequence stratigraphic distribution of organic-rich rocks (abstract): AAPG Bulletin, v. 75, p. 557.

Cunningham, K., and Newell, K., 1991, Sequence stratigraphic and reservoir studies of the Misener formation at Zenith field, south-central Kansas: *in* W. Watney, A. Walton, C. Caldwell, and M. DuBois, workshop organizers, Midcontinent core workshop: integrated studies of petroleum reservoirs in the Midcontinent: Midcontinent Section, AAPG, p. 81-82.

Demaison, G., and Moore, G., 1980, Anoxic environments and oil source bed genesis: AAPG Bulletin, v. 64, p. 1179-1209.

Doveton, J., 1986, Log analysis of subsurface geology: John Wiley & Sons, New York, 273 p.

Dow, W., 1974, Application of oil-correlation and source-rock data to exploration in Williston basin: AAPG Bulletin, v. 58, p. 1253-1262.

Ellison, S., 1950, Subsurface Woodford black shale, west Texas and southeast New Mexico: Bureau of Economic Geology Report of Investigation 7, 17 p.

Ettensohn, F., 1985, Controls on the development of the Catskill Delta complex basin-facies, *in* D. Woodrow and W. Sevon, eds., The Catskill Delta: Geological Society of America Special Paper 201, p. 65-77.

Ettensohn, F., and Elam, T., 1985, Defining the nature and location of a Late Devonian–Early Mississippian pycnocline in eastern Kentucky: Geological Society of America Bulletin, v. 96, p. 1313-1321.

Fertl, W., and Chilingarian, G., 1990, Hydrocarbon resource evaluation in the Woodford Shale using well logs: Journal of Petroleum Science and Engineering, v. 4, p. 347-357.

Fritz, R., 1991, Bakken Shale and Austin Chalk, *in* R. Fritz, M. Horn, and S. Joshi, eds., Geological aspects of horizontal drilling: AAPG Continuing Education Course Note 33, p. 91-147.

Goebel, E., 1968, Undifferentiated Silurian and Devonian, *in* D. Zeller, ed., The stratigraphic succession in Kansas: Kansas Geological Survey Bulletin 189, p. 15-17.

Hass, W., and Huddle, J., 1965, Late Devonian and Early Mississippian age of the Woodford Shale in Oklahoma, as determined from conodonts: United States Geological Survey Professional Paper 525-D, p. D125-D132.

Heckel, P., and Witzke, B., 1979, Devonian world

palaeogeography determined from distribution of carbonates and related lithic palaeoclimatic indicators, *in* M. House, C. Scrutton, and M. Bassett, eds., The Devonian System: Special Papers in Palaeontology 23, p. 99-123.

Hester, T., Sahl, H., and Schmoker, J., 1988, Cross sections based on gamma-ray, density, and resistivity logs showing stratigraphic units of the Woodford Shale, Anadarko basin, Oklahoma: United States Geological Survey Miscellaneous Field Studies Map 2054, 2 plates.

Hester, T., Schmoker, J., and Sahl, H., 1989, Regional distributional trends and organic-carbon content of the Woodford Shale, Anadarko basin, Oklahoma, based on gamma-ray, density, and resistivity logs (abstract): Geological Society of America Abstracts with Programs, v. 21, p. 14.

Hester, T., Schmoker, J., and Sahl, H., 1990, Log-derived regional source-rock characteristics of the Woodford Shale, Anadarko basin, Oklahoma: United States Geological Survey Bulletin 1866-D, 38 p.

Hilpman, P., 1967, Devonian stratigraphy in Kansas: a progress report: Tulsa Geological Society Digest, v. 35, p. 88-98.

Johnson, K., 1989, Geologic evolution of the Anadarko basin, *in* K. Johnson, ed., Anadarko basin symposium, 1988: Oklahoma Geological Survey Circular 90, p. 3-12.

Jones, R., and Demaison, G., 1980, Organic facies—stratigraphic concept and exploration tool (abstract): AAPG Bulletin, v. 64, p. 729.

Kominz, M., and Bond, G., 1991, Unusually large subsidence and sea-level events during middle Paleozoic time: new evidence supporting mantle convection models for supercontinent assembly: Geology, v. 19, p. 56-60.

Lambert, M., 1982, Vitrinite reflectance of Woodford Shale in Anadarko basin, Oklahoma (abstract): American Association of Petroleum Geologists Bulletin, v. 66, p. 591-592.

Lambert, M., 1992, Internal stratigraphy of the Chattanooga Shale in Kansas and Oklahoma, *in* K. Johnson and B. Cardott, eds., Source rocks in the southern Midcontinent: 1990 symposium: Oklahoma Geological Survey Circular 93, p. 94-103.

Lambert, M., in press, Revised Upper Devonian and Lower Mississippian stratigraphic nomenclature in Kansas, *in* D. Baars, ed., Revision of stratigraphic nomenclature in Kansas: Kansas Geological Survey Bulletin 230.

Lee, W., 1940, Subsurface Mississippian rocks of Kansas: Kansas Geological Survey Bulletin 33, 114 p.

Lee, W., 1956, Stratigraphy and structural development of the Salina basin area: Kansas Geological Survey Bulletin 121, 167 p.

Manger, W., 1985, Devonian–Lower Mississippian lithostratigraphy, northwestern Arkansas: southcentral section: Geological Society of America guidebook, 25 p.

McGhee, G., Jr., and Bayer, U., 1985, The local signature of sea-level change, *in* U. Bayer and A. Seilacher, eds., Sedimentation and evolutionary cycles, p. 98-112.

Meissner, F., 1978, Petroleum geology of the Bakken Formation, Williston basin, North Dakota and Montana, *in* D. Estelle and R. Miller, eds., The economic geology of the Williston basin, International Williston basin symposium: Montana Geological Society, p. 207-227.

Merriam, D., 1963, The geologic history of Kansas: Kansas Geogical Survey Bulletin 162, 317 p.

Mitchum, R., Jr., and Van Wagoner, J., 1991, High frequency sequences and their stacking patterns: sequence-stratigraphic evidence of high-frequency eustatic cycles: Sedimentary Geology, v. 70, p. 131-160.

Moore, R., Frye, J., Jewett, J., Lee, W., and O'Connor, H., 1951, The Kansas rock column: Kansas Geological Survey Bulletin 89, 132 p.

Newell, K., 1987, Salina basin sub-Chattanooga subcrop map: Kansas Geological Survey Open-File Report 87-4, 1 plate.

Newell, K., 1989, Salina basin: distribution of Upper Devonian–Lower Mississippian Misener Sandstone (superimposed on sub-Chattanooga subcrop map): Kansas Geological Survey Open File Report 89-18, 1 plate.

Over, D., and Barrick, J., 1990, The Devonian/Carboniferous boundary in the Woodford Shale, Lawrence uplift, southcentral Oklahoma, *in* S. Ritter, ed., Early to Middle Paleozoic conodont biostratigraphy of the Arbuckle Mountains, southern Oklahoma: Oklahoma Geological Survey Guidebook 27, p. 63-73.

Potter, P., Maynard, J., and Pryor, W., 1980, Sedimentology of shale: Springer-Verlag, New York, 306 p.

Reed, E., 1946, Boice Shale, new Mississippian subsurface formation in southeast Nebraska: AAPG Bulletin, v. 30, p. 348-352.

Roberts, C., 1988, Laminated black shale–chert cyclicity in the Woodford Formation (Upper Devonian of southern Mid-continent): unpublished Masters of Science thesis, University of Texas at Dallas, 85 p.

Schmoker, J., 1980, Organic content of Devonian shale in western Appalachian basin: AAPG Bulletin, v. 64, p. 2156-2165.

Schmoker, J., 1981, Determination of organic-matter content of Appalachian Devonian shales from gamma-ray logs: AAPG Bulletin, v. 65, p. 1285-1298.

Sloss, L., 1963, Sequences in the cratonic interior of North America: Geological Society of America Bulletin, v. 74, p. 93-114.

Sloss, L., 1988, Tectonic evolution of the craton in Phanerozoic time, *in* L. Sloss, ed., Sedimentary cover—North American Craton; U.S.: The Geology of North America, v. D-2: Geological Society of America, p. 25-51.

Streel, M., Fairon-Demaret, M., and Loboziak, S.,

1990, Givetian–Frasnian phytogeography of Euramerica and western Gondwana based on miospore distribution, *in* W. McKerrow and C. Scotese, eds., Palaeozoic palaeogeography and biogeography: Geological Society Memoir 12, p. 291-296.

Sullivan, K., 1985, Organic facies variation of the Woodford Shale in western Oklahoma: Shale Shaker, v. 35, p. 76-89.

Tissot, B., and Welte, D., 1984, Petroleum formation and occurrence: Springer-Verlag, New York, 699 p.

Tourtelot, H., 1979, Black shale—its deposition and diagenesis: Clays and Clay Minerals, v. 27, p. 313-321.

Twenhofel, W., 1939, Environments of origin of black shales: AAPG Bulletin, v. 23, p. 1178-1198.

Urban, J., 1960, Microfossils of the Woodford Shale (Devonian) of Oklahoma: unpublished Masters of Science thesis, University of Oklahoma, 77 p.

Vail, P., Mitchum, R., Jr., and Thompson, S., III, 1977, Seismic stratigraphy and global changes of sea level, part 4: global cycles of relative changes of sea level, *in* C. Payton, ed., Seismic stratigraphy—applications to hydrocarbon exploration: AAPG Memoir 26, p. 83-97.

Veevers, J., and Powell, C., 1987, Late Paleozoic glacial episodes in Gondwanaland reflected in transgressive–regressive depositional sequences in Euramerica: Geological Society of America Bulletin, v. 98, 475-487.

Ver Wiebe, W., 1946, Kinderhook dolomite of Sedgwick County, Kansas: AAPG Bulletin, v. 30, p. 1747-1755.

Waples, D., 1985, Geochemistry in petroleum exploration: International Human Resources Development Corporation, Boston, 232 p.

Williams, J., 1974, Characterization of oil types in Williston basin: AAPG Bulletin, v. 58, p. 1243-1252.

Wilson, M., and Berendsen, P., 1988, Chattanooga Shale isopach: Kansas Geological Survey Open-File Report 88-6, 1 plate.

Witzke, B., and Heckel, P., 1988, Paleoclimatic indicators and inferred Devonian paleolatitudes of Euramerica, *in* N. McMillan, A. Embry, and D. Glass, eds., Devonian of the world: proceedings of the second international symposium on the Devonian system, v. 3: Canadian Society of Petroleum Geologists, p. 49-63.

Woodrow, D., Fletcher, F., and Ahrnsbrak, W., 1973, Paleogeography and paleoclimate at the deposition sites of the Devonian Catskill and Old Red Sandstone Facies: Geological Society of America Bulletin, v. 84, p. 3051-3064.

Wright, W., 1963, Woodford: source of 10% of Permian basin oil reserves: Oil & Gas Journal, v. 61, p. 188-190.

Chapter 12

Source Quality Variations Tied to Sequence Development in the Monterey and Associated Formations, Southwestern California

Kevin M. Bohacs
Exxon Production Research Company
Houston, Texas, USA

ABSTRACT

The occurrence and hydrocarbon potential of the source rocks in the Monterey Formation (Miocene) are strongly controlled by stratigraphic and geographic position. The major shifts in the geochemical properties of the rocks occur at sequence boundaries or downlap surfaces; there are systematic variations in source quality within the sequences and sequence sets. Source richness (total organic carbon content) tends to be moderate in the lowstand systems tracts, increasing to a maximum around the mid-sequence downlap surface, and relatively low in the highstand systems tracts. Source quality in this setting is a function of the organic-sulfur content of the kerogen; quality is generally higher in the lowstand systems tracts and lower in the highstand systems tracts. Both richness and quality vary strongly at the sequence-set scale: the thin depositional sequences in the transgressive sequence set have the largest total organic carbon contents and lowest quality (highest sulfur) rocks.

The major source potential is in specific mud rock lithofacies (phosphatic shales and carbonaceous marls). These source facies are diachronous across the basins of Santa Maria and Santa Barbara Channel. The time-transgressive nature of these lithofacies units across the various subbasins make it essential to employ the physical surfaces used by sequence stratigraphy to properly correlate among outcrops and wells and to decipher the distribution of source potential. A sequence stratigraphic framework allows comparison of the geochemical measurements of time-equivalent rocks; this portrays genetic relations that can reveal depositional processes and enable construction of predictive models.

INTRODUCTION

Complexly interbedded, diverse lithologies of the Monterey Formation (Miocene) form the source, seal, and reservoir of its hydrocarbon system. Predicting the arrangement of the components of this hydrocarbon system requires an understanding of the control that basin evolution, tectonics, oceanography (including sea level change), and sediment input had on the distribution of the various lithologies. The best way to decipher the effects of these controls and the resultant lithofacies distribution is within a sequence-stratigraphic framework. Sequence stratigraphy uses the three-dimensional array of surfaces and bedding to reveal genetically related packages of rocks. This approach allows the recognition and mapping of coeval depositional environments and lithofacies and enables the construction of predictive models based on understanding the depositional processes. Having this actualistic framework makes it possible to predict source quality variations from standard exploration data such as regional geology, reflection seismic lines, and well logs.

This chapter presents the sequence stratigraphic and source distribution results of a large-scale investigation of the Miocene depositional system of southern California conducted by numerous workers at Exxon Production Research Company and Exxon Company, USA; a more detailed report of our stratigraphic work can be found in Bohacs, 1990. Much of this study built on the many excellent investigations of the larger scale aspects of the Monterey Formation contained in the literature (including, but not limited to: Arends and Blake, 1986; Barron, 1986; Bramlette, 1946; Grivetti, 1982; Isaacs, 1980, 1981, 1983; Pisciotto and Garrison, 1981; Stanley and Surdam, 1984) and in unpublished company reports.

Our observations of two key outcrop sections illustrate the attributes of depositional sequences in relatively deep marine environments and the relation of the depositional sequences to variations in rock properties. This work extends the application of sequence stratigraphy to deeper-marine environs and establishes criteria for constructing sequence frameworks in slope and deep-basinal mud rocks using outcrop, core, well-log, biostratigraphic, and seismic data. This approach to investigating fine-grained rocks and predicting their properties has significant potential beyond the scope of this study.

GENERAL GEOLOGIC SETTING

The Monterey Formation contains rocks deposited during the Miocene in a series of borderland basins off the coast of California (Figure 1). These extensional basins varied in size, topography, and tectonic history; they were formed by the interaction of a spreading ridge (between the Farallon and Pacific plates) with the western edge of the North American plate (Pisciotto and Garrison, 1981). This study concentrated on the Santa Barbara Channel and Santa Maria basins (Figure 2). The development of the area during the Tertiary is diagrammed in Figure 3; in broad terms, starting in the Oligocene from shallow shoreline and paralic clastic environs, the area experienced a long-term deepening that culminated in the Miocene with the deep-basinal deposition of the dominantly biogenic Monterey Formation. Subsequent gradual shallowing was accelerated by uplift and greatly increased input of terrigenous clastics during the Pliocene.

The depositional systems in the area record persistent low-oxygen conditions in generally deep-water depths (mostly bathyal to abyssal); sediment supply rates varied significantly both laterally and temporally. The energy level at the bottom also varied widely: sedimentary structures range from thin, evenly laminated shales to submarine scours, slumps, and slides with 9-m-long blocks. There is ample evidence of sediment transport along the bottom during deposition and of reworking of deposited sediment by currents (Garrison and Ramirez, 1989; Bohacs, 1990; Schwalbach and Bohacs, 1991). These basins did not experience a deathly quiet, even snow of plankton through quiescent waters.

The input of organically produced material to the Monterey system was large due to persistent and generally intense upwelling (Isaacs, 1980; Calvert, 1966). The dominant type of organically produced material varied widely among biogenic silica, carbonate, and phosphate. High levels of productivity throughout the system were an essential attribute of the Monterey facies development, but do not appear to be the major control on variations in source quality. The estimated range of overall organic productivity varied by little more than an order of magnitude (Figure 4). This indicates that variations in productivity alone were insufficient to generate the range of source-potential variations (e.g., total organic carbon [TOC] content = 0–26 wt.%; hydrogen index [HI] = < 10 to > 600 mg hydrocarbon/g organic carbon; see Figures 8 and 27). Our investigations indicate that the variations in source quality are mostly controlled by shifting conditions of deposition and preservation, such as changing rates of terrigenous sediment supply and changing burial rates, rather than changes in primary organic input. Also, as the system evolved, the varying interaction of the oceanography with local topography may have locally intensified upwelling, thereby focusing organic productivity and accumulation.

DATABASE AND METHODOLOGY

We constructed the sequence stratigraphic framework and source rock model using data from outcrop sections totaling 1066 m (3500 ft), 77 wells, 7 cores, and 4312 line-kilometers of reflection seismic lines. The outcrops studied in detail are near Point Pedernales and Naples Beach, both in Santa Barbara County (Figure 2). The interval investigated ranged in age from Saucesian to Delmontian (approximately

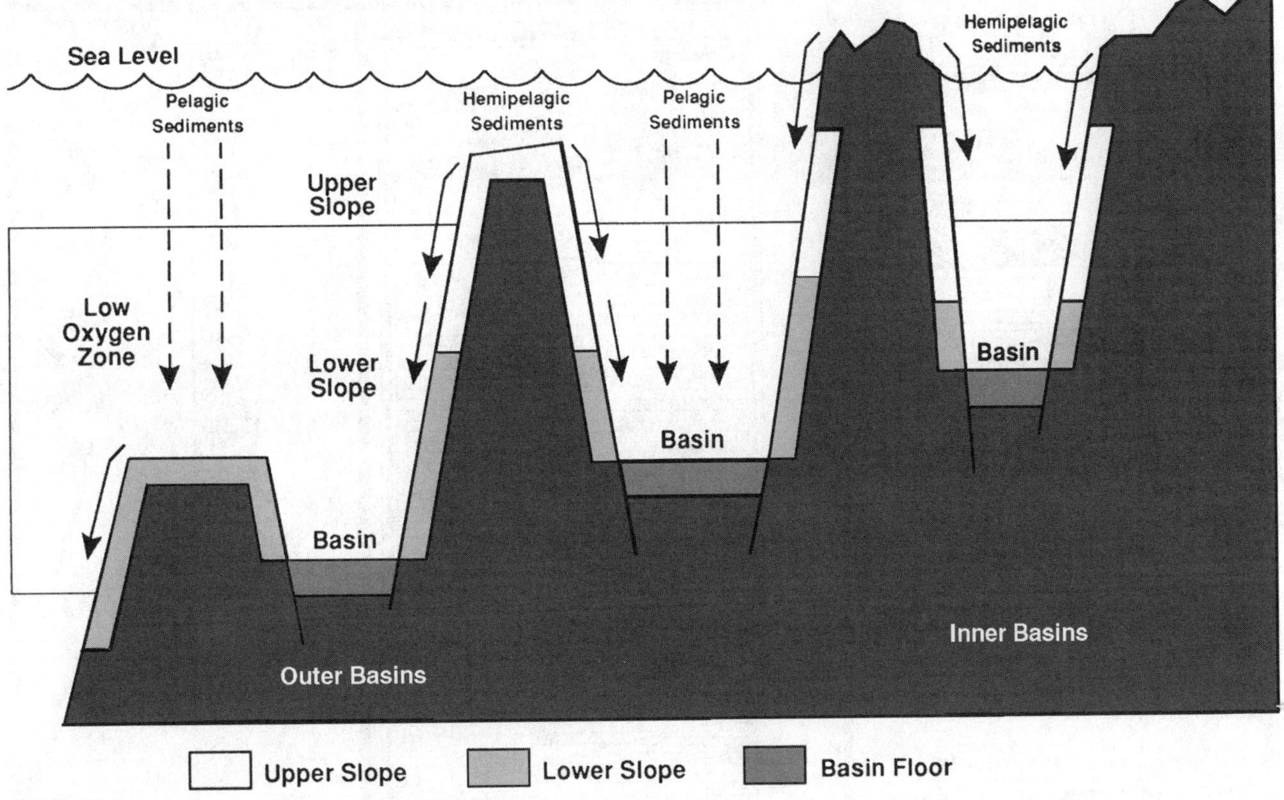

Figure 1. Schematic diagram of borderland basin setting. The varied topography and sedimentological influences on the depositional environments of the Monterey Formation are shown schematically on an onshore–offshore transect (after Gorsline and Emery, 1959).

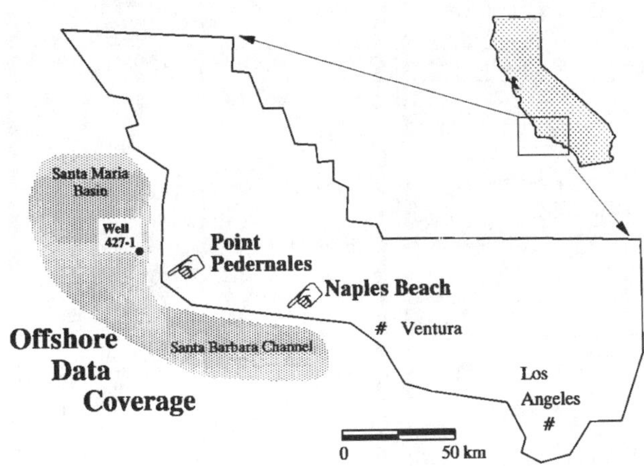

Figure 2. Location of outcrops and subsurface data used in this investigation (after Bohacs, 1990).

17.5 to 5.5 Ma) and contained portions of the underlying Rincon Shale and overlying Sisquoc Shale that bracket the Monterey Formation. The outcrop sections were described at a scale of 2.5 cm = 3 m for lithology, bedding, stratal stacking, and structural features. Natural gamma-ray spectra were measured with a portable spectrometer at a stratigraphic spacing of 1.5 meters; this revealed the relation of the gamma-ray spectra to lithology and assisted correlation to well logs offshore (Figure 5). Samples were taken at a spacing of about 1.5 m, and all were analyzed for bulk geochemical properties (total organic carbon, hydrogen index, oxygen index, total sulfur), 11 major element oxides, and 14 trace elements. Samples selected on the basis of stratigraphic position and bulk chemical properties were extracted and examined for molecular organic facies variations.

All samples were screened for microfossils with varied results. The Naples Beach section returned plentiful, well-preserved foraminifera and diatoms; the Point Pedernales section yielded few foraminifera and diatoms (due to its advanced state of silica diagenesis), but abundant and fairly well preserved dinoflagellates and other palynomorphs. The microfossils revealed important information concerning both the age zonation and climatic-environmental conditions. The outcrops were tied to a grid of well-log cross sections and seismic lines in the offshore Santa Maria and Santa Barbara Channel basins.

CONSTRUCTION OF SEQUENCE-STRATIGRAPHIC FRAMEWORK

Our approach boils down to looking at the geological record as composed of rock packages bounded by physical surfaces formed by distinct events. The

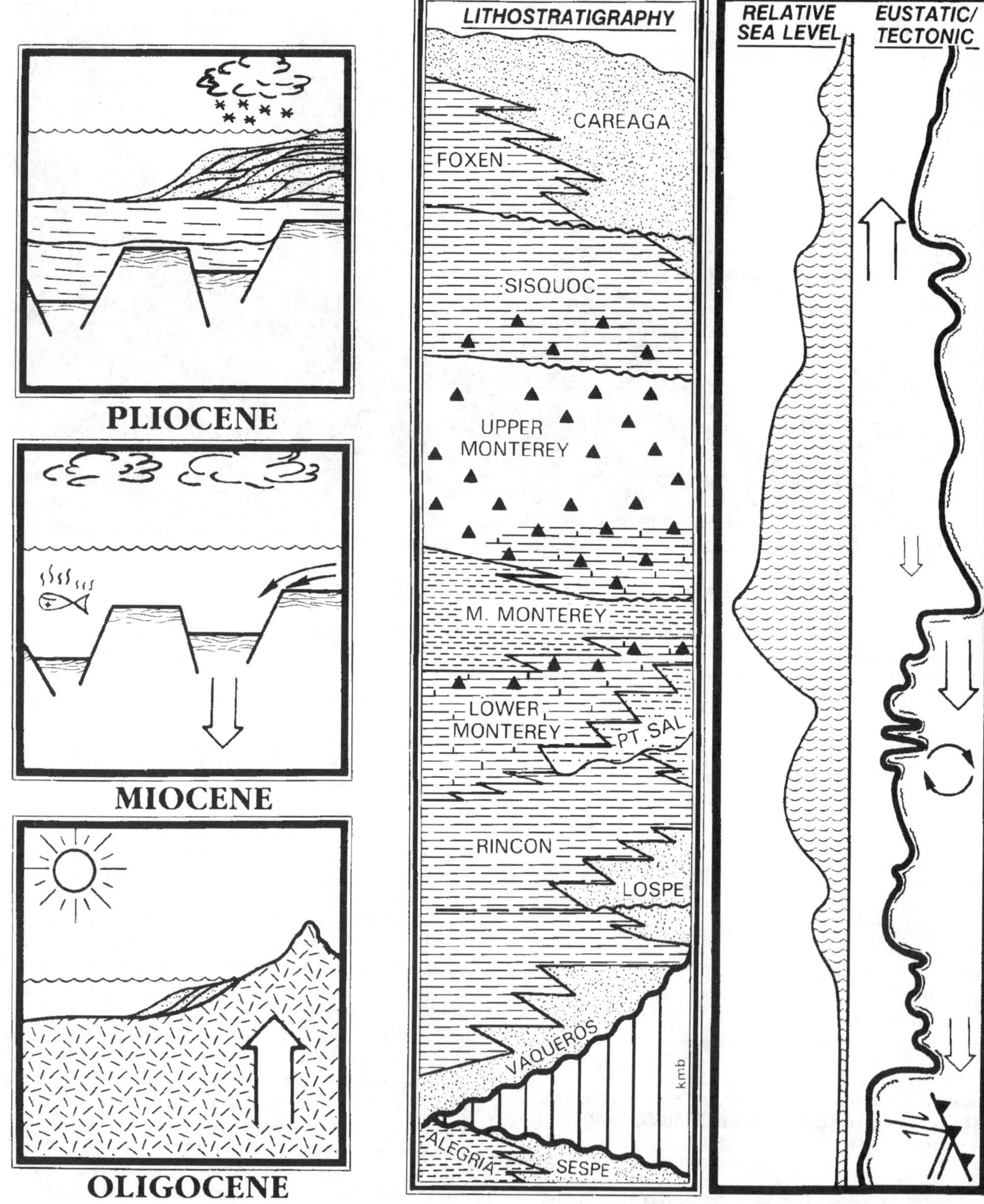

Figure 3. General geological history of southwestern California, Oligocene to Pliocene. The lithostratigraphy of this portion of the Tertiary (center column) is the result of the interaction of changing sea level, basin tectonics, sediment supply, climate, and oceanography. The right column portrays eustatic sea level and major tectonic events: the collision of the Mendicino triple junction, rotation of the Santa Barbara Channel plate, and episodes of subsidence and erosion. Their net effect is shown in the relative sea level curve.

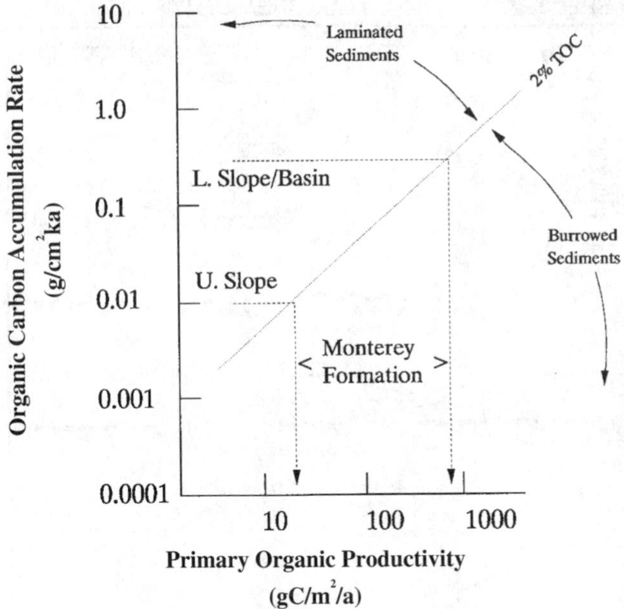

Figure 4. Organic productivity in the Monterey system. The rates of primary organic productivity estimated from organic-carbon accumulation rates varied by little more than one order of magnitude. Such variations were probably insufficient to generate the major variations in source potential. (Primary productivity rates estimated using the relations of Bralower and Thierstein, 1987.)

strata between the physical surfaces are observed to stack in repeated, and therefore predictable, patterns. The physical surfaces are preserved sediment–fluid interfaces and thus represent time lines (Mitchum et al., 1977; Schoch, 1989); hence, they bound genetically related packages of rocks and yield insights into the relation of depositional processes to resultant lithofacies packages.

The stratal stacking patterns result from interactions among eustatic sea level, basin tectonics, and sediment supply. It is difficult or impossible to deconvolve the individual influence of each of these factors (e.g., Kendall and Lerche, 1988). However, the sedimentary system is controlled by the net combination of the three factors (accommodation), so deciphering the individual influences is not essential in constructing the local sequence stratigraphic framework. Sediments respond mainly to the physical energy levels in the system relative to their grain size, with only a slight influence of flow depth (Southard, 1971). The physical energy levels (and often oxygenation states) do change in repeated patterns that can be used to decipher the sequence stratigraphy within each basin. Only as a final step, to extrapolate beyond the local area or to construct a global model, is it necessary to assign the relative influences of tectonics, eustatic sea level, and sediment supply.

The Monterey Formation was deposited as the distal portions of a number of depositional sequences, a region not shown on most conventional sequence-stratigraphic models (the right-hand third of Figure 6). In this region, the expression of depositional sequences is more subtle, but still distinct and recognizable. Here, the changes in bottom-energy levels, oxygenation state, and sediment input still leave distinctive imprints on the surfaces, stacking patterns of beds and lithofacies, and on larger scale stratal geometries (Zelt, 1985; Miskell-Gearhart, 1989). Even deep-basinal environments record the effects of changing sea level through physical and chemical processes. Changing sea level modifies oceanic and atmospheric circulation patterns and climate. Lowered sea level can intensify circulation in parts of a basin and can intensify coastal upwelling (Barron and Baldauf, 1989). Rearranged land drainages change the location, amount, and proximity of terrigenous clastic input. Changing amounts and intensities of freshwater runoff may affect the vertical density stratification of the ocean waters. Modifying areas of erosion and deposition in both land and oceanic realms affects the input and distribution of nutrients. The changing nutrient budget and circulation patterns affect the amount and type of biologic productivity. Moving the location of the shoreline and ocean currents changes the location, bottom area, and strength of the oceanic oxygen minimum zone. Garrison and Ramirez (1989) discussed similar physical effects in their work on the coarse-grained portions of the Monterey Formation.

We constructed the sequence stratigraphic framework for this distal marine setting following the standard flow of analyses and interpretation: (1) examine the lithologic and stratal characteristics of the rocks, (2) group the lithologic and stratal characteristics into facies packages, (3) relate the stacking patterns of facies packages to depositional environment, and (4) use the stacking patterns and lateral relations of rock packages from the different depositional environments to interpret the sequence stratigraphy.

Facies

Facies were defined based on bedding, visual and chemical lithology, sedimentary structures, total organic carbon (TOC), total sulfur content, chemical-oxide analyses, natural gamma-ray spectra, and various combinations of these parameters (Th/U, Al_2O_3/TOC, etc.). The lithofacies defined by outcrop observations correspond well with the geochemical facies defined by the gamma-ray spectra and chemical analyses. This indicates that the chemical properties of these rocks do indeed covary with the physical parameters. Geochemical properties also show a strong correspondence to stratigraphic position: the major shifts in geochemical properties occur at sequence boundaries and downlap surfaces (Figure 7).

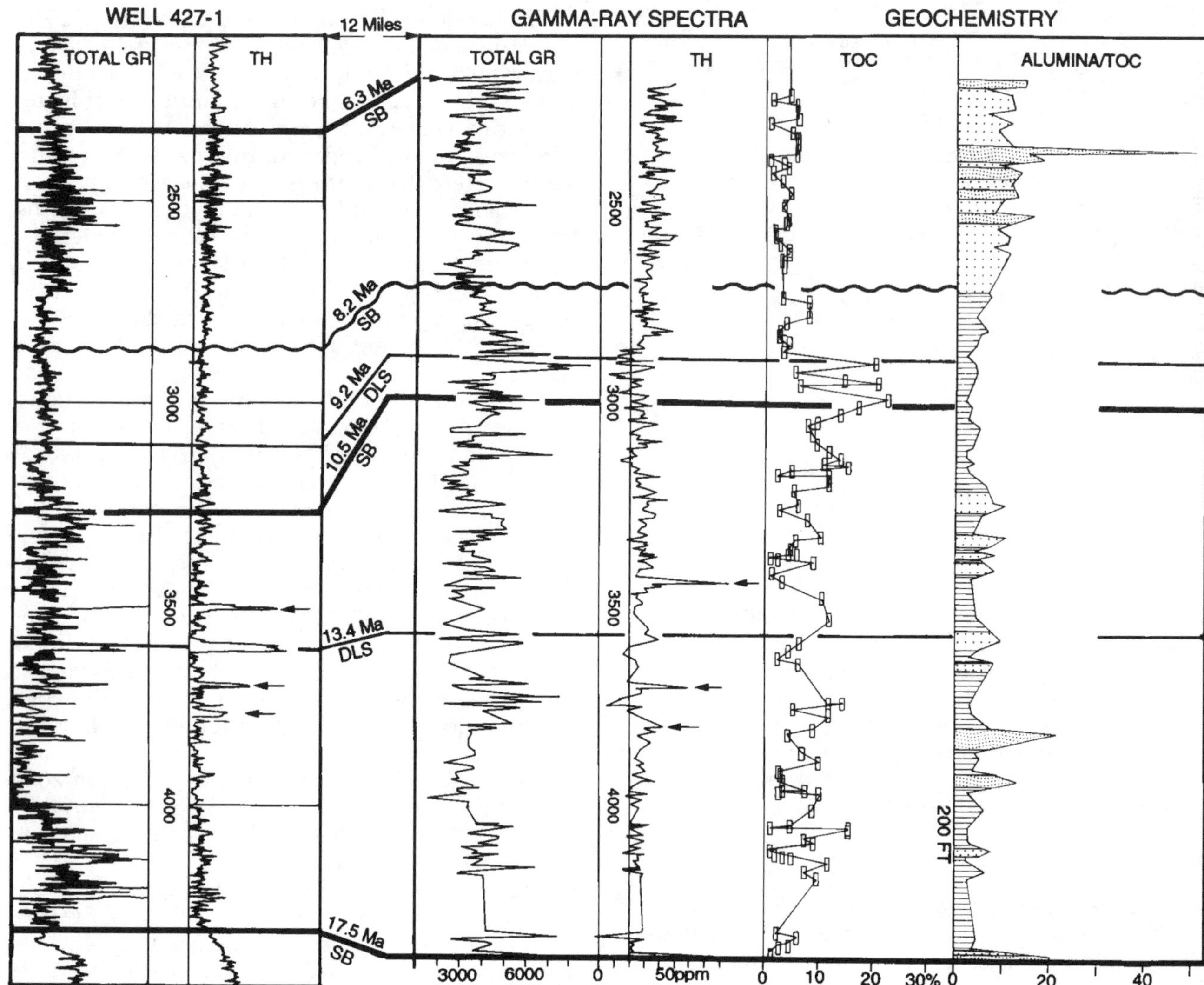

Figure 5. Correlation of outcrop to subsurface. Illustrated is the tie of the Point Pedernales outcrop section to well 427-1, approximately 20 km (12 mi) offshore using the spectral gamma-ray character. Note the four peaks in thorium adjacent to the 13.4 Ma downlap surface; these correspond to volcanic ash beds in the outcrop.

Two major facies were recognized using these parameters: hemipelagic and pelagic. The amount and relative importance of terrigenous influence distinguishes these two deep-marine facies; their differences are best revealed by their physical aspects (Table 1). The facies have characteristic ranges of all the properties listed above (Figures 8–12) although they are but end members in a lithologic/chemical continuum (Figure 13). The full range of variation and detailed stratification is portrayed in the outcrop descriptions (Figures 14, 15).

The two facies possess distinctive well-log signatures. Hemipelagic facies are characterized by overall high gamma-ray activity, thin beds, and numerous interbeds of relatively high and moderately low gamma-ray activity (bottom right, Figure 16). The well-log pattern of the pelagic facies (bottom left, Figure 16) shows a more thickly bedded pattern and overall low gamma-ray activity; thin, moderately high gamma-ray intervals are interbedded with relatively thick low gamma-ray zones. The differences in gamma-ray activity results from the relative proportions of biogenic and detrital lithologies in each facies. The pelagic facies contains abundant components with low radioactivity (diatoms, radiolarians, calcareous plankton). The hemipelagic facies tends to be enriched in clays, phosphates, and organic matter, all of which tend to be radioactive due to their uranium and potassium content.

Depositional Environment

The facies stack in distinctive manners and proportions according to their depositional environments. Each depositional environment is identified by a characteristic stacking of pelagic and hemipelagic

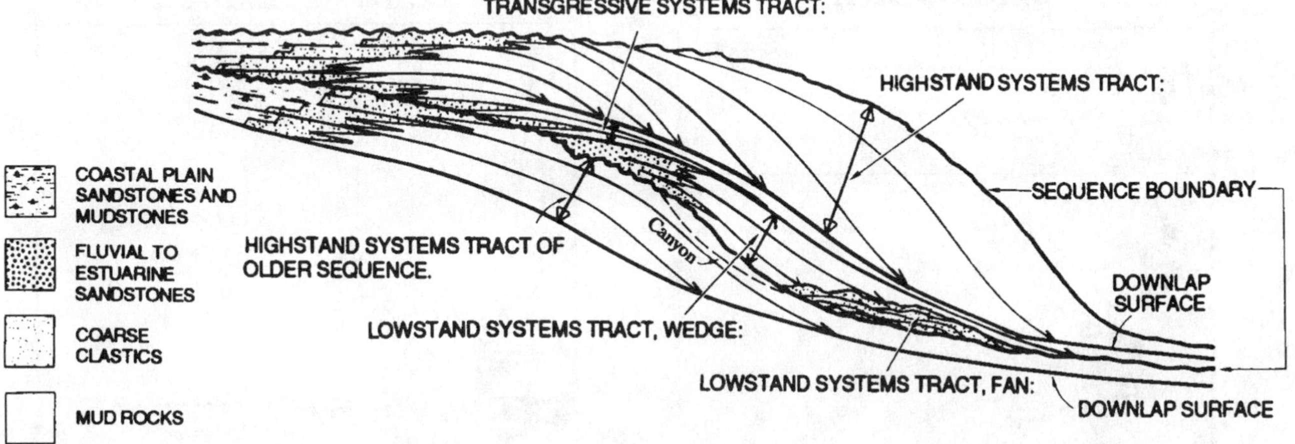

Figure 6. Physical stratigraphy of an idealized sequence (after Van Wagoner et al., 1988). The internal architecture of a depositional sequence and its component systems tracts comprise parasequence sets, parasequences, and their bounding surfaces. A depositional sequence is "a relatively conformable succession of genetically related strata bounded by unconformities and their correlative conformities" (Mitchum, 1977). A parasequence is a relatively conformable succession of genetically related beds or bedsets bounded by marine flooding surfaces and their correlative conformities (Van Wagoner et al., 1988). Variations from this idealized sequence are caused by regional and local factors that affect accommodation, including subsidence, eustatic sea level, sediment supply, and basin morphology. Thus, different systems tracts differ in thickness, areal extent, geometry, and degree of preservation. The Monterey and associated formations were deposited in positions generally in the basinal portions of the diagram, where most of the surfaces and strata are conformable (after Bohacs, 1990).

facies; it is analogous to defining environments within a deltaic complex based on the stacking of sandstone and mudstone facies (Figure 17). The stacking patterns may be recognized in outcrop from the bedding-surface arrangements, laminae geometries, and lithologic changes, and in the subsurface on various well logs. The characteristic distribution of gamma-ray activity for each environment is shown in Figure 18: basin-floor depositional environments tend to have a large proportion of pelagic facies; upper-slope environments tend to have a large proportion of hemipelagic facies. The determination of a depositional environment for any one map point also requires a three-dimensional perspective gained from mapping paleotopography, regional paleogeography, and isopach thickness of the stratigraphic interval on a palinspastic base map, as well as estimates of open-ocean conditions (surface currents, undercurrents, upwelling zones). The interpretation of the depositional environment therefore includes input at all scales of observation, from cores, outcrops, well logs, and seismic data to plate tectonics.

Combining the lithologic and stratal characteristics of the rocks revealed from outcrop and well logs with the paleogeography and paleotopographic setting derived from seismic and regional studies, we defined three depositional environments: bank top—upper slope, lower slope, and basin floor. (Note that the terms "basin" and "slope" are topographic features of this borderland basin and do not necessarily relate to the continental slope and basin floor physiographic features developed on simple Atlantic-type margins.) Similar environmental settings have been outlined by Pisciotto and Garrison, (1981). Table 2 lists the general attributes of each depositional environment. These environments are subject to particular modes of deposition, preservation, and early burial that have an important impact on the ultimate organic and inorganic chemistry of the rocks. The organic matter especially is very sensitive to not only the depositional process (how the organic matter gets transported to the bottom) but also to the preservational environment (processes that occur within the first few centimeters of burial). Hence, the term "depositional environment" throughout this chapter refers to the deposition of organic matter on the sea floor, its very early burial, and subsequent preservation.

Sequence Stratigraphy

The stacking of the depositional environments, along with the physical attributes of the surfaces separating the environments, reveal the sequence stratigraphic framework. We analyzed and correlated the Miocene section from the sequence-set to parasequence scale; the key surfaces that reveal geochemically significant rock packages are the sequence boundaries and the mid-sequence downlap surfaces. Especially in these deep-water environments, no sin-

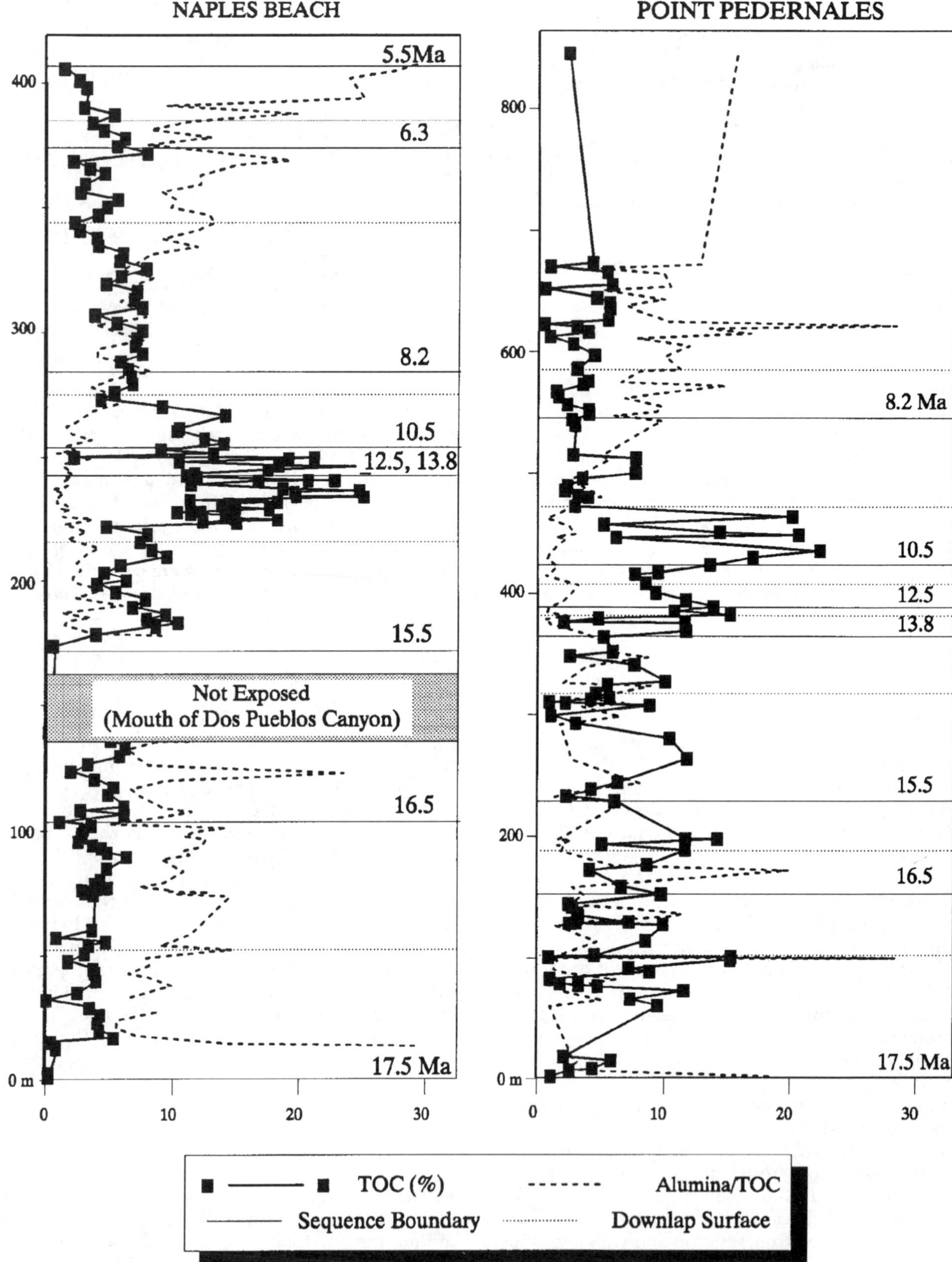

Figure 7. Variation of source richness and quality with stratigraphic position. Major shifts in geochemical parameters tend to occur at sequence boundaries and downlap surfaces. Ages of sequence boundaries are labeled in millions of years. (After Bohacs, 1990; positions of surfaces at Naples Beach have been changed to agree with additional age and well data.)

Table 1. Key attributes of the hemipelagic and pelagic facies.

Attribute*	Hemipelagic Facies	Pelagic Facies
Bedding surfaces	Sharp, planar, and curved Discontinuous wavy, non-parallel Wavy, non-parallel	Sharp and gradational; wavy, curved Wavy, parallel Planar, parallel (in mudrocks)
Lamina geometries	Discontinuous, wavy, parallel Wavy, non-parallel Planar, parallel (only in sandstones and siltstones)	Discontinuous, wavy, non-parallel Curved, non-parallel (continuous and discontinuous) Discontinuous, wavy, parallel
Lithology**	Porcelanite/mudstone (10–30 cm/10–30 cm) Porcelanite/clay shale (10–30 cm/2–4 cm) Mudstone (5–25 cm) Clay shale (5–15 cm; calcareous at Naples Beach) Dolomite/mudstone (10–30 cm), (plus dolomite replacing mudstone) Dolomite grainstone (M-F grained, 5–25 cm) Siliceous shale (10–20 cm) Siliceous mudstone (10–30 cm) Chert/porcelanite (interbedded at 2–4 cm to 5–15 cm scale as "tiger stripe") Siltstone (5–10 cm) Sandstone (VF, F, M grained; 5–25 cm; sublitharenite) Conglomerate: Clast supported; clasts less than or equal to 40 cm; see "slump zones" below	Porcelanite/phosphatic shale (1–5 cm) Phosphatic shale (calcareous) (2–15 cm) Porcelanite/dolomite/phosphatic shale (5–10 cm) Porcelanite/dolomite (20–40 cm) Chert/porcelanite ("tiger stripe")(2–5 cm) Phosphatic grainstones (VF grained) (0.2–1.5 cm) Phosphatic shale/dolomite (5–10 cm)
Concretions	Chert (2–5 cm; rounded to flattened) Phosphate (3 populations: 5–10 cm, 1–2 cm, 0.05–0.1 cm; mostly rounded, some flattened or irregular; generally concentrated along bedding surfaces)	Phosphate (2 populations: 1–2 cm, 0.1–0.5 cm; irregular, flattened, scattered; small blebs, 0.05–0.1 cm, some aligned along beds, some scattered)
Minor Constituents	Phosphatic shale (5–15 cm) Phosphatic grainstone (VC–M grained; 1–3 cm)	Limestone (VF grained) Clay shale (1–15 cm; some partially replaced by dolomite) Siltstone (1–5 cm) Siliceous shale (with phosphatic concretions)
Sedimentary Structures		
Primary	Scours (common) Load structures Graded beds: normal and inverted (5–25 cm)	Graded beds (only in thin siltstones and phosphatic grainstones) Scours (sparse, generally at base of graded beds)
Secondary	Convolute bedding Swirled, flowed bedding (water-escape structures) Microfaults Bouma sequences: T_{ab}, thinning, fining-upward bedsets in clay shales (5–15 cm) Slump zones: 2–30 m thick; blocks (2–6 m across) of interbedded dolomite and clay shale	Convolute lamination Swirled, flowed bedding Microfaults
Burrows	Sparse, inclined to vertical: *Planolites* *Terebellina* {Two 50-cm intervals near base of Point Pedernales section may be burrow-churned.}	None observed

* All attributes are noted in order of decreasing occurrence within each category.

** Bed thickness is indicated in centimeters. A slash between two lithology types indicates interbedding at a fine scale.

gle feature is completely diagnostic of a sequence boundary or downlap surface. The criteria used to pick these surfaces rely on an association of rock properties, the stratal stacking pattern of the lithofacies on either side of the surface, systematic changes in lamina and bed geometries, and the local expression of the surface. Criteria for recognizing these surfaces in deep-basinal mud rocks are listed in Table 3.

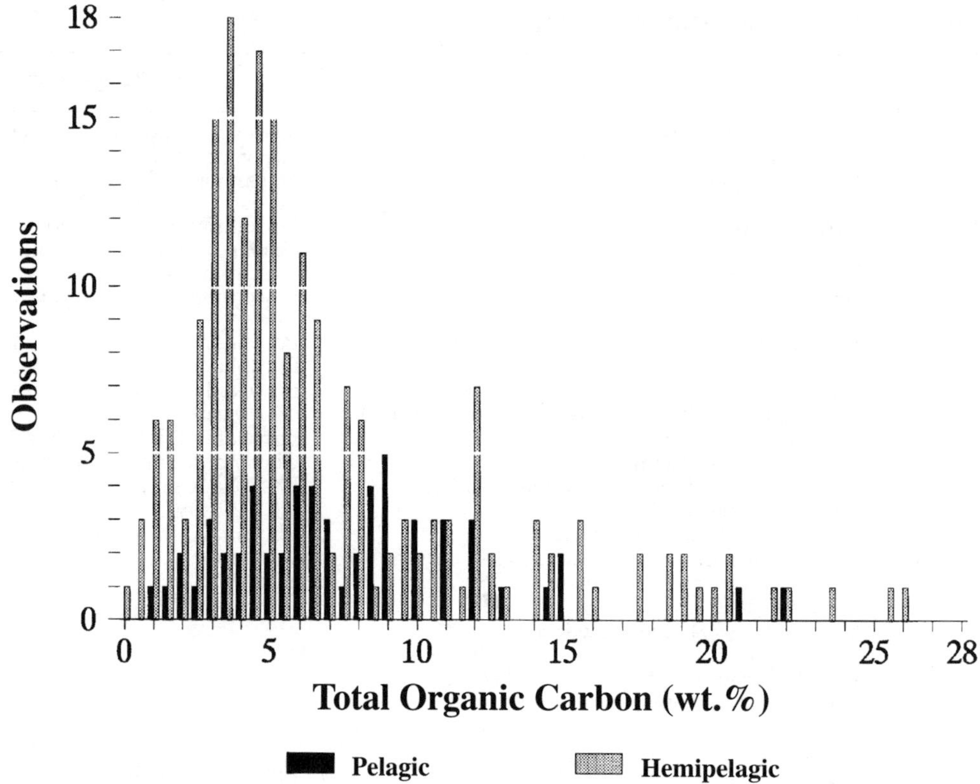

Figure 8. Distribution of total organic carbon contents of the hemipelagic and pelagic facies. Hemipelagic facies: mean TOC = 6.6%, mode = 3.5%. Pelagic facies: mean TOC = 7.4%, primary mode = 9% and secondary mode at 6%.

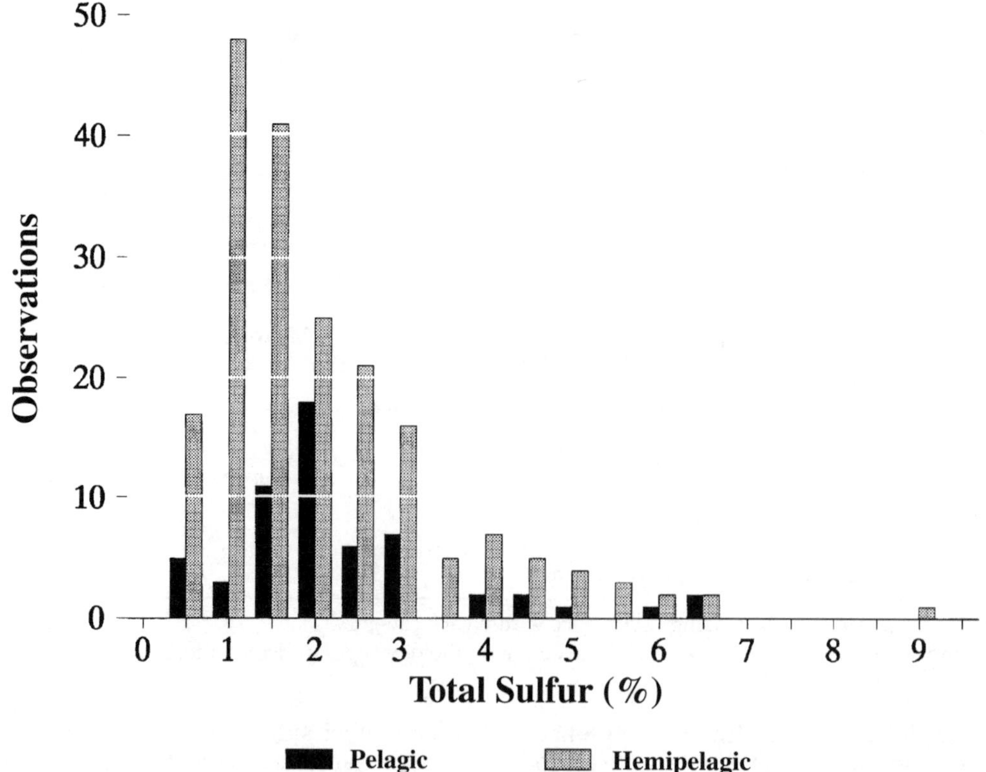

Figure 9. Distribution of total sulfur contents of the hemipelagic and pelagic facies. Hemipelagic facies: mean sulfur content = 1.81%, mode = 1.0%. Pelagic facies: mean sulfur content = 2.1%, mode = 2%.

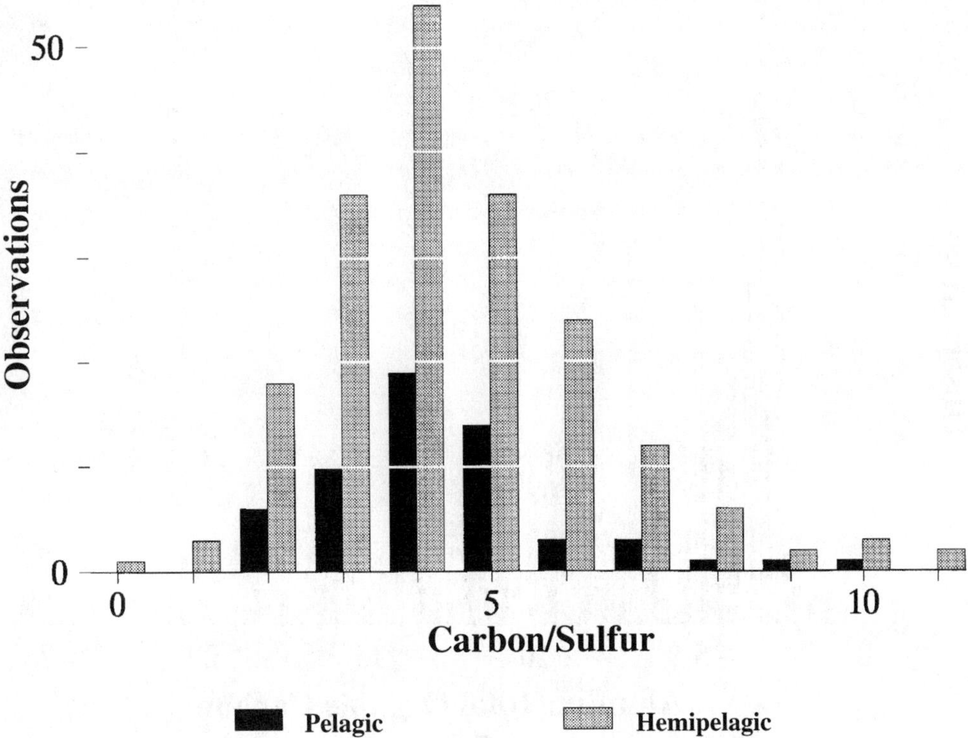

Figure 10. Distribution of organic carbon/sulfur ratios of the hemipelagic and pelagic facies. Hemipelagic facies: mean TOC/S = 4.11, mode = 4.0, standard deviation = 3.07. Pelagic facies: mean TOC/S = 3.85, mode = 4, standard deviation = 1.57.

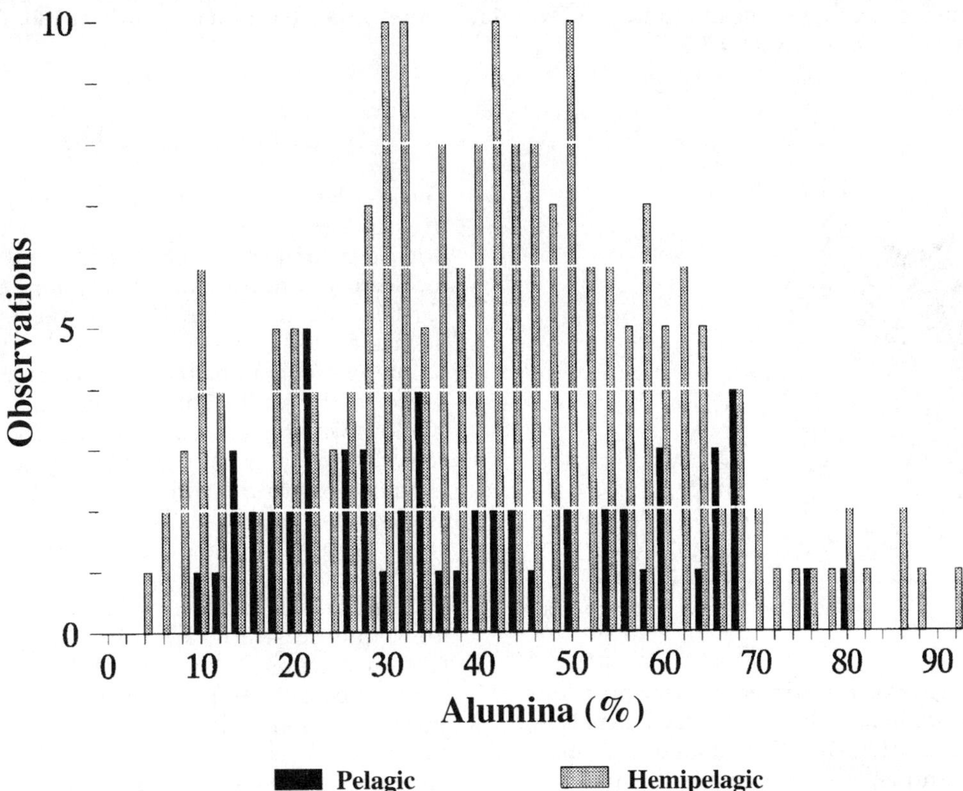

Figure 11. Distribution of alumina in the hemipelagic and pelagic facies. Alumina content is an indication of the terrigenous clay content of the rocks. Hemipelagic facies: mean alumina content = 40.67%, mode = 32% (secondary modes at 40%, 50%), standard deviation = 18.55%. Pelagic facies: mean alumina content = 38.51%, mode = 22% (secondary modes at 34%, 68%), standard deviation = 19.01%.

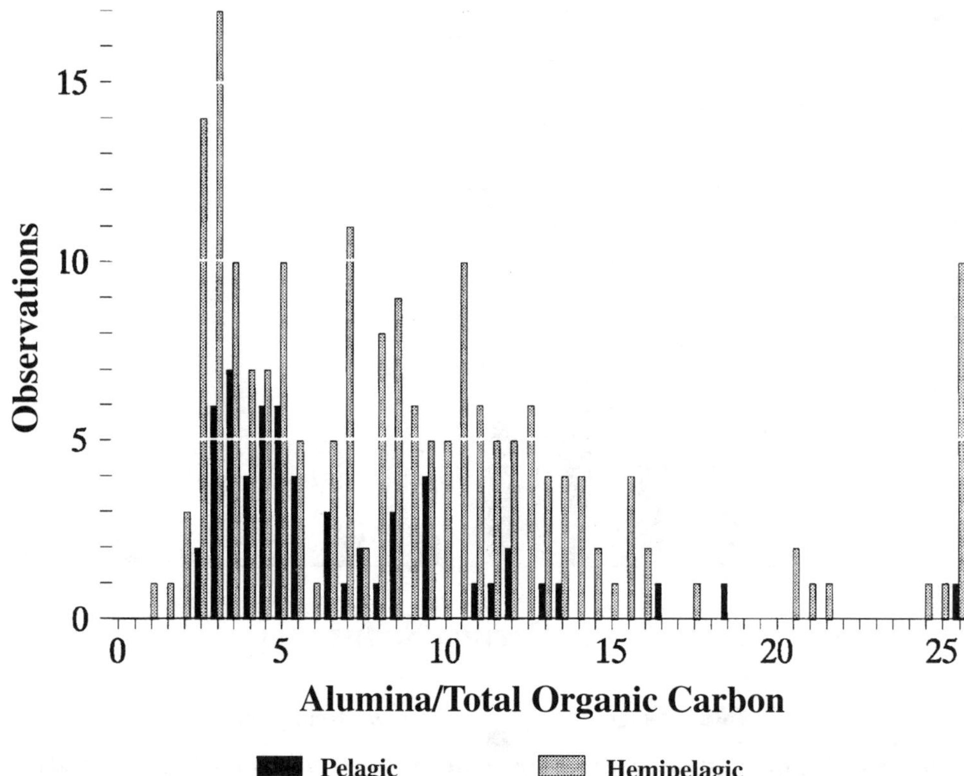

Figure 12. Distribution of alumina-to-total organic carbon ratios of the hemipelagic and pelagic facies. This ratio indicates the quality of the source rock according to its sulfur partitioning between inorganic and organic species. Hemipelagic facies: mean alumina/TOC = 13.36, mode = 3.0 (secondary mode at 7.0). Pelagic facies: mean alumina/TOC = 6.96, mode = 3.5.

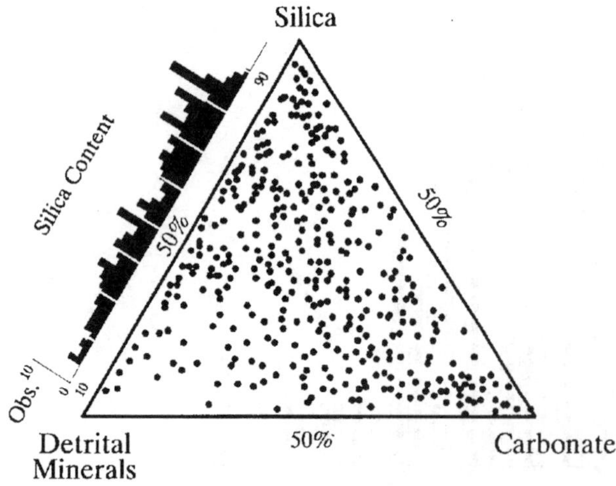

Figure 13. Wide range of sedimentary compositions among individual beds in the Monterey Formation, south–central coastal basins. Each data point represents a single sample; the histogram is the marginal distribution of samples containing less than 1% carbonate minerals (including apatite). All compositions are calculated disregarding organic matter content. (Data mostly from Isaacs, 1986.)

These criteria contrast with the "conventional" attributes of sequence boundaries and mid-sequence downlap surfaces, such as subaerial erosion, deep local erosion and stratal truncation, and basinward shift in coastal onlap (Van Wagoner et al., 1988) because of the distal deep-water setting of the Monterey. Although changing accommodation drives the formation of the entire sequence in both deep and shallow water, the local response and record differs according to the local depositional environment. This is analogous to the differences in the expression of depositional sequences between clastic- and carbonate-dominated systems (Sarg, 1988; Van Wagoner et al., 1988).

One can only make a preliminary interpretation of the sequence stratigraphy from any single vertical section, because very different three-dimensional stratal geometries may have very similar vertical expressions depending on position in the basin and completeness of sequences. A relatively broad four-dimensional perspective (space and time) is necessary to confirm the preliminary interpretation. In this study, we confirmed the identification of the key surfaces in outcrop in two ways: (1) by tying the outcrops to the integrated well-log/seismic/biostratigraphic grid through the outcrop gamma-ray spectral profiles (see Figure 5), and (2) by the biofacies analyses, especially of dinoflagellates. Figures 19 and 20 (taken from a grid of detailed well-log cross sections) illustrate the characteristic geometries associated with sequence boundaries and downlap surfaces.

As a check on the position of the system tracts boundaries determined from physical evidence, we plotted the frequency of occurrence of paleoecologically significant species of dinoflagellates against the physical stratigraphy (Figure 21; see also Table 4). The abundance of shallower water genera corresponds well with the sequence boundaries picked based on physical evidence. The two occurrences of reworked forms correspond to the 16.5 and 10.5 Ma sequence boundaries—the time of two of the largest sea level falls in the Miocene according to Haq et al. (1987). Downlap surfaces picked in outcrop are generally marked by a local maximum in the abundances of deep-water indicator species. This integration of physical stratigraphy and biofacies is the ideal way that sequence stratigraphy works: bringing together all the data that attest to the depositional conditions of the rocks.

FACIES CHANGES WITHIN A SEQUENCE STRATIGRAPHIC FRAMEWORK

The physical and chemical attributes of the Monterey Formation vary systematically with position within the sequences and sequence sets. Within each sequence, the parameters of the depositional environment that control source potential vary systematically with changing accommodation; these controls include the type and amount of organic production, bottom-energy levels, oxygenation levels, and the amount and distribution of terrigenous material. These variations are seen at many different scales: from parasequence sets, through systems tracts and sequences, to sequence sets. Not all systems tracts nor sequences in the Monterey look alike, but the nature of the changes in vertical section is analogous.

The importance of a chronostratigraphic framework to understanding the variations in source potential is illustrated in Figure 22. This diagram illustrates how the mud rock-dominated portion of a depositional sequence, commonly mapped as a single formation, ranges widely in age and detailed depositional setting. Significant variations in mud rock properties and source potential can occur within a single sequence of a "black shale" formation (e.g., Mowry Shale—Bohacs, [unpublished data, 1985], Miskell-Gearhart, 1989; Tropic Shale—Zelt, 1985; Pierre Shale—Loutit et al., 1990); additionally, a "black-shale" formation may contain many sequences (Bohacs and Isaksen, 1991). Considering the geochemical measurements in light of the sequence-stratigraphic framework allows comparison of time-equivalent rocks; this portrays genetic relations that can reveal depositional processes and enable construction of predictive models.

The physical aspects of systems tracts in the Monterey Formation show good concordance of features, although the detailed lithology may vary widely. The interval between the sequence boundary and mid-sequence downlap surface (lowstand and transgressive systems tracts) exhibits maximum bottom energy and terrigenous input at the base, gradually decreasing upward to a minimum at the mid-sequence downlap surface. Bedding planes are mostly sharp, planar to curved; lamina geometries are most commonly discontinuous, wavy nonparallel at the base, overlain by discontinuous, wavy parallel, and ultimately wavy and planar parallel (bedding terminology after Campbell, 1967). Primary sedimentary structures include scours (common to abundant), graded beds (5–25 cm thick), and T_{ab} and T_{de} turbidites (Bouma sequences a–b and d–e; Bouma, 1962). The lowstand also commonly contains the deposits of submarine slides, slumps, and debris flows (further discussion in Bohacs, 1990). Secondary sedimentary structures that are common in the lowstand systems tract are load structures, convolute, swirled, and flowed bedding, microfaults, and vein structures (see Garrison and Ramirez, 1989).

The highstand systems tract (mid-sequence downlap surface to overlying sequence boundary) generally contains minimal (background) terrigenous input. Bedding planes in the highstand systems tracts are both sharp and gradational, wavy and curved; lamina geometries are generally planar parallel and wavy parallel, with more common discontinuous, wavy parallel in the upper portion of the systems tract. Primary sedimentary structures other than planar parallel laminations are not common; graded beds (sandy siltstone to mudstone) above broad scours (1–2 cm deep by 1–2 m wide) occur sparsely. Secondary sedimentary structures are sparse in occurrence and include convolute, swirled, and flowed bedding.

No trace fossils were observed in practically all of the interval, most probably due to anoxic bottom waters. Sand-filled *Planolites* and *Terebellina* are sparsely concentrated in one 50-cm-thick interval of a lowstand at Naples Beach. These burrows descend from the base of turbidite beds contained within a thick sedimentary slumped zone, indicating transport downslope from a more oxygenated area.

Siliceous sponge spicules are ubiquitous from the 17.5 Ma sequence boundary to the top of the section. Lowstands tend to contain abundant spicules, generally aligned in laminae; the transgressive and highstand systems tracts contain sparse to common spicules dispersed in the rock. The spicules are often broken and scattered; we observed few complete axons. The presence of sponge spicules strongly indicates advective transport of material along the bottom from more proximal, more highly oxygenated areas. Figure 23 shows the present-day abundance of

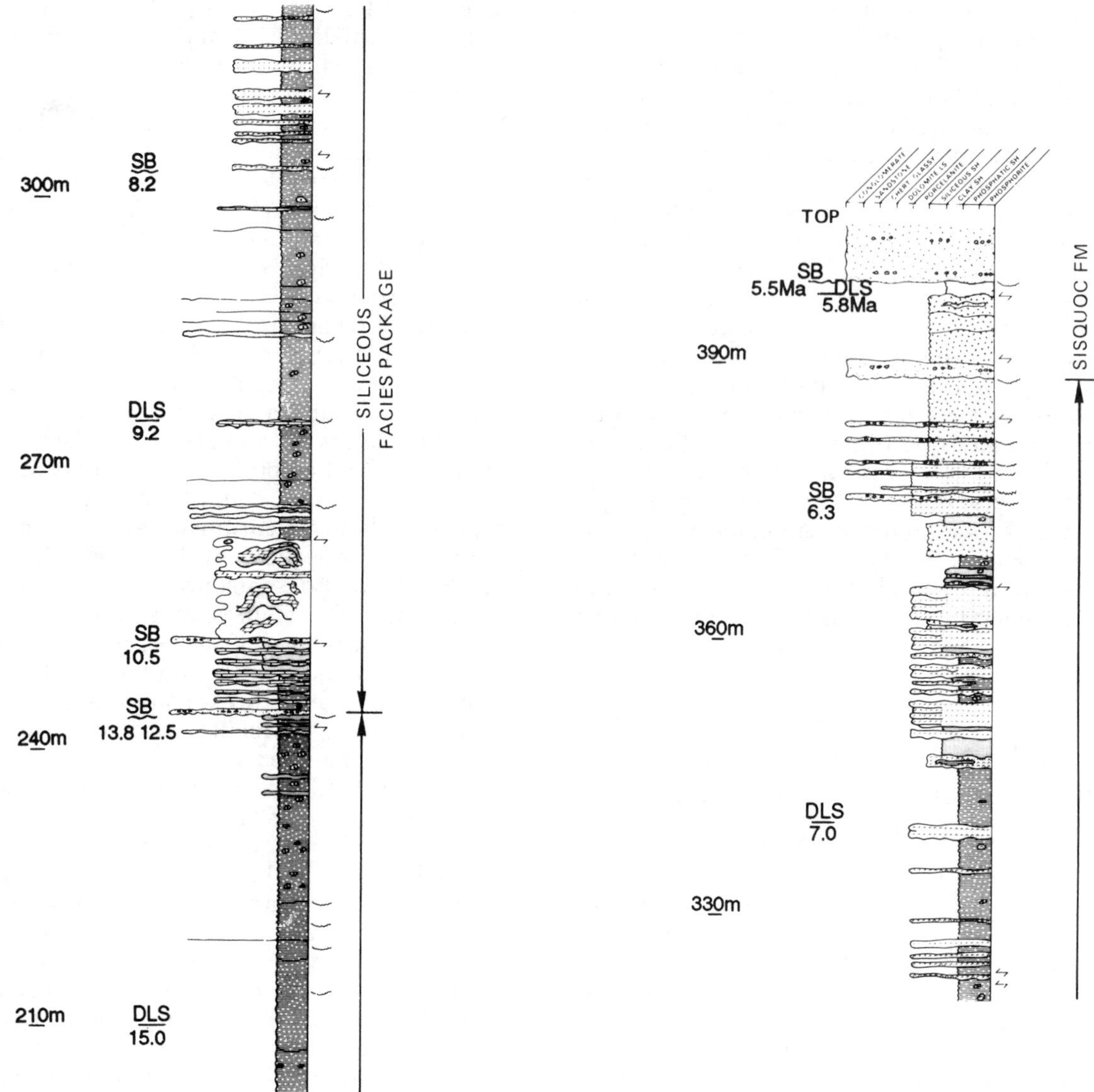

Figure 14. Columnar description of the Naples Beach outcrop (Santa Barbara County, California). Lithology is represented by outcrop profile and pattern; key bedding planes shown on right-hand side. Ages of sequence boundaries (SB) and downlap surfaces (DLS) in millions of years. (After Bohacs, 1990; positions of most surfaces between 16.5 and 9.2 Ma have been adjusted to agree with additional age and well data.)

siliceous sponges under a strong upwelling zone, where they are restricted to the more shallow, well-circulated regions. Most sponge physiologists indicate that these forms require an aerobic environment: ". . . the oxygen consumption [in sponges] has been found to be approximately the same as in other marine invertebrates, for example, 0.15 cc of consumed oxygen per hour per cc of sponge . . ." (De Laubenfels, 1955, p. E32). The presence of bottom currents able to carry and disseminate these and other clay- to silt-sized particles is an important factor in the energy regime of the Monterey environment and bears directly on the transport and preservation of organic matter. The terrigenous clays carried in by these episodic bottom currents strongly control the quality of the source rock (as described in the following section).

Lithology within a particular depositional sequence is a function of geographic, bathymetric, and stratigraphic position within the sequence (Sangree and Widmier, 1977; Van Wagoner et al., 1988). In a marine clastic-dominated depositional sequence, the lithologies in the nearshore regions will be mostly coarse clastics and mud rocks will dominate the deeper offshore regions. Hence, the lithologic expression of any particular systems tract varies according to the depositional systems that occupied the specific site under consideration. What does remain consistent are the trends in lithology and bedding within the sequence (the stacking patterns). The

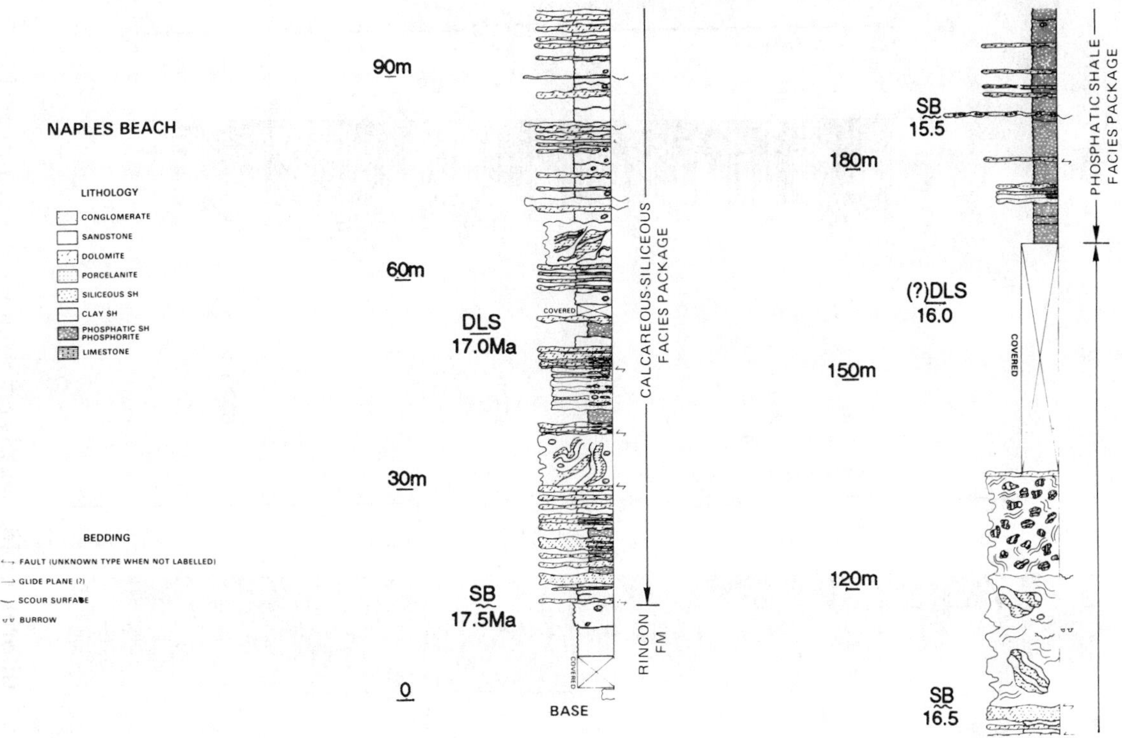

Figure 14 (continued).

lithologies and organic content of the Monterey Formation resulted from complex interactions of shifting terrigenous input, different kinds of organic productivity (siliceous vs. calcareous), and changing preservational conditions in the water and sediment column. The interaction of these processes is a function of not only relative sea level (systems tract), but also position within the basin. Each depositional environment (upper slope, lower slope, basin floor) still has characteristic proportions of hemipelagic and pelagic facies, but exactly which depositional environments stack in a particular systems tract is dependent on the geomorphic position and depositional history of a particular site. Thus, each systems tract does not have a generalized characteristic lithology, but systems tracts throughout the basin do show equivalent secular trends of lithology change within sequences and sequence sets.

This equivalent secular change is illustrated in Figure 24; it shows the 10.5 Ma sequence at Naples Beach (upper-slope setting) and Point Pedernales (lower-slope and basin-floor setting). A first-order difference between the outcrops is the thickness of the section; Point Pedernales is 10 m thicker than Naples Beach, even though Point Pedernales is the more distal location. This illustrates an essential attribute of the Monterey system that contrasts with "normal" sequence stratigraphy: due to the large biogenic input, depositional sequences thicken in an offshore direction.

The section above the sequence boundary in both locations shows evidence of high bottom energy levels and sediment reworking: at Point Pedernales, high rates of biogenic silica deposition and unstable bottom conditions generated ten or more slump zones (1–7 m thick; Figure 25) of thinly interbedded chert and porcelanite. The large amount of biogenic material in the lowstand at Point Pedernales accumulated in the lower slope environment and subsequently slid into its final resting place; the predominance of biogenic siliceous sediments and their relative instability most probably resulted from invigorated circulation and elevated siliceous productivity during the lowstand (Barron and Baldauf, 1989). At Naples Beach, a 30-cm-thick bed of reworked phosphatic concretions (≤ 5 cm) rests upon the sequence boundary and is overlain by two thick slump zones (3–6 m) of dolomite and phosphatic shale and three graded beds of siltstone and very fine grained sandstone. The larger terrigenous component at Naples Beach is due to its more proximal location.

The transgressive systems tracts at both sections record lower bottom-energy levels and more continuous sedimentation: the more distal Point Pedernales section in porcelanite with phosphatic shale interbeds, and Naples Beach in phosphatic mudstones and grainstones with a 60-cm-thick limestone bedset at the downlap surface.

The highstand at Point Pedernales accumulated porcelanite and phosphatic shale with a subtle

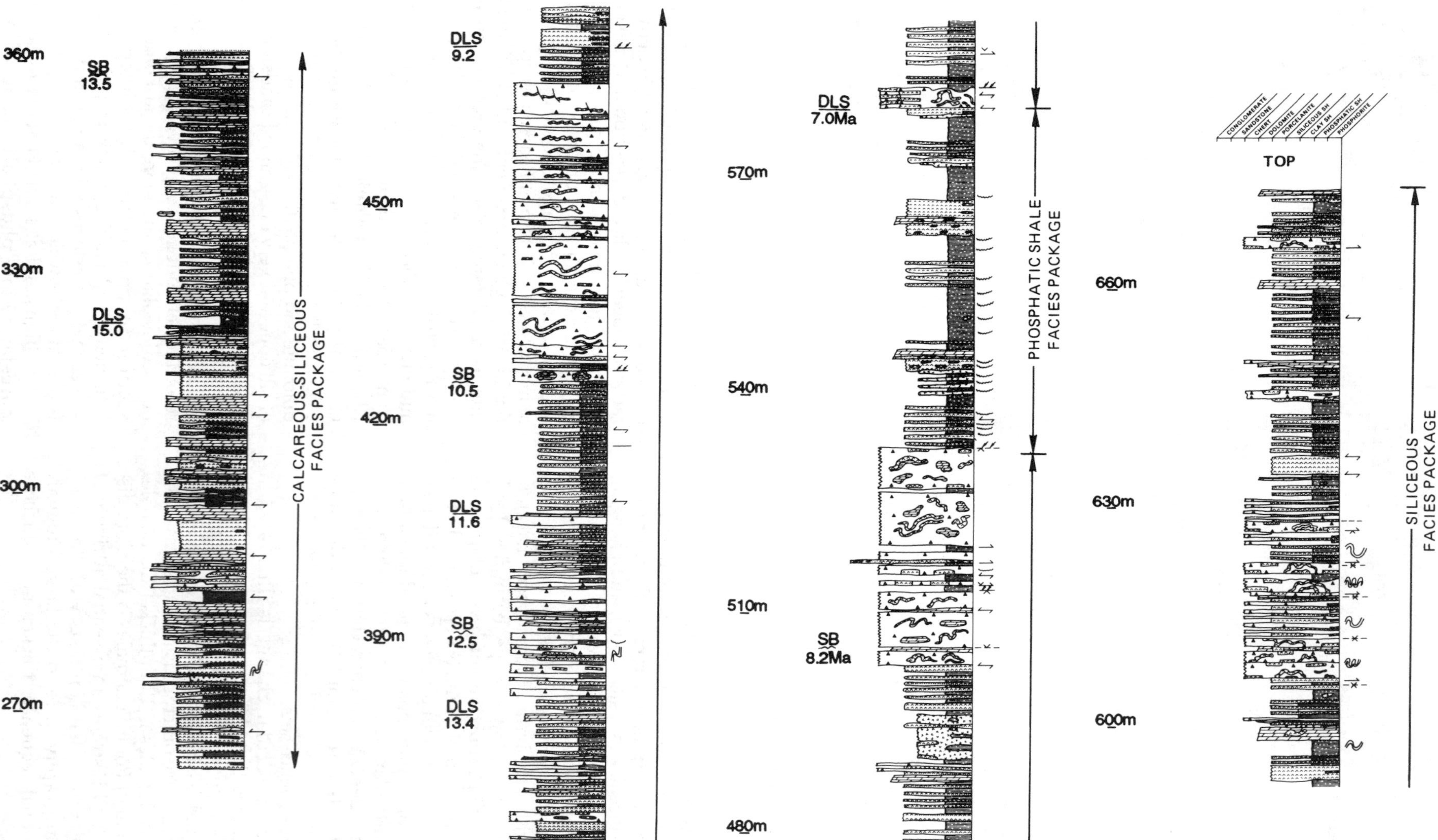

Figure 15. Columnar description of the Point Pedernales outcrop (Santa Barbara County, California). Lithology is represented by outcrop profile and pattern; key bedding planes shown on right-hand side. Ages of sequence boundaries (SB) and downlap surfaces (DLS) in millions of years. (After Bohacs, 1990.)

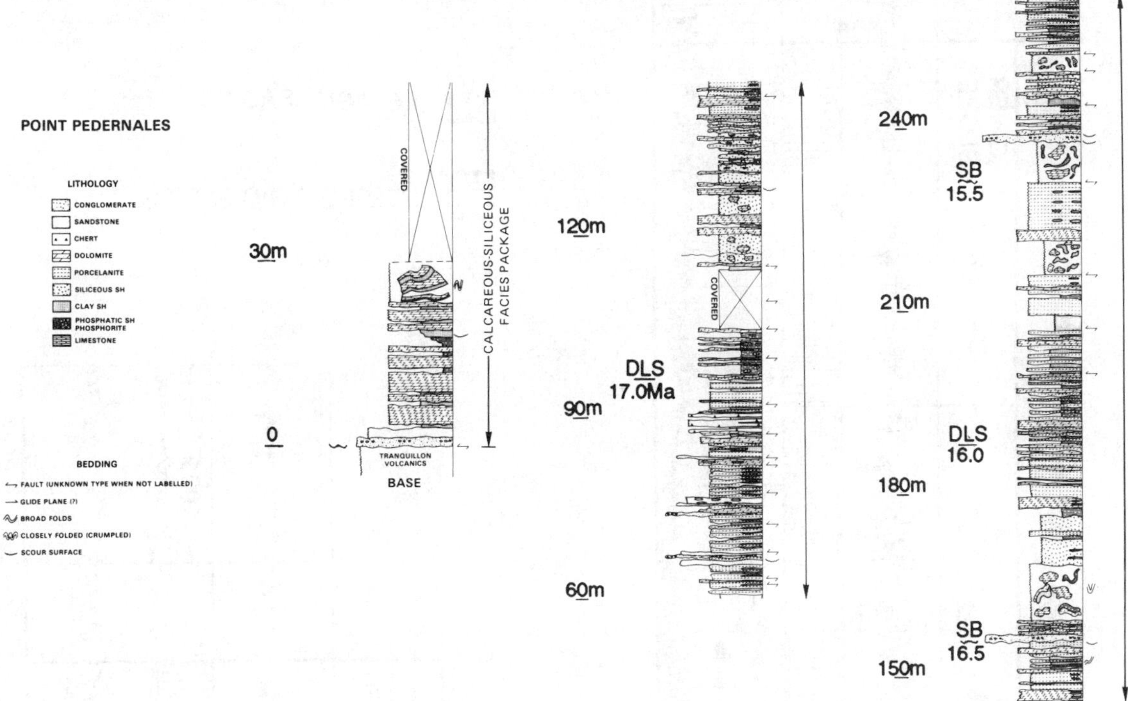

Figure 15 (continued).

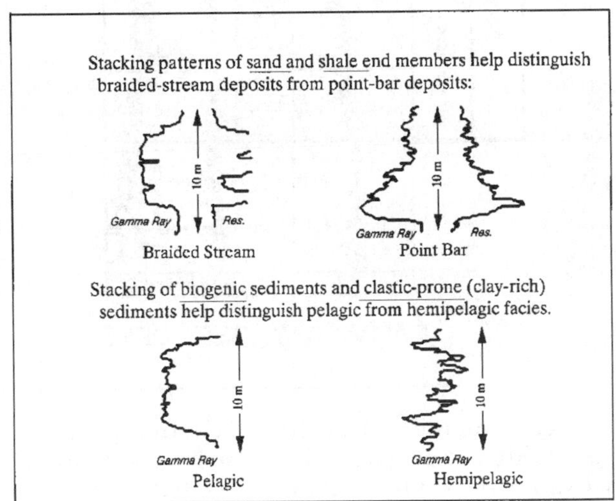

Figure 16. Stacking of strata indicates facies. Just as one defines fluvial facies based on the stacking of sandstones and mudrocks, so can one interpret the pelagic and hemipelagic facies of basinal mudrocks. (From Bohacs, 1990.)

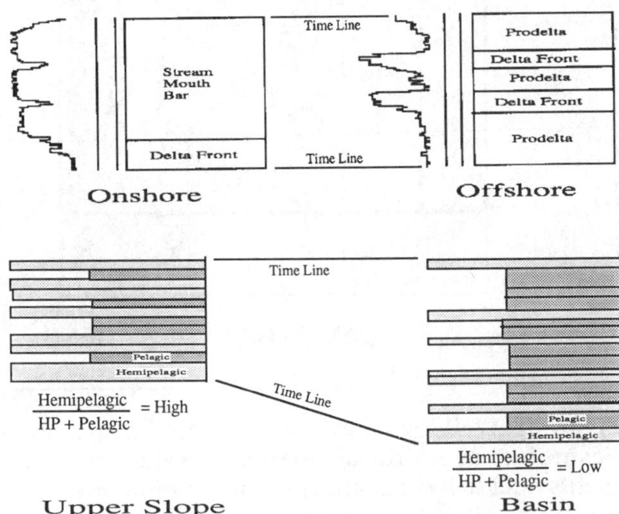

Figure 17. Stacking of subenvironments indicates the overall depositional setting. Just as the portions of a river-dominated shoreline system stack in characteristic associations, so does the stacking of hemipelagic and pelagic facies indicate the depositional settings within the Monterey Formation. (From Bohacs, 1990.)

decrease upward in biogenic silica content (due to a relative increase of phosphatic-shale input over biogenic silica production). The highstand at Naples Beach is marked by thick phosphatic mudstone and grainstone bedsets at the base, with gradually more scours and thin siltstone beds (5–15 cm) upwards towards the overlying sequence boundary, portraying its more proximal position. Thus, the 10.5 Ma

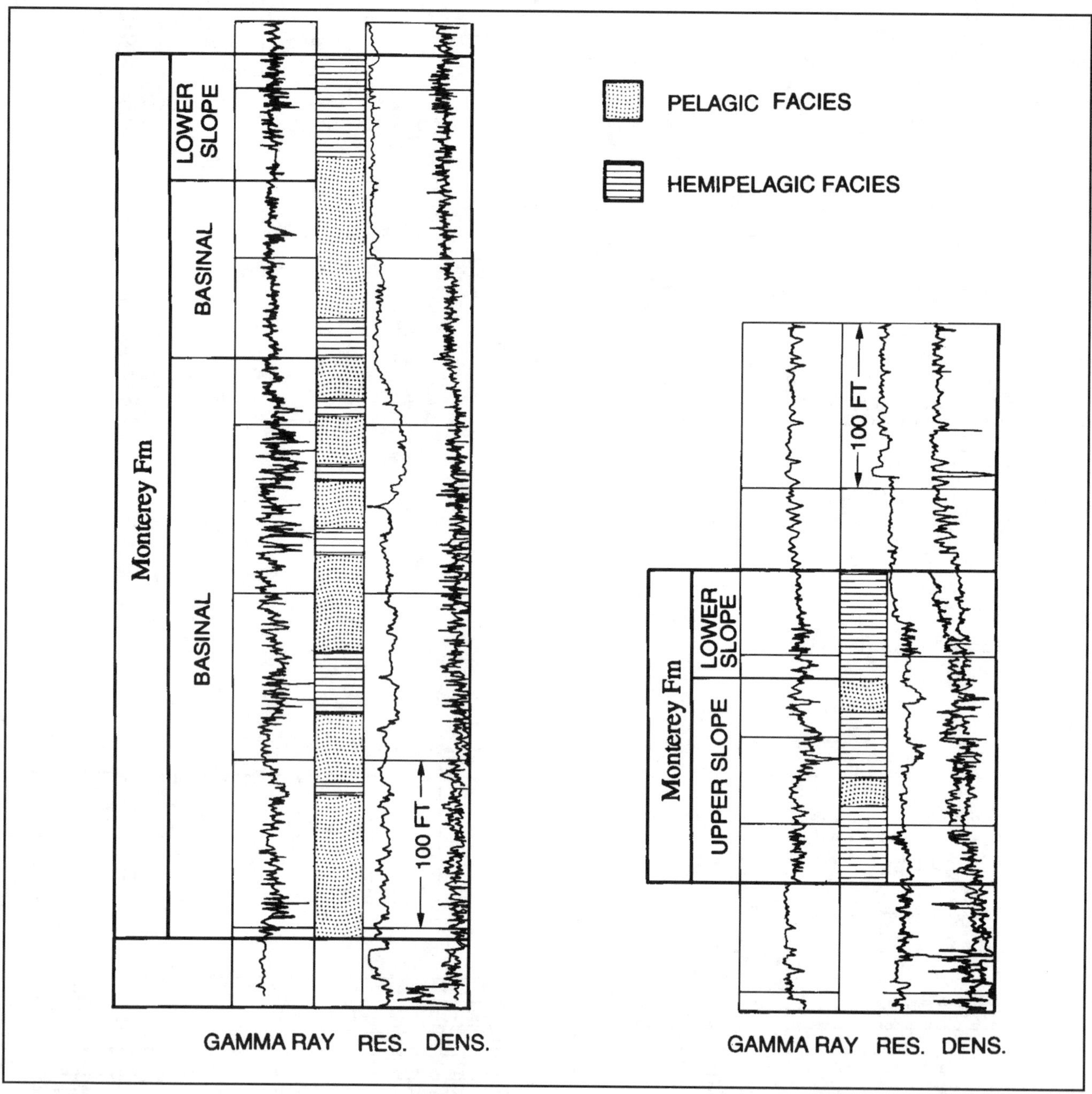

Figure 18. Well logs showing the vertical distribution of interpreted Monterey depositional environments. Basin-floor depositional environments tend to have a large proportion of pelagic facies; upper-slope environments tend to have a large proportion of hemipelagic facies. (RES. = resistivity; DENS. = density.)

sequence at both locations shows similar trends in bedding and recorded energy levels; the lithology changes are disparate but consonant with the paleogeographic position of the two outcrops.

SOURCE QUALITY VARIATIONS

The quality of the source rocks of the Monterey Formation is a function of the richness (TOC), oil-proneness (hydrogen content), and sulfur content of its kerogen. The richness and sulfur content vary significantly although the composition (oil proneness) of the primary organic input was probably relatively constant: most of the rock samples (89.5%) have hydrogen indices greater than 400 (Figure 26) and the molecular signature of all the oils studied vary little, suggesting that the contributing organic matter is similar (Figure 27). The molecular compositions are dominated by compounds characteristic of diatoms. Most (90 to 95%) of the organic matter is amorphous

Table 2. Generalized attributions of Monterey depositional environments.

Environment	Paleotopographic position	Gamma-ray activity	Isopach thickness	Paleogeographic position
Bank Top	On top of highs, center of highs	Thick zone of very high gamma-ray activity	Less than 30 m	On interbasin highs
Upper Slope	To the sides of interbasin highs	Generally thick zones of high gamma-ray activity with thin interbeds of moderately low gamma-ray activity	30–60 m	Around the edges of interbasin highs
Lower Slope	Midway between interbasin highs and intrabasin deeps	Relatively equal portions of moderate to thinly bedded high gamma-ray and low gamma-ray bedset	60–140 m	Along the lower edges of interbasin highs, midway between the interbasin high and basin deeps; best developed in the inner tier of basins
Basin Floor	In intrabasin deeps or local maxima in isopach thickness	Relatively thick bedsets of low gamma-ray activity interbedded with thin bedsets of high gamma-ray activity	Generally greater than 140 m	In center of interbasin deeps

under microscopic examination and there is no terrigenous organic matter apparent. Much of the variation in richness is controlled by changing depositional conditions. For instance, the richness decreases where the organic carbon content is diluted by increased biogenic silica deposition. The sulfur content of the rocks also varies significantly as a function of depositional environment, controlled by the amount of terrigenous clastic input and burial rate (discussed below). We observed that the sulfur content of the kerogen is a major control on density, viscosity, and sulfur content of the oils associated with these source rocks (both extracted and in reservoirs) (see also Baskin and Peters, 1992).

The major shifts in richness and sulfur content occur at sequence boundaries or downlap surfaces. That is, these through-going physical surfaces bound packages of rocks (sequences and sequence sets) that have distinct chemistries. The total organic carbon distribution shown in Figure 7 demonstrates these major stratigraphic packages and their attendant geochemistry. At the sequence scale, TOC content tends to be moderate in the lowstands, increasing to the downlap surface, and relatively low in the highstands. The thin sequences in the transgressive sequence set contain the largest TOC contents. The TOC contents of each systems tract results from the balance among the different types of organic productivity, and their deposition and preservation. Thus, although the lowstand systems tracts were probably deposited under intensified upwelling, they have only moderate TOC contents because the abundant biogenic silica also being produced diluted the organic matter deposited with it. The slower rates of net sedimentation in the transgressive portions of the sequence set tended to concentrate the organic matter.

The stratigraphic distribution of the ratio of alumina to total organic carbon is used as an indicator of hydrocarbon source rock quality (Figure 7). This ratio indicates the amount of sulfur that is free in the geochemical system to be bound in the kerogen, which in

Table 3. sequence stratigraphic criteria.

The criteria for the interpretation of a sequence boundary in outcrops of deep-basinal mudrocks are:

1. Evidence of increased circulation above the surface:
   - increased proportion of hemipelagic facies in proximal subbasins.
   - increased proportion of biogenic silica in distal subbasins.
2. Evidence of increased bottom-energy levels above the surface:
   - a larger proportion of the section above the surface contains coarser clastics, sandstone turbidites, slumps, and slides.
   - lamina geometries above the surface tend to be wavy, non-parallel; the bed surfaces are typically wavy.
   - bedset thicknesses increase above the surface.
3. Local relief and erosion are generally present on the surface.
4. The proportion of shallower-water and reworked dinoflagellates increases to a local maximum above the surface.

The criteria for the recognition of a downlap surface in outcrops of deep-basinal mudrocks are:

1. An increased proportion of pelagic facies in the interval adjacent to the surface.
2. A secular change in the dominant lithology across the surface (e.g., from calcareous to siliceous rocks, phosphatic shale to clay shale).
3. Evidence of decreased bottom-energy levels and terrigenous sediment input:
   - a larger proportion of phosphatic concretions and phosphatic shale is present.
   - lamina geometries tend to be planar, parallel, and continuous, and are often rather diffusely expressed; bed surfaces typically are planar.
4. Little or no evidence of significant erosion is present on the surface (but some local reworking not uncommon).
5. The proportion of deeper-water dinoflagellates increases to a local maximum at or near the surface; there is no evidence of reworked individuals.

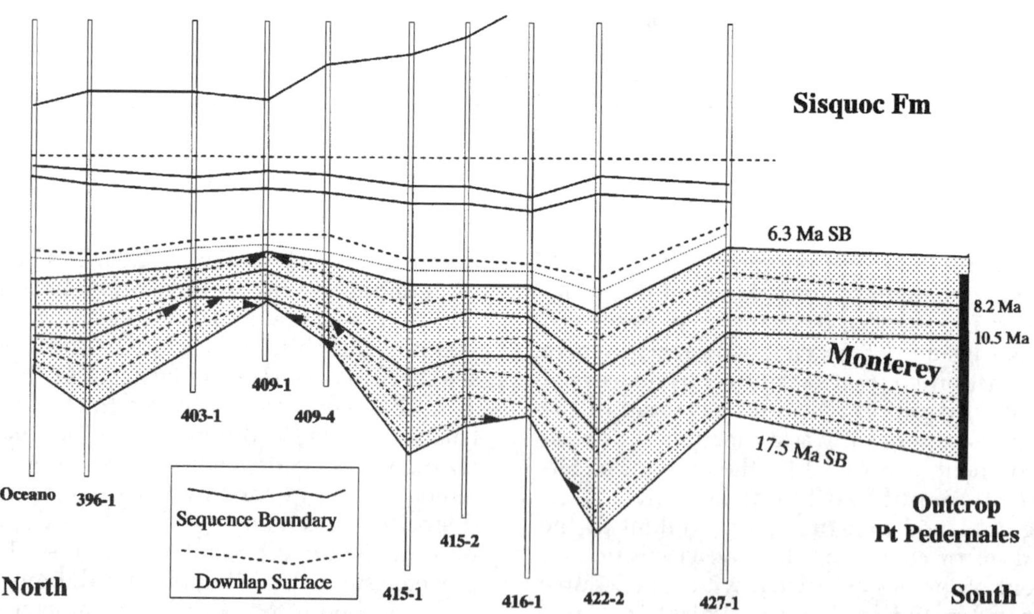

Figure 19. Schematic north–south cross section of the Miocene section, offshore southwestern California. This schematic section is taken from a grid of well-log cross sections that were tied with extensive seismic and biostratigraphic data. Illustrated are the characteristic geometries of depositional sequences in the Monterey Formation (large-scale truncation under sequence boundaries, regional onlap of downlap surfaces, depositional thinning onto topography). Not all sequence boundaries are shown below the 10.5 Ma sequence boundary: within this interval the sequence boundaries are generally conformable with the downlap surfaces (at the scale of this schematic cross section) and show the same large-scale geometries as their associated downlap surfaces. Also shown is the correlation to outcrops at Point Pedernales. Section is approximately 40 km (24 mi) long.

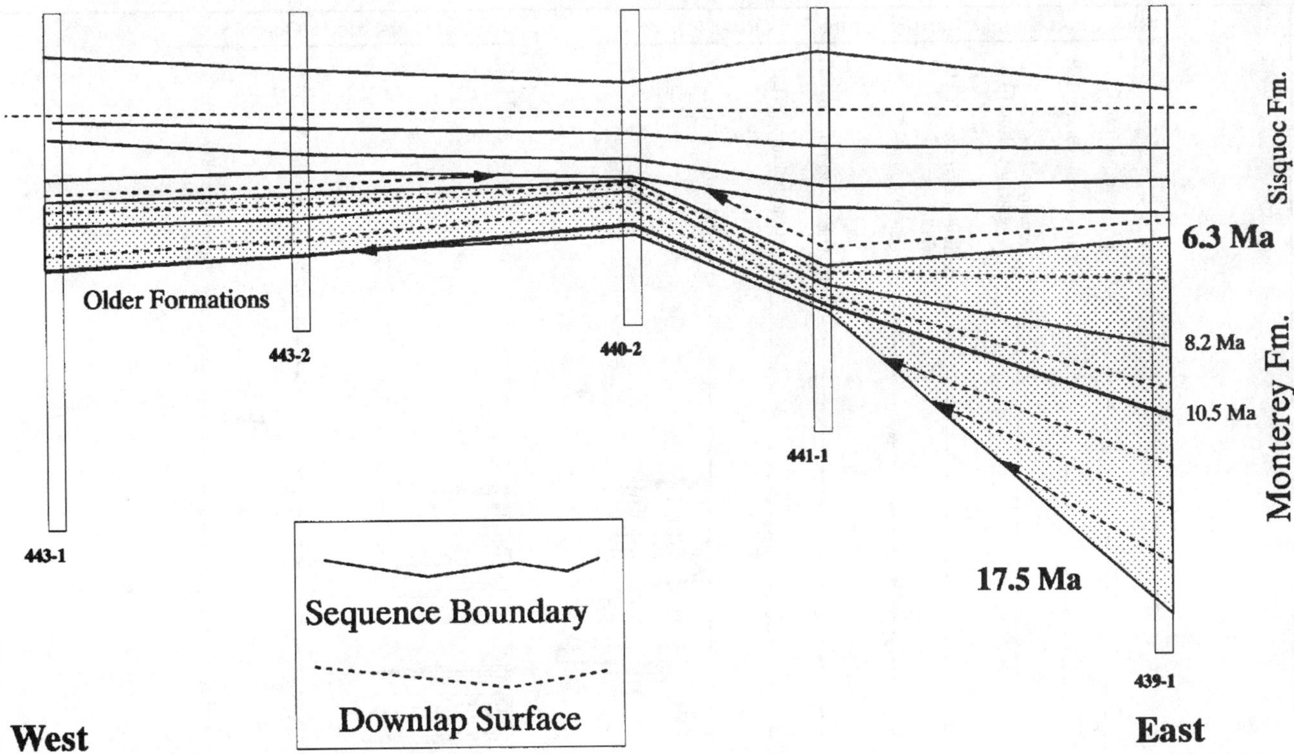

Figure 20. Schematic east–west cross section of the Miocene section, offshore southwestern California. This schematic section is taken from a grid of well-log cross sections that were tied with extensive seismic and biostratigraphic data. Illustrated are the characteristic geometries of depositional sequences in the Monterey Formation (large-scale truncation under sequence boundaries, regional onlap of downlap surfaces, depositional thinning onto topography). Not all sequence boundaries are shown below the 10.5 Ma sequence boundary: within this interval the sequence boundaries are generally conformable with the downlap surfaces (at the scale of this schematic cross section) and show the same large-scale geometries as their associated downlap surfaces. Section is approximately 25 km (15 mi) long.

turn influences the quality of any oil generated. The content of alumina is a function of the amount of clay minerals that come into the depositional environment as clastics; this mineral detritus also contains iron and other metals. The iron is key because it combines with an equivalent amount of sulfur to make pyrite, thereby removing the sulfur from the organic-geochemical system and improving the quality of the kerogen (Raiswell and Berner, 1986; Mankiewicz, personal communication, 1987). Hence the alumina/TOC ratio portrays not only the richness but also the quality of a source that was deposited in these deep-basinal environments. At Naples Beach (Figure 7) the section between the 15.5 Ma sequence boundary and 10.5 Ma sequence boundary has very low ratios of alumina to TOC (generally less than 7.5), indicating a sulfur-rich oil source. Up section from the 10.5 Ma sequence boundary toward the sequence boundary at 6.3 Ma, the detrital content increases and the hydrocarbon source rock quality improves. The lower section, below the 15.5 Ma sequence boundary to the 17.5 Ma sequence boundary, contains two sequences with low to intermediate source rock quality in the zones around the downlap surfaces and higher quality in the lowstand systems tracts. Thus the lowstand systems tracts contain higher quality source rocks due to their larger content of mineral detritus; although these rocks are not the richest, they are most prone to generate lower sulfur (higher quality) oil.

The sulfur content is also influenced by the burial rate of organic matter: the longer the organic matter remains in the zone of sulfate reduction in the sediment, connected to a source of seawater sulfate, the greater the sulfur content of the kerogen (Raiswell and Berner, 1986). Thus, the thinner sequences in the transgressive portion of the sequence set with lower rates of sediment accumulation have the largest sulfur content. So, the major depositional controls on source rock quality (input of terrigenous clastics [i.e., Fe] and burial rate) are both strongly influenced by position within the sequence and sequence set. Portraying these parameters within a sequence strati-

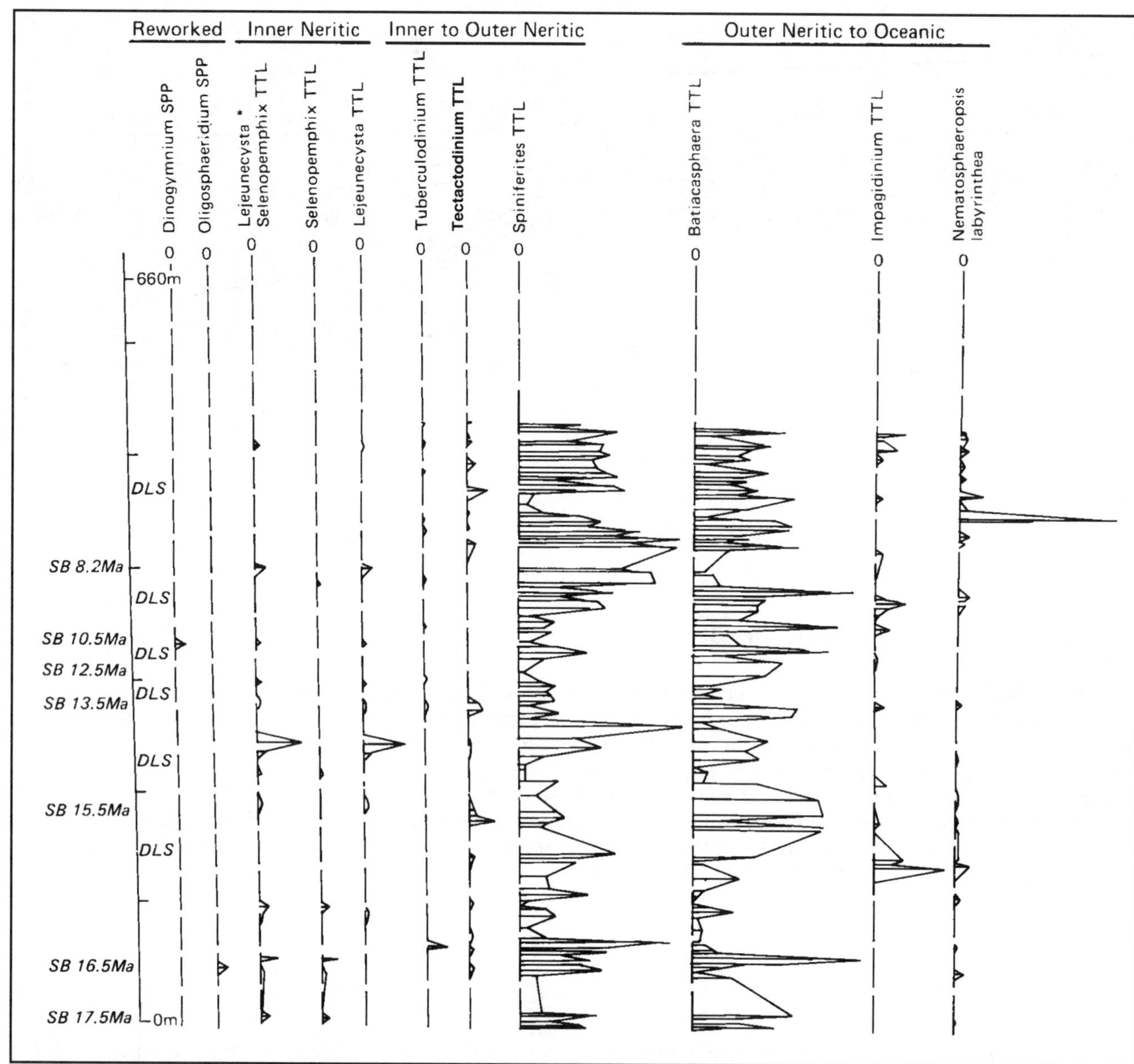

Figure 21. Biofacies indices of sequence stratigraphy, Point Pedernales outcrop. The frequencies of useful environmental indicator dinoflagellates plotted against depth. Positions of sequence boundaries (SB) and downlap surfaces (DLS) are chosen from the physical stratigraphy of the outcrop. Four classes of indicator genera are shown: inner neritic, inner to outer neritic, inner neritic to oceanic, and reworked. The palaeoecology of these genera was interpreted by W. Gregory of Exxon Company, USA based on investigations of the ecology of present-day and Neogene dinoflagellates (Williams and Bujak, 1977; Harland, 1983; Wrenn and Kokinos, 1986). Table 4 shows his categories of paleoecologically significant species.

graphic framework makes it possible to map these attributes and to compare genetically related rocks around the basin.

SUMMARY AND CONCLUSIONS

To derive predictive models of depositional facies and source potential, it is essential to compare genetically related rocks. Sequence stratigraphy provides the chronostratigraphic framework that reveals coeval rocks, genetic relations, and detailed timing of deposition. The sequence stratigraphic approach also recognizes a fundamental aspect of the rock record: sediment accumulation is punctuated by discontinuities at many scales. Even the fine-grained rocks of the Monterey Formation do not form a simple layer-cake

Table 4. Paleoecology of selected Miocene dinoflagellates.*

Inner Neritic	Inner Neritic to Oceanic
Selenopemphix spp. *Lejeunecysta* spp.	*Lingulodinium machaerophorum* *Cleistosphaeridium diversispinosum* *Spiniferites mirabilis* *Operculodinium centrocarpum*
Inner to Outer Neritic	**Reworked**
Spiniferites suite *Multispinula* spp. *Operculodinium israelianum* *Polysphaeridium zoharyi* *Tuberculodinium vancampoae* *Tectactodinium* spp. *Impagidium* suite *Hystrichosphaeropsis obscura* *Nematosphaeropsis* spp.	*Oligosphaeridium* spp. *Dinogymnium* spp.

* After Harland (1983) and Wrenn and Kokinos (1986)

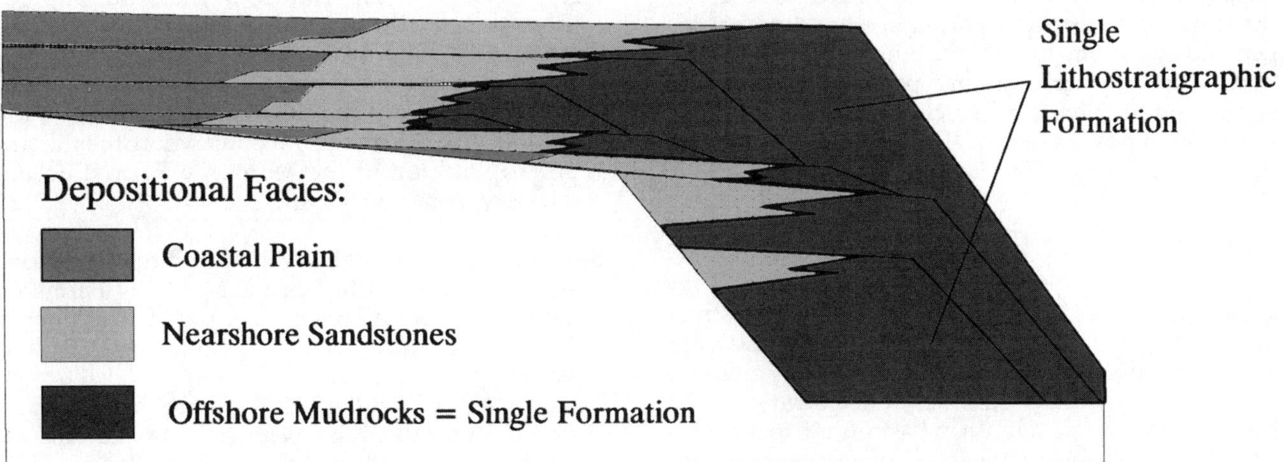

Figure 22. Lithostratigraphy contrasted with chronostratigraphy. The mudrock portion of a sequence is often mapped as a single formation. This diagram illustrates the wide range in age (and potentially depositional environment) covered by the mudrocks of a single sequence, each parasequence boundary (———) represents a time line. Recognizing this allows the collection and comparison of genetically related samples to enable the construction of a depositional model of geochemical variations.

stratigraphy, but record an active setting with significant erosion and non-deposition as well as deposition. As Ager (1981) said: ". . . remember a child's definition of a net as a lot of holes tied together with string. The stratigraphical record is a lot of holes tied together with sediment." Consequently, the geochemical properties of the Monterey Formation do not necessarily change gradually, but have significant shifts at sequence boundaries and downlap surfaces.

By integrating many scales of observation, from seismic through outcrop to thin section and molecular analyses, this study provided fundamental insights into the nature of the Monterey depositional system, its stratal patterns, and geochemical character. The correspondence of the geochemical facies with the depositional environment indicates that the chemical properties of the Monterey Formation covary with its physical parameters. This covariation, along with the

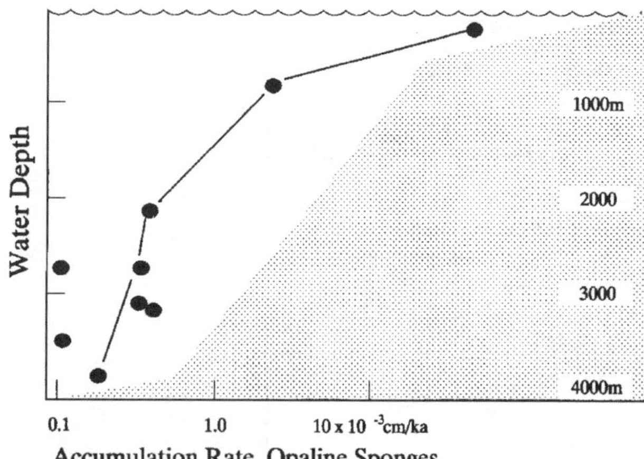

Figure 23. Accumulation rate of opaline sponges in Holocene sediments as a function of water depth. This relation, from core data from the upwelling region off northwest Africa, shows that siliceous sponges thrive in relatively shallow water. This agrees with observations that sponges require oxic environments. Siliceous sponge spicules are common to abundant in many portions of the Monterey Formation that were deposited under dysoxic to anoxic conditions. This occurrence indicates that these spicules were transported in from a more oxygenated environment and it supports the observations of the physical stratigraphy that also point to a significant near-bottom advective component in the sediment-supply system of the Monterey Formation. (Data from Bojé and Tomszak, 1978, p. 270.)

geological framework, makes it possible to construct models for the prediction of rock properties from standard exploration data. This study also established criteria for constructing a sequence stratigraphic framework in slope and deep-basin mud rocks. This extends our working knowledge of sequence stratigraphy; these criteria need to be tested on similar settings to confirm their utility and to allow the extension of the results of this investigation.

ACKNOWLEDGMENTS

This work is part of a large-scale investigation of the Miocene of southwestern California conducted as a joint project of Exxon Production Research Company (EPR) and Exxon Company, USA under the able leadership of Wendy Burgis. Andrew Scott, Ron Kleist, and Glenn Buckley of Exxon USA provided management support. I am grateful to them and to the members of the project team: J. A. Farre, G. G. Gray, K. E. Green, P. S. Koch, P. J. Mankiewicz (EPR), W. A. Gregory (then at Louisiana State University), V. K. Hohensee, G. D. Kidd, G. S. Kompanik, J. R. Schwalbach, M. L. Smith, and L. S. Wall (Exxon USA). Yeoperson services in the field were also provided by L. Hood, G. H. Isaksen, and M. McCoy. J. Rubenstone (Lamont-Doherty Geological Observatory) and T. S. Loutit (EPR) were so kind as to provide strontium-isotope data. I thank Exxon Company, USA and Exxon Production Research Company for permission to publish this material.

My understanding and focus on the mysteries of the Monterey have been sharpened and honed by many long and intensive discussions with the friends of these mudrocks, most notably Jon Schwalbach, Caroline Isaacs, and Margaret Keller. I also appreciate the comments and queries of those who have reviewed this work in its previous incarnations as poster and oral presentations. The fine paleontologists of Union Oil—Gregg Blake, Mary Lou Cotton-Thornton, and Bob Arends, among others—have been most generous in sharing their insight and expertise. Shiraz Kahn and John Barron kindly shared their age data. V. Robison and L. Liro reviewed the manuscript and provided helpful suggestions. All problems with interpretations of the myriad of data are, of course, my own.

REFERENCES CITED

Ager, D.V., 1981, The Nature of the Stratigraphical Record, 2nd edition: London, Macmillan.

Arends, R.G., G.H. Blake, 1986, Biostratigraphy and paleoecology of the Naples Bluff coastal section based on diatoms and benthic foraminifera, *in* R.E. Casey, J.A. Barron, Siliceous Microfossils and Microplankton of the Monterey Formation and Modern Analogs: Pacific Section, SEPM, p. 121-136.

Barron, J.A., 1986, Paleoceanographic and tectonic controls on deposition of the Monterey Formation and related siliceous rocks in California: Palaeogeography, Palaeoclimatology, Palaeoecology, v. 53, p. 27-45.

Barron, J.A., J.G. Baldauf, 1989, Tertiary cooling steps and paleoproductivity as reflected by diatoms and biosiliceous sediments, *in* W.H. Berger, V.S. Smetacek, G. Wefer, Productivity of the Ocean: Present and Past: New York, J. Wiley and Sons, p. 341-354.

Baskin, D.K., K.E. Peters, 1992, Early generation characteristics of a sulfur-rich Monterey kerogen: American Association of Petroleum Geologists Bulletin, v. 76, p. 1-13.

Bohacs, K.M., 1990, Sequence stratigraphy of the Monterey Formation, Santa Barbara County: Integration of physical, chemical, and biofacies data from outcrop and subsurface, *in* M.A. Keller and M.K. McGowen, Miocene and Oligocene petroleum reservoirs of the Santa Maria and Santa Barbara–Ventura basins, California: San Francisco, SEPM Core Workshop No. 14, p. 139-201.

Bohacs, K.M., G.H. Isaksen, 1991, Source quality variations tied to sequence development: integration of physical and chemical aspects, Lower to Middle Triassic, western Barents Sea: American Asso-

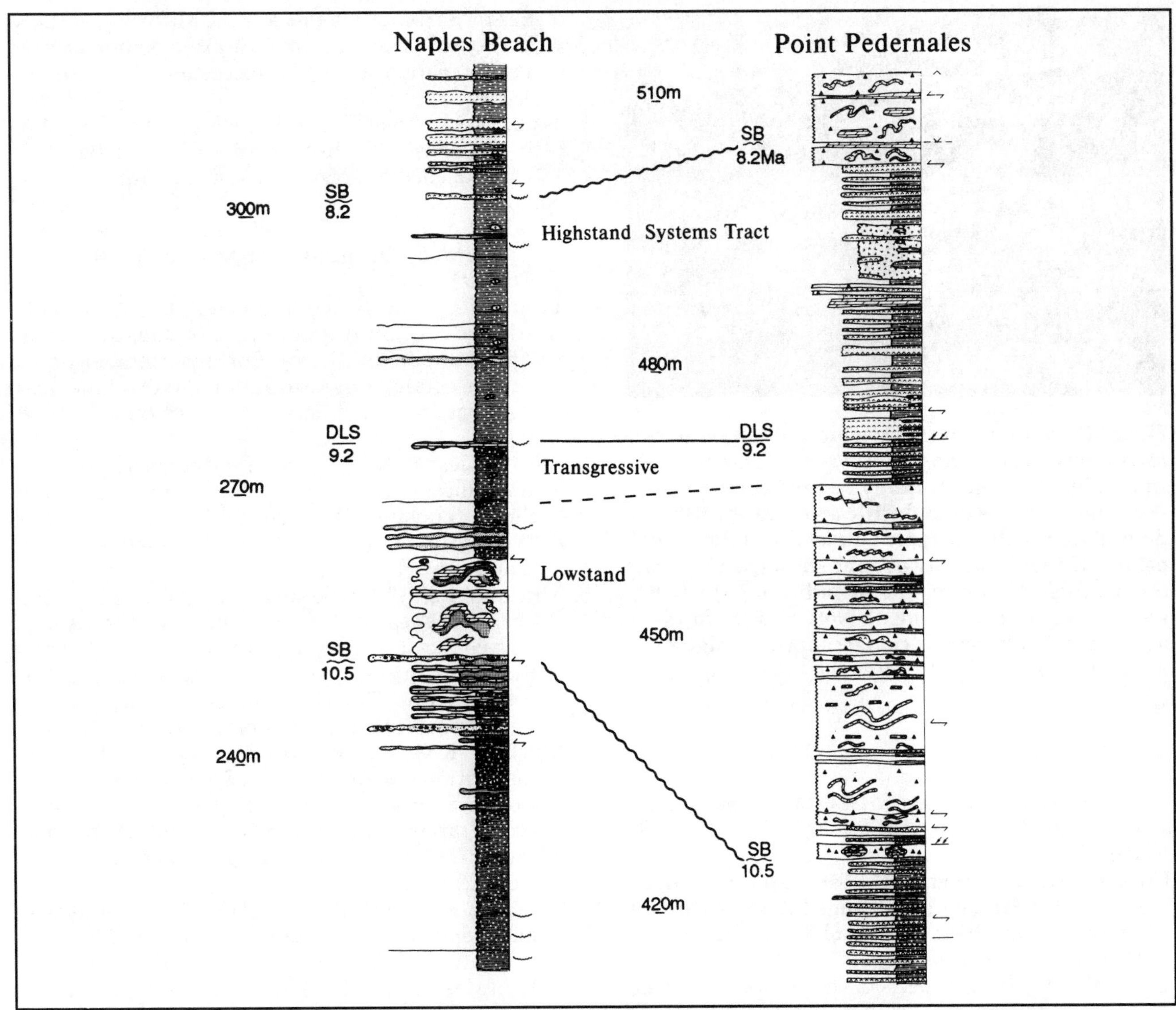

Figure 24. Variation in lithology of age-equivalent systems tracts, 10.5 Ma depositional sequence. The systems tracts at Naples Beach and Point Pedernales show similar trends in physical aspects; the different lithologic expressions are a function of the paleogeographic position of each outcrop. The more distal Point Pedernales is dominated by biogenic sedimentation; the more proximal Naples Beach section demonstrates larger terrigenous influence.

ciation of Petroleum Geologists Bulletin, v. 75, p. 544.

Bojé, R., M. Tomszak, 1978, Up-welling ecosystems: New York, Springer-Verlag.

Bouma, A.H., 1962, Sedimentology of some Flysch deposits: a graphic approach to facies interpretation: Amsterdam, Elsevier, 168 p.

Bralower, T.J., H.R. Thierstein, 1987, Organic carbon and metal accumulation rates in Holocene and mid-Cretaceous sediments: palaeoceanographic significance, *in* J. Brooks and A.J. Fleet, Marine Petroleum Source Rocks: Oxford, Geological Society Special Publication No. 26, p. 345-369.

Bramlette, M.N., 1946, The Monterey Formation and the origin of its siliceous rocks: United States Geological Survey Professional Paper 212, 57 p.

Calvert, S.E., 1966, Origin of diatom-rich varved sediments from the Gulf of California: Journal of Geology, v. 74, p. 546-565.

Campbell, C.V., 1967, Lamina, laminaset, bed, bedset: Sedimentology, v. 8, p. 7-26.

De Laubenfels, 1955, Porifera, *in* R.C. Moore, ed., Treatise on Invertebrate Paleontology, Part E: p. E32.

Garrison, R.E., P.C. Ramirez, 1989, Conglomerates and breccias in the Monterey Formation and related units as reflections of basin margin history, *in* I.P. Colburn, Conglomerates in basin analysis; a

Figure 25. Slump/slide zone typical of lowstand systems tracts in distal positions, Point Pedernales. Internally, this zone of thinly bedded chert and porcelanite shows both ductile and semi-brittle deformation with consistent offset down the paleoslope. Upper boundary of zone shows sedimentary onlap across the top (pointers), indicating synsedimentary age of emplacement. Note also conformable contact at base; this contact shows no sign of penetrative shear. (Field notebook is 19 cm long.)

symposium dedicated to A.O. Woodford: Field Trip Guidebook—Pacific Section, SEPM, 62, p. 189-206.

Gorsline, D.S., K.O. Emery, 1959, Turbidity current deposits in San Pedro and Santa Monica basins, off southern California: Geological Society of America Bulletin, v. 70, p. 279-290.

Grivetti, M.C., 1982, Aspects of stratigraphy, diagenesis, and deformation in the Monterey Formation near Santa Maria-Lompoc, California: M.A. Thesis, University of California, Santa Barbara, California.

Haq, B.U., J. Hardenbol, P.R. Vail, 1987, Chronology of fluctuating sea levels since the Triassic: Science v. 235, p. 1156-1167.

Harland, R., 1983, Quaternary dinoflagellate cysts from hole 548 and 549A, Goban Spur (Deep Sea Drilling Project leg 80), *in* P.D. de Graciansky, C.W. Poag, et al., Initial Reports of the Deep Sea Drilling Program, v. 80, p. 761-766.

Hornafius, J.S., 1991, Facies analysis of the Monterey Formation in the northern Santa Barbara Channel: American Association of Petroleum Geologists Bulletin, v. 75, p. 894-909.

Isaacs, C.M., 1980, Diagenesis in the Monterey Formation examined laterally along the coast near Santa Barbara, California: Ph.D. Dissertation, Stanford University, Stanford, California.

Isaacs, C.M., 1981, Field characterization of rocks in the Monterey Formation along the coast near Santa Barbara, California: AAPG Field Guide 4, p. 9-24.

Isaacs, C.M., 1983, Compositional variation and sequence in the Miocene Monterey Formation, Santa Barbara coastal area, California, *in* D.K. Larue, R.J. Steel, Cenozoic Marine Sedimentation, Pacific margin, USA: Pacific section SEPM, p. 117-132.

Isaacs, C.M., 1986, Guide to the Monterey Formation in the coastal California area, Ventura-San Luis Obispo: Pacific Section AAPG v. 52, 91 p.

Kendall, C.G.St.C., and I. Lerche, 1988, The rise and fall of eustacy, *in* C.K. Wilgus, et al., Sea-Level Changes: an Integrated Approach: SEPM Special Publication 42, p. 3-18.

Loutit, T.S., A.E. Bence, J. Smale, J.D. Shane, 1990, Anatomy of an organic-rich condensed section: The Campanian Sharon Springs Member of the Pierre Shale, Powder River Basin: American Association of Petroleum Geologists Bulletin, v. 74, p. 708.

Miskell-Gearhart, K.J., 1989, Productivity, preservation, and cyclic sedimentation within the Mowry Shale depositional sequence, Lower Cretaceous, Western Interior Seaway: Ph. D. dissertation, Rice University, Houston, 225 p.

Mitchum, R.M., 1977, Seismic stratigraphy and global changes of sea level, Part 1: Glossary of terms used in seismic stratigraphy, *in* C.E. Payton, Seismic Stratigraphy—Applications to Hydrocarbon Exploration: Tulsa, American Association of Petroleum Geologists Memoir 26, p. 205-212.

Mitchum, R.M., P.R. Vail, S. Thompson, III, 1977, The depositional sequence as a basic unit for stratigraphic analysis, *in* C.E. Payton, Seismic Stratigraphy—applications to hydrocarbon exploration: Tulsa, American Association of Petroleum Geologists Memoir 26, p. 53-62.

Pisciotto, K.A., and R.E. Garrison, 1981, Lithofacies and depositional environments of the Monterey Formation, California, *in* R.E. Garrison and R.G. Douglas, eds. The Monterey Formation and Related Siliceous Rocks of California: Pacific Section SEPM, p. 97-122.

Raiswell, R., and R.A. Berner, 1986, Pyrite and organic matter in normal marine shales: Geochimica et Cosmochimica Acta, v. 50, p. 1967-1976.

Sangree, J.B. and J.M. Widmier, 1977, Seismic interpretation of clastic facies. *in* C.E. Payton, Seismic Stratigraphy—applications to hydrocarbon exploration: Tulsa, American Association of Petroleum Geologists Memoir 26, p. 165-184.

Sarg, J.F., 1988, Carbonate sequence stratigraphy, *in* C.K. Wilgus, B.S. Hastings, H. Posamentier, J. Van Wagoner, C.A. Ross, and C.G.St.C. Kendall, Sea-level Changes: An Integrated Approach: Tulsa, SEPM Special Publication 42, p. 155–181.

Schoch, R.M., 1989, Stratigraphy: Principles and Methods: New York, Van Nostrand Reinhold, 375 p.

Schwalbach, J.R. and K.M. Bohacs, 1991, Depositional sequences in continental margin settings: Hemipelagic and pelagic facies of the Monterey Formation, California: SEPM 1991 Theme Meeting, Portland.

Southard, J.B., 1971, Representation of bed configura-

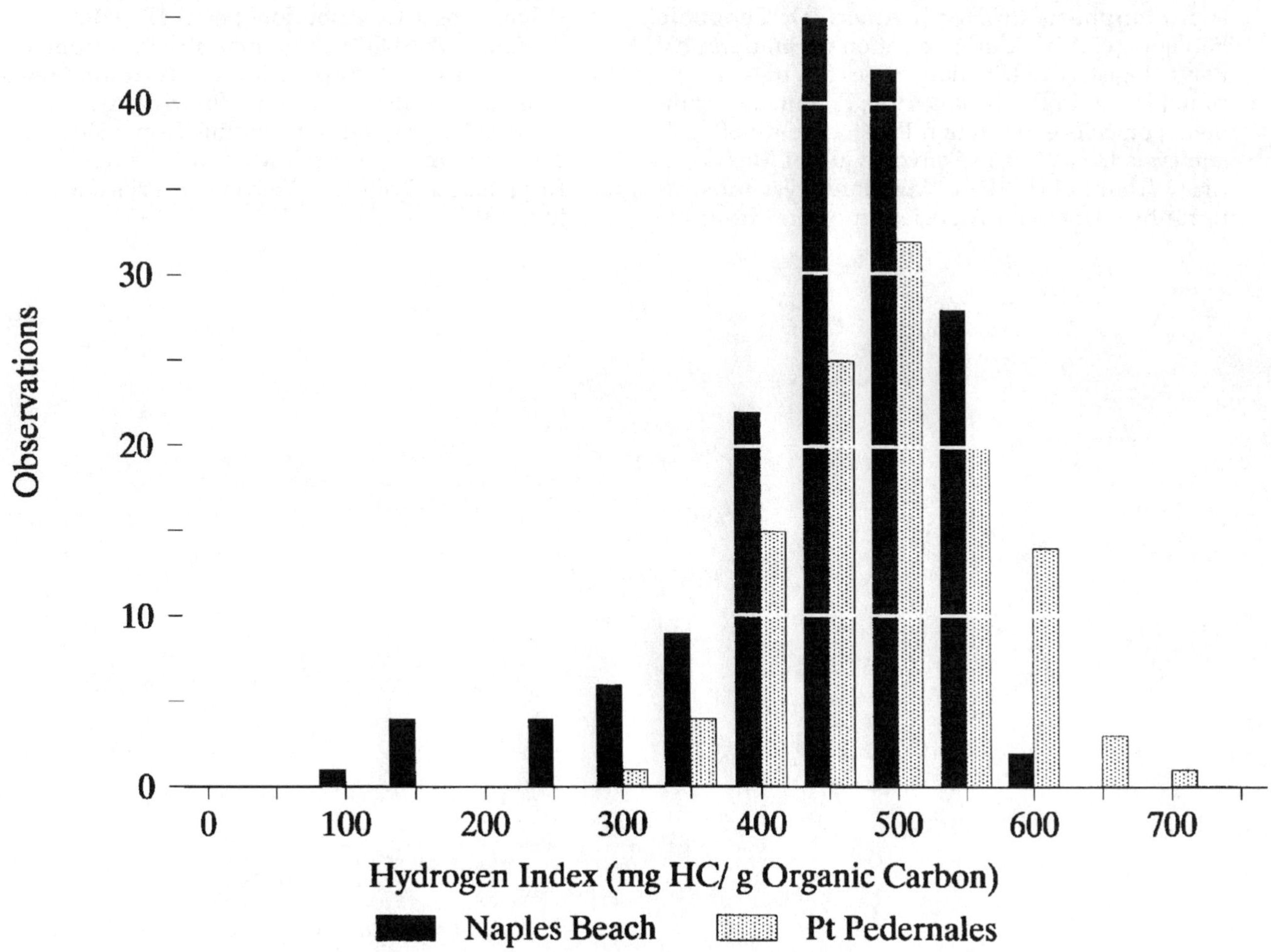

Figure 26. Distribution of hydrogen indices of outcrop samples of the Monterey Formation from Naples Beach and Point Pedernales. Most of the rock samples (89.5%) have hydrogen indices greater than 400 indicating that the primary organic input was probably relatively constant.

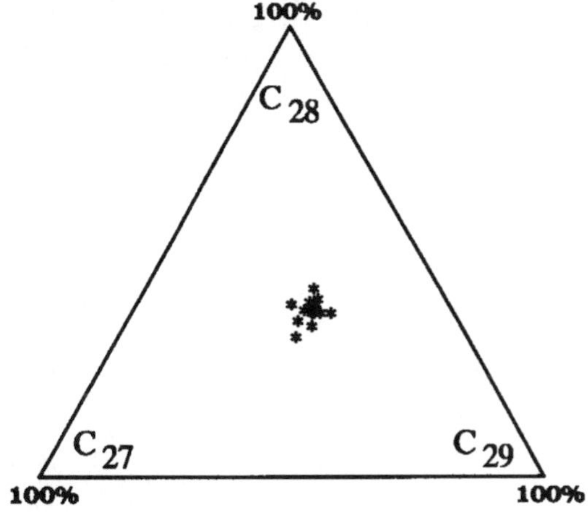

Figure 27. Distribution of steranes in Monterey oils from offshore Santa Maria and Santa Barbara Channel basins. The tight grouping of the sterane distribution suggests that the primary organic matter contribution is similar for all oils.

tions in depth-velocity-size diagrams: Journal of Sedimentary Petrology, v. 41, p. 903-915.

Stanley, K.M. and R.C. Surdam, 1984, The role of wrench fault tectonics and relative change of sea level on deposition of upper Miocene Pliocene Pismo Formation, Pismo Syncline, California, *in* R.C. Surdam, A Guidebook to the Stratigraphic, Tectonic, Thermal, and Diagenetic History of the Monterey Formation, Pismo and Huasna Basin, California: SEPM San Jose Guidebook no. 2, 94 p.

Van Wagoner, J.C., H.W. Posamentier, R.M. Mitchum, P.R. Vail, J.F. Sarg, T.S. Loutit and J. Hardenbol, 1988, An overview of the fundamentals of sequence stratigraphy and key definitions, *in* C.K. Wilgus, B.S. Hastings, H. Posamentier, J. Van Wagoner, C.A. Ross, and C.G.St.C. Kendall, Sea-level Changes: An Integrated Approach: Tulsa, SEPM Special Publication 42, p. 39-46.

Williams, G.L., J.P. Bujak, 1977, Cenozoic palynostratigraphy of offshore eastern Canada, *in* W.C. Elsik, Contributions to Stratigraphic Palynology (with emphasis on North America), Cenozoic Palynology: American Association of Stratigraphic Palynologists, Contributions Series 5A, p. 14-47.

Wrenn, J.H., and J.P. Kokinos, 1986, Preliminary comments on Miocene through Pleistocene dinoflagellate cysts from DeSoto Canyon, Gulf of Mexico, *in* J.H. Wrenn, et al., Dinoflagellate Cyst Biostratigraphy: American Association of Stratigraphic Palynologists, Contributions Series 17, p. 169-225.

Zelt, F.B., 1985, Natural gamma-ray spectrometry, lithofacies, and depositional environments of selected Upper Cretaceous marine mudrocks, western United States, including Tropic Shale, and Tununk member of Mancos Shale: Ph.D. Dissertation, Princeton University, Princeton, N.J., June 1985.

Chapter 13

Paleoceanographic Interpretation of Variations in the Sulfur Isotopic Compositions and Mn/Fe Ratios in the Miocene Monterey Formation, Santa Maria Basin, California

Doreen A. Zaback
Lisa M. Pratt
Department of Geological Sciences
Indiana University
Bloomington, Indiana, USA

ABSTRACT

Variations in sulfur isotopic compositions of pyrite and Mn/Fe ratios for core samples from the Monterey Formation, Santa Maria Basin, correspond to sea level fluctuations associated with periods of glacial expansion in the Antarctic and changes in the oceanic thermal gradient during the Miocene. Elevated ratios of Mn/Fe occur in both the lower calcareous and middle phosphatic lithofacies during a period when sea level was relatively high. A change to more positive sulfur isotopic compositions of pyrite at or near the transition from the middle phosphatic to upper siliceous lithofacies is associated with a major drop in sea level and stabilization of cold bottom waters in the Pacific Ocean. An inverse relationship between titanium concentration and organic carbon content suggests both the elevated ratios of Mn/Fe and enriched isotopic values of pyrite occur within a condensed section. The occurrence of these geochemical signals within a condensed section may be helpful in identifying chemically reactive sedimentary strata and potential source rocks in frontier basins.

In addition to fluctuations of sea level and oceanic circulation patterns, intra-basin variations, such as intensity of bottom currents, rates of bacterial sulfate reduction relative to rates of sulfate throughput in sediments, and thickness of the oxygen-minimum zone, influence manganese and sulfur reactions. These intra-basin processes resulted in differences in cycling of sulfur and sequestering of manganese within the Santa Maria basin. Specifically, the highest Mn/Fe ratios are observed near the flanks of the basin, whereas pyrite sulfur is enriched toward the center of basin.

INTRODUCTION

Deposition of three distinct stratigraphic intervals in the Monterey Formation (lower calcareous, middle phosphatic, and upper siliceous facies) was driven by second- and third-order sea level fluctuations and by onset of increased glaciation during the Miocene (Isaacs, 1989). Although these factors are known to affect physical characteristics of rock units, changing circulation patterns and energy of bottom currents associated with sea level fluctuations and glaciation also influence the position and intensity of the oxygen-minimum zone within the water column and movement of redox boundaries within the sediment column. For instance, widespread anoxic events can be associated with transgressive phases of sea level cycles (Schlanger and Jenkyns, 1976) and epicontinental enrichment of manganese is documented during the Cenomanian–Turonian transgression (Schlanger et al., 1987; Pratt et al., 1991). Numerous other examples of enrichment of manganese in shallow-marine basins are cited by Force and Cannon (1988). Because sulfur and manganese are sensitive to redox processes, concentrations of reduced sulfur and manganese in conjunction with isotopic composition of reduced sulfur within a basin should reflect paleoceanographic and paleoclimatic changes.

Manganates (Mn-oxides) and sulfate act as electron acceptors in the dysaerobic and anaerobic oxidation of organic matter, respectively (Claypool and Kaplan, 1974; Jørgensen, 1983; Erhlich, 1987). Manganate and sulfate reduction occur in spatially separated zones of bacterial diagenesis and, therefore, distributions of sulfur and manganese species can be used to infer positions of redox boundaries (Figure 1). Oxidation of HS^- to elemental sulfur (S^0) and reduction of Mn^{4+} to Mn^{2+} occurs at the O_2–H_2S boundary as a consequence of abiological MnO_2–H_2S redox reactions (Burdige and Nealson, 1986; Sorensen and Jørgensen, 1987) or by various sulfur-oxidizing bacteria (Jørgensen, 1983). Ideally, elevated concentrations of elemental sulfur and reduced manganese species should indicate the position of the O_2–H_2S redox boundary. Concentrations of elemental sulfur are often low in organic–carbon-rich rocks, however, making it an unsuitable redox-boundary indicator. Pyrite, a major sulfur species in marine organic-rich rocks, is most likely to form near the O_2–H_2S boundary where moderate Eh conditions exist. (Eh represents the reducing potential of a solution and is expressed in volts.) The following reactions of Goldhaber and Kaplan (1974) balanced for electron flow show the sequence of pyrite formation:

$$8H_2S + 6FeO \bullet OH \rightarrow 5FeS + S^0 + Fe^{2+} \rightarrow 2Fe_3S_4$$

$$2S^0 + Fe_3S_4 \rightarrow 3FeS_2.$$

Isotopic composition of hydrogen sulfide and subsequently pyrite, is controlled both by the kinetic isotope effect associated with bacterial sulfate reduction (approximately 60‰; Chambers and Trudinger, 1979; Goldhaber and Kaplan, 1980) and by the extent to which the pore-water sulfate reservoir is exhausted.

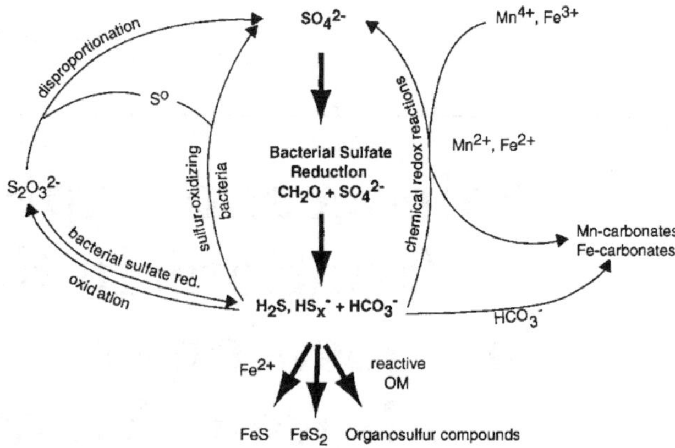

Figure 1. Generalized view of the sulfur cycle and its interaction with manganese in the geosphere. No specific depths are inferred for the reactions. The complexity of the cycle will depend on the environment.

Consumption or renewal of the sulfate reservoir is determined by the rate of bacterial sulfate reduction relative to sulfate transport. Reactivity of organic matter, bottom-water oxygen content, and transport of sulfate into sediment by advection and/or diffusion (Goldhaber and Kaplan, 1975, 1980) are major controls on the early diagenetic cycling of sulfur. With increasing depth, reactivity of organic matter will decrease (due to the extended length of time the organic matter has been exposed to oxidizers) and diffusion of oxygen and sulfate will be inhibited. Therefore, the position of the O_2–H_2S boundary relative to the sediment–water interface has a major influence on the isotopic value of pyrite sulfur.

In this chapter we present stable sulfur isotope data for pyrite and concentration data for total reduced sulfur, total organic carbon, manganese, total iron, titanium, and phosphorus for samples from the early to late Miocene Monterey and late Miocene to early Pliocene Sisquoc formations in the Santa Maria basin. Elevated manganese/iron ratios coincide with a major global transgression during the early to mid-Miocene. The sulfur isotopic enrichment of pyrite corresponds to a period of falling sea level during the middle Miocene and occurs throughout the basin. This chapter focuses on factors controlling variations in manganese concentration and in isotopic composition of pyrite sulfur during deposition of the Monterey Formation in the Santa Maria basin.

GEOLOGICAL SETTING

Lithology

The lower lithofacies of the Monterey Formation is calcareous and is composed predominantly of foraminiferal and coccolith mudstones and shales with minor interbedded cross-laminated sandy lenses, and concretionary carbonate (Garrison et al., 1979). The middle phosphatic facies is principally composed of phosphatized foraminiferal and coccolith mudstones and shales (Pisciotto and Garrison, 1981). Phosphatic rocks occur as laterally persistent layers of cryptocrystalline carbonate–fluoroapatite that are up to several centimeters thick (Isaacs, 1983). The major components of the upper siliceous facies are biogenic and diagenetic silica and clay (Pisciotto and Garrison, 1981). Strata in this facies alternate between packages of laminated and massive strata.

Climatic–Oceanic Influences

The Monterey Formation of the Santa Maria basin accumulated by hemipelagic and pelagic sedimentation in a subsiding silled basin effectively starved from large-scale terrigenous clastic input (Isaacs, 1989; Bohacs, 1990). Monterey sediments were affected by numerous factors including fluctuations in redox boundaries, coastal upwelling, active tectonics, and paleoclimate. Changes in composition, sedimentary structures, and faunal distribution in sediments of Neogene Pacific basins appear to be related to variation in atmospheric–oceanic circulation, the onset of Antarctic glacial expansion, and an increased rate of spreading at the East Pacific Rise beginning 25 to 22 Ma (see Figure 2) (Ingle, 1981; Woodruff, 1985). Deposition of the lower calcareous and middle phosphatic lithofacies coincides with periods of high sea level and occurrences of benthonic foraminifera depleted in ^{18}O. Relatively high sea level and increased accommodation space caused the upper portion of the lower calcareous and the middle phosphatic lithofacies to have characteristics typical of condensed sections (Baum and Vail, 1988; Loutit et al., 1988). Oxygen isotopic data and uniform depth distribution of benthic foraminifera suggest that the lower calcareous facies was deposited in a relatively warm, poorly oxygenated ocean (Ingle, 1981; Woodruff, 1985). The change from a calcareous to a phosphatic facies at about 15 Ma is considered to reflect the onset of coastal upwelling. During the mid-Miocene (16 to 12 Ma) movement of benthic fauna into shallower bathymetric zones and an oxygen isotopic shift from values of approximately 1.0‰ to 2.5‰ (vs. the Peedee belemnite) in tests of benthic foraminifera collected from three sites of the Deep Sea Drilling Project record an influx of cooler bottom waters into the California coastal basins and an increased latitudinal temperature gradient associated with glacial expansion (Woodruff, 1985). Benthic foraminifera associated with modern poorly oxygenated environments are more common at shallow depths after 12 Ma (Woodruff, 1985). The onset of predominantly biogenic-siliceous sedimentation is an indication of intense coastal upwelling and is inferred to record the effects of increased atmospheric and oceanic circulation (Pisciotto and Garrison, 1981).

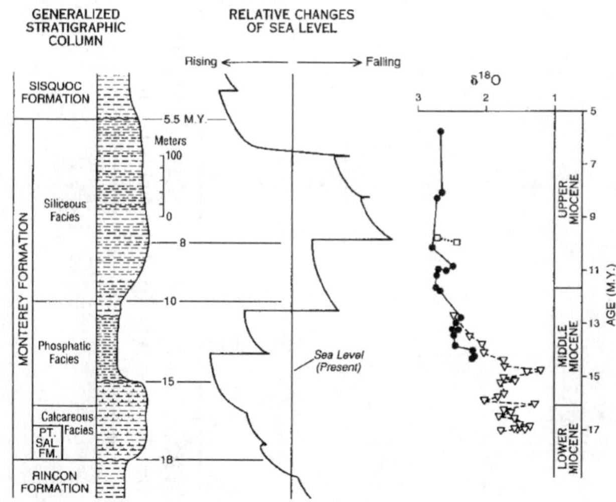

Figure 2. Comparison of stratigraphy of the Miocene Monterey Formation, relative sea level changes, and oxygen isotopic composition of benthonic foraminifera (*Cibicidoides*) recorded during the Miocene. Stratigraphic column modified from Isaacs, 1983, and Bohacs, 1990. Sea level curve modified from Vail et al., 1977. Oxygen isotope record from Barrera et al., 1985, for DSDP sites 173 (squares), 470 (closed circles), and 495 (inverted triangles).

GEOCHEMICAL METHODS

Whole rock samples from seven cores within the Santa Maria basin were analyzed for total sulfur, silica, calcium, magnesium, manganese, titanium, phosphorus, and iron concentrations and isotopic composition of pyrite sulfur. The weight percent of total sulfur was obtained using a LECO C/S-244 analyzer. Silica, calcium, magnesium, total iron and manganese concentrations were measured in duplicate on mixtures of rock powder and lithium tetraborate/lithium metaborate that were fused to allow mineral decomposition. Phosphorus (as P_2O_5) concentrations were determined by dissolution of whole rock samples in a mixture of concentrated nitric acid, 72% perchloric acid, and 49% hydrofluoric acid. Sample solutions were analyzed on a Jarrell Ash 975 Plasma Atomcomp emission spectrometer. Pyrite sulfur was obtained by sequential chemical separation of sulfur species following the methods described in Zaback and Pratt (1992). Sulfur dioxide was analyzed on a Nuclide 6-60 mass spectrometer. Isotope ratios are reported in standard per mil values relative to Canon Diablo Troilite.

RESULTS

Data on sulfur isotope ratios and on concentrations of sulfur species, total organic carbon, titanium, manganese, iron, silica, calcium, and magnesium are shown in Table 1 grouped according to core identifi-

Table 1. Geochemical data for samples from individual cores.

Sample	Depth (ft.)	Formation/ Lithofacies	TS (wt.%)	TOC (wt.%)	$\delta^{34}S$ pyrite	Pyrite-S*	Kerogen-S*	MnO_2 (wt.%)	Fe_2O_3 (wt.%)	TiO_2 (wt.%)	SiO_2 (wt.%)	P_2O_5 (wt.%)	CaO (wt.%)	MgO (wt.%)
Inboard Flank														
Bradley Lands 1														
Mont-10	4080	Sisquoc	1.13	2.20	-6.01	0.44	0.32	n.d.**	n.d.	n.d.	n.d.	0.36	n.d.	n.d.
Mont-11	4181	Mont./upper sil	0.89	2.59	8.37	0.27	0.34	0.010	1.16	0.15	71.50	0.29	4.57	2.33
Mont-12	4797	Mont./mid. phos	2.40	9.83	8.92	0.14	0.40	0.017	1.33	0.19	32.10	2.00	29.60	0.88
Mont-13	5005	Mont./low. calc	1.56	7.66	5.95	0.37	0.51	0.155	1.08	0.12	10.80	0.65	23.10	13.40
Basin Center														
Los Nietos SST-25														
Mont-1	8434	Mont./upper sil	1.88	4.66	9.96	0.34	0.34	0.020	2.69	0.39	75.80	0.19	0.74	0.82
Mont-2	8587	Mont./mid. phos	1.64	2.99	15.61	0.67	0.07	0.014	1.87	0.26	77.80	0.38	1.86	1.12
Mont-3	8696	Mont./mid. phos	1.65	5.24	11.98	0.28	0.41	0.023	1.72	0.23	27.50	3.81	29.90	2.83
Mont-4	9053	Mont./low. calc	2.74	2.00	11.95	0.32	0.45	0.038	3.53	0.39	45.20	1.02	17.50	3.19
Union Dome														
Mont-15	9773	Sisquoc	0.78	0.71	n.d.	n.d.	n.d.	0.030	3.89	0.46	71.40	0.33	1.12	1.64
Mont-16	10546	Mont./upper sil	1.50	2.69	n.d.	n.d.	n.d.	0.021	2.73	0.54	68.70	n.d.	3.52	1.59
Mont-17	10767	Mont./upper sil	1.25	2.93	n.d.	n.d.	n.d.	0.025	2.92	0.57	70.30	n.d.	1.23	0.95
Mont-18	10980	Mont./low. calc	2.18	2.61	n.d.	n.d.	n.d.	0.017	3.14	0.50	70.10	0.52	1.52	1.04
Mont-19	11349	Mont./low. calc	1.00	0.71	n.d.	n.d.	n.d.	0.045	3.71	0.52	67.40	0.26	1.61	1.95

Outboard Flank														
Union Harris #A-1														
Mont-28	4586	Sisquoc	1.06	1.59	-5.14	0.71	0.06	0.035	3.69	0.33	74.8	n.d.	2.00	2.20
Mont-29	5088	Mont./upper sil	2.07	4.75	2.45	0.64	0.13	0.017	2.35	0.31	70.4	n.d.	2.61	1.51
Mont-30	5328	Mont./upper sil	1.66	3.42	8.00	0.41	0.13	0.016	2.56	0.36	72.2	n.d.	0.67	0.88
Longitudinal Axis														
Security Fee #39														
Mont-9	5785	Mont./upper sil	0.64	1.25	11.83	0.32	0.25	0.009	1.14	0.09	89.4	0.19	0.48	0.26
Mont-5	5751	Mont./upper sil	1.94	3.44	n.d.	n.d.	n.d.	0.016	2.02	0.31	75.7	0.63	1.36	0.67
Mont-6	5778	Mont./upper sil	2.09	3.24	n.d.	n.d.	n.d.	0.020	2.51	0.36	71.8	1.60	3.34	1.27
Mont-7	6616	Mont./mid. phos	0.89	2.74	n.d.	n.d.	n.d.	0.008	0.90	0.06	85.0	0.19	2.28	1.35
Mont-8	6648	Mont./mid. phos	1.17	3.32	17.58	0.26	0.31	0.018	1.04	0.14	22.9	1.38	24.90	13.30
Los Nietos Leroy														
Mont-40	4135	Sisquoc	0.78	0.44	-4.02	0.86	0.04	0.021	2.27	0.33	n.d.	0.48	2.03	0.97
Mont-38	4602	Monterey	2.48	7.15	7.89	0.13	0.41	0.020	2.41	0.34	n.d.	0.41	0.91	0.73
Mont-39	5102	Monterey	2.32	1.45	6.31	n.d.	n.d.	0.083	3.14	0.36	n.d.	0.52	17.30	9.31
Golco														
Mont-26	6379	Mont./upper sil	2.69	4.11	12.60	n.d.	n.d.	0.030	3.75	0.56	59.70	0.46	1.49	1.45
Mont-25a	7357	Mont./mid. phos	1.83	2.54	2.30	0.61	0.19	0.015	2.16	0.25	68.10	2.06	5.01	1.20
Mont-25	7443	Mont./low. calc	0.57	1.42	n.d.	n.d.	n.d.	0.015	0.39	0.04	11.60	n.d.	27.30	17.20
Mont-23	7447	Mont./low. calc	0.48	1.84	n.d.	n.d.	n.d.	0.010	0.35	0.04	33.80	n.d.	20.40	12.40
Mont-24	7447	Mont./low. calc	0.64	2.96	-1.15	0.27	0.24	0.012	0.77	0.11	57.10	n.d.	11.60	6.48

* expressed as proportion of total sulfur
** n.d. = not determined.

cation. Data pertaining to sulfur species is reported as the actual concentrations obtained and has not been normalized. Incomplete recovery of sulfur species is attributed mainly to loss of pyrite sulfur as SO_2 during dissolution of pyrite sulfur (Zaback, 1992). Testing of the pyrite dissolution procedure with an isotopically known pyrite standard indicates that isotopic fractionation does not occur at this step (Zaback, 1992). Figure 3 illustrates core locations and lines of cross section that are described in this study. Cross sections are used to show relationship between position in the basin and sulfur isotopic, Mn/Fe, and titanium data. A positive isotopic excursion in pyrite sulfur occurs at or near the boundary between the middle phosphatic and upper siliceous lithofacies and is generally located stratigraphically above an enrichment in manganese (Figures 4, 5). Samples from the Union Bradley Lands 1 and Union Harris cores, which are proximal to the flanks of the present-day basin, have similar ranges of sulfur isotope values (Union Bradley, –6.01 to +8.92‰; Union Harris, –5.14 to +8.00‰). In contrast, samples from the Los Nietos SST-25 core, near the center of the basin, are consistently enriched in ^{34}S (+9.96 to +15.61‰) through the entire Monterey section. Isotopic values of pyrite from cores along cross section B–B′ (–4.02 to +17.58‰) are similar to values recorded for samples near the flanks of cross section A–A′(–6.01 to +8.92‰).

Manganese abundances range from 0.009 to 0.16 wt.% with an average of 0.029 wt.%; total iron abundances range from 0.35 to 3.89 wt.% with an average of 2.16 wt.% (Table 1). Overall, Mn/Fe ratios are higher and more variable in the shallower cores (0.055 ± 0.077, inboard flank; 0.015 ± 0.010, longitudinal axis), than in the deep basinal cores (0.009 ± 0.002).

Titanium concentrations, measured as titanium oxide, range from approximately 0.04 to 0.57 wt.% with an average of 0.30 wt.%. Samples from the Union Bradley Lands 1 and Security Fee cores have the lowest titanium concentrations (0.06 to 0.36 wt.%), while the highest concentrations of titanium (0.23 to 0.57 wt.%) occur in samples from the Union Dome and Los Nietos SST-25 cores (Figure 4). Titanium is more abundant in the upper siliceous facies than in the middle phosphatic and lower calcareous lithofacies (Zaback and Pratt, 1992).

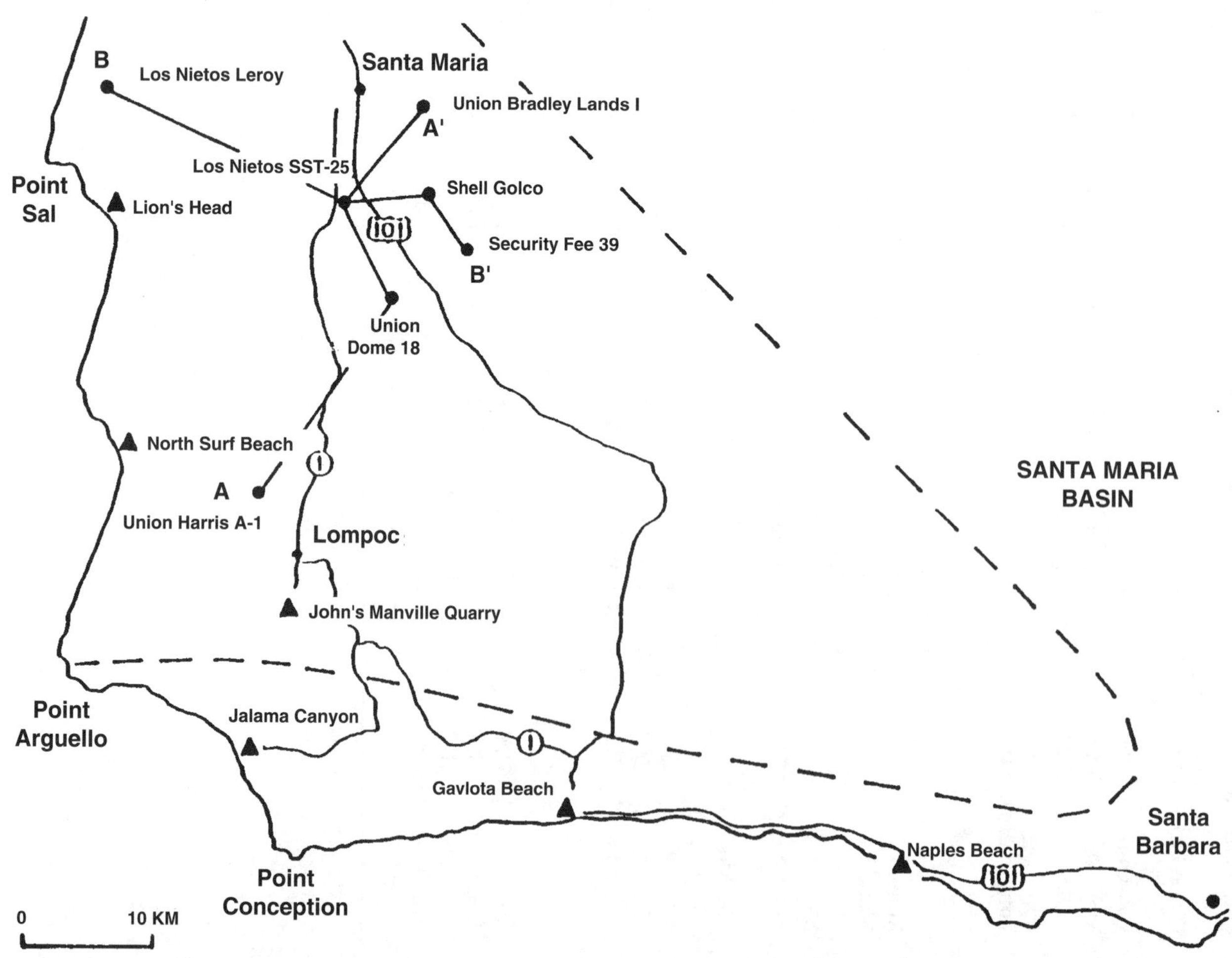

Figure 3. Location of cores (triangles) and cross sections A–A′, B–B′ discussed in text.

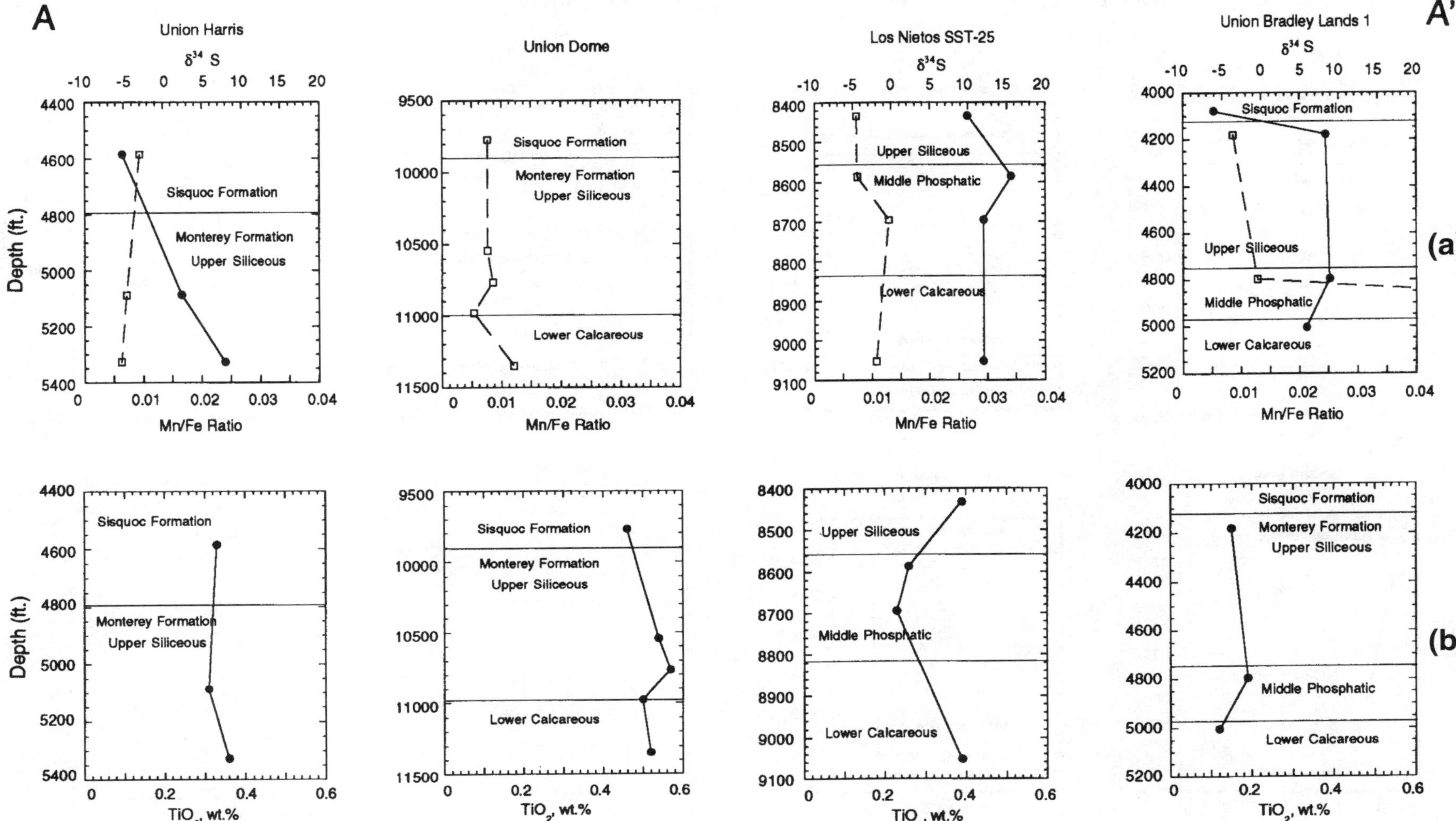

Figure 4. Changes in sulfur isotopic composition of pyrite (closed circles) and Mn/Fe ratios (open squares), (A), and titanium concentration, (B), vs. depth for core along cross sections A–A´.

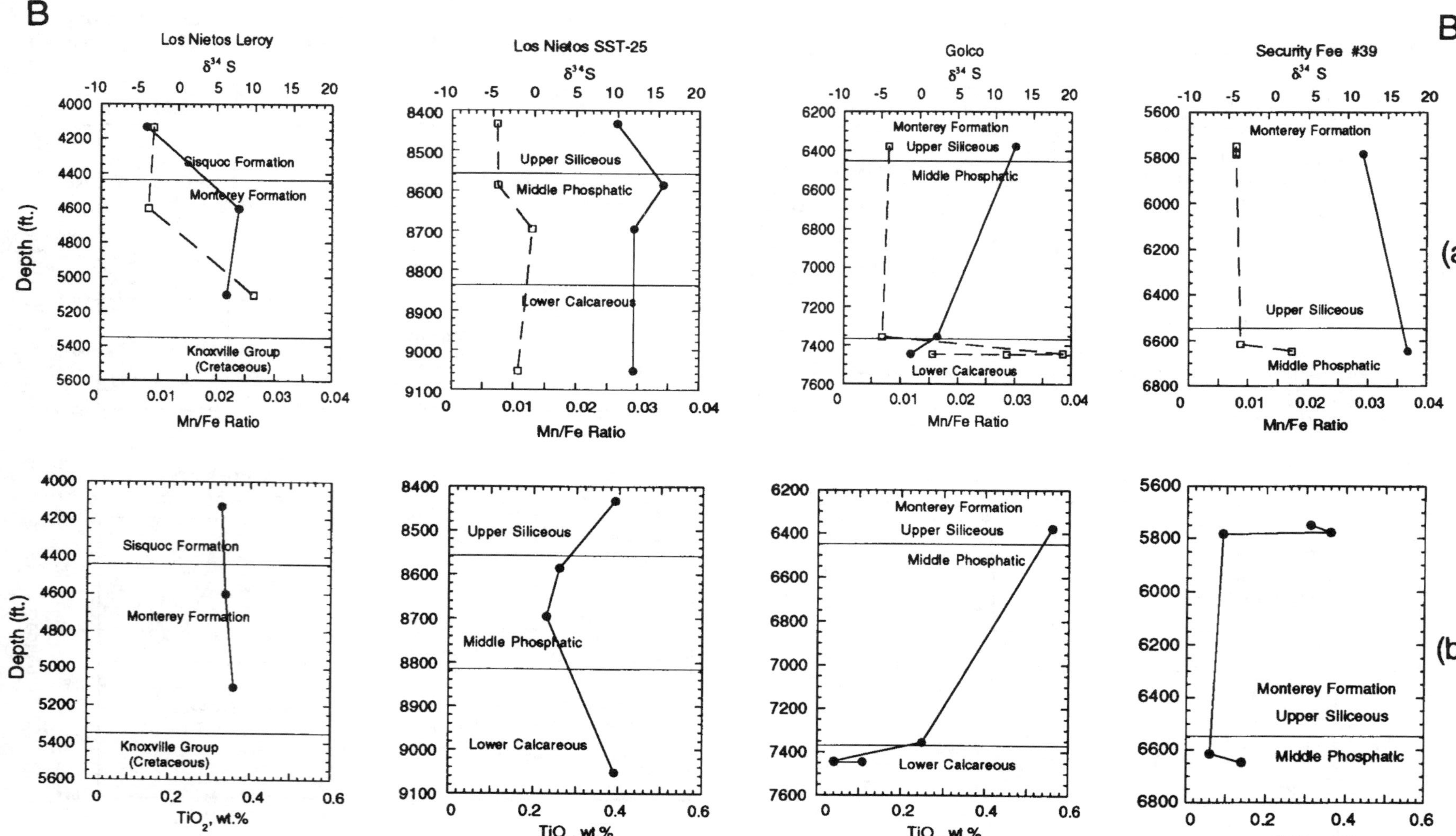

Figure 5. Changes in sulfur isotopic composition of pyrite (closed circles) Mn/Fe ratios (open squares), (A), and titanium concentration, (B), vs. depth for cores along cross sections and B–B′.

Concentrations of calcium and magnesium are only weakly correlated ($r^2 = 0.461$; Figure 6), probably due to the occurrence of calcium in both carbonate and apatite minerals. Elimination of samples Mont-3 and Mont-12, both of which are from the middle phosphatic lithofacies (apatite- and calcite-rich), results in a strengthened correlation ($r^2 = 0.900$). Two groups of samples are evident when manganese is plotted vs. magnesium (Figure 7). Group I is characterized by increasing concentrations of manganese with increasing concentrations of magnesium. Group II is characterized by near-constant manganese concentrations with increasing magnesium concentrations. Three of the samples in Group II (Mont-23, Mont-24, and Mont-25) contain well-crystallized vuggy dolomites. Manganese and phosphorus concentrations are not correlated (Figure 8) in the samples analyzed.

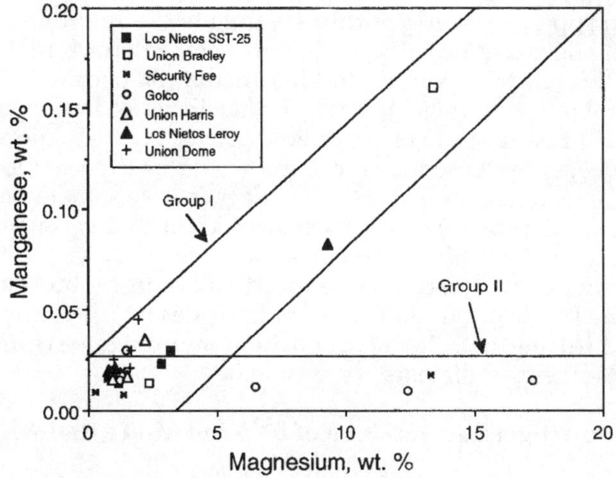

Figure 7. Distribution of manganese relative to magnesium for Monterey core samples. Group I is inferred to represent early diagenetic, Mn-bearing dolomite. Group II represents samples containing vugular dolomite or apatite.

DISCUSSION

Manganese–Sulfur Relationships

The stratigraphic position of increasing Mn/Fe ratios relative to the more positive values of pyrite sulfur is intriguing in view of potential coupling between manganese and sulfur cycles. While the sampling frequency relative to depth is low, similar relationships are observed in each of the seven cores from the Santa Maria basin analyzed. For this reason, the repetition of the geochemical patterns, especially at lithologic boundaries, is believed to be significant and warrants discussion.

Reduction of manganates has been shown to occur via biological (Burdige and Nealson, 1985; Chandramohan et al., 1987; Ehrlich, 1987) and chemical (Emerson et al., 1979; Burdige and Nealson, 1986) reactions. Biological reduction of manganates has been documented in aerobic (Trimble and Erhlich, 1968), dysaerobic (Reeburgh, 1983; Erhlich, 1987), and anaerobic (Nealson, 1983; Burdige and Nealson, 1985) environments due to heterotrophic mineralization of organic carbon and lithotrophic oxidation (oxidation of inorganic compounds for energy gain) of hydrogen sulfide (Jørgensen, 1987). Because reduction of manganates appears to occur within a wide range of diagenetic zones, the mode and depth of reduction may have an effect on the efficiency of mobilization and

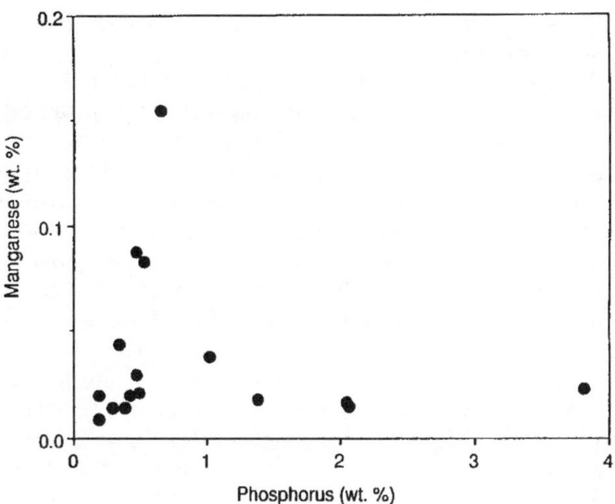

Figure 8. Cross-plot showing the lack of correlation between manganese and phosphorus concentration.

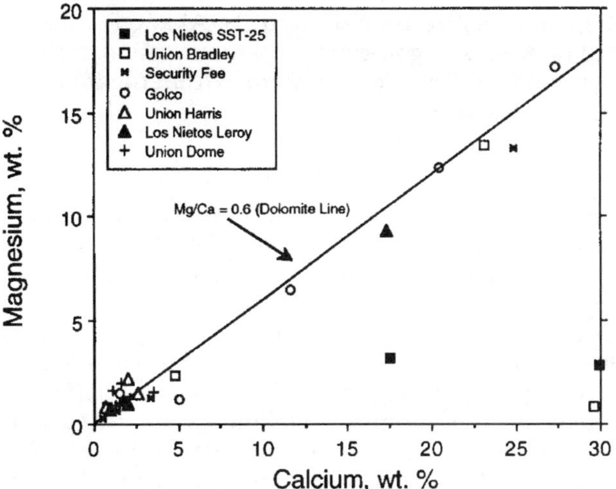

Figure 6. Cross-plot of magnesium and calcium concentration for Monterey samples. Line is representative of the Mg/Ca weight ratio of dolomite. Points plotting away from the line represent samples from the middle phosphatic zone.

trapping of Mn^{2+} within sediments. Reduction of manganates near the sediment–water interface may facilitate transport of Mn^{2+} into an oxygen-poor water column and, thus, decrease the likelihood of trapping of Mn at the site of reduction. Conversely, reduction deeper in the sediment column would be expected to favor retention of Mn^{2+} as a diagenetic carbonate or sulfide phase. The data are insufficient to determine the mode of Mn-oxide reduction, however, given the high concentration of reduced sulfur in Monterey rocks, chemical reduction of Mn-oxides by upwardly diffusing H_2S during deposition and diagenesis of Monterey sediments is probable.

Stratigraphic Variation of $\delta^{34}S$ and Mn/Fe Ratios

An enrichment of manganese in the lower calcareous–middle phosphatic zone could be attributed to an increase in the primary input of manganese. However, because manganese concentrations are poorly correlated with titanium ($r^2 = 0.025$) or total iron ($r^2 = 0.051$) concentrations, increases in Mn/Fe ratios, therefore, cannot be explained by increased terrigenous clastic input during deposition of Monterey sediments. The most likely explanation is chemical partitioning of manganese relative to iron in the sediments.

Burdige and Nealson (1986) found reduction of MnO_2 to be controlled by bacterial sulfide production. High rates of sulfate reduction and sulfide production would, at first, be expected to result in reduction and mobilization of Mn, isotopic enrichment of pyrite, and an increase in phosphorus concentration. However, as mentioned earlier, the rate of sulfate reduction relative to sulfate transport has a significant influence on the isotopic composition of reduced sulfur species (Zaback et al., 1993). In other words, pyrite sulfur can be isotopically depleted at high rates of sulfate reduction as long as sulfate transport is high as well (i.e., under open system conditions). Whereas diagenetic chemical reactions link the sulfur and manganese cycles, physical phenomena such as diffusion and advection across the sediment–water interface may result in a decoupling of manganese concentration and isotopic expression of reduced sulfur species in marine sediments.

Influences of Reactive Organic Matter

The stratigraphic position of elevated ratios of Mn/Fe and ^{34}S enrichment of pyrite may be related to partitioning of sulfur species (pyrite vs. kerogen) as a function of organic carbon content within the Monterey. A decrease in concentration of pyrite sulfur coupled with an increase in concentration of kerogen sulfur relative to increasing organic carbon content is observed in Figure 9. While the correlation coefficient relating concentration of kerogen sulfur to organic carbon content is low, the trend nonetheless suggests that organic matter is a significant sulfur sink at organic carbon concentrations above 5 wt.%. In addition, kerogen sulfur is isotopically enriched relative to pyrite sulfur and pyrite sulfur shows a progressive isotopic enrichment with increasing organic carbon content (Zaback, 1992). Considering the stratigraphic position of the isotopic enrichment together with higher concentrations of organic carbon in the middle phosphatic lithofacies, it is reasonable to postulate that deposition of large amounts of reactive organic matter expand the zone of bacterial sulfate reduction within the sediments. Increased rates of bacterial sulfate reduction and higher trapping efficiency of H_2S by organic carbon deeper in the sediment column would decrease the amount of H_2S diffusing upward for reduction of Mn^{4+} to Mn^{2+}.

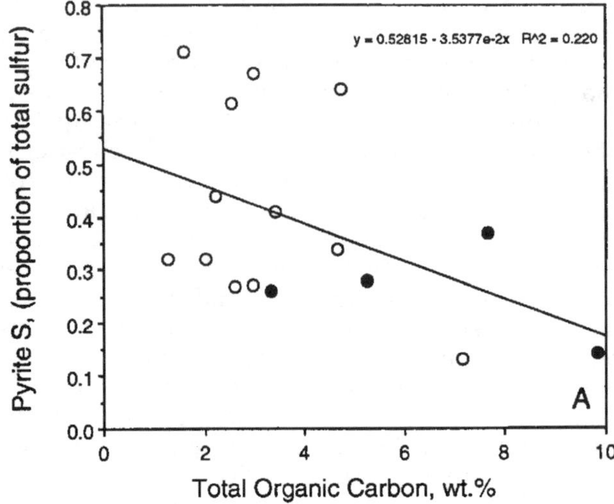

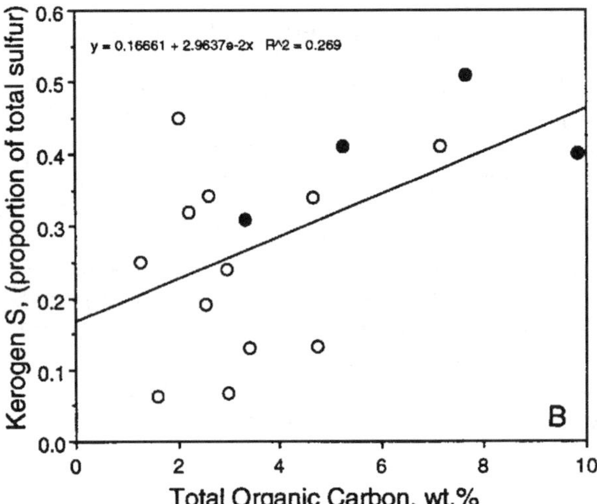

Figure 9. Distribution of pyrite S (A) and kerogen S (B) relative to organic carbon concentration. Filled symbols indicate data of samples from the middle phosphatic zone.

Changes in Rates of Sedimentation

It has been proposed that high sedimentation rates enhance deposition of well-preserved organic matter and result in sediments characterized by high organic carbon content (Toth and Lerman, 1977; Demaison and Moore, 1980) and elevated rates of sulfate reduction (Goldhaber and Kaplan, 1975; Canfield, 1991). Alternatively, suggestions have been made that high sedimentation rates dilute sediments with clastics and result in sediments with low organic carbon contents (Demaison and Moore, 1980). Because of the effect on sulfate reduction rates, sedimentation rates may also be expected to affect the isotopic composition of reduced sulfur species as well. For example, rapid burial of reactive organic matter below the sediment–water interface would sustain sulfate reduction at depths where transport of sulfate would become impaired. Therefore, high rates of sedimentation might be expected to correspond to isotopic enrichment of pyrite.

Average sedimentation rates (calculated from stratigraphic thicknesses and ages of strata) for the three major lithofacies of the Monterey Formation varied as follows: lower calcareous, 38 m/m.y.; middle phosphatic, 35 m/m.y.; and upper siliceous, 53 m/m.y. (Isaacs, 1989; Bohacs, 1990). The Sisquoc Formation, which overlies the Monterey Formation, was deposited at an average rate of 700 m/m.y. (Compton, 1991). In the samples analyzed for this study, the highest organic carbon contents and the most enriched isotopic values of pyrite are found in the phosphatic lithofacies, the zone with the lowest sedimentation rate. The most depleted sulfur isotopic values occur in the Sisquoc Formation and are associated with the highest rate of sedimentation. While the amounts of organic carbon oxidized during bacterial sulfate reduction are approximately equal for the Monterey and Sisquoc formations, this amount of carbon represents a larger proportion of deposited carbon in the Sisquoc (Zaback et al., 1993). Increased clastic input to the Sisquoc likely had a dilution effect on organic carbon content and resulted in termination of sulfate reduction before transport of sulfate was inhibited (Zaback et al., 1993).

The association between sedimentation rate and isotopic values of pyrite found in the Santa Maria basin should not be generally applied to all marine black shales. For instance, within the New Albany Shale (where sedimentation rates are calculated to be an order of magnitude less than the Monterey and sulfur isotopic values are generally lighter), the more isotopically enriched pyrite occurs in members where sedimentation rates are highest (Beier, 1988; Zaback et al., 1993). It is likely, therefore, that amount and reactivity of organic matter, depth of sulfate reduction, and pyrite formation within the sediment, in addition to sedimentation rate, control the sulfur isotopic value of pyrite.

Diagenetic Precipitation of Dolomite vs. Apatite

The stratigraphic relationship between the Mn/Fe and sulfur isotopic variations may also be associated with the change from dolomite-dominated to apatite-dominated diagenetic mineral precipitation. The absence of a positive trend between manganese and phosphorus concentrations (Figure 8) suggests manganese exclusion from apatite and strengthens the hypothesis that manganese is present as a Ca,Mg,Mn carbonate. Increased rates of anaerobic mineralization of organic matter are required to provide the elevated concentrations of phosphate necessary for pore waters to become supersaturated with respect to apatite (Gulbrandsen, 1969; Burnett, 1977; Baturin, 1982). Such conditions, however, also favor precipitation of dolomite by increasing carbonate alkalinity (Baker and Kastner, 1981; Compton, 1988) and possibly by decreasing the sulfate concentration of pore waters (Baker and Kastner, 1981). In order to explain the observed predominance of apatite minerals over carbonate minerals in sediments with equivalent amounts of CaO, precipitation of carbonates must have been inhibited in the middle phosphatic lithofacies.

Various theories have been proposed to explain early diagenetic formation of dolomite and apatite. Ion exchange reactions involving clay minerals have been proposed as a mechanism to deplete pore waters of Mg^{2+} in anoxic marine sediments (Drever, 1971; Baturin, 1982). Removal of Mg^{2+} from pore water would inhibit dolomite precipitation and kinetically favor apatite formation. The ability of Mg^{2+} to be transported into sediments may play a role in diagenetic mineral formation. Because dolomite forms at deeper levels than apatite (Kastner et al., 1984), a switch from dolomite-dominated to apatite-dominated phases may be related to depth of the O_2–H_2S boundary and to the amount of seawater that moved through the sediments. The relative sulfur isotopic enrichment of pyrite near the phosphatic–siliceous boundary suggests that mass transport may have controlled sulfate reduction at that time. During such transport-limited conditions, precipitation of apatite near the sediment–water interface might be favored.

Alternatively, the shift from a dolomite to an apatite diagenetic phase may be independent of Mg^{2+} concentration and, instead, related to pH and concentration of HPO_4^{2-} (Stumm and Morgan, 1981). Increased primary production associated with intensification of upwelling at approximately 12 Ma, may have promoted deposition of a more nutrient-rich organic matter which would result in increased rates of sulfate reduction and elevated concentrations of HPO_4^{2-} in the pore water. The net result of these factors would likely favor precipitation of apatite over dolomite. The above arguments are speculative, but do illustrate the need to understand organic–inorganic reactions and effects of diffusion during early diagenesis.

Effects of Paleoceanographic Changes

Although elevated Mn/Fe ratios do not occur within a single stratigraphic layer in the Santa Maria basin, they do occur within lower- to mid-Miocene strata deposited during a second-order rise in sea level defined by Vail et al., 1977 (Figure 2) and is coincident with the mid-Miocene maximum in $\delta^{13}C$ values of foraminifera (Vincent and Berger, 1985). The increased ratios of Mn/Fe in the Monterey may be an example of high sea level affecting the reducing potential of sediments and, in this regard, is similar to the geochemistry of other condensed sections (Loutit et al., 1988).

Variations from negative to positive isotopic compositions of reduced sulfur species within a single rock formation are not well documented in the geological literature. In the Monterey samples analyzed in this study, coincidence of pyrite displaying the greatest enrichment of ^{34}S with the onset of falling sea level at approximately 10 Ma suggests that the transport of sulfate within marine sediments and sea level change are linked on a regional, if not global, scale. At the onset of these changes, upwelling was likely intensified and resulted in increased primary productivity and preservation of organic matter at a time when terrigenous input was relatively low (Zaback and Pratt, 1992). Deposition of sediments containing high contents of reactive organic matter, in turn, caused rates of bacterial sulfate reduction to overwhelm rates of sulfate transport within the sediments (i.e., approach closed system conditions), leading to positive isotopic values of pyrite which, in some cases, approach the inferred isotopic value of Miocene seawater sulfate.

Relationship Between Excursions and Intrabasinal Processes

Manganese/Iron Variations

The increase in Mn/Fe ratios generally occur stratigraphically lower than the increase in sulfur isotopic values in each core, and this enrichment of Mn is more pronounced along the flanks of the basin (Figures 4, 5). The relationship between magnitude of Mn enrichment and its geographic position within the basin suggests that Mn^{2+} was mobilized, transported, and concentrated in a carbonate phase toward the basin flanks during deposition of Monterey sediments. Displacement of bottom water by turbidity currents in the Santa Barbara Basin has been postulated as the cause of anomalies in oxygen and nitrate content within the water column (Sholkovitz and Soutar, 1975) and diffusion of Mn^{2+} from sediments into the bottom water in coastal California basins has been reported (Martin and Knauer, 1984; Shiller et al., 1985). Evidence of turbidity flows in the Santa Maria basin then offers a means of transporting Mn^{2+} in a poorly oxygenated water column from the basinal site of reduction to the site of trapping on the flanks.

Titanium Concentrations

Distributions of clay-controlled elements such as titanium, aluminum, and iron are often used to infer proximity to detrital sources (Brumsack, 1983; Bralower and Thierstein, 1987) and may be helpful in determining direction of detrital transport within a basin. While iron is often considered to be diagenetically mobile relative to titanium (Wintsch et al., 1991), correlations between titanium and aluminum, and titanium and iron (r^2 = 0.96 and 0.92, respectively) throughout the basin (Zaback and Pratt, 1992) are believed to indicate that these elements had a consistent provenance. It may be argued that changes in concentration of titanium within a basin more accurately reflect grain-size distribution or segregation of clays (i.e., illite, smectite, and chlorite) (Chamley, 1989) than direction and relative amount of clastic input. However, no changes in grain size were detected using visual comparison of thin-sections for three samples from the basin flanks and three samples from the basin center. Uniform grain size throughout the basin is not surprising because the Santa Maria was one of the more outboard of the California Miocene borderland basins. Thus, titanium content is probably an accurate indicator of relative amounts of detrital input to different parts of the Santa Maria basin.

On the assumption that titanium concentrations represent terrigenous clastic input, two inferences regarding direction of transport are possible. First, higher overall concentrations near the center of the basin (Union Dome and Los Nietos SST-25 cores) and along the longitudinal axis coupled with low concentrations of titanium along the northeast or inboard flank (Union Bradley Lands 1 core) suggest terrigenous clastic influx originated from the northwest and continued along the longitudinal axis of the basin. Second, distribution of titanium may reflect transverse distribution of sediment along submarine canyons. Under such circumstances, detritus could bypass the upper slope of the basin and occur within distal submarine fan deposits toward the center of the basin. Both types of dispersal probably characterized delivery of sediments to the Santa Maria basin.

Sulfur Isotopic Enrichment of Pyrite in Center vs. Flanks of Basin

The occurrence of turbidity flows may also be invoked to explain the relationship between the magnitude of the sulfur isotopic change and depositional environment. Physical disruption or agitation of slope sediments would increase the advective transport of sulfate into the sediments. Increased transport of sulfate would, in effect, increase the sulfate reservoir and be expressed in strata as a relative depletion of ^{34}S in reduced sulfur species. Such a mechanism may explain why the flanks are more depleted in ^{34}S relative to the center of the basin. The continuous nature of laminae in strata inferred to represent the center of the basin is evidence of weakened bottom currents. Minimal agitation of sediment would have little effect on sulfate throughput and would result in more complete reduction of the sulfate reservoir (given equivalent rates of bacterial sulfate reduction in both slope and basinal environments), thus explaining the more enriched $\delta^{34}S$ values of pyrite from the center of the basin.

Description of the Oxygen-Minimum Zone

Expansion of the oxygen-minimum zone is necessary to reduce and mobilize large amounts of manganese to locations where manganese can be concentrated in the sediment (Pratt et al., 1991, and references therein). In the Monterey, relatively high Mn/Fe ratios near the flanks of the basin indicate manganese was more extensively concentrated there than in the center of the basin. In addition, the laminated structure of the basinal strata indicate the oxygen-minimum zone extended from the slope to the floor of the basin (Pisciotto and Garrison, 1981) or that bottom waters were depleted in oxygen, at least during deposition of the lower calcareous and middle phosphatic lithofacies. The elevated concentrations of manganese near the flanks relative to concentrations of manganese in the center probably marks the boundary between the oxygenated water column and the oxygen-minimum zone in the Santa Maria basin.

Evidence of Condensed Section

Following Loutit et al. (1988), condensed sections are relatively thin, marine, stratigraphic units consisting of pelagic and hemipelagic sediments deposited at times of maximum regional transgression. Rising sea level causes terrigenous clastic material to be trapped landward and causes sedimentation rates farther offshore to decline (Baum and Vail, 1988). In areas of strong upwelling (e.g., California borderland basins) however, this concept needs modification. A decrease in terrigenous clastic input may not be linked to a decrease in overall rates of sediment accumulation due to the importance of biogenic sediment. Thus, condensed sections deposited in upwelling regimes can be thicker than condensed sections originating on platforms and continental shelves lacking high primary productivity.

Overall, samples from the upper siliceous lithofacies have the highest average concentrations of elements derived solely from terrigenous clastics (titanium, 0.31 ± 0.20; aluminum, 6.07 ± 3.84), whereas samples from the middle phosphatic lithofacies average the lowest concentrations of titanium (0.17 ± 0.07) and aluminum (3.33 ± 1.27) (Zaback and Pratt, 1992). Samples from the lower calcareous and middle phosphatic lithofacies (which were deposited during a maximum regional transgression of a second-order sea level rise; see Figure 2) show an increase in organic carbon content with decreasing titanium concentration (Figure 10A). These results suggests that the lower calcareous and middle phosphatic lithofacies represent a condensed section deposited in a deep-water, continental-margin basin. At times of limited input of continentally derived detritus and high primary productivity, reactive organic matter will be abundant in marine sediments. Deposition of large amounts of reactive organic matter enhances anaerobic degradation processes and causes sediments to become highly reducing. For this reason, geochemical signals, such as elevated ratios of Mn/Fe and enrichment of ^{34}S in reduced sulfur species, are enhanced in condensed sections. In addition, condensed sections of this type should be expected to yield petroleum source rocks of excellent quality but elevated sulfur content. While there is debate regarding the predominant petroleum-generating unit within the Monterey Formation, Orr (1986) offers evidence that the middle phosphatic lithofacies is the major source of petroleum in and around the Santa Maria basin area.

Samples from the upper siliceous lithofacies of the Monterey Formation and the overlying Sisquoc Formation show no correlation between titanium and organic carbon content (Figure 10B). Within the upper siliceous lithofacies, rocks alternate between laminated and massive strata. Such an alternation is explained by deposition of beds during times when bottom waters were oxygen-poor (laminated strata) or oxygen-rich (massive strata) (Isaacs, 1989).

Plots of titanium vs. silica (reported as oxides) aid in explaining the relationships observed between titanium and organic carbon for the lithofacies of the Monterey Formation. For the lower calcareous and middle phosphatic lithofacies, titanium and silica contents (while low relative to the upper siliceous lithofacies) are positively correlated (r^2 = 0.775 and 0.659, respectively), indicating that silica was derived mainly from clastic input during deposition of these

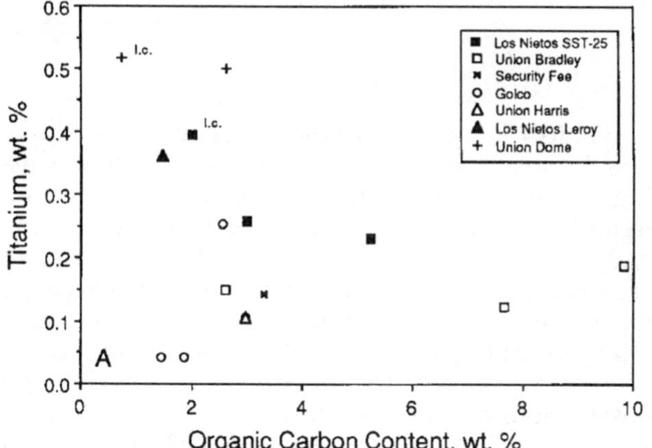

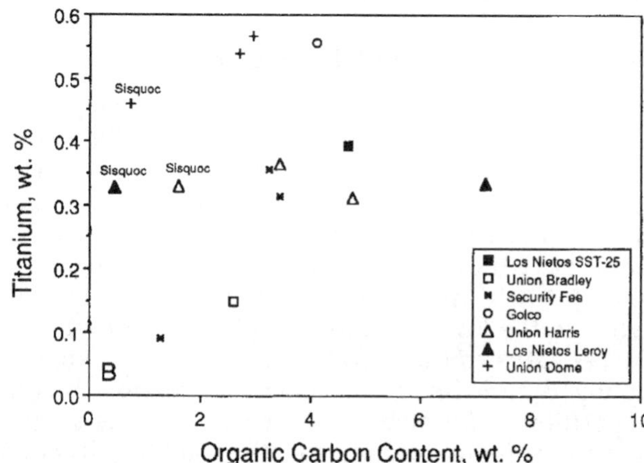

Figure 10. Relationship between titanium concentration and organic carbon content in a condensed marine section (A) in comparison to stratigraphic units receiving increased terrigenous clastic input (B).

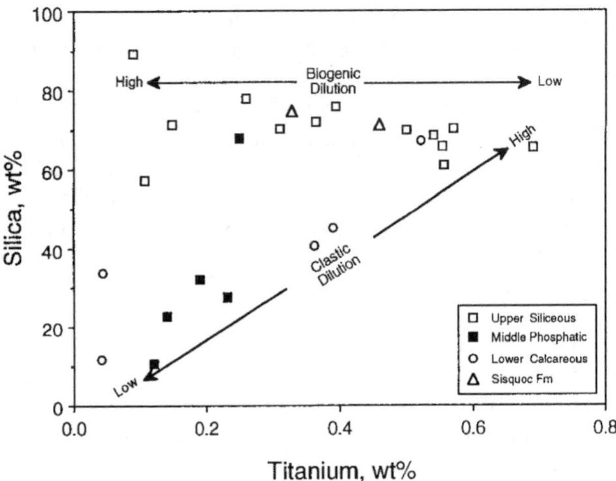

Figure 11. Trends between concentrations of silica and titanium for samples from the lower calcareous, middle phosphatic, and upper siliceous lithofacies of the Monterey and the overlying Sisquoc Formation. The two trends observed are a result of increasing terrigenous clastic input and biogenic dilution by diatoms.

two units (Figure 11). For the upper siliceous lithofacies, however, silica is nearly invariant with increasing concentrations of titanium (Figure 11). Such a trend is explained by periods of enhanced upwelling and preservation of biogenic silica. Thus, the lack of correlation between titanium and organic carbon is most likely attributed to dilution of the sediments with biogenic silica during deposition of the upper siliceous lithofacies. This hypothesis is supported by observations of Isaacs (1989) wherein laminated strata were found to contain a greater abundance of diatoms associated with upwelling relative to the massive strata of the upper siliceous unit.

CONCLUSION

Documentation of the fluctuations of Mn/Fe ratios and sulfur isotopic composition of pyrite in a condensed section of the Monterey Formation of the Santa Maria basin illustrates some of the effects that sea level changes and rates of terrigenous clastic input have on the diagenetic cycling of manganese and sulfur in marine basins. Elevation of Mn/Fe ratios in the lower calcareous lithofacies offers evidence that chemical partitioning of the two elements occurred at a time of marine transgression during deposition of Monterey sediments. Near the transition from the middle phosphatic to upper siliceous lithofacies, the initiation of falling sea level and introduction of cold, bottom waters into the basin likely intensified upwelling and primary productivity. Because detrital clastic input remained limited while deposition of organic matter was high, organic-rich sediments resulted. Deposition of organic-rich sediments promoted high rates of sulfate reduction at and below the sediment–water interface resulting in a shift toward more positive sulfur isotopic values of pyrite. Therefore, in deep-water, continental-margin basins, units deposited during times of low sedimentation rate may contain the greatest abundance of reactive organic matter.

The stratigraphic position and the spatial relationships of variations of Mn/Fe ratios and sulfur isotopic composition of pyrite provide information regarding the differences in the chemical nature of strata in the Monterey and Sisquoc formations. The occurrence of sulfur and manganese geochemical signals within known productive source rocks and their association with sea level changes, may aid in defining economically important frontier basins that were affected by sea level changes similar to those recorded in the Santa Maria basin. The data may also be helpful in better predicting the location of optimal source rocks within these basins.

Ancillary data such as concentrations of titanium and aluminum are useful in explaining the location of differences in intensity of sulfur and manganese signals within the Santa Maria basin. These data suggest that during a single transgressive event, local basin phenomena, such as intensity and direction of bottom-currents, may affect the degree to which Mn/Fe ratios and sulfur isotopic values of pyrite increase within chemically reactive sediments.

ACKNOWLEDGMENTS

This manuscript benefited from discussions with G. B. Hieshima, L. J. Suttner, and A. Basu and reviews by J. S. Compton and V. D. Robison. Financial support was provided by the National Science Foundation (Grant #EAR-8816371 to L. M. Pratt), AAPG Student Grant-In-Aid (1988), Sigma Xi Student Grant-In-Aid (1988), and Mobil Research and Development Corporation. A. Kornacki of Shell Oil Company arranged for the sampling of cores. S. A. Studley and M. Gilstrap are thanked for their technical assistance.

REFERENCES CITED

Baker, P.A., and M. Kastner, 1981, Constraints on the formation of sedimentary dolomite: Science, v. 213, p. 214-216.

Barrera, E., G. Keller, and S.M. Savin, 1985, Evolution of the Miocene ocean in the eastern North Pacific as inferred from oxygen and carbon isotopic ratios of foraminifera, *in* J.P. Kennett, ed., The Miocene Ocean: Paleoceanography and Biogeography: Geological Society of America, Memoir 163, p. 83-103.

Baturin, G.N., 1982, Phosphorites on the sea floor, origin, composition and distribution: Developments in Sedimentology, v. 33, Elsevier Sci. Public., Amsterdam, 343 p.

Baum, G.R., and P.R. Vail, 1988, Sequence stratigraphic concepts applied to Paleogene outcrops, Gulf and Atlantic Basins, *in* C.K. Wilgus, B.S. Hastings, C.A. Ross, H. Posamentier, and C.G.St.C. Kendall, eds., Sea-Level Changes: An Integrated Approach: SEPM Special Publication No. 42, p. 309-327.

Beier, J.A., 1988, Sulfide, phosphate, and minor element enrichment in the New Albany (Devonian–Mississippian) of Southern Indiana [Ph.D. thesis]. Indiana University, Bloomington, 141 p.

Bohacs, K.M., 1990, Sequence stratigraphy of the Monterey Formation, Santa Barbara County: Integration of physical, chemical, and biofacies data from outcrop and subsurface, *in* M.A. Keller and M.K. McGowen, eds., Miocene and Oligocene Petroleum Reservoirs of the Santa Maria and Santa Barbara–Ventura Basins, California: SEPM Core Workshop No. 14, p. 139-200.

Bralower T.J., and H.R. Thierstein, 1987, Organic carbon and metal accumulation rates in Holocene and mid-Cretaceous sediments: palaeoceanographic significance, *in* J. Brooks and A.J. Fleet, eds., Marine Petroleum Source Rocks: Geological Society Special Publication 26, p. 345-369.

Brumsack, H.J., 1983, A note on Cretaceous black shales and recent sediments from oxygen deficient environments: paleoceanographic implications, *in* E. Suess and J. Thiede, eds., Coastal Upwelling; its sediment record. Part I.: Plenum Press, New York, p. 471-484.

Burdige, D.J., and K.H. Nealson, 1985, Microbial manganese reduction by enrichment cultures from coastal marine sediments: Applied Environmental Microbiology, v. 50, p. 491-497.

Burdige, D.J., and K.H. Nealson, 1986, Chemical and microbiological studies of sulfide-mediated manganese reduction: Geomicrobiology Journal, v. 4, p. 361-387.

Burnett, W.C., 1977, Geochemistry and origin of phosphorite deposits from off Peru and Chile: Geological Society of America Bulletin, v. 88, p. 813-823.

Canfield, D.E., 1991, Sulfate reduction in deep-sea sediments: American Journal of Science, v. 291, p. 177-188.

Chamley, H., 1989, Clay Sedimentology: Springer-Verlag, Berlin, 623 p.

Chandramohan, D., P.A. Lokabharathi, Shanta Nair, and S.G.P. Matondkar, 1987, Bacteriology of ferromanganese nodules from the Indian Ocean: Geomicrobiology Journal, v. 5, p. 17-31.

Chambers, L.A. and P.A. Trudinger, 1979, Microbiological fractionation of stable sulfur isotopes: a review and critique: Geomicrobiology Journal, v. 1, p. 249-293.

Claypool, G.E., and I.R. Kaplan, 1974, The origin and distribution of methane in marine sediments, *in* I.R. Kaplan, ed., Natural Gases in Marine Sediments, Plenum Press, p. 99-139.

Compton, J.S., 1988, Degree of supersaturation and precipitation of organogenic dolomite: Geology, v. 16, p. 318-321.

Compton, J.S., 1991, Porosity reduction and burial history of siliceous rocks from the Monterey and Sisquoc Formations, Point Pedernales area, California: Geological Society of America Bulletin, v. 103, p. 625-636.

Demaison, G.J., and G.T. Moore, 1980, Anoxic environments and oil source bed genesis: AAPG Bulletin, v. 64, p. 1179-1209.

Drever, J.I., 1971, Magnesium–iron replacement in clay minerals in anoxic marine sediments: Science, v. 172, p. 1334-1336.

Ehrlich, H.L., 1987, Manganese oxide reduction as a form of anaerobic respiration: Geomicrobiology Journal, v. 5, p. 423-431.

Emerson, S., R.E. Cranston, P.S. Liss, 1979, Redox species in a reducing fjord: equilibrium and kinetic considerations: Deep-Sea Research, v. 26A, p. 859-878.

Force, E.R., and W.F. Cannon, 1988, Depositional model for shallow-marine manganese deposits around black shale basins: Economic Geology, v. 83, p. 93-117.

Garrison, R.E., R.G. Stanley, and L.J. Horan, 1979, Middle Miocene sedimentation on the southwestern edge of the Lockwood High, Monterey County, California, *in* S.A. Graham, ed., Tertiary and Quaternary geology of the Salinas Valley and Santa Lucia Range, Monterey County, California, SEPM Field Guide #4, p. 51-65.

Goldhaber, M.B., and I.R. Kaplan, 1974, The sulfur cycle, *in* E.D. Goldberg, ed., The Sea: Wiley-Interscience, New York, v. 5, p. 569-655.

Goldhaber, M.B., and I.R. Kaplan, 1975, Controls and consequences of sulfate reduction rates in recent marine sediments: Soil Science, v. 119, p. 42-55.

Goldhaber, M.B., and I.R. Kaplan, 1980, Mechanisms of sulfur incorporation and isotope fractionation during early diagenesis in sediments of the Gulf of California: Marine Chemistry, v. 9, p. 95-143.

Gulbrandsen, R.A., 1969, Physical and chemical factors in the formation of marine apatite: Economic Geology, v. 64, p. 365-382.

Ingle, J.C., 1981, Origin of Neogene diatomites around the North Pacific rim, *in* R.E. Garrison and R.G. Douglas, eds., The Monterey Formation and related siliceous rocks of California: Pacific Section SEPM Special Publication, p. 159-179.

Isaacs, C.M., 1983, Compositional variation and sequence in the Miocene Monterey Formation, Santa Barbara coastal area, California, *in* D.K. Larue and R.J. Steel, eds., Cenozoic Marine Sedimentation, Pacific margin, U.S.A.: Pacific Section SEPM, p. 117-132.

Isaacs, C.M., 1989, Field trip guide to deposition and diagenesis of the Monterey Formation, Santa Barbara and Santa Maria areas, California: USGS Open-File Report 84-98, 91p.

Jørgensen, B.B., 1983, Processes at the sediment–water interface: *in* Bolin and Cook, eds., The Major

Biogeochemical Cycles and Their Interactions, SCOPE, p. 477-515.

Jørgensen, B.B., 1987, Ecology of the sulphur cycle: oxidative pathways in sediments, *in* J.A. Cole and S. Ferguson, eds., The Nitrogen and Sulphur Cycles: Society for General Microbiology Sumposium 42. Society for General Microbiology Limited, Cambridge University Press, p. 31-63.

Kastner, M., K. Mertz, D. Hollander, and R. Garrison, 1984, The association of dolomite–phosphorite–chert: causes and possible diagenetic sequences, *in* R.E. Garrison, M. Kastner, and D.H. Zenger, eds., Dolomites of the Monterey Formation and Other Organic-rich Units: Pacific Section SEPM, v. 41, p. 75-86.

Loutit, T.S., J. Hardenbol, P.R. Vail, and G.R. Baum, 1988, Condensed sections: the key to age determination and correlation of continental margin sequences, *in* C.K. Wilgus, B.S. Hastings, C.A. Ross, H. Posamentier, and C.G.St.C. Kendall, eds., Sea-Level Changes: An Integrated Approach: SEPM Special Publication No. 42, p. 183-213.

Martin, J.H., and G.A. Knauer, 1984, VERTEX: Manganese transport through oxygen minima: Earth and Planetary Science Letters, v. 67, p. 35-47.

Nealson, K.H., 1983, The microbial manganese cycle, *in* W. Krumbein, ed., Microbial Biochemistry: Blackwell Scientific, Oxford, p. 191-222.

Orr, W.L., 1986, Kerogen/asphaltene/sulfur relationships in sulfur-rich Monterey oils: Organic Geochemistry, v. 10, p. 499-516.

Pisciotto, K.A., and R.E. Garrison, 1981, Lithofacies and depositional environments of the Monterey Formation, California, *in* R.E. Garrison and R.G. Douglas, eds., The Monterey Formation and Related Siliceous Rocks of California: Pacific Section, SEPM Sp. Publ., p. 97-122.

Pratt, L.M., E.R. Force, and B. Pomerol, 1991, Coupled manganese and carbon-isotopic events in marine carbonates at the Cenomanian–Turonian boundary: Journal of Sedimentary Petrology, v. 61, p. 370-383.

Reeburgh, W.S., 1983, Rates of biogeochemical processes in anoxic sediments: Annual Review of Earth and Planetary Science, v. 11, p. 269-298.

Schlanger, S.O., and H.C. Jenkyns, 1976, Cretaceous ocean anoxic events: causes and consequences: Geologie en Mijnbouw, v. 55, p. 179-184.

Schlanger, S.O., M.A. Arthur, H.C. Jenkyns, and P.A. Scholle, 1987, The Cenomanian–Turonian oceanic anoxic event, I. Stratigraphy and distribution of organic carbon-rich beds and the marine ^{13}C excursion, *in* J. Brooks and A. Fleet, eds., Marine Petroleum Source Rocks: Geological Society of London Special Publication, no. 2, p. 371-399.

Shiller, A.M., J.M. Gieskes, and N.B. Price, 1985, Particulate iron and manganese in the Santa Barbara Basin, California: Geochimica et Cosmochimica Acta, v. 49, p. 1239-1249.

Sholkovitz, E., and A. Soutar, 1975, Changes in the composition of the bottom water of the Santa Barbara Basin: effect of turbidity currents: Deep-Sea Research, v. 22, p. 13-21.

Sorensen, J., and B.B. Jørgensen, 1987, Early diagenesis in sediments from Danish coastal waters: microbial activity of Mn-Fe-S geochemistry: Geochimica et Cosmochimica Acta, v. 51, p. 1583-1590.

Stumm, W., and J.J. Morgan, 1981, Aquatic Chemistry, 2nd edition: John Wiley and Sons, Inc., New York, 780 p.

Toth, D.J., and A. Lerman, 1977, Organic matter reactivity and sedimentation rates in the ocean: American Journal of Science, v. 277, p. 465-485.

Trimble, R.B., and H.L. Ehrlich, 1968, Bacteriology of manganese nodules. III. Reduction of MnO_2 by two strains of nodule bacteria. Applied Microbiology, v. 16, p. 695-702.

Vail, P.R., R.M. Mitchum, Jr., and S. Thompson, III, 1977, Seismic stratigraphy and global changes of sea level, part four: global cycles of relative changes of sea level: AAPG Memoir 26, p. 83-98.

Vincent, E., and W.H. Berger, 1985, Carbon dioxide and polar cooling in the Miocene: The Monterey hypothesis: *in* E.T. Sundquist and W.S. Broecker, eds., Natural variations in carbon dioxide and the carbon cycle: American Geophysical Union, Washington, D.C., Geophysical Monograph 32, p. 455-468.

Wintsch, R.P., C.M. Kvale, and H.J. Kisch, 1991, Open-system, constant-volume development of slaty cleavage, and strain-induced replacement reactions in the Martinsburg Formation, Lehigh Gap, Pennsylvania: Geological Society of America Bulletin, v. 103, p. 916-927.

Woodruff, F., 1985, Changes in Miocene deep-sea benthic foraminiferal distribution in the Pacific Ocean—relationship to paleoceanography, *in* J.P. Kennett, ed., The Miocene Ocean—Paleoceanography and Biogeochemistry: Geological Society of America Memoir 163, p. 131-175.

Zaback, D.A., 1992, Geochemical and stable isotopic study of the C-S-Fe-Mn-P system in the Miocene Monterey Formation, Santa Maria Basin, California [Ph.D. thesis]: Bloomington, Indiana, Indiana University, 192 p.

Zaback, D.A., and L.M. Pratt, 1992, Isotopic composition and speciation of sulfur in the Miocene Monterey Formation: re-evaluation of sulfur reactions during early diagenesis in marine environments: Geochimica et Cosmochimica Acta, v. 56, p. 763-774.

Zaback, D.A., L.M. Pratt, and J.M. Hayes, 1993, Transport and reduction of sulfate and immobilization of sulfide in marine black shales: Geology, v. 21, 141-144.

Chapter 14

Sequence Stratigraphic Significance of Organic Matter Variations: Example from the Upper Cretaceous Mancos Shale of the San Juan Basin, New Mexico

Mark A. Pasley
Greg W. Riley
Amoco Production Company
Houston, Texas, USA

Dag Nummedal
Department of Geology and Geophysics
Louisiana State University
Baton Rouge, Louisiana, USA

ABSTRACT

Study of a portion of the Upper Cretaceous Mancos Shale in the San Juan Basin of New Mexico shows that organic facies deposited on the shelf change noticeably at surfaces that have sequence stratigraphic significance. Shelf sediments below transgressive surfaces contain abundant, well-preserved terrestrial organic matter (phytoclasts) whereas sediments above transgressive surfaces contain sparse and highly degraded phytoclasts and more hydrogen-rich organic matter. Shelf sediments associated with the maximum flooding surface typically contain the least terrestrial organic matter. These results indicate that the type and preservation of organic matter is related to both the rate of terrigenous sediment supply to the shelf and the bottom water oxygen conditions present on the shelf.

Variations in the amount and type of organic matter (organic facies) preserved in shelf sediments are predictable within a sequence stratigraphic framework. Each systems tract has a distinctive depositional style that affects the amount of terrigenous sediment influx to the shelf and, consequently, the type and preservation of organic matter that is deposited on the shelf. Fine-grained marine sediments in transgressive systems tracts possess high total organic carbon and yield relatively high amounts of hydrocarbons during pyrolysis. Petrographically, this organic matter is composed primarily of amorphous nonstructured protistoclasts. Phytoclasts in the transgressive systems tract are highly degraded. In contrast, progradational marine depositional systems of both the lowstand and highstand systems tracts contain less total organic carbon and less pyrolyzable hydrocarbons. Petrographic analysis of organic matter in these rocks reveals abundant macerals of terrestrial origin. Phytoclasts are especially well preserved in the lowstand systems tract.

Integration of data from the characterization of organic matter with sedimentologic and regional stratigraphic information provides greater precision in locating surfaces that bound systems tracts within the depositional sequence. An example of this approach is presented for a part of the Upper Cretaceous Mancos Shale of the San Juan Basin, New Mexico. Organic matter data not only improve systems tract identification in fine-grained, basinward facies but also demonstrate that the predictive capabilities of sequence stratigraphy are applicable to marine petroleum source rocks. These results indicate that optimum source rock potential is found in the transgressive systems tract below the condensed section facies that contains the downlap surface.

INTRODUCTION

Data from organic geochemistry and organic petrology have proven invaluable in the evaluation of petroleum source potential and thermal maturity of rocks in sedimentary basins. However, these data are also useful in sedimentologic and stratigraphic interpretations. The concept of organic facies is important in this regard. Organic facies, as defined by Jones (1987), are mappable stratigraphic units that are distinguished by the composition of their constituent organic matter. Several studies have demonstrated that organic facies deposited in shelf environments during regional transgression of the shoreline differ markedly from those deposited in shelf sediments associated with regression (Habib and Miller, 1989; Leckie et al., 1990; Pasley and Hazel, 1990; Davies et al., 1991; Harput and Gokcen, 1991; Pasley, 1991; Pasley et al., 1991; Wignall, 1991b; Posamentier and Chamberlin, 1993). In a study of Upper Cretaceous shelf sediments in the Atlantic Coastal Plain, Habib and Miller (1989) demonstrated that transgressive sediments contained a large number of dinoflagellate species and an organic facies dominated by amorphous debris. Regressive sediments, in contrast, possessed an organic facies of terrestrially derived particulate organic matter (vascular tissue and inertinite). Comparison of the organic matter from different stratigraphic intervals of the Cretaceous Mancos Shale in the San Juan Basin by Pasley et al. (1991) revealed that transgressive offshore marine sediments contain higher organic carbon, more hydrogen-rich organic matter, and less terrestrial organic matter than the underlying regressive deposits. Recent work has examined the connection between sequence stratigraphy and the deposition of organic matter and petroleum source rocks (Leckie et al., 1990; Pasley and Hazel, 1990; Stefani and Burchell, 1990; Bohacs and Isaksen, 1991; Creaney et al., 1991; Gorin and Steffen, 1991; Pasley, 1991; Pasley et al., 1991; Wignall, 1991a; Posamentier and Chamberlin, 1993).

Sequence stratigraphic principles have begun to alleviate many of the problems associated with classical approaches to lithostratigraphy, especially those pertaining to time–rock relationships. This advantage is possible because sequence stratigraphy emphasizes the establishment of a regional chronostratigraphic framework based, in part, on physical stratigraphic surfaces. Key sequence stratigraphic surfaces are thought to have chronostratigraphic significance because all rocks above them are younger than all rocks below. Each of the depositional systems tracts in a depositional sequence exhibits a characteristic parasequence stacking pattern that develops in response to changing rates of generation of accommodation space (Jervey, 1988; Van Wagoner et al., 1988, 1990). Stacking of parasequences is initially progradational (or forestepping) and becomes more aggradational in the lowstand systems tract. The transgressive systems tract is characterized by an overall retrogradational (or backstepping) stacking pattern and the subsequent highstand systems tract is characterized by a stacking of parasequences that is initially aggradational but becomes dominantly progradational. Recognition of these stacking patterns on well logs (especially gamma ray logs) provides the foundation for regional subsurface sequence stratigraphic interpretations at scales below the limit of seismic resolution (Vail, 1990; Vail and Wornardt, 1990; Van Wagoner et al., 1990).

Objectives

The overall objective of this chapter is to integrate data from the characterization of organic matter with concepts from sequence stratigraphy. Towards this goal, three specific objectives are addressed. First, the relationship between organic matter characteristics and position within the depositional sequence is examined. Second, the utility of this relationship to an integrated sequence stratigraphic analysis is demonstrated. Third, the implications of this relationship to the prediction of marine petroleum source rocks are discussed.

Geologic Setting

Samples from the San Juan Basin, located in northwestern New Mexico and southwestern Colorado (Figure 1), were used to study the relationship between organic matter deposition and the sequence stratigraphy of offshore marine sediments. The basin accumulated more than 1500 m of Upper Cretaceous, predominantly clastic marine sediments while it was part of the Western Interior Seaway. These sediments were deposited during multiple transgressive and regressive cycles (Molenaar, 1983; Kauffman, 1977, 1985).

Fine-grained, marine sediments from the Upper Cretaceous Mancos Shale were sampled at four locations (Figure 2). Conventional cores were obtained from the Bisti oil field (location 1) and from two wells located in a basinward position: Union Texas Petroleum Company (UTPC) Angel Peak B#37, location 3; and Newsom A3E (A20), location 4. Outcrop samples were collected at the Hogback oil field (location 2). Samples obtained from cores at Bisti oil field, from the UTPC Newsom A3E (A20) well, and from the UTPC Angel Peak B#37 well were studied in detail and the stratigraphic intervals represented by these cores are shown in the generalized lithostratigraphic diagram of Figure 3. These rocks range in age from latest Cenomanian to earliest Santonian (Figure 4) and were deposited in shallow marine environments (Molenaar, 1983; Nummedal et al., 1989b; Riley, 1993). Well logs, lithostratigraphy, and sedimentological observations for the Angel Peak and Newsom cores are presented in Figures 5 and 6.

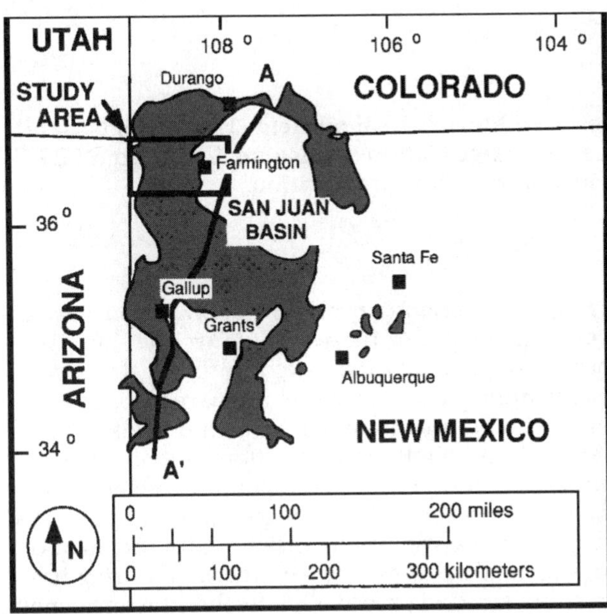

Figure 1. Location map for the San Juan Basin of New Mexico and Colorado, USA. Cretaceous outcrops denoted by shaded areas and study area of Figure 2 shown as rectangle in northwest corner of New Mexico. Line of section A–A′ represents approximate position of schematic diagrams in Figures 3 and 4. Figure adapted from Molenaar (1983).

RESULTS

Organic Matter Characterization

The organic matter in the Mancos Shale was characterized using both transmitted white light petrography and programmed pyrolysis (Rock-Eval). Results from the Bisti oil field have been described previously by Pasley et al. (1991) and are summarized in Figure 7. Results for the Angel Peak and Newsom cores are presented in Figures 8 and 9. The organic matter in the Newsom and Angel Peak cores is thermally mature with respect to liquid hydrocarbon generation (T_{max} = 442–452°C) whereas organic matter in the Bisti samples is only marginally mature (T_{max} = 430°C, R_o = 0.60%; Pasley et al., 1991).

Parameters from programmed pyrolysis include: amount of hydrocarbons generated by pyrolytic breakdown of the kerogen (S_2 in mg HC/g rock), hydrogen index (HI in mg HC/g C_{org}), and total organic carbon (TOC in wt.%) (see Espitalié et al., 1977, and Peters, 1986, for more detailed explanation). Petrographic analysis provides a measure of the amount of terrestrial organic matter (% TOM) and the amount of degradation that these phytoclasts have experienced (degradational index, or D). The percent terrestrial organic matter is derived from point count data and is a measure of the relative proportion of terrestrially derived organic matter to the total organic matter (Pasley et al., 1991). The degradational index (D) is a ratio of the amount of well preserved phytoclasts in a sample to the amount of highly degraded (or amorphous) phytoclasts present in that same sample and is, therefore, numerically high when the terrestrial organic matter is well preserved and low when it is highly degraded (Hart, 1986; Pasley and Hazel, 1990; Pasley, 1991; Pasley et al., 1991). These organic matter data are shown with grain size, ichnofabric (intensity of bioturbation on an ordinal scale where 1 = no bioturbation and 5 = total bioturbation; see Droser and Bottjer, 1986), and lithostratigraphy for the Angel Peak and Newsom cores in Figures 8 and 9.

Sequence Stratigraphy

A sequence stratigraphic interpretation for a portion of the Upper Cretaceous strata in the San Juan

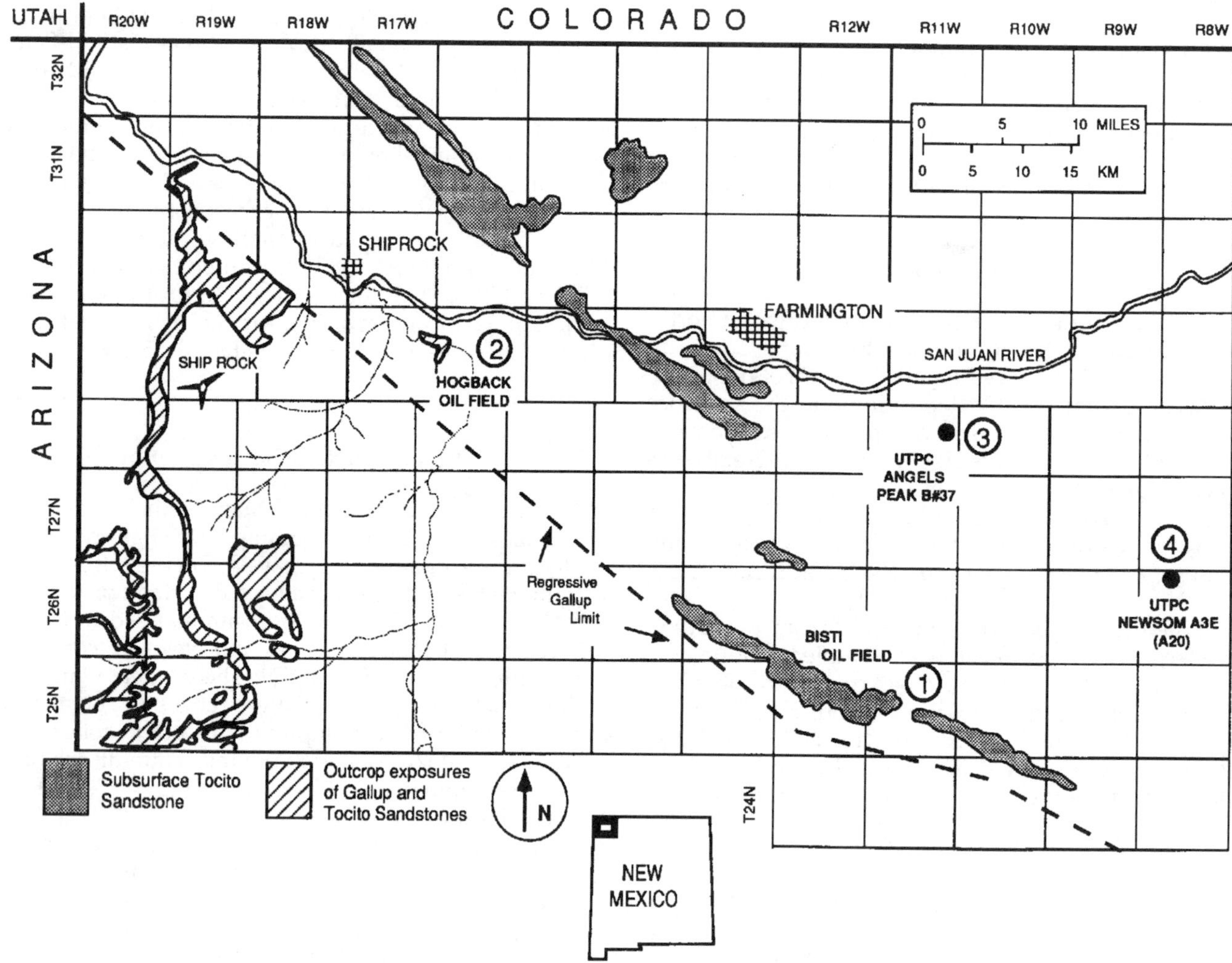

Figure 2. Location map of study area in the San Juan Basin of New Mexico. Bisti oil field (1) and Hogback oil field (2) are located close to the most basinward extent of the regressive Gallup Sandstone. Cores from UTPC Angel Peak B#37 (3) and Newsom A3E (A20) (4) are located in a more basinward position.

Basin of New Mexico is shown for the Angel Peak and Newsom wells in Figures 10 and 11, respectively. This interpretation is the result of integration of data from the characterization of organic matter (Figures 8, 9) with well log cross sections and core and outcrop studies (Molenaar, 1983, 1989; Bergsohn, 1988; Nummedal et al., 1988, 1989a,b; Nummedal, 1990; Pasley, 1991; Riley, 1993; R. Bottjer, written communication, 1987). The organic matter data, which were used in this interpretation, are plotted against the sequence stratigraphy for this portion of the Mancos Shale in Figures 12 and 13.

DISCUSSION

Organic Matter Variations Associated With Transgressions

Organic matter variations that can be recognized in shelf sediments are largely the result of changing terrigenous sediment supply to the shelf. During seaward movement of the shoreline (regression), sediments are generally deposited in delta, strandplain, and prodelta environments and are subsequently redistributed and transported to the shelf as storm deposits and shelf plumes. Therefore considerable amounts of terrigenous sediment are delivered to the shelf environments during regression and both TOC and HI in the regressive shelf sediments are low. Depositional systems active on the shelf during transgression, however, receive relatively little terrigenous sediment because most of this material is trapped in estuarine and related coastal environments (Swift, 1976; Nummedal and Swift, 1987; Jervey, 1988; Posamentier and Vail, 1988; Roberts and Coleman, 1988). Modern estuaries are known to be efficient sediment traps (McCave, 1973; Boyle et al., 1977; Nichols and Biggs, 1985) and trapping of terrestrial organic matter in modern estuarine environments has been documented in the Amazon (Showers and Angle,

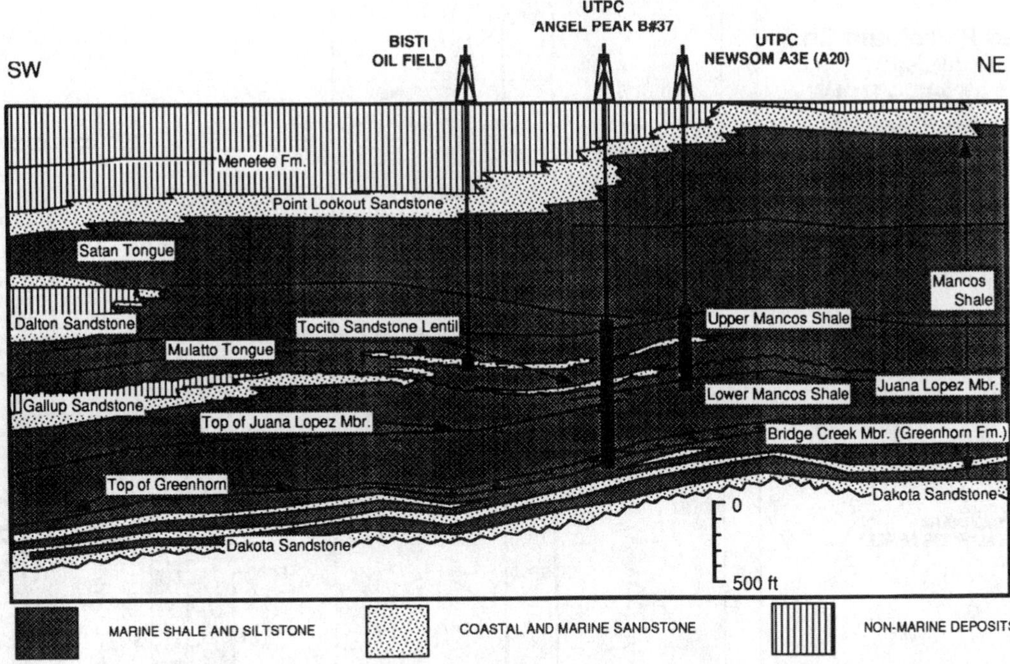

Figure 3. Generalized lithostratigraphic diagram of Upper Cretaceous strata in San Juan Basin. Approximate position of cores, which are out of the section (Figure 1), are denoted by dark vertical boxes. Figure based on regional work of Molenaar (1983).

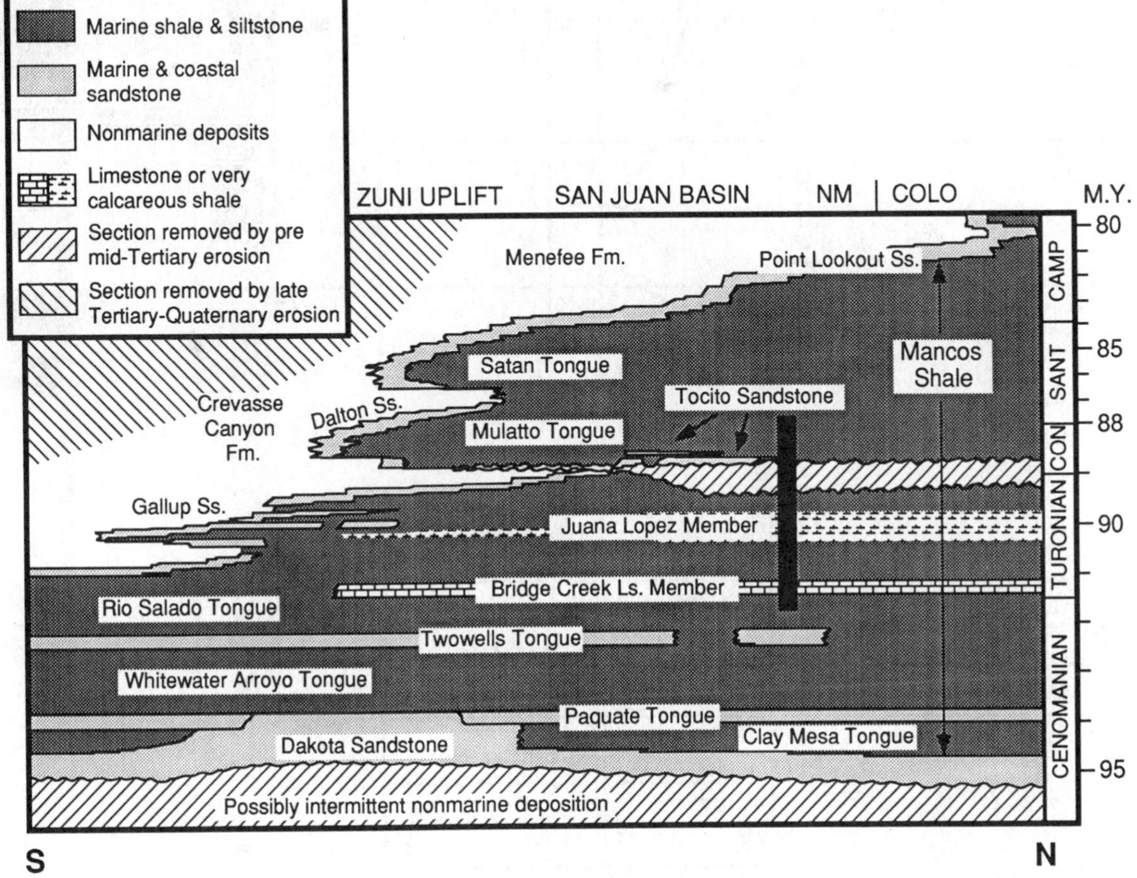

Figure 4. Chronostratigraphy of Late Cretaceous deposits in the San Juan Basin. Vertical bar approximates interval of study. Approximate line of section is A–A´ in Figure 1. Figure adapted from Molenaar (1983).

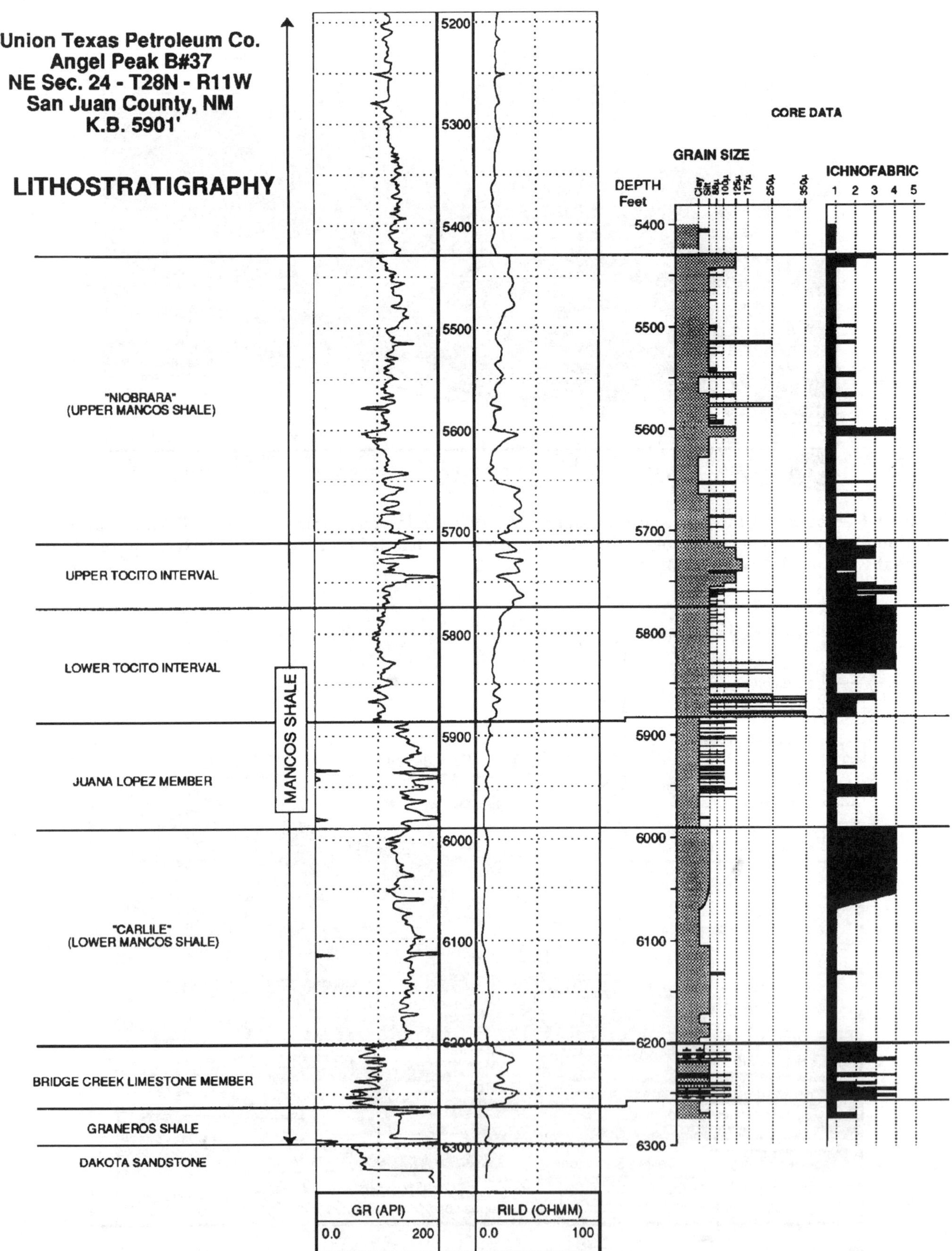

Figure 5. Lithostratigraphy for the UTPC Angel Peak B#37 well. Well log and core data (grain size and ichnofabric) are shown. Stratigraphic information was compiled from several sources (e.g., McCubbin, 1969; Molenaar, 1983, 1989; Pasley, 1991; Riley, 1993, and R. Bottjer, written communication, 1987).

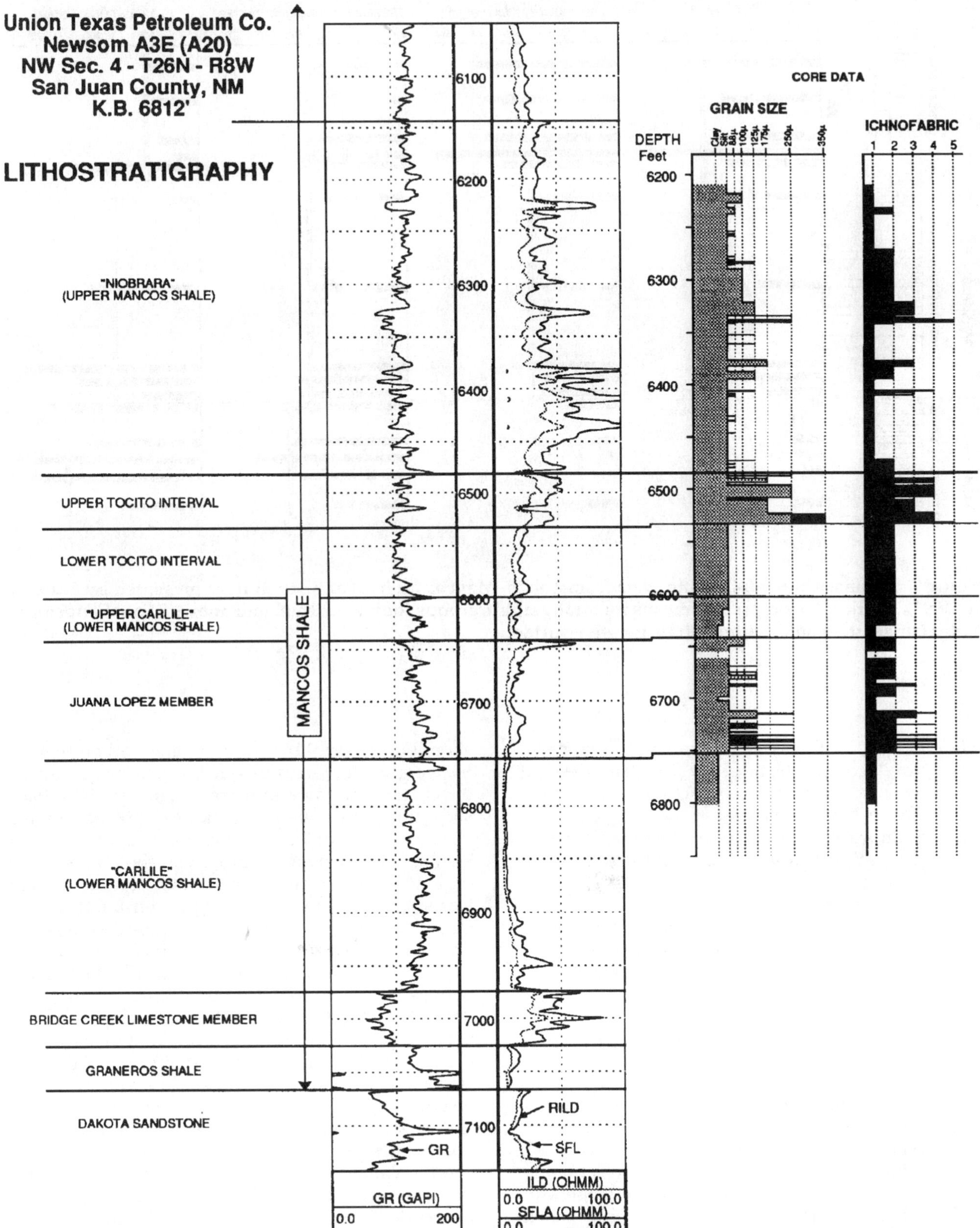

Figure 6. Lithostratigraphy, well log, and core data for UTPC Newsom A3E (A20) well.

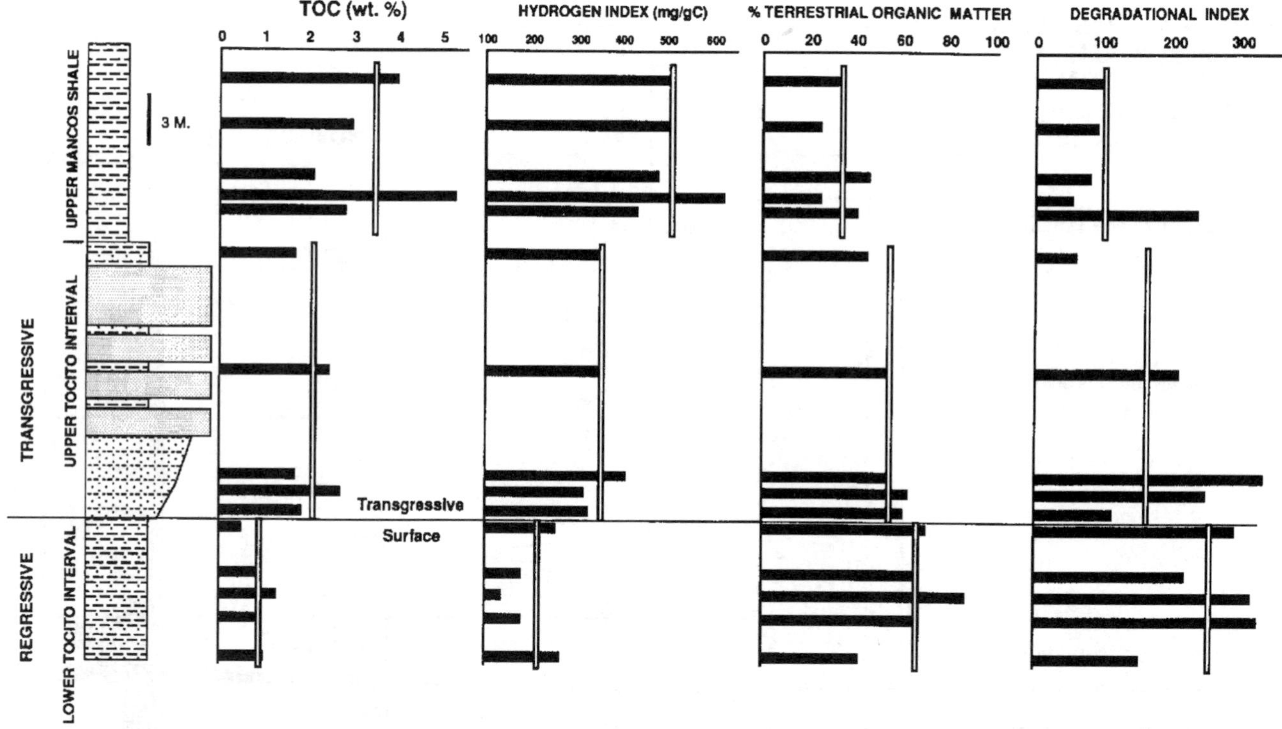

Figure 7. Summary of organic matter data from Tocito–Mancos interval at Bisti oil field presented by Pasley et al. (1991). Vertical bars represent means for stratigraphic groups (below, within, and above Tocito Sandstone). Note change in organic matter at transgressive surface.

1986), the Bay of Biscay (Fontugne and Jouanneau, 1987), Nova Scotia (LeBlanc et al., 1989), and the Gulf of St. Lawrence (Lucotte et al., 1991).

The Tocito–Mancos interval of the San Juan Basin provides an excellent example of the organic matter variations on the shelf associated with transgression of the shoreline (Figure 7) (Pasley et al., 1991). Sediments below the transgressive surface are generally low in TOC and HI. The overlying transgressive sediments, however, contain more total organic carbon and more marine organic matter (higher HI and lower TOM).

Organic Matter and Depositional Systems Tracts

As mentioned earlier, each depositional systems tract within a depositional sequence exhibits a characteristic parasequence stacking pattern (Van Wagoner et al., 1988, 1990). This concept provides the link between the nature of organic matter deposition and depositional systems tract because the type and preservation of organic matter deposited on the shelf is directly associated with parasequence stacking. When each successive parasequence progrades to a more basinward position (forestepping), a position on the shelf represented by a single outcrop or borehole will observe an increase in the relative amount of terrestrial organic matter upward because that position is becoming progressively more proximal to the shoreline (Figure 14A). Total organic carbon (percent by weight) decreases upward as sediment dilution effects become more important. Conversely, a backstepping parasequence stacking pattern, which is characteristic of the transgressive systems tract, results in the delivery of progressively less terrestrial organic matter (relative to the total organic content) to a single site on the shelf (Figure 14B). In the backstepping packages, sediment dilution becomes progressively less important and TOC increases upward.

Therefore, it may be inferred that the relationship between organic matter type and preservation and depositional systems tract reflects the relative importance of progradational depositional systems in each systems tract. Shallow marine sediments in the lowstand and highstand systems tracts are influenced heavily by progradation (hence their aggradational and forestepping stacking patterns) and contain predominantly well preserved phytoclasts (high % TOM and D). Values for TOC and HI are generally lower in these sediments. Autochthonous organic matter is proportionately more common in the transgressive systems tract because progressively less terrestrial organic matter is delivered to the shelf (Figure 14B). In addition, fine-grained facies in the transgressive systems tract contain more hydrogen-rich organic matter and may be more prone to anoxia (high TOC and HI; see discussion by Wignall, 1991b). The change in organic matter characteristics within a

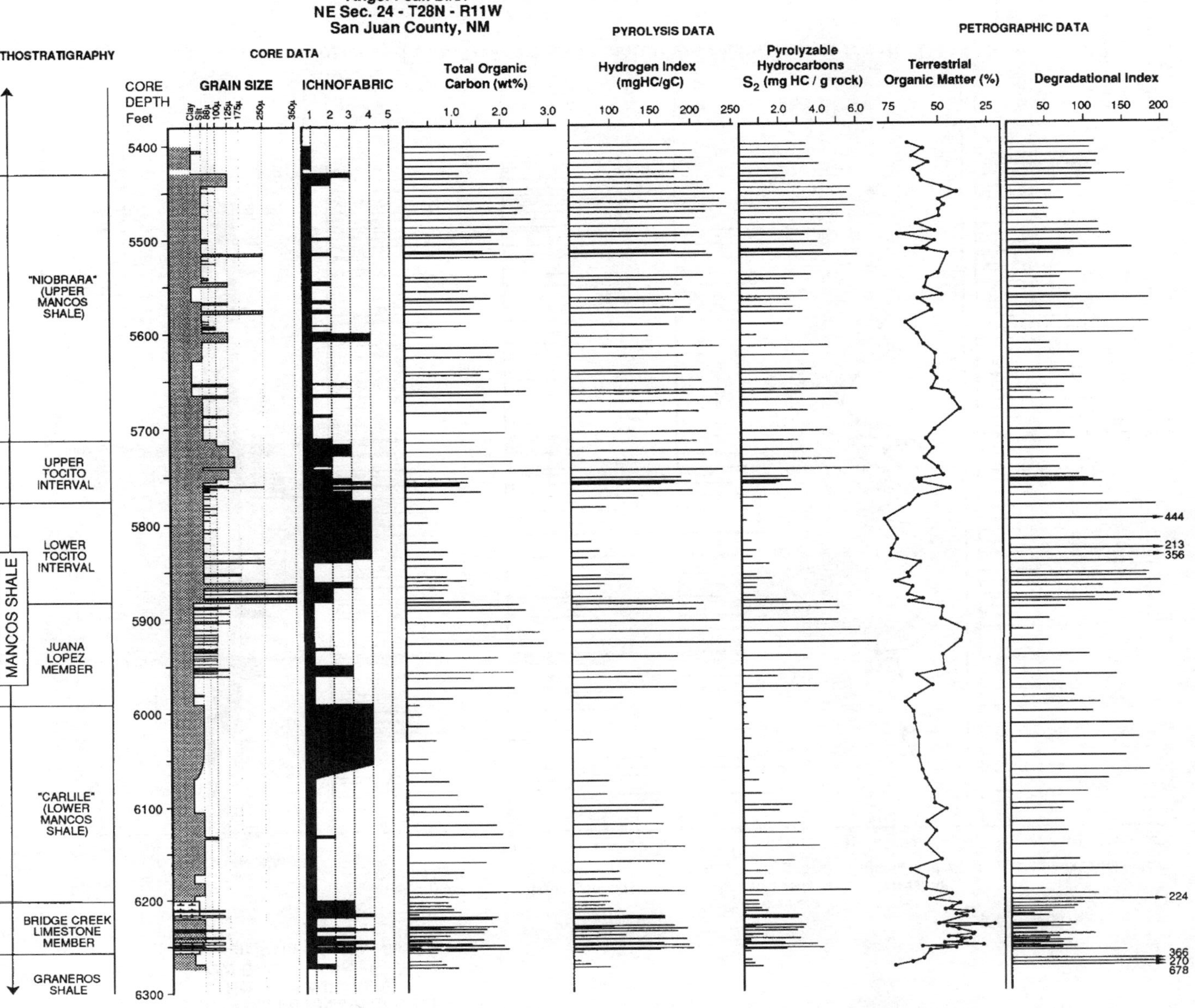

Figure 8. Lithostratigraphy, grain size, ichnofabric, and data from organic matter characterization for the UTPC Angel Peak B#37 well.

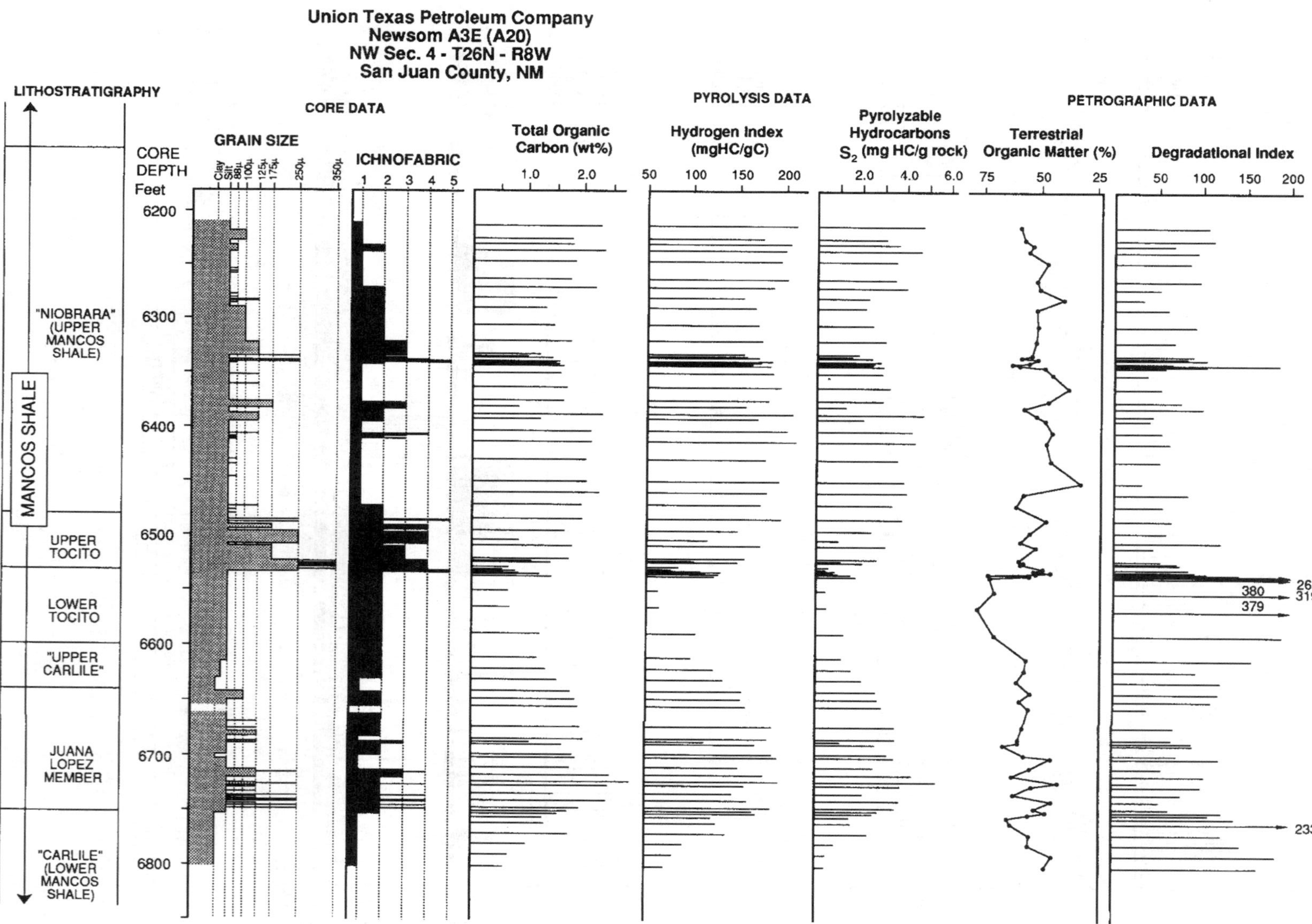

Figure 9. Lithostratigraphy, grain size, ichnofabric, and data from organic matter characterization for the UTPC Newsom A3E (A20) well.

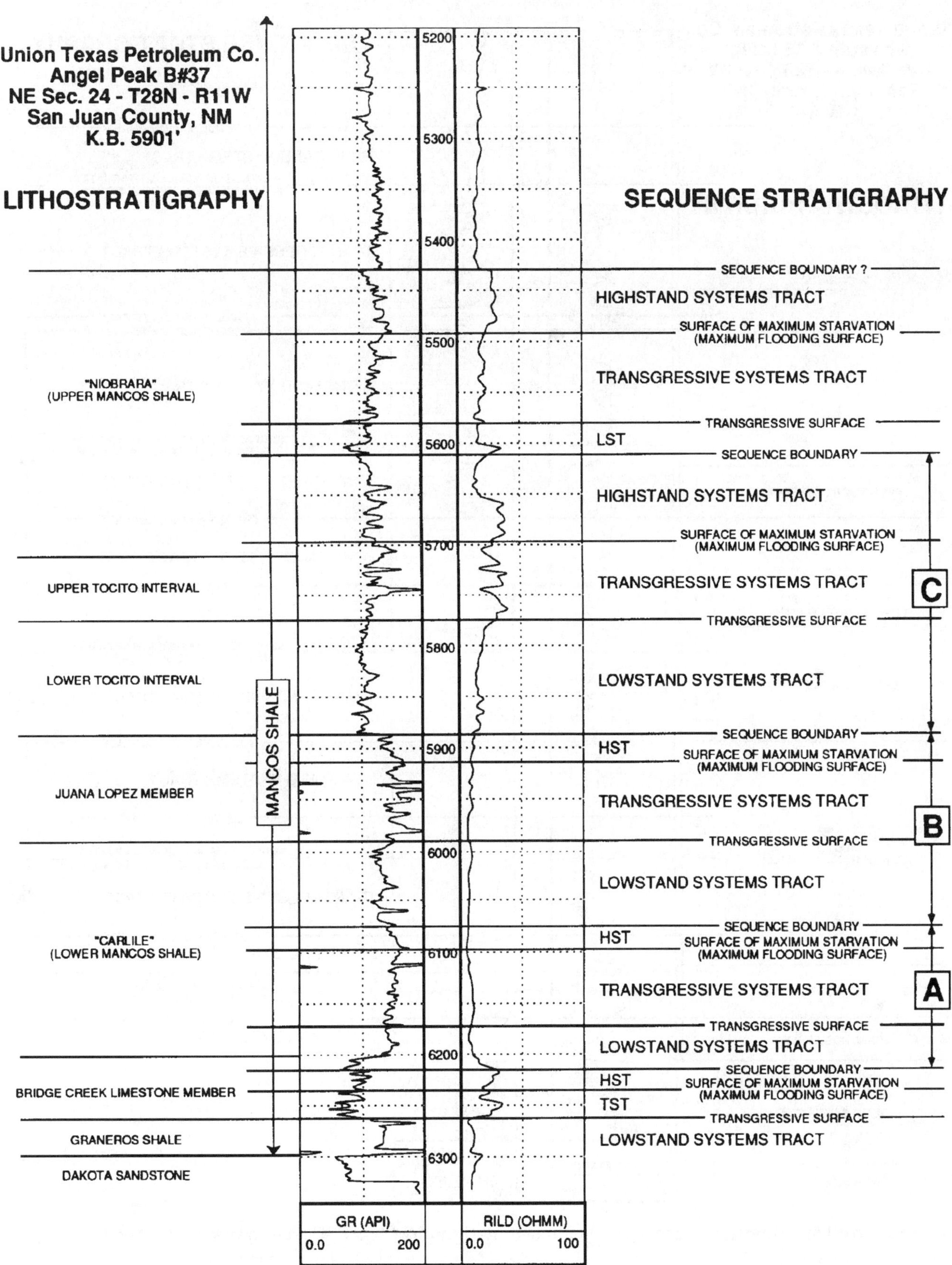

Figure 10. Well log, lithostratigraphy, and sequence stratigraphy for UTPC Angel Peak B#37 well. Sequences A, B, and C labeled for reference. Sequence C is correlative to section at Bisti oil field (Figure 7).

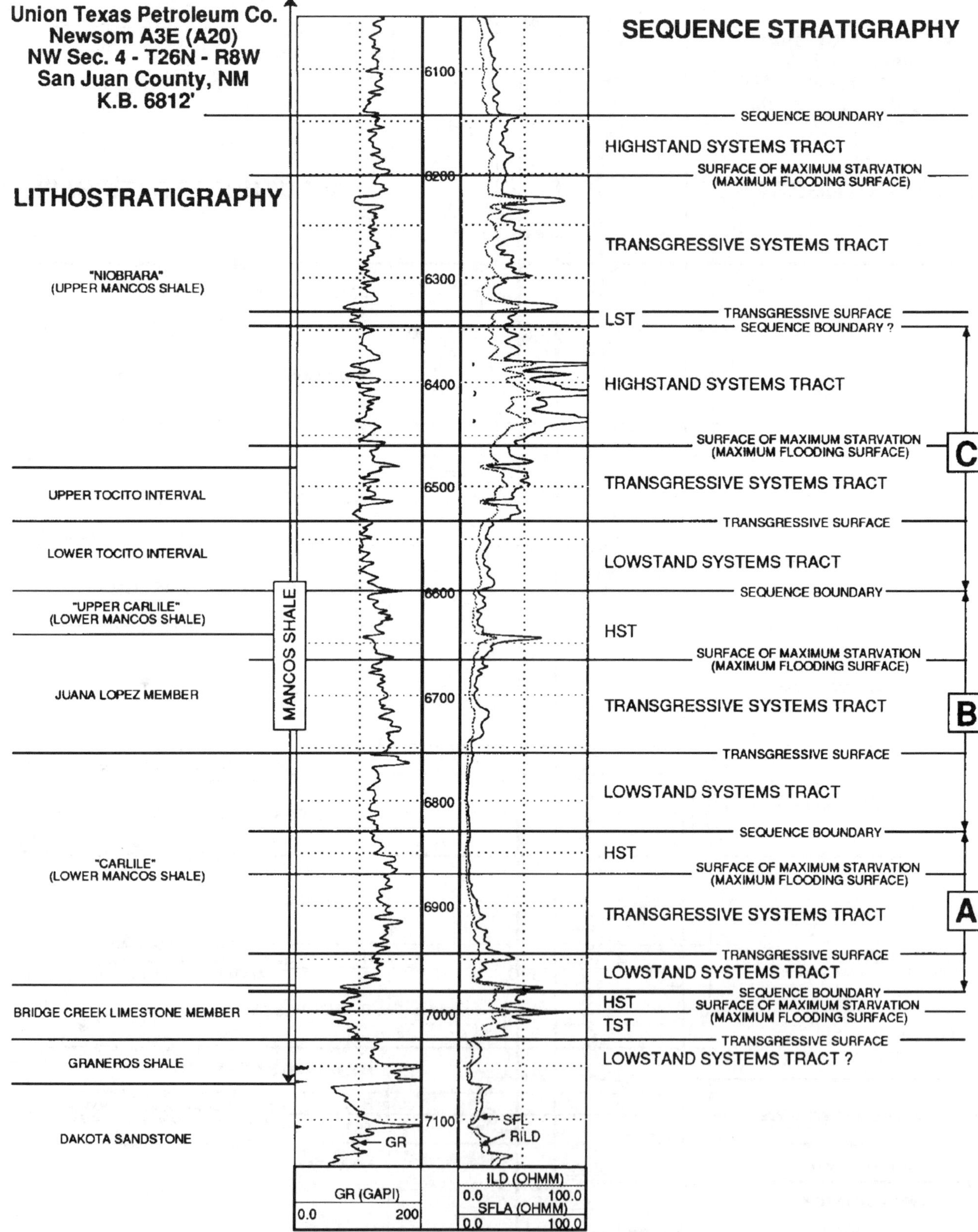

Figure 11. Well log, lithostratigraphy, and sequence stratigraphy for UTPC Newsom A3E (A20) well.

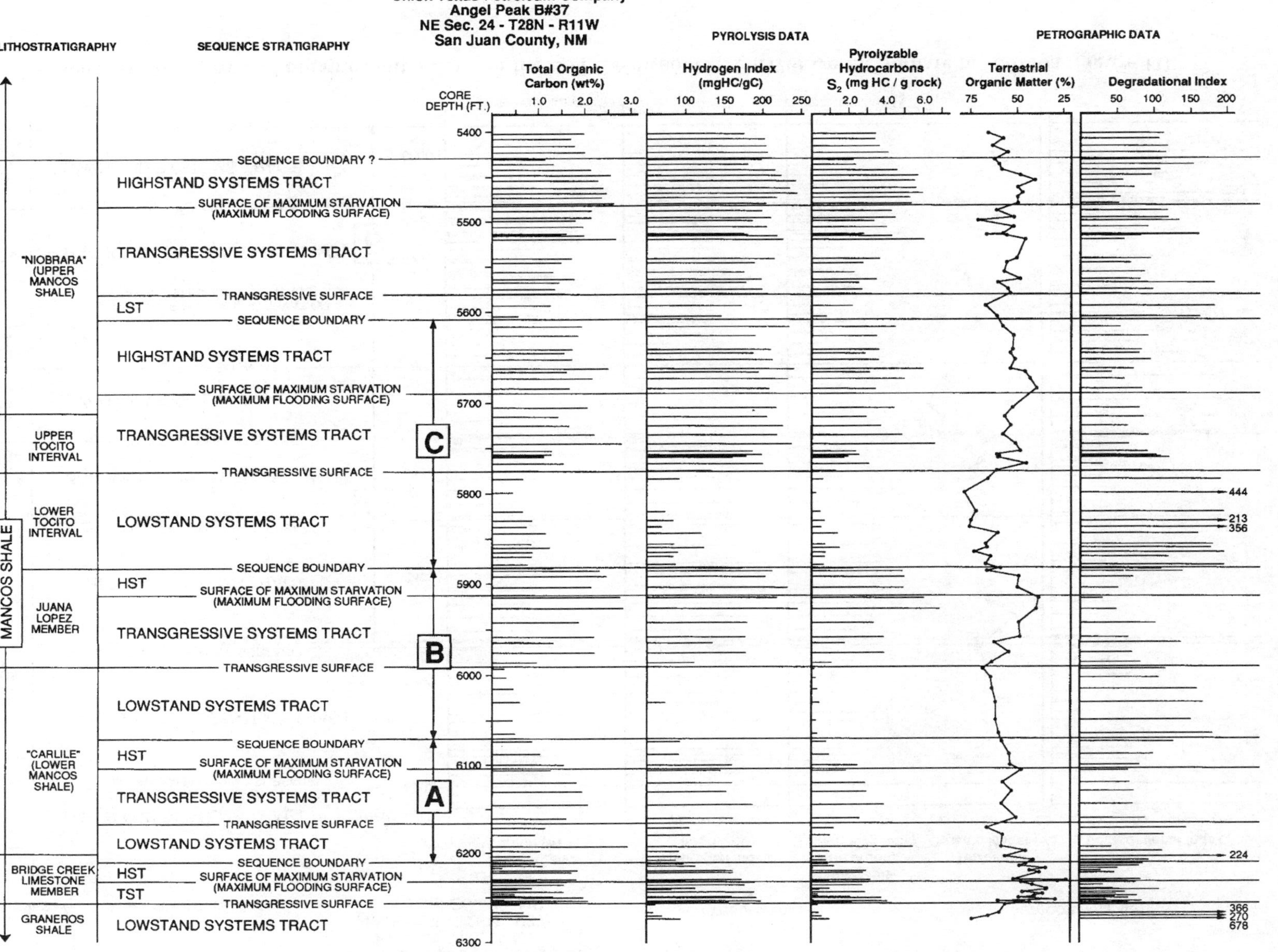

Figure 12. Sequence stratigraphy and organic matter data for the UTPC Angel Peak B#37 well. Compare to well log (Figure 10). Sequence C is correlative to section at Bisti oil field (Figure 7).

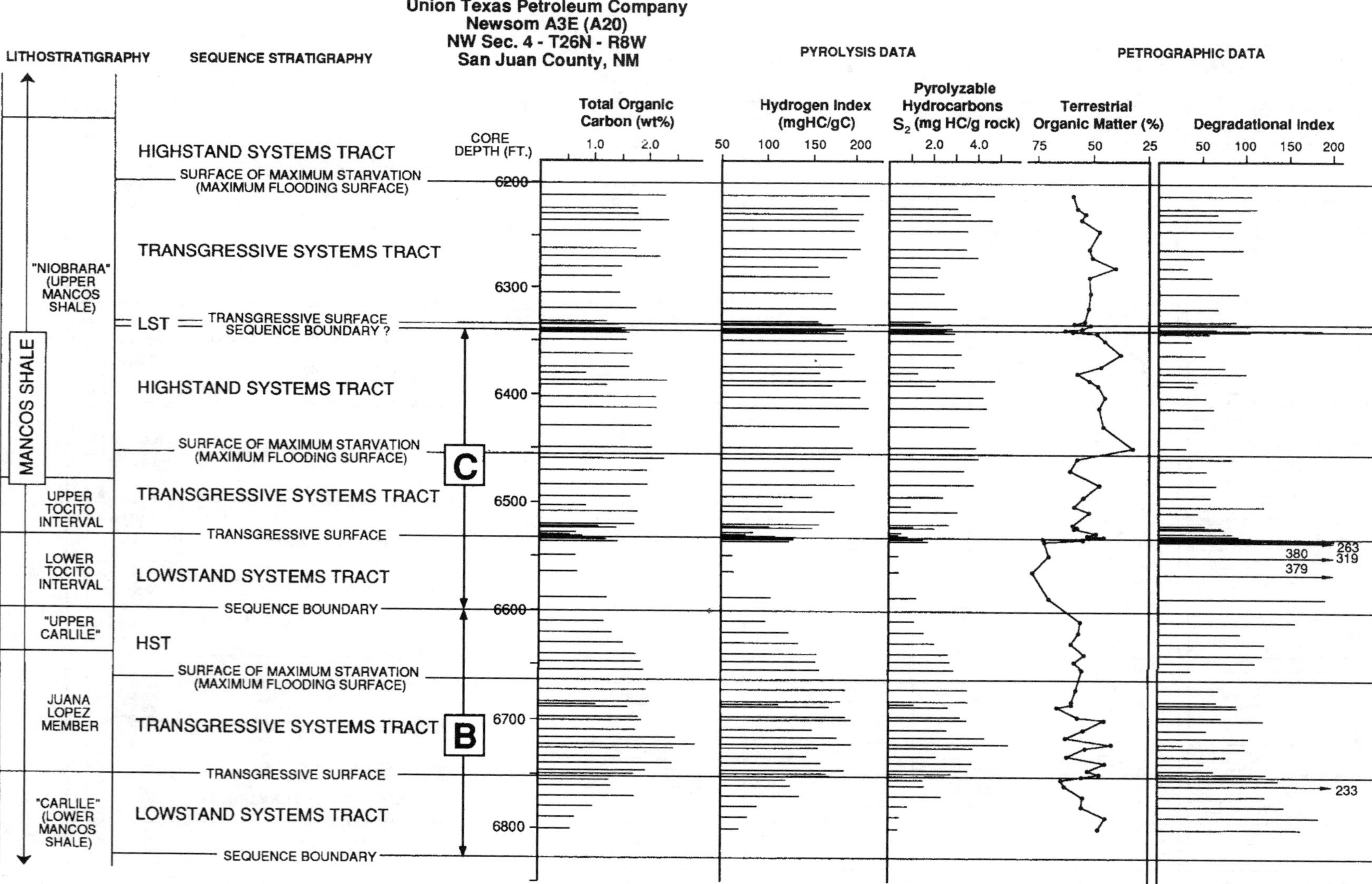

Figure 13. Sequence stratigraphy and organic matter data for the UTPC Newsom A3E (A20) well. Compare to well log (Figure 11).

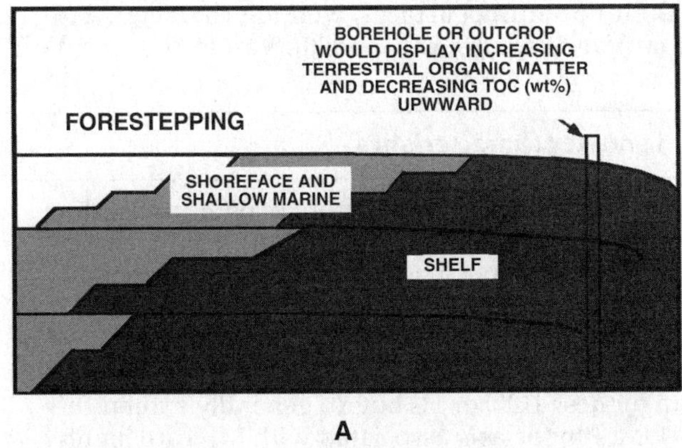

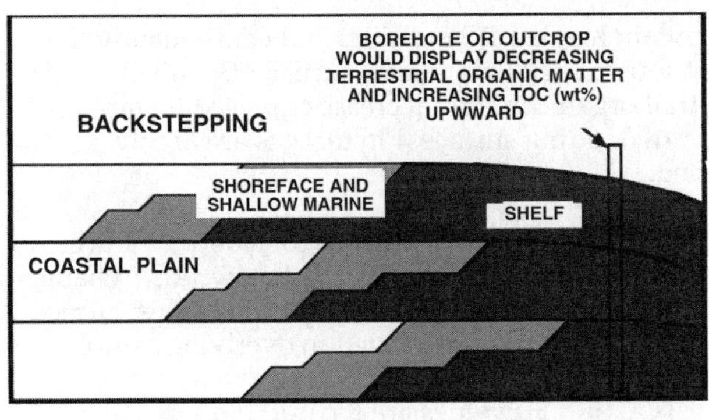

Figure 14. Relationship between parasequence stacking and type of organic matter delivered to shelf environment. Figure adapted from Van Wagoner et al. (1988, 1990). (A)—forestepping (progradational) stacking results in increasing terrestrial organic matter and decreasing TOC upward at single location on the shelf (denoted by vertical box) as that position becomes more proximal. (B)—backstepping pattern results in decreasing terrestrial organic matter and increasing TOC upward as the terrigenous sediment is trapped in estuarine and coastal environments and position on shelf becomes progressively more distal.

depositional sequence is shown schematically in Figure 15. Both the organic richness and organic facies change predictably in a sequence stratigraphic framework (see also Pasley and Hazel, 1990; Bohacs and Isaksen, 1991; Creaney et al., 1991; Gorin and Steffen, 1991; Pasley et al., 1991).

Implications for Sequence Stratigraphic Interpretations

The results of this study demonstrate that data from organic matter characterization are useful in sequence stratigraphic interpretations (see also Pasley and Hazel, 1990; Posamentier and Chamberlin, 1993). This methodology not only employs commonly discussed sequence stratigraphic techniques with well logs and cores (e.g., Vail, 1987, 1990; Van Wagoner et al., 1990; Vail and Wornardt, 1990) but also uses the relationship between organic matter deposition, transgressions of the shoreline, and parasequence stacking to enhance the resolution in locating surfaces that bound depositional systems tracts in the fine-grained, offshore parts of a depositional sequence. Integration of well log correlations, core descriptions, regional stratigraphic relationships, and organic matter data into sequence stratigraphy is possible because each systems tract in a depositional sequence has a distinctive depositional style that affects the rate of terrigenous sediment supply to the shelf and, consequently, the type and preservation of organic matter. These concepts are summarized in Table 1 and Figure 15.

It is suggested that use of the ideas presented in Table 1 improves precision in the location of transgressive surfaces and surfaces of maximum starvation (or downlap surfaces), especially in fine-grained intervals where conventional (i.e., physical) lithologic changes associated with these surfaces are often equivocal. Delineation of these surfaces is important to sequence stratigraphic studies of well log, core, and outcrop data that are concerned with intervals below the limit of seismic resolution (Nummedal and Swift, 1987; Nummedal et al., 1989a; Van Wagoner et al., 1990). In addition, the use of organic matter data in sequence stratigraphic interpretations of fine-grained strata improves the understanding of sedimentary basin fill as a whole because these strata generally contain good biostratigraphic information and a more continuous record of deposition. The use of data from the characterization of organic matter in a sequence stratigraphic interpretation for the Upper Cretaceous Mancos Shale (Figures 10–13) has provided results that are generally corroborated by well log correlations (Molenaar, 1989; Pasley, 1991; Riley, 1993; R. Bottjer, written communication, 1987) and do not significantly conflict with regional stratigraphic information (McCubbin, 1969; Hook and Cobban, 1980; Molenaar, 1983; Glenister and Kauffman, 1985; Kauffman, 1985; Nummedal, 1990; J. Stein, personal communication, 1991).

Specifically, well log cross sections and observations from core descriptions are used to correlate packages of similar stacking patterns. Transgressive systems tracts, for example, are identified in the well logs as intervals with a backstepping parasequence stacking pattern. In core and outcrop, lithologic aspects such as a decrease in grain size, change in ichnofabric, or an identifiable disconformity are important in identification of transgressive surfaces. However, the combination of log, core, and outcrop data commonly provides more than one plausible alternative for the location of the transgressive surface that forms the lower boundary of the transgressive systems tract. This is especially true for sequence stratigraphic interpretations of fine-grained intervals. Precision in these interpretations is improved when transgressive surfaces are identified as a change in

Table 1. Summary of relationship between systems tracts, their bounding surfaces, well log character, and organic matter characteristics. Well log signatures taken from Van Wagoner et al. (1990), Vail (1990), and Vail and Wornardt (1990). See also Figure 15.

	Log character	Organic matter characteristics
Highstand Systems Tract	Forestepping (progradational)	TOC and HI decrease upward from maximum flooding surface. Increase in both amount of terrestrial organic matter (TOM) and its preservation. (Degradational index increases).
Maximum Flooding Surface	Generally highest gamma ray count	Samples associated with this surface do not necessarily contain highest TOC or HI but do generally exhibit the least TOM. Phytoclasts associated with MFS are highly degraded (Degradational index at a minimum).
Transgressive Systems Tract	Backstepping (retrogradational)	May contain highest TOC and HI and consequently the highest source rock potential (if sufficiently thick). Terrestrial organic matter decreases upward toward maximum flooding surface. Phytoclasts are highly degraded.
Transgressive Surface	Change from aggradational below to backstepping above	Samples above this surface contain considerably lower amounts of TOM than those directly below. Also, change in degradational index is pronounced across this surface as phytoclasts are more degraded in overlying samples.
Lowstand Systems Tract	Aggradational to forestepping (progradational)	Lowest TOC and HI values are recorded in samples from this sytems tract. Terrestrial organic matter is abundant and well preserved (degradational index high).
Sequence Boundary	Change from forestepping below to aggradational above	Most difficult surface to recognize using organic matter characteristics. General increase in degradational index as phytoclasts become better preserved in LST above the boundary than in the HST below.

well log stacking pattern (from forestepping/aggradational to backstepping) that coincides with the lithologic characteristics described above, decreases in percent terrestrial organic matter (TOM) and degradational index (D), and increases in total organic carbon (TOC) and hydrogen index (HI). Examples of the utility of this integrated approach can be seen in the location of the transgressive surfaces at the base of the transgressive systems tracts in Sequence C (Angel Peak—5775 ft or 1760 m, Figures 10 and 12; Newsom—6533 ft or 1991 m, Figures 11 and 13) and in Sequence A (Angel Peak—6167 ft or 1880 m, Figures 10 and 12; Newsom—6942 ft or 2116 m, Figure 11). The precise location of these surfaces is confirmed because they are recognizable in the core, correlatable on the logs, and are marked by an increase in TOC, HI, and S_2 and a decrease in TOM and D (Figures 12, 13).

Surfaces of maximum starvation are identified on well log cross sections as correlatable intervals of relatively high gamma ray counts that separate backstepping stacking patterns from forestepping ones. Lithologically, surfaces of maximum starvation are commonly characterized by the presence of concretions (calcareous, pyrite-rich, or septarian), fossil-rich zones, phosphatized shell fragments, and/or changes in ichnofabric. Data from maceral analysis provide additional information in locating these surfaces as minimums in both TOM and D values (Pasley and Hazel, 1990). The surface of maximum starvation that forms the boundary between the highstand and transgressive systems tracts in Sequence B (Angel Peak core—5917 ft or 1803.5 m, Figures 8, 10, and 12; Newsom core—6663 ft or 2031 m, Figures 9, 11, and 13) exhibits these characteristics.

It should be noted, however, that organic matter from the highstand and lowstand systems tracts can appear similar (Figure 15), a result not entirely surprising when one considers that these systems tracts contain predominantly regressive sediments and share similar stacking patterns. In general, an increase in the relative proportion of terrestrial organic matter

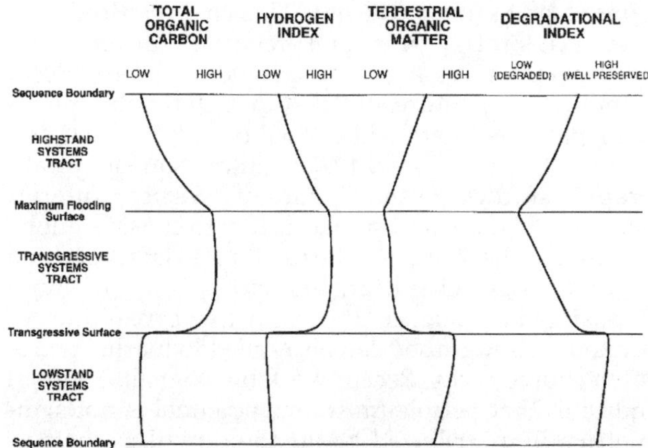

Figure 15. Schematic diagram showing relationship between organic matter characteristics and position with the depositional sequence.

(% TOM) and an increase in its preservation (increase in D) can be recognized at sequence boundaries. It should be emphasized, however, that identification of sequence boundaries within fine-grained, basinward strata requires appreciable amounts of regional stratigraphic information. It is important to remember that organic matter changes that occur on the shelf are related to transgressions and regressions of the shoreline and not necessarily to the subtle downward shifts in coastal onlap that are associated with sequence boundaries in the Upper Cretaceous marine succession of the San Juan Basin (see discussion by Christie-Blick, 1991).

Implications to Petroleum Source Rock Prediction

Previous studies have suggested that the Mancos Shale interval contains the major petroleum source rocks for the San Juan Basin (Ross, 1980; Rice, 1983). Results presented by Pasley et al. (1991) have shown that the Upper Mancos Shale above the Tocito Sandstone does indeed contain appreciable source potential. Data from the Newsom and Angel Peak cores (Figures 8, 9, 12, 13) provide additional information concerning the petroleum source potential of a larger portion of the Mancos Shale. The lower Tocito interval in both the Newsom and Angel Peak cores and at Bisti oil field (lowstand systems tract of Sequence C) possesses no appreciable source potential. The same is true for the Carlile interval of the lower Mancos Shale directly below the Juana Lopez Member (lowstand systems tract of Sequence B; Figures 12, 13). These shales are low in TOC and HI and the most common macerals are of terrestrial origin. However, the Carlile interval above the Bridge Creek Member (between 6100–6150 ft, or 1859–1874.5 m, in the Angel Peak core), the lower part of the Juana Lopez Member, and the upper Mancos Shale directly above the Tocito Sandstone (Mulatto Tongue equivalent) possess excellent source potential (given their present thermal maturity). These transgressive, marine shales contain sufficient organic matter that is hydrogen-rich and may have already generated significant amounts of hydrocarbons in the Angel Peak and Newsom cores.

One of the appealing aspects of sequence stratigraphy is that it provides a framework for predictions concerning the nature of sedimentary basin fill (e.g., Vail, 1987; Jervey, 1988; Posamentier and Vail, 1988; Vail and Wornardt, 1990). Whereas these predictive capabilities have been applied to reservoir facies, information concerning detailed source rock predictions in a sequence stratigraphic framework is lacking. Previous workers have suggested that, in a given depositional sequence, the condensed section facies associated with the downlap surface should contain the best marine source rocks (e.g., Vail, 1987). However, the intervals listed above (the Carlile above the Bridge Creek, the lower Juana Lopez, and the Mulatto Tongue equivalent of the upper Mancos) possess considerable source potential and are not exclusively condensed sections. These units were deposited as offshore marine sediments in the transgressive systems tracts of Sequences A, B, and C (Figures 12, 13). The surfaces of maximum starvation (also known as downlap or maximum flooding surfaces) in each of these sequences are generally found at the top of the intervals containing the optimum source potential. In some cases, potential source sediments directly overlie the transgressive surface that forms the lower boundary of the transgressive systems tract. Values for TOC and HI (and, as a consequence, source potential) generally decrease as one moves upward from the surface of maximum starvation into the highstand systems tract (see Figure 15). Studies on Cretaceous rocks in Canada have also found meaningful source potential in transgressive shales below the condensed section associated with the downlap surface and only a short distance above the transgressive surface (Leckie et al., 1990; Posamentier and Chamberlin, 1993). Models proposed by Wignall (1991a,b) indicate that organic-rich shales are most common in the transgressive systems tract and are not symmetrically distributed about the surface of maximum starvation. These results, in conjunction with the findings of the present study, suggest that the best marine petroleum source rocks are found in the offshore, fine-grained intervals of the transgressive systems tract below the condensed section facies that contains the downlap surface.

The relationship between organic matter characteristics and position within a depositional sequence (Figure 15) implies that correlations based on sequence stratigraphy can enhance petroleum source rock prediction. Data from the Mancos Shale suggest that the transgressive systems tract of a sequence is most likely to contain marine petroleum source rocks. In addition, some depositional sequences are more likely to have better source potential than others and these sequences can be correlated regionally. Lateral variations in source potential are observed within the transgressive systems tract of a particular sequence. Differences within the transgressive systems tract of Sequence C at Bisti, Angel Peak, and Newsom (Figures 7, 12, 13) are thought to be related to local variations that affect the depositional environment. It has been suggested that differential subsidence related to basement faults (Four Corners Lineament; Stevenson and Baars, 1986) had a direct effect on Late Cretaceous deposition in this portion of the Western Interior Seaway (Huffman and Taylor, 1991; Riley, 1993). This structural movement may explain some of the lateral variability in petroleum source potential within the Mancos Shale. Similar conclusions concerning the relationship between local basin morphology and spatial distribution of organic-rich deposits within transgressive systems tracts have been documented for the lower Kimmeridgian of the Tethys seaway (Baudin et al., 1992).

CONCLUSIONS

Integration of data from the characterization of organic matter with sedimentologic and stratigraphic observations has shown that organic facies deposited in shallow marine environments change abruptly and predictably across surfaces associated with transgressions of the shoreline. This change reflects the sudden reduction in the amount of terrigenous sediment that reaches the shallow marine environments during transgression. This finding has special relevance to the lithogenetic approach employed in sequence stratigraphy. Because the rate of terrigenous sediment supply to the marine environment varies with changing systems tracts, each systems tract in a sequence has a characteristic parasequence stacking pattern. Therefore, a direct relationship between type and preservation of organic matter and position within the depositional sequence is expected.

The predictable relationship between organic matter type and preservation and position within the depositional sequence facilitates an integrated approach to sequence stratigraphy. The addition of data from organic matter characterization to those commonly used in sequence stratigraphy (well log correlations, core and outcrop descriptions, biostratigraphy, and regional stratigraphic relationships) provides greater resolution in the location of critical surfaces that bound depositional systems tracts within the depositional sequence. Use of organic matter data in fine-grained intervals is important because these strata often contain the best biostratigraphic information and the most complete sedimentary record. Therefore, organic matter data, which are commonly gathered for petroleum source rock evaluation, should also be used in the broader context of sedimentary basin analysis.

An example of this approach has been presented for the Upper Cretaceous Mancos Shale of the San Juan Basin. Observations from the characterization of organic matter have been used in conjunction with core descriptions and well log correlations (parasequence stacking patterns). This has resulted in a sequence stratigraphic interpretation that improves on previous work and, at the same time, provides a framework for future study. Although the interpretations presented here will undoubtedly be refined and corrected, it is suggested that future sequence stratigraphic studies on the Cretaceous Western Interior should take these and similar data into consideration.

Finally, because a relationship exists between organic matter characteristics and position within a depositional sequence, the predictive capabilities of sequence stratigraphy can be applied to marine petroleum source rocks. Recent work by Wignall (1991a,b) indicates that petroleum source potential is not symmetrically distributed about the downlap surface. Results presented here agree with this assertion and have shown that optimum source potential is found in the transgressive systems tract below the condensed section (and associated downlap surface). In some cases, appreciable source potential has been recognized in shales directly above the transgressive surface that forms the lower boundary of the transgressive systems tract.

ACKNOWLEDGMENTS

This chapter represents a portion of the senior author's Ph.D. dissertation in the Department of Geology and Geophysics at Louisiana State University (LSU). Assistance by George F. Hart, Joseph E. Hazel, Jeffrey S. Hanor, Ray E. Ferrell, and William A. Gregory of LSU is acknowledged. The authors are grateful to Amoco Production Company for permission to study the Angel Peak and Newsom cores, and to Shell Western Exploration and Production Incorporated for permission to use the cores from Bisti oil field. Financial support from Amoco Production Company, Shell Oil Company Foundation, AAPG (Lewis Austin Weeks Grant), and the New Mexico Bureau of Mines and Mineral Resources is gratefully acknowledged. Stratigraphic expertise by C. M. "K" Molenaar (USGS) and Rich Bottjer and Jeff Stein (Amoco) was especially helpful, as was technical support from both Ron Snelling and Elizabeth Chinn at LSU. The authors thank Amoco Production Company for assistance with the preparation of this chapter.

REFERENCES CITED

Baudin, F., F. Cecca, É. Fourcade, and J. Azéma, 1992, Paléoenvironnements et faciès organiques du Kimméridgien inférieur de la Téthys: C.R. Académie des Sciences, t. 314, Série II, p. 373-379.

Bergsohn, I., 1988, Lithofacies architecture of the Tocito Sandstone, northwest New Mexico: unpublished M.S. thesis, Department of Geology and Geophysics, Louisiana State University, Baton Rouge, 170 p.

Bohacs, K.M., and G.H. Isaksen, 1991, Source quality variations tied to sequence development: integration of physical and chemical aspects, Lower to Middle Triassic, western Barents Sea (abs.): AAPG Bulletin, v. 75, p. 544.

Boyle, E.A., J.M. Edmond, and E.R. Sholkovitz, 1977, The mechanism of iron removal in estuaries: Geochimica et Cosmochimica Acta, v. 41, p. 1313-1324.

Christie-Blick, N., 1991, Onlap, offlap, and the origin of unconformity-bounded depositional sequences: Marine Geology, v. 97, p. 35-56.

Creaney, S., Q.R. Passey, and J. Allan, 1991, Use of well logs and core data to assess the sequence stratigraphic distribution of organic-rich rocks (abs.): AAPG Bulletin, v. 75, p. 557.

Davies, J.R., A. McNestry, and R.A. Waters, 1991, Palaeoenvironments and palynofacies of a pulsed transgression: the late Devonian and early Dinantian (Lower Carboniferous) rocks of south-east Wales: Geological Magazine, v. 128, p. 355-380.

Droser, M.L., and D.J. Bottjer, 1986, A semiquantitative field classification of ichnofacies: Journal of Sedimentary Petrology, v. 56, p. 558-559.

Espitalié, J., J. Madec, B. Tissot, J. Mennig, and P. Leplat, 1977, Source rock characterization method for petroleum exploration: Proceedings, 1977 Offshore Technology Conference, v. 3, p. 439-443.

Fontugne, M.R., and J.-M. Jouanneau, 1987, Modulation of the particulate organic carbon flux to the ocean by a macrotidal estuary: evidence from measurements of carbon isotopes in organic matter from the Gironde system: Estuarine, Coastal and Shelf Science, v. 24, p. 377-387.

Glenister, L.M., and E.G. Kauffman, 1985, High resolution stratigraphy and depositional history of the Greenhorn regressive hemicyclothem, Rock Canyon Anticline, Pueblo, Colorado, *in* L.M. Pratt, E.G. Kauffman, and F.B. Zelt, eds., Fine-grained Deposits and Biofacies of the Cretaceous Western Interior Seaway: Evidence of Cyclic Sedimentary Processes: SEPM Field Trip Guidebook No. 4, 1985 Midyear Meeting, p. 170-183.

Gorin, G.E., and D. Steffen, 1991, Organic facies as a tool for recording eustatic variations in marine fine-grained carbonates—example of the Berriasian stratotype at Berrias (Ardéche, SE France): Palaeogeography, Palaeoclimatology, Palaeoecology, v. 85, p. 303-320.

Habib, D., and J.A. Miller, 1989, Dinoflagellate species and organic facies evidence of marine transgression and regression in the Atlantic Coastal Plain: Palaeogeography, Palaeoclimatology, Palaeoecology, v. 74, p. 23-47.

Harput, O.B., and S.L. Gokcen, 1991, Application of the organic facies method in the Thrace Basin: Sedimentary Geology, v. 72, p. 171-187.

Hart, G.F., 1986, Origin and classification of organic matter in clastic systems: Palynology, v. 10, p. 1-23.

Hook, S.C., and W.A. Cobban, 1980, Reinterpretation of type section of Juana Lopez Member of Mancos Shale: New Mexico Geology, v. 2, no. 2, p. 17-22.

Huffman, A.C., Jr., and D.J. Taylor, 1991, Basement fault control on the occurrence and development of San Juan Basin energy resources: Abstracts with Program, GSA Rocky Mountain Section, v. 23, p. 34.

Jervey, M.T., 1988, Quantitative geological modeling of siliciclastic rock sequences and their seismic expression, *in* C.K. Wilgus, B.S. Hastings, C.G.St.C. Kendall, H.W. Posamentier, C.A. Ross, and J.C. Van Wagoner, eds., Sea-Level Changes: An Integrated Approach: Tulsa, SEPM Special Publication No. 42, p. 47-69.

Jones, R.W., 1987, Organic facies, *in* J. Brooks and D. Welte, eds., Advances in Petroleum Geochemistry, Volume 2: London, Academic Press, p. 1-90.

Kauffman, E.G., 1977, Geological and biological overview: Western Interior basin: Mountain Geologist, v. 14, p. 75-100.

Kauffman, E.G., 1985, Cretaceous evolution of the Western Interior Basin of the United States, *in* L.M. Pratt, E.G. Kauffman, and F.B. Zelt, eds., Fine-grained Deposits and Biofacies of the Cretaceous Western Interior Seaway: Evidence of Cyclic Sedimentary Processes: SEPM Field Trip Guidebook No. 4, 1985 Midyear Meeting, p. iv-xiii.

LeBlanc, C.G., R.A. Bourbonniere, H.P. Schwarcz, and M.J. Risk, 1989, Carbon isotopes and fatty acids analysis of the sediments of Negro Harbor, Nova Scotia, Canada: Estuarine, Coastal and Shelf Science, v. 28, p. 261-276.

Leckie, D.A., C. Singh, F. Goodarzi, and J.H. Wall, 1990, Organic-rich, radioactive marine shale: a case study of a shallow-water condensed section, Cretaceous Shaftsbury Formation, Alberta, Canada: Journal of Sedimentary Petrology, v. 60, p. 101-117.

Lucotte, M., C. Hillaire-Marcel, and P. Louchouarn, 1991, First-order organic carbon budget in the St. Lawrence lower estuary from ^{13}C data: Estuarine, Coastal and Shelf Science, v. 32, p. 297-312.

McCave, I.N., 1973, Mud in the North Sea, *in* E.D. Goldberg, ed., North Sea Science: Cambridge, Mass., The MIT Press, p. 75-100.

McCubbin, D.G., 1969, Cretaceous strike-valley sandstone reservoirs, northwestern New Mexico: AAPG Bulletin, v. 53, p. 2114-2140.

Molenaar, C.M., 1983, Major depositional cycles and regional correlations of Upper Cretaceous rocks, southern Colorado Plateau and adjacent areas, *in* M.W. Reynolds and E.D. Dolly, eds., Mesozoic Paleogeography of West-Central United States: Rocky Mountain Section of SEPM, p. 201-224.

Molenaar, C.M., 1989, Stratigraphic cross sections of Upper Cretaceous rocks across San Juan Basin, northwestern New Mexico and southwestern Colorado (abs.): AAPG Bulletin, v. 73, p. 1167.

Nichols, M.M., and R.B. Biggs, 1985, Estuaries, *in* R.A. Davis, Jr., ed., Coastal Sedimentary Environments, 2nd Edition: New York, Springer-Verlag, p. 77-186.

Nummedal, D., 1990, Sequence stratigraphic analysis of Upper Turonian and Coniacian strata in the San

Juan Basin, New Mexico, U.S.A., *in* R.N. Ginsburg and B. Beaudoin, eds., Cretaceous Resources, Events, and Rhythms: Background and Plans for Research: Dordrecht, The Netherlands, Kluwer Academic Publishers, p. 33-46.

Nummedal, D., and D.J.P. Swift, 1987, Transgressive stratigraphy at sequence–bounding unconformities: some principles derived from Holocene and Cretaceous examples, *in* D. Nummedal, O.H. Pilkey, and J.D. Howard, eds., Sea Level Fluctuation and Coastal Evolution: SEPM Special Publication No. 41, p. 242-260.

Nummedal, D., D.J.P. Swift, and B.M. Kofron, 1989a, Sequence stratigraphic interpretation of Coniacian strata in the San Juan Basin, New Mexico, *in* R.A. Morton and D. Nummedal, eds., Shelf Sedimentation, Shelf Sequences and Related Hydrocarbon Accumulation: Gulf Coast Section SEPM Foundation Seventh Annual Research Conference Proceedings, p. 175-202.

Nummedal, D., R. Wright, R.D. Cole, and R.R. Remy, 1989b, Cretaceous shelf sandstones and shelf depositional sequences, Western Interior Basin, Utah, Colorado, and New Mexico: Washington, American Geophysical Union, 28th International Geological Congress, Field Trip Guidebook T119, 87 p.

Nummedal, D., R. Wright, and D.J.P. Swift, 1988, Sequence stratigraphy of Upper Cretaceous strata of the San Juan Basin, New Mexico: Field Trip Guidebook, March 1988, AAPG/SEPM 73rd Annual Meeting, Houston, Texas, 216 p.

Pasley, M.A., 1991, Organic matter variation within depositional sequences: stratigraphic significance and implication to petroleum source rock predicition: unpublished Ph.D. dissertation, Department of Geology and Geophysics, Louisiana State University, 148 p.

Pasley, M.A., W.A. Gregory, and G.F. Hart, 1991, Organic matter variations in transgressive and regressive shales: Organic Geochemistry, v. 17, p. 483-509.

Pasley, M.A., and J.E. Hazel, 1990, Use of organic petrology and graphic correlation of biostratigraphic data in sequence stratigraphic interpretations: example from the Eocene–Oligocene boundary section, St. Stephens Quarry, Alabama: Gulf Coast Association of Geological Societies Transactions, v. 40, p. 661-683.

Peters, K.E., 1986, Guidelines for evaluating petroleum source rocks using programmed pyrolysis: AAPG Bulletin, v. 70, p. 318-329.

Posamentier, H.W., and C.J. Chamberlin, 1993, Sequence stratigraphic analysis of Viking Formation lowstand beach at Joarcam Field, Alberta, Canada, *in* H.W. Posamentier, C.P. Summerhayes, B.U. Haq, and G.P. Allen, eds., Stratigraphy and Facies Associations in a Sequence Stratigraphic Framework: International Association of Sedimentologists, Special Publication 18, p. 469-485.

Posamentier, H.W., and P.R. Vail, 1988, Eustatic controls on clastic deposition II-sequence and systems tracts models, *in* C.K. Wilgus, B.S. Hastings, C.G.St.C. Kendall, H.W. Posamentier, C.A. Ross, and J.C. Van Wagoner, eds., Sea-Level Changes: An Integrated Approach: SEPM Special Publication No. 42, p. 125-154.

Rice, D.D., 1983, Relation of natural gas composition to thermal maturity and source rock type in San Juan Basin, northwestern New Mexico and Southwestern Colorado: AAPG Bulletin, v. 67, p. 1199-1218.

Riley, G.W., 1993, Origin of a coarse-grained shallow marine sandstone complex: the Coniacian Tocito Sandstone, northwestern New Mexico: unpublished Ph.D. dissertation, Department of Geology and Geophysics, Louisiana State University, 342 p.

Roberts, H.H., and J.M. Coleman, 1988, Lithofacies characteristics of shallow expanded and condensed sections of the Louisiana distal shelf and upper slope: Gulf Coast Association of Geological Societies Transactions, v. 38, p. 291-301.

Ross, L.M., 1980, Geochemical correlation of San Juan Basin oils—a study: Oil and Gas Journal, v. 78, no. 44, p. 102-110.

Showers, W.J., and D.G. Angle, 1986, Stable isotopic characterization of organic carbon accumulation on the Amazon continental shelf: Continental Shelf Research, v. 6, p. 227-244.

Stevenson, G.M., and D.L. Baars, 1986, The Paradox: a pull-apart basin of Pennsylvanian age, *in* J.A. Peterson, ed., Paleotectonics and Sedimentation in the Rocky Mountain Region, U.S.: AAPG Memoir 41, p. 513-539.

Stefani, M., and M. Burchell, 1990, Upper Triassic (Rhaetic) argillaceous sequences in northern Italy: depositional dynamics and source potential, *in* A.Y. Huc, ed., Deposition of Organic Facies: Tulsa, AAPG Studies in Geology 30, p. 93-106.

Swift, D.J.P., 1976, Continental shelf sedimentation, *in* D.J. Stanley and D.J.P. Swift, eds., Marine Sediment Transport and Environmental Management: New York, Wiley, p. 311-350.

Vail, P.R., 1987, Seismic stratigraphy interpretation using sequence stratigraphy, part one: seismic stratigraphy interpretation procedure, *in* A.W. Bally, ed., Atlas of Seismic Stratigraphy: AAPG Studies in Geology, v. 27, p. 1-10.

Vail, P.R., 1990, Fundamentals of sequence stratigraphy: Sequence Stratigraphy Workbook, Part 4, Short Course Notes—1990 Annual Meeting Gulf Coast Association of Geological Societies, unpaginated.

Vail, P.R., and W.W. Wornardt, 1990, Well log-seismic stratigraphy: an integrated tool for the 90's, *in* J.M. Armentrout, J.M. Coleman, W.E. Galloway, and P.R. Vail, eds., Sequence Stratigraphy as an Exploration Tool—Concepts and Practices in the Gulf Coast: Eleventh Annual Research Conference, Gulf Coast Section SEPM Foundation, Program and Extended Abstracts, p. 379-388.

Van Wagoner, J.C., R.M. Mitchum, K.M. Campion, and V. Rahmanian, 1990, Siliciclastic sequence stratigraphy in well logs, cores, and outcrops: concepts for high-resolution correlation of time and facies, AAPG Methods in Exploration Series No. 7, 55 p.

Van Wagoner, J.C., H.W. Posamentier, R.M. Mitchum, P.R. Vail, J.F. Sarg, T.S. Loutit, and J. Hardenbol, 1988, An overview of the fundamentals of sequence stratigraphy and key definitions, *in* C.K. Wilgus, B.S. Hastings, C.G.St.C. Kendall, H.W. Posamentier, C.A. Ross, J.C. Van Wagoner, eds., Sea-Level Changes: An Integrated Approach: SEPM Special Publication No. 42, p. 39-45.

Wignall, P.B., 1991a, A model for transgressive black shales in a sequence stratigraphic framework (abs.): AAPG Bulletin, v. 75, p. 693.

Wignall, P.B., 1991b, Model for transgressive black shales?: Geology, v. 19, p. 167-170.

Index

Middlebury College
0 00 02 0616832 0